Paul P. Wang, Da Ruan, Etienne E. Kerre (Eds.)

Fuzzy Logic

Studies in Fuzziness and Soft Computing, Volume 215

Editor-in-chief
Prof. Janusz Kacprzyk
Systems Research Institute
Polish Academy of Sciences
ul. Newelska 6
01-447 Warsaw
Poland
E-mail: kacprzyk@ibspan.waw.pl

Further volumes of this series can be found on our homepage: springer.com

Vol. 199. Zhong Li
Fuzzy Chaotic Systems, 2006
ISBN 978-3-540-33220-6

Vol. 200. Kai Michels, Frank Klawonn, Rudolf Kruse, Andreas Nürnberger
Fuzzy Control, 2006
ISBN 978-3-540-31765-4

Vol. 201. Cengiz Kahraman (Ed.)
Fuzzy Applications in Industrial Engineering, 2006
ISBN 978-3-540-33516-0

Vol. 202. Patrick Doherty, Witold Łukaszewicz, Andrzej Skowron, Andrzej Szałas
Knowledge Representation Techniques: A Rough Set Approach, 2006
ISBN 978-3-540-33518-4

Vol. 203. Gloria Bordogna, Giuseppe Psaila (Eds.)
Flexible Databases Supporting Imprecision and Uncertainty, 2006
ISBN 978-3-540-33288-6

Vol. 204. Zongmin Ma (Ed.)
Soft Computing in Ontologies and Semantic Web, 2006
ISBN 978-3-540-33472-9

Vol. 205. Mika Sato-Ilic, Lakhmi C. Jain
Innovations in Fuzzy Clustering, 2006
ISBN 978-3-540-34356-1

Vol. 206. A. Sengupta (Ed.)
Chaos, Nonlinearity, Complexity, 2006
ISBN 978-3-540-31756-2

Vol. 207. Isabelle Guyon, Steve Gunn, Masoud Nikravesh, Lotfi A. Zadeh (Eds.)
Feature Extraction, 2006
ISBN 978-3-540-35487-1

Vol. 208. Oscar Castillo, Patricia Melin, Janusz Kacprzyk, Witold Pedrycz (Eds.)
Hybrid Intelligent Systems, 2007
ISBN 978-3-540-37419-0

Vol. 209. Alexander Mehler, Reinhard Köhler
Aspects of Automatic Text Analysis, 2007
ISBN 978-3-540-37520-3

Vol. 210. Mike Nachtegael, Dietrich Van der Weken, Etienne E. Kerre, Wilfried Philips (Eds.)
Soft Computing in Image Processing, 2007
ISBN 978-3-540-38232-4

Vol. 211. Alexander Gegov
Complexity Management in Fuzzy Systems, 2007
ISBN 978-3-540-38883-8

Vol. 212. Elisabeth Rakus-Andersson
Fuzzy and Rough Techniques in Medical Diagnosis and Medication, 2007
ISBN 978-3-540-49707-3

Vol. 213. Peter Lucas, José A. Gámez, Antonio Salmerón (Eds.)
Advances in Probabilistic Graphical Models, 2007
ISBN 978-3-540-68994-2

Vol. 214. Irina Georgescu
Fuzzy Choice Functions, 2007
ISBN 978-3-540-68997-3

Vol. 215. Paul P. Wang, Da Ruan, Etienne E. Kerre (Eds.)
Fuzzy Logic, 2007
ISBN 978-3-540-71257-2

Paul P. Wang
Da Ruan
Etienne E. Kerre
(Eds.)

Fuzzy Logic

A Spectrum of Theoretical & Practical Issues

With 108 Figures and 28 Tables

Professor Dr. Paul P. Wang
Department of Electrical &
Computer Engineering
Pratt School of Engineering
Duke University
Science Drive, West Campus
27708 Durham, N. C.
USA
e-mail: ppw@ee.duke.edu

Professor Dr. Etienne E. Kerre
Department of Applied Mathematics &
Computer Science
Ghent University
Krijgslaan 281 (S9), 9000 Gent
Belgium
e-mail: etienne.kerre@ugent.be

Professor Dr. Da Ruan
The Belgian Nuclear Research
Centre (SCK•CEN)
Boeretang 200, 2400 Mol
Belgium
e-mail: druan@sckcen.be

and

Department of Applied Mathematics &
Computer Science
Ghent University
Krijgslaan 281 (S9), 9000 Gent
Belgium
e-mail: da.ruan@ugent.be

ISSN print edition: 1434-9922
ISSN electronic edition: 1860-0808
ISBN 978-3-642-09033-2 e-ISBN 978-3-540-71258-9

Springer is a part of Springer Science+Business Media
springer.com

Softcover reprint of the hardcover 1st edition 2007

Cover design: WMX Design

Contents

Foreword

In order to properly characterize the content of this book, it is important to clarify first the intended meaning of its title *Fuzzy Logic.* This clarification is needed since the term "fuzzy logic," as currently used in the literature, is viewed either in a *narrow sense* or in a *broad sense.* In the narrow sense, fuzzy logic is viewed as an area devoted to the formal development, in a unified way, of the various logical systems of many-valued logic. It is concerned with formalizing *syntactic aspects* (based on the notion of *proof*) and *semantic aspects* (based on the notion of *truth*) of the various logical calculi. In order to be acceptable, each of these logical calculi must be *sound* (provability implies truth) and *complete* (truth implies provability). The most representative publication of fuzzy logic in this sense is, in my opinion, the classic book by Peter Hajek [1].

When the term "fuzzy logic" is viewed in the broad sense, it refers to an extensive agenda whose primary aim is to utilize the apparatus of fuzzy set theory for developing sound concepts, principles, and methods for representing and dealing with knowledge expressed by statements in natural language. Although work in fuzzy logic in the broad sense is not directly concerned with the issues that are investigated under fuzzy logic in the narrow sense, the importance of the latter is that it provides the former with solid theoretical foundations.

After examining the content of this book, it is easy to conclude that its title, *Fuzzy Logic,* refers to fuzzy logic in the broad sense. This is consistent, by and large, with the usual meaning of the term "fuzzy logic" in the literature. Indeed, most papers and books that use the term "fuzzy logic" in their titles or in their lists of keywords are, in fact, dealing with issues of fuzzy logic in the broad sense. Literature devoted to fuzzy logic in the narrow sense, which is lately referred to as *mathematical fuzzy logic* or *formal fuzzy logic,* is only a small fraction of the overall literature on fuzzy logic, and it is primarily literature shared by a relatively small community of researchers working in this area. However, in spite of this rather small visibility of fuzzy logic in the narrow sense, its importance for building foundations for fuzzy logic in the broad sense and for developing well-founded wholesale fuzzy mathematics [2] can hardly be exaggerated.

The publication of this book is very timely and, as I see it, its significance is twofold. On the one hand, it is a testimonial of the current maturity of fuzzy logic (in both senses), which has been achieved within a relatively short period of four decades. On the other hand, the book is an invaluable resource for newcomers interested in utilizing in their own domain areas the many powerful concepts, principles, and methods of fuzzy logic that are now available.

As is well known, the history of fuzzy logic began with the simple idea conceived by Lotfi Zadeh in the early 1960s that classes of objects that do not have precisely defined criteria of membership are prevalent when we deal with real world. To allow mathematics to represent such classes, Zadeh introduced in 1965 in his famous seminal paper [3] the concept of a *fuzzy set*, which abandoned the requirement of classical set theory that each object either belongs or does not belong to a given set. This concept, which on the surface looks very simple, represented in fact an extremely radical intrusion into mathematics, as it was understood at that time, with profound consequences, some of which are examined in this book.

The significance of the simple idea of abandoning the sharp boundaries of classical sets was initially recognized by only a very small group of mostly young researchers. Outside this group, the idea was largely ignored or, worse, it was ridiculed or even attacked, sometimes quite viciously. The new area of fuzzy set theory was initially highly vulnerable since it did not have any defendable foundations or any convincing applications. It took some time to make visible advances in both these areas. Again, it was initially a small group of researchers who made these crucial advances. These were talented people who were able to see earlier than most of their peers the tremendous significance and potential of the deceptively simple idea of sets with unsharp boundaries. It is very appropriate to view them as pioneers who started a grand paradigm shift from classical logic and mathematics to their fuzzy counterparts [4]. This paradigm shift is still ongoing and this book is in some sense an indicator of how far this paradigm shift has progressed.

After reading this book and comparing the various issues discussed in its chapters with the simple idea of a set with unsharp boundaries suggested in 1965 by Zadeh [3], one can hardly fail to recognize that the idea, regardless of or, perhaps, due to its simplicity, has proven to be an extremely profound idea. Considering the huge body of theoretical results and the fascinating spectrum of applications that are described in this book, which all emerged over the period of about forty years from such a simple idea, one can hardly be left unimpressed. Let me mention just some of the most impressive developments in this area that are well captured in this book.

In his seminal paper [3], Zadeh introduced a particular type of fuzzy sets, which are now usually referred to as *standard fuzzy sets*. These are fuzzy sets in which degrees of membership are represented by numbers in the unit interval [0,1]. A number of other types of fuzzy sets are now recognized. They include *interval-valued fuzzy sets, fuzzy sets of type 2 and higher types, fuzzy sets of level 2 and higher levels, L-fuzzy sets, intuitionistic fuzzy sets,* and *rough fuzzy sets* [5]. By and large, these types of fuzzy sets have already been well researched and have found utility in some applications.

Distinct types of fuzzy sets imply distinct types of fuzzy set theories. Moreover, further distinctions within each type result from the choice of operations on the respective type of fuzzy sets. For standard fuzzy sets, the classes of justifiable operations of intersection, union, and complementation have been extensively studied. Distinct choices from the three classes result in distinct algebraic structures and thus in distinct theories based on standard fuzzy sets. In addition, averaging operation of various kinds, which have no meaning in classical set theory, are perfectly meaningful in fuzzy set theories.

A lot of important results have been obtained over the last forty years in the area of *fuzzy systems.* These are systems whose variables, usually referred to as *linguistic variables*, range over states that are fuzzy sets. Representing states of variables by appropriate fuzzy sets, which is called a *granulation*, is a generalization of the classical notion of quantization. The primary aim of fuzzy systems is to deal mathematically with knowledge expressed by statements in natural language. Closely associated with fuzzy systems is the area of *approximate reasoning*, which has been a subject of intensive research that has produced a large body of very impressive results. Fuzzy systems are also closely connected with the area of *fuzzy relations*, which is now quite well developed. In this area, the concept of *fuzzy relation equations* is particularly important for dealing with the various inverse problems associated with fuzzy systems.

An important area, which emerged from the simple idea of a standard fuzzy set, is *fuzzy mathematical analysis.* Its roots are in the concepts of a *fuzzy interval*, which is a generalization of the classical concept of an interval of real numbers, and *fuzzy arithmetic*, which is a generalization of classical interval arithmetic. Employing these two concepts, further investigations into fuzzy counterparts of the various concepts of classical mathematical analysis (continuity, differentiation, integration, etc.) led, for example, to the powerful tool of *fuzzy differential equations.*

Mathematical analysis is just one area of classical mathematics that has been generalized to its fuzzy counterpart or, using the more common term, which has been *fuzzified.* By now, almost all areas of mathematics have been fuzzified: topology, geometry, various algebraic structures, graphs, hypergraphs, finite-state automata, cellular automata, formal languages, etc. While these many results of fuzzification are quite impressive and useful in selected applications, they are based, by and large, on fuzzifying only some aspects of each of these areas of mathematics. The previously mentioned research program outlined by Behounek and Cintula [2], which is based on fuzzy logic in the narrow sense, seems to be the right approach to developing fuzzy mathematics in a comprehensive way, so that all that can be fuzzified in the various areas of mathematics will actually be fuzzified.

I could go on and on to mention many additional results contained in the rapidly growing literature on fuzzy logic in the two senses, but I consider the above mentioned results sufficient for illustrating the profound implications that resulted from the simple idea of a set with unsharp boundaries that was conceived some forty years ago by Lotfi Zadeh. As I previously mentioned, the timely publication of this book is a testimonial of these implications. The editors should be congratulated on selecting a representative blend of theoretical and application oriented topics and inviting the right authors to cover them. It is not surprising that quite a number of these authors were forty years ago among the pioneers who had the talents to recognize the significance of the emerging new area.

References

1. P. Hajek, *Metamathematics of Fuzzy Logic* (Kluwer, Boston, 1998).
2. L. Behounek and P. Cintula, From fuzzy logic to fuzzy mathematics: a methodological manifesto. *Fuzzy Sets and Systems* **157** (2006) 642-646.
3. L. A. Zadeh, Fuzzy sets. *Information and Control* **8** (1965) 338-353.
4. G. J. Klir, From classical sets to fuzzy sets: a grand paradigm shift. In: P. P. Wang, editor, *Advances in Fuzzy Theory and Technology,* Vol. 2 (Duke Univ. Press, Durham, NC, 1994) pp. 3-47.
5. G. J. Klir, Foundations of fuzzy set theory and fuzzy logic: a historical overview. *Intern. J. of General Systems* **30** (2001) 91-132.

October 2006

George J. Klir
Binghamton University (SUNY)

Preface

It is definitely not our intention to publish yet another edited volume on fuzzy logic as there are numerous volumes already available in the literatures. Honestly, we not only wish to produce a volume, which is substantially different from what is already available, but also to devote an urgently needed one!

Inspired by George Klir, who has contributed a lot in bringing to public's attention the widely applicability of fuzzy logic to new frontiers, we feel we should join in and extend this task covering even more application areas. This is what we have envisioned, a newly designed edited volume that is very focused on four specific objectives:

(1) A collection for someone who would like to find out what is fuzzy logic? Why fuzzy logic? Where can fuzzy logic methodologies make a significant contribution?
(2) A volume for someone who is already a member of the fuzzy logic research community and who wants to gain some frank opinions on the cutting-edge research problems. What directions may yield the greatest advantages via future research efforts?
(3) A collection for the program directors of funding agencies to decide where to put money for profitable returns? The kind of information would help to determine a possible maximum yield?
(4) For scientific and mathematical philosophers, this volume will provide as an accumulated knowledge base as to the state-of-the-art of fuzzy logic research as they evaluate the accuracy of their forecasting.

Obviously, this project represents a snap shoot of the active on-going research. Indeed we realize the enormous boom of fuzzy logic based papers, theoretical as well as practical. Some figures to illustrate this boom: up to now more than 47,000 papers with "fuzzy" in the title have been counted in the INSPEC database and MATHSCINET reports about 14,000 mathematical papers with "fuzzy" in the title. The practical impact of fuzzy logic can be seen from the huge numbers of fuzzy-logic related patents included in Japan

(4,800) and US (1,700) with applications in almost all domains ranging from industrial and quality control over automobile industrial and steel manufacturing to financial engineering and economics. It may be advisable that we do this at least once every decade.

To accomplish our much focused objectives, three of us went out to solicit the leaders of the fuzzy logic research community with diversity mix in the back of our mind. The results, we must say, have been a more than satisfactory yields. This is particularly rewarding because the researchers asked were very enthusiastically supportive of our efforts.

Even though we are gratified with the response, we did not stop there! We then speared head for filling the gaps of those research areas not covered by the crops. In particular, two PhD dissertations from two leading U.S. universities, one in geology and the other in micro-biological informatics, are of some significance.

In this way we reinforced the well-known fact that fuzzy logic is not only important in those known application areas, but it also breaks the ground of the unknown territories. The value of this volume is further enhanced by the excellent 'foreword' written specifically by George Klir. Readers are strongly urged to read George's foreword before embarking the details of Chapter one entitled, appropriately, 'Why fuzzy logic?'

At very outset, we feel strongly to make more observations, commentary, and suggestiveness or even down right controversy. To this end, we created an introductory chapter as Chapter one in the beginning of the book. By doing so, we have conserved our preface to a normal length of an edited volume.

Chapter one will be devoted to some detailed introduction to each of the ten covered technical areas. It also serves to link all eighteen chapters as a cohesive volume. We wish our readers to appreciate the fact that it is not always possible to twist a researcher's arm to write a specific chapter in order to fit in our need of serving our research community, primarily due to a researcher's busy scheduling. To single out only one example, fuzzy logic has been widely used in solving economical problems over the past two decades and the most qualified author is not available during this period of time and we must wait!

Finally, we must thank all the participating authors for their contributions. For a large number of reviewers who also contributed their valuable time unselfishly. To George Klir, whose foreword has made this volume much more worthy!

December 2006

Paul P. Wang, Duke University, NC, USA
Dạ Ruan, The Belgian Nuclear Research Centre (SCK•CEN) and Ghent University, Belgium
Etienne E. Kerre, Ghent University, Belgium

List of Contributors

Piero Baraldi
Politecnico di Milano Dipartimento di Ingegneria Nucleare Via Ponzio 34/3 20133 Milan Italy
piero.baraldi@polimi.it

Louis R. Bartek
Department of Geological Sciences University of North Carolina at Chapel Hill Mitchell Hall Campus Box 3315 Chapel Hill NC 27599-3315
bartek@email.unc.edu

Mustapha Baziz
IRIT-UPS 118 route de Narbonne - 31062 Toulouse Cedex 9
baziz@irit.fr

Jorge S. Benítez-Read
National Nuclear Research Institute of Mexico (ININ) Toluca Institute of Technology (ITT) CINVESTAV Fac. de Ing. UAEM México
jsbr@nuclear.inin.mx

Bernadette Bouchon-Meunier
Université Pierre et Marie Curie-Paris6 CNRS UMR 7606 LIP6 8 rue du Capitaine Scott Paris F-75015 France
Bernadette.Bouchon-Meunier@lip6.fr

Mohand Boughanem
IRIT-UPS 118 route de Narbonne - 31062 Toulouse Cedex 9
bougha@irit.fr

Yingjun Cao
Department of Mathematics & Computer Science North Carolina Central University 1801 Fayetteville Street Durham NC 27707
ycao@nccu.edu

Bernard De Baets
Dept. of Applied Mathematics Biometrics and Process Control
Ghent University
Coupure links 653
B-9000 Gent Belgium
Bernard.DeBaets@ugent.be

Robert V. Demicco
Department of Geological Sciences and Environmental Studies Binghamton University
PO Box 6000 Binghamton
NY 13902-6000
demicco@binghamton.edu

Marcin Detyniecki
Université Pierre et Marie Curie-Paris6 CNRS UMR 7606 LIP6
8 rue du Capitaine Scott Paris
F-75015 France
marcin.detyniecki@lip6.fr

Tharam Dillon
Faculty of Information Technology
University of Technology Sydney
POBox 123 Broadway NSW 2007
Australia
tharam@it.uts.edu.au

János Fodor
John von Neumann Faculty of Informatics Budapest Tech Bécsi út 96/b H-1034 Budapest Hungary
Fodor@bmf.hu

Daniel Gómez
Faculty of Statistics Complutense University Madrid 28040 Spain
dagomez@estad.ucm.es

Siegfried Gottwald
Universität Leipzig Institut für Logik und Wissenschaftstheorie
Beethovenstr. 15 04107 Leipzig
Germany
gottwald@uni-leipzig.de

Sergio Guadarrama
Departamento de Inteligencia Artificial
Universidad Politécnica de Madrid
28660 Boadilla del Monte. Madrid
Spain
sguada@dia.fi.upm.es

J. Humberto Pérez-Cruz
Toluca Institute of Technology (ITT)
CINVESTAV Fac. de Ing.
UAEM México

Etienne E. Kerre
Department of Applied Mathematics and Computer Science Ghent University Krijgslaan 281 (S9) 9000 Gent Belgium
Etienne.Kerre@UGent.be

Marie-Jeanne Lesot
Université Pierre et Marie Curie-Paris6 CNRS UMR 7606 LIP6
8 rue du Capitaine Scott Paris
F-75015 France
marie.jeanne.lesot@lip6.fr

Zhong Li
Faculty of Electrical and Computer Engineering FernUniversität in Hagen 58084 Hagen Germany
zhong.li@fernuni-hagen.de

Victoria López
Computer Engineering
Antonio de Nebrija University
Madrid 28040 Spain
mlopez@nebrija.es

Yannick Loiseau
IRIT-UPS 118 route de Narbonne - 31062 Toulouse Cedex 9
loiseau@irit.fr

Jie Lu
Faculty of Information Technology
University of Technology Sydney
POBox 123 Broadway NSW 2007
Australia
jielu@it.uts.edu.au

André Maïsseu
Professeur IAE de ¨Paris
Université de Paris
1 "Panthéon-La Sorbonne"
France
andre.maisseu@wanadoo.fr

Benoît Maïsseu
Ecole Supérieure d'Electricité "SUPELEC" Gif sur Yvette France
benoit.maisseu@renault.com

Michael Margaliot
School of Electrical Engineering
Tel Aviv University
Tel Aviv 69978
Israel
michaelm@eng.tau.ac.il

Christophe Marsala
Université Pierre et Marie
Curie-Paris6
CNRS UMR 7606 LIP6
8 rue du Capitaine
Scott Paris F-75015 France
christophe.marsala@lip6.fr

Javier Montero
Faculty of Mathematics Complutense
University
Madrid 28040
Spain
monty@mat.ucm.es

Vilém Novák
University of Ostrava
Institute for Research
and Applications of
Fuzzy Modeling 30.
dubna 22 701 03 Ostrava 1
Czech Republic
and Institute of Information
and Automation Theory Academy of
Sciences of the Czech Republic
Pod vodárenskou věží 4
186 02 Praha
8 Czech Republic
Vilem.Novak@osu.cz

Ibrahim Ozkan
Department of Economics Hacettepe
University Beytepe Ankara
Turkey
ibrahim.ozkan@utoronto.ca

Witold Pedrycz
Department of Electrical & Computer Engineering University of
Alberta
Edmonton T6R 2G7
Canada
and Systems Research Institute
Polish Academy of Sciences
Warsaw
Poland
pedrycz@ece.ualberta.ca

Irina Perfilieva
University of Ostrava Institute for
Research and Applications of Fuzzy
Modeling 30.
dubna 22 701 03 Ostrava
1 Czech Republic
Irina.Perfilieva@osu.cz

Henri Prade
IRIT-UPS 118 route de Narbonne -
31062 Toulouse Cedex 9
prade@irit.fr

Ana Pradera
Departamento de Informática
Estadística y Telemática
Universidad Rey Juan Carlos 28933
Móstoles. Madrid. Spain
ana.pradera@ urjc.es

Eloy Renedo
Departamento de Inteligencia
Artificial
Universidad Politécnica de Madrid
28660 Boadilla del Monte.
Madrid. Spain
erenedo@dia.fi.upm.es

Maria Rifqi
Université Pierre et Marie Curie-
Paris6 CNRS UMR 7606 LIP6 8 rue
du Capitaine
Scott Paris F-75015 France
maria.rifqi@lip6.fr

Da Ruan
Belgian Nuclear Research Centre (SCK•CEN)
Boeretang 200 2400 Mol Belgium
druan@sckcen.be

Alade Tokuta
Department of Mathematics & Computer Science North Carolina Central University
1801 Fayetteville Street
Durham NC 27707
atokuta@nccu.edu

Enric Trillas
Departamento de Inteligencia Artificial
Universidad Politécnica de Madrid
28660 Boadilla del Monte. Madrid Spain
etrillas@fi.upm.es

I.B. Turksen
Department of Industrial Engineering
TOBB ETU Sogutozu Ankara Turkey
bturksen@etu.edu.tr

Paul P. Wang
Department of Electrical & Computer Engineering Pratt School of Engineering Duke University
Durham NC 27708 USA
ppw@ee. duke.edu

Jeffrey D. Warren
North Carolina Division of Coastal Management 1638 Mail Service Center
Raleigh NC 27699-1638
jeff.warren@ncmail.net

Guangquan Zhang
Faculty of Information Technology
University of Technology Sydney
POBox 123 Broadway NSW 2007 Australia
zhangg@it.uts.edu.au

Xu Zhang
Key Laboratory of Systems and Control Academy of Mathematics and
Systems Sciences Academia Sinica Beijing 100080
and
Yangtze Center of Mathematics
Sichuan University
Chengdu 610064
China

Enrico Zio
Politecnico di Milano Dipartimento di Ingegneria
Nucleare Via Ponzio 34/3
20133 Milan Italy
enrico.zio@polimi.it

Why Fuzzy Logic? – A Spectrum of Theoretical and Pragmatics Issues

Paul P. Wang, Da Ruan, and Etienne E. Kerre

Abstract. This chapter briefly summarizes the contents of the whole book. The research survey presents a snap shot of the state-of-the-art of the active on-going research of the current maturity of fuzzy logic. In addition, the editors trace the origin and the evolution of fuzzy logic, in which the vagueness philosophy of William James is highly relevant. Finally, the editors comment, critical review and recommend the strategy for the future growth of this young theory and technology. The most significant part of this book project is to emerge the society of uncertainty. This volume covers all the grounds of the past, present and the future so far as fuzzy logic is concerned.

1 Introduction

According to Professor Sir G. Elliot Smith, Invention is not an isolated phenomenon in the history of civilization [1]. Even the simplest advance represents the interweaving of many threads of knowledge that took centuries or thousands of years to spin. The enormous complexity of the process and the fact that a progressive development is built on the foundations of the accumulated knowledge of the whole world of civilization are fatal to the common opinion that significant inventions can be made independently. A case in point, the subject matter of this volume dealing with fuzzy logic is no exception. Hence at the very outset, we would like to claim that despite of the enormous literature available at present on fuzzy logic, it just started the study of its very foundations and that important application field such as the highly relevant discipline of psychological research and its relations is of particularly lacking.

Historically, the probability and statistics researchers have persistently raising such question as to the problems solvable by fuzzy logic can also be solved by probability and statistics. Such a misinterpretation and misunderstanding are indeed unfortunate. The irony is that even for probability and statistics' development and evolution it has taken them nearly 350 years to answer some fundamental questions.

In fuzzy logic research community, some feel somewhat frustrated that the speed of the development in fuzzy logic is not fast enough. To find out at least some partial answers to this frustration, through a recent research in mathematical history, Wang has concluded that the development and the evolution of fuzzy logic as compared with that of the probability and statistics actually is quite normal. [2]. The study finds out that the methodology of probability and statistics only became a 'mainstream in the sense of viable economics' after 350 years of evolution. The correct status of fuzzy logic research, at present, is called 'mainstream in the sense of viable research activities.'

2 Theoretical Issues

Pradera et al. analyze some of the main issues involved in the construction of fuzzy set theories in Chap. 2. Based upon the arguments of this chapter, even though we do have well agreed about t-norms and t-conorms standard solutions, it is still necessary to investigate some alternative fuzzy connectives and their properties.

In Chap. 3 Fodor and De Baets have investigated the so-called uninorms as a generalization of the t-norms and t-conorms since they allow for an arbitrary element of [0, 1] as a neutral element.

The works of the above two chapters reflect the foundations of the mathematics of uncertainty that has yet to be cast in stone! This also says that most mathematics of uncertainty, especially fuzzy logic is very much an active research area.

Fuzzy relations have always been the main concept and hence they play an instrumental role in the world of mathematics of uncertainty because they soften the rigid requirement and very tight constraints in traditional mathematics. Pedrycz has shown to us in Chap. 4 that the problem of structural interpolation and approximation has been realized in the framework of fuzzy relations. The key issue of this chapter reflects the essence of the mathematics of uncertainty of which the ultimate answers may not be unique in final evaluations.

In everyday life, chaos is a state of complete disorder in which the normal laws of cause and effect seem to be abandoned. The intriguing thing about the phenomenon scientists call chaos is that it involves completely unpredictable behavior which follows from absolutely predictable deterministic laws [3]. The connection between fuzzy logic and chaos theory has been known to fuzzy researchers for quite sometime now and Li and Zhang have examined this phenomenon further in Chap. 5. The interesting aspect of this issue is that fuzzy logic naturally relates both to chaos phenomena as well as human reasoning. Hence, it adds another dimension so far as the business of information processing is concerned.

More importantly, this thesis also launches the curiosity into an orbit that we now know that there is no unique way to generate a chaos. In fact, Chap. 12

will discuss the generation of a new type of chaos by the so-called "Mutable Fuzzy Logic Networks."

It is difficult not to mention that in the world of information processing, chaos exists not only all over in nature, it is also a visible phenomenon in animal's brain and the regulating activities of the Genome networks [4].

The last theoretical issue, but not the least, can be found in Ozkan and Turksen's Chap. 6. Semantics issues have been brought into this chapter. It is true that we have known very well that we are in our very existence in this world full of fuzziness and uncertainty of many kinds, the next question to be raised is that how one should and can do a best job to measure it? Chap. 6 does suggest a pragmatic way to do just that! On the other hand, one should also take notice that the very foundations of such measurement can be argued as being fuzzy!

Mathematics of fuzziness, one important example of the rigorous mathematics of uncertainty has been flourished from the very early evolution of the fuzzy set theory introduced in [5]. The whole theory has been built upon rigorous classical mathematics. It has to be a joy for genuine mathematicians that they got the opportunity to work on the generalization of this classical mathematics, although some establishments may think this is not necessary. Nevertheless, a community of fine mathematicians was established and with moderate expansion over the years. The philosophical arguments among mathematicians are, however, more troublesome because not all mathematicians have agreed to softening the mathematics. Softening means softening in the basic hypothesis and modified axioms. Very often, the softened mathematics means the final answer may not be unique. A mathematician believing that all answers must be unique will have difficult time to accept such results. On the other hand, the real world is full of uncertainty and vagueness. Should one push these philosophical arguments further, to the end, one must answer the bottom line question "Then, what is mathematics good for? If it does not solve real problems? It is clear that this debate will most likely to continue for many years to come! During JCIS 2005, the Joint Conference on Information Sciences, more than 100 researchers participated the debate and indeed has concluded that the nomenclature of "Mathematics of Uncertainty" appropriately reflects the state of the affair that a major paradigm has been shifted. The significance of such a shift is that the new mathematics, the mathematics of uncertainty, will have done whole lots of good such that all academic units will be benefited from using the mathematics of uncertainty.

The major publication of the SMU, Society for Mathematics of Uncertainty, was launched to serve for the movement! New Mathematics & Natural Computing has been designed as a long lasting vehicle for this particular movement.

So far as the fuzzy mathematics is concerned, we believe that fuzzy mathematicians should go on the business of concentrated effort to develop fuzzy mathematics, entirely free and independent of the application domains. The main job for fuzzy mathematicians is to search for the truth, nothing but the truth! An excellent survey and reviewing article about the historical development of the fuzzy mathematics has been written by Kerre and Mordeson [6].

Before we leave the Sect. I on theoretical issues, we must bring up a very important issue which is not represented by the chapters of this volume. This issue, the study of the evolution and origin of the mathematics of uncertainty is certainly of paramount importance. Unfortunately, the philosophy has not been embraced by most members of the fuzzy logic research community. We feel contrarily, we must study its evolutionary process as we said this at the very beginning of the Sect. I.

As it turns out, the philosophical bases of the mathematics of uncertainty is the principle of vagueness. For many years, the philosopher and founder of American psychology, William James, is the champion of the principle of vagueness [7]. The vagueness has subsequently been embraced by a group of excellent philosophers who have done wonderful research works [8]. The communications among philosophers and mathematicians deserves much more attention. We will not devote more pages to discuss this important desired improvement, but we feel strongly that the historical perspective is of immense importance.

3 Pragmatics Issues

As odd as it sounds, it is the pragmatics issues that will reveal the strength or weakness of a theory. This section serves to explore the issues associated with different domains of the application areas of fuzzy logic.

3.1 Mathematical Modeling

Appropriately, the first application area of fuzzy logic deals with mathematical modeling. Chap. 7, as it turns out, is a good example of using fuzzy logic as a vehicle for mathematical modeling. Margaliot correctly argues that human observers very naturally use verbal descriptions and explanations of various systems. Specifically, he illustrates two examples on mimicking a natural behavior -the field of ethology-; one has to do with the territorial behavior of fish and the other has to do with the orientation to light of a flatworm.

As a matter of fact, there are plenty of examples before Margaliot's thesis making the use of fuzzy logic as a means of modeling. During 1980–1990's, the applications of fuzzy logic in mathematical modeling have been wide-spread.

The underlining problem is this: should one find a differential equation to model the behavior of a dynamic system, one should never hesitate to do just that. Why? The solution of a dynamic system, based upon the theory of differential equations, will be unique for a given set of initial conditions as long as the system is linear. The reason we reach at using fuzzy logic is due to the fact the dynamic system will have no solution at all, unless we use fuzzy logic. Most likely, fuzzy logic is used because the system is nonlinear and to have a suboptimal solution is much better than no solution at all!

As expected, this kind of solution still faces a serious challenge of uncertainty in the sense that such a solution may not be unique! This is caused by the process of approximation. Even in reasoning, that is the reason why we call it approximate reasoning.

Surprisingly, the rule-based or natural language based approach, luckily, added yet another dimension to the solution because now you have gotten a humane touch. Only as hindsight, we realize that the modeling using fuzzy logic has potential of solving so many problems in social and biological sciences.

3.2 Natural Language

The next pragmatics issue we single out to be examined is the problem of the natural language. Novak's work in Chap. 8 primarily alludes to the calculus of fuzzy logic rather than the modeling of the semantics. Nevertheless, fuzzy logic has given the rise of the hope that the modeling and handling of the vagueness of the natural language might eventually have some reasonable solutions.

Number one issue in language research has been how one can cope with the vagueness of a language. This indeed is a huge important issue as explained by William James as his reinstatement of the vague [7]. James stated in his principle "language works against our perception of the truth." He argues that we all take language too much for granted.

We all assume that each word has one meaning and that when the word is used in n numbers of sentences, the meaning is the same. As a matter of fact, a language is not an objective copy of the reality. Semantics issue has always been a very difficult problem to tackle.

Probing further the second issue would be the question of how to model the language quantitatively, this issue, as it turns out to be much of the research initiated by Zadeh's so-called Computing with Words.

Decades of research in the design of expert systems ended up with the question how to solve the problem in linguistics. We are basically stalked because of our inability of making major progress in linguistics research. That is the reason why we said at the very outset that the natural language problem is a huge problem without a trace of exaggerating.

Fuzzy logic may not solve all the problems; at least it gives us hope. The next question would then be that researchers must have the will and the resources to tackle this huge problem.

3.3 Data Mining and Knowledge Engineering

This is a topic of our current interest among computer scientists and many fields of intensive applications. The collections of Chap. 9–11, as it turns out, have more to do with theory rather than some specific applications. Perfilieva's contribution in Chap. 9 has been the development of some analytical theory of fuzzy IF-THEN rules where all three familiar rules of inference can work properly. The theory and more technical know-how as needed in order to design knowledge engineering systems have also been tackled by her.

The advantage of using fuzzy logic in ontology-based information is explored by Baziz et al. in Chap. 10. The main advantage of using fuzzy logic, as advocated by the authors is its capability of performing some uncertainty measurements, which is normally lacking when one has to deal with natural language. There is no surprise here for the arguments.

The last of the group, Chap. 11 by Bouchon-Meunier et al. focuses on the real world applications of fuzzy logic for information retrieval and data mining. Another real world application in this chapter covers a wide range of medical, educational, chemical, and multimedia. Once again, the authors were able to conclude that a marked advantage is possible due to the application of fuzzy logic.

To sum up this trio, one reveals the research in data mining and knowledge engineering has a similar goal, namely, ultimately, we got to have effective expert systems. However, both data mining and knowledge engineering are undoubtly the basic building blocks for an expert system. We will never leave the fundamentals behind. Yet, the expert systems will never be successfully built unless we solve the linguistics problem first. It is interesting to observe the link of this application area to the last one when we discussed the problem of natural language. As a matter of fact, the pragmatics philosophy of William James, at least in America, has deep special meaning involving so many fundamental areas of knowledge [8].

Even we focus only on the short term issue of data mining technology as applied to two very hot applications nowadays, economics and security; we leave this topic with the impressions that this indeed is a technology which is completely necessary.

3.4 Bioinformatics

There are several features which make this volume distinctly different from the other volumes published by the fuzzy research community. This one, Bioinformatics, has been missing in previous ones.

Fuzzy logic has been utilized in the genome regulation networking research before by invoking its 'approximate reasoning' capability, but not in more fundamental issues of uncovering the mechanism of how they do regulates. The theory first proposed by Stuart Kauffman more than three decades ago remains to be the most accepted model in gene regulation networks. Even though the Boolean network is more general than the cellular automata theory as introduced by John von Neumann as a model for biological research [9]. Yet, the Boolean network, especially the (N,K) model is only capable of providing the symbolic information. Cao et al.'s Chap. 12 is most likely a significant contribution in mathematical modeling of biology, because the model of 'Mutable

Fuzzy Networks' by far is the most complete mathematical model in the sense it absorbs all the essential features of the genome regulation networks.

Historically speaking, it was Bertrand Russell who unified the philosophical logic and the rigorous mathematics into the elegant mathematical logic that is known today [10]. It is not surprising that the biological modeling finds the right tool in Mutable Fuzzy Logic as first proposed by Wang. We cannot resist the temptation here to remind our readers that William James' pragmatics principle and the principle of vagueness has many dimensions in applications [11], hence, biology is yet to be another example of his powerful principle.

The principle of vagueness is a philosophical notion and the mathematics of uncertainty is a tool, which will help to modeling and manipulating the vagueness. It is our strong belief that two major academical areas ultimately will benefit more than other areas such as social sciences and biological sciences.

Before we leave the application area of bioinformatics, a discussion in the area of funding policy is in order; the super computing cost of this particular research project has an expenditure of reaching seven figures in terms of the U.S. Dollars. This is also served as an example that one stand out feature of the mathematical biological research, we are afraid, is the phenomenon of being too complex! A research project, especially dealing with the problem of complexity necessarily requires adequate funding. Unfortunately, this inadequate funding has always been a handicap in the fuzzy logic research community.

Readers who are interested in Chap. 12 may consult the Chap. 9 of the book by Martin A. Nowak entitled "Evolutionary Dynamics—Exploring the equations of life" [12]. Perhaps use the Chap. 9 of the book [12] as a prerequisite matter, especially the spatial chaos, dynamic fractals and evolutionary kaleidoscopes.

3.5 Fuzzy Logic Control

The general mathematical problem of fuzzy control is identified by Gottwald in Chap. 13 as an interpolation problem. The interpolation is definitely needed because this is the only way to solve the nonlinear problem, which has been abundant in control engineering problems. However, as it will become clearer later, the more important problem in control systems design is the more general rule based formulation in fuzzy logic and furthermore, which does not limit to only the IF ... THEN ... rules. More attractive property of fuzzy logic in control, as it turns out, is the 'approximate reasoning' power.

No surprisingly, Gottwald is addressing some general sufficient conditions for the solvability of the interpolation problem, as well as similar conditions for suitably modified data. It is very nice for him to treat these issues in a very rigorously manner.

Chapter 14 by Benitez et al. presents an innovative design by integrating state variables (a state vector) the first order numerical integration for state gain estimation, and fuzzy logic to attain a suitable power regulation in the nuclear reactors.

The robustness characteristic and properties of fuzzy logic is, once again, demonstrated in this design of workable control policy or algorithms. However, the overall picture that deals with the role of fuzzy logic cannot be completed without further supplementary explanation which we will offer some comments in the following paragraphs.

The evolution of the fuzzy logic control itself, as it turns out, is a fascination story on its own right. It has been recognized for a long time that fuzzy logic is useful to solving control problems. Zadeh and Bellman further stated that the usefulness stands from two fundamental characteristics of fuzzy logic; one is the modeling of nonlinear dynamics and the other is the 'approximate reasoning' capability. In fact, the very first wave of the success of fuzzy logic came from the Japanese researchers 'applications of the theory in solving consumer electronics applications in the 1970's. Quite large specific applications include rice cooker, washing machine, vacuum cleaner, camera focusing, and the like. This was the beginning of a surge of many applications in solving control problems up to today that all cars in the world are utilizing this technology for one kind of application or the others. This very technology is one of the main reasons responsible for the birth of so-called intelligent control, even the autonomous control.

Clearly, two stand-out phenomena have been acutely observed; the approximate reasoning capability and the nonlinear modeling of the dynamics. The irony is that all these applications in the automobile industry have never used the nomenclature of fuzzy logic, but the artificial intelligence instead. Unfortunately, there is a somewhat negative impact due to the lack of far reaching vision and the technology has been hyped to solve nearly all unsolved control problems. It may be possible that the pioneering applications in Japan in particular have been carried out by control engineers who were prone to have such belief carried over from the success of consumer applications. At the height of such wave of surging, it has been not uncommon to have several hundreds of papers in fuzzy control of which, unfortunately, many of them, as it turns out, to reinventing the wheels. One example would be the fuzzy controllers designed were comparable to the performance of the conventional PID controllers. This phenomenon could have been avoided if we do have a power society to assume some leadership in pointing out a clear direction where the technology should have been headed!

This misled hype did have adverse impact, at least in several academic programs in leading research countries. Nevertheless, a lesson learned may have a very positive impact in the sense that it serves as a wake up call for the establishment of some pragmatics and functional society for the community. This will be reflected in our concluding remarks.

3.6 Pattern Recognition

Next we will turn our attention to a topic which possesses huge potential for our future research, pattern recognition. Fuzzy logic is a natural fit as an important tool for pattern recognition. In fact, one of the very first applications, as early as 1969, was pointed out by Bellman and Zadeh in Information Sciences journal.

Zio et al. have demonstrated yet another application of fuzzy logic in pattern recognition in Chap. 14. Their problem is identification in complex engineering systems as related to faults and anomalities, needless to say, is a natural fit! This chapter, in addition, re-enforces our belief that pattern recognition is instrumental to the research of data mining and knowledge engineering described in previous sections.

More fundamentally, pattern recognition is closely related to the study of human brain functions – one of the most important endeavors in this new century. Pattern recognition is very useful to perceptions. Furthermore, perceptions are an act of creation that involves many distributed brain regions and discovering the network interactions among these regions is important for understanding a range of issues of several basic sciences. It is safe to predict that the pattern recognition is also the most important contribution of the computational intelligence to the progress of these basic sciences.

3.7 Decision Making

Pattern recognition is also intimately related to the decision making issues. Decision making has huge implications in the world of economics and business applications. Our ability to make decisions is needed everywhere at anytime. The usefulness of the mathematics of uncertainty is fully demonstrated beyond any doubt in this especially important area.

Montero et al. said it very well in Chap. 16 that the true objective of fuzzy decision making models should focus their attention on decision processes itself. While the probability theory can properly model crisp acts, this is not the case for fuzzy decisions. Zhang et al., on the other hand, look at a more complicated structure on decision making in Chap. 17. Their arguments are: when a decision maker at the upper level (the leader) attempts to optimize an objective, the decision maker at the lower level (the follower) tries to find an optimized strategy according to each of the possible decisions made by the leader. To achieve their objectives, they suggest a bi-level programming that provides a means of supporting two levels of cooperative decision making.

It is not necessary to remind our readers that the research areas in decision making are indeed huge and we can only barely show two aspects of the issues in decision making problems as two samples of pressing issues.

3.8 Geology

Warren contributes immensely to this volume by summarizing his PhD dissertation from University of North Carolina in Chap. 18. Since nonlinear phenomena exist in every field of research, it is not surprising that geological problems will use fuzzy logic. It is nice to see the research projects of which the main thrust in solving geological problem are emerging more in the Western nations; it is not a surprise at all in Chinese language literatures. In fact, one can discover many seminal works in this geological research in China for many years.

Warren's work covers a wide range of problems essential to geological research. It is indeed rewarding for this volume to be able to include Chap. 18 that summarizes a wide spectrum of potentially useful research findings.

3.9 Management and Operations Research

The last area, certainly not the least, to be discussed in this volume is Chap. 19 by Maisseu and Maisseu in which they describe how to manage the complexity and diversity of the parameters influencing consumer decisions. Note that fuzzy logic plays a multiple role in this application. To say the very least, the authors' idea is very interesting in demonstrating the applicability of the theory yet in some virgin territory of which the problem solving tools and the problems are a natural fit.

The degree of enthusiasm of the fuzzy logic research in many academic disciplines has been a real riding in roller coasters, but one area to be singled out is the embrace of the technology by industrial engineering and operations research. It has never experienced any declinations. In fact, this academic remains as the brightest academic departments so far as the ongoing research activities in the fuzzy logic research.

We have indeed presented in this volume a wide spectrum of the applications of the fuzzy logic theory. Even for these ten areas of applications, they themselves are intertwined with various patterns of the mixture. We have stressed the importance of the foundations and the pragmatics philosophy of William James. To apply this powerful theory to solving the real world problems, we must conclude that the concept of vagueness is a reality that we must cope with it!

We use Jonathan Lawry's quotes to reemphasize this basic concept from his book entitled "Modeling and Reasoning with Vague Concepts." Jonathan uses label semantics to state his findings as follows [13]:

Vagueness is central to the flexibility and robustness of natural language descriptions. Vague concepts are robust to the imprecision of our perceptions, while still allowing us to convey useful, and sometimes vital, information. The study of vagueness in Artificial Intelligence (AI) is therefore motivated by the desire to incorporate this robustness and flexibility into intelligent computer systems. Such a goal, however, requires a formal model of vague concepts that will allow us to quantify and manipulate the uncertainty resulting from their use as a means of passing information between autonomous agents.

4 Concluding Remarks

As we said in the very beginning, this project has provided an excellent opportunity for us to take a snap shoot on the state of the research activities concerning the fuzzy logic research community. The collections in this volume, however, do not represent even the average of the current state of research activities, only a snap shoot.

Nevertheless, the collections will be proved to be invaluable in some other objectives such as an exercise of designing a strategically plan for rapid future growth of the technology. In other words, how can the fuzzy logic research community utilize their very limited and scare resource in order to have optimal growth under many constraints?

This last session serves to summarize our findings for a strategical plan. One way, as we decided to do this, is state in a very concise manner the following conclusions:

(1) Society for Mathematics of Uncertainty, which was found one year ago with the collective wisdom of more than one hundred chartered members in Salt Lake City, can be the core society which will glue us together. Even though some may not agree, but we have the past experience and models which did not really work very well so far. For a period of six years, the Japanese government invested fairly sizable funds to promote the fuzzy logic research. This event has been extremely exciting; unfortunately, it has only a short life span. Many other opportunities coming and go for several countries eventually proved to yield similar results due to the inevitable nationalists' feelings and pride involved. Hence what we have hoped for is a wake-up call that we really need a global organization which will step out such a pitfall. It becomes clear to us that a fresh approach is badly needed. This brings out the second major point.

(2) Fuzzy Logic Foundation: If indeed a major funding from some government appears to be unlikely, then the best approach is the creation of an FLF, Fuzzy Logic Foundation on global basis. Perhaps the time has finally arrived with a surge of the establishments of various private foundations via United Nations or other meaningful organizations.
We believe it is a good idea for the call for an FLF, Fuzzy Logic Foundation on global basis.

(3) In order to accomplish the above objective, great deals of the preparation works are needed. SMU, Society for Mathematics of Uncertainty, can assume such a role to perform the needy tasks. Our final paragraph will be entirely devoted to discuss what tasks lay ahead and we will leave the lists of tasks needed as a forum of public debate within or out of fuzzy logic community.

(4) Tasks lie ahead: As expected, the tasks which are urgently needed would be too numerous to be listed here. What we hope to accomplish is to provide a spark-plug that would jump start the formation process.

We believe strongly that the concept of the mathematics of uncertainty has far reaching impact in all academic endeavors simply because this is exactly what we would need in order to develop any academic disciplines which will solve all the difficult problems such as social sciences and biological sciences.

As to what specific strategically plan that will work well with the current constraints, we believe this is the topic of all our community's concern and should have appropriate forum for debate and discussions.

Nevertheless, we shall leave the discussion here with some suggested examples concerning the tasks that lie ahead. We will now conclude our discussion in chapter one by providing some sample examples concerning the future tasks required.

- Recognizing the huge benefits that can be derived from the research of the mathematics of uncertainty; we must support the young scholars who are currently active in this research field. They will need the voice which will make sure their welfare are guaranteed, such as tenure and retaintion, promotion and job opportunities for their students, etc.
- The production of children books with the concepts of the mathematics of uncertainty, continuing education, public exhibitions, etc.
- To speed-up the processes which will enable for the fuzzy logic technology to become a mainstream in active research, more importantly, to become a main stream in the social and economical sciences.

Acknowledgments

This research survey is partially supported by the bilateral China-Flanders project BIL 011S1105 at Ghent University.

References

1. Smith GE (1946) In the beginning, the origin of civilization. The Thinkers Library, No.29, London: Watts & Co., Second Edition, page 2.
2. Wang PP (2006) The evolution of the fuzzy logic, Report to the Executive Council of SMU, Society for Mathematics of Uncertainty. Dec. 1, 2006, in press.
3. Gribbin J (1999) The Little Book of Science, Barne and Noble Books, New York, page 8.
4. Cao Y and Wang PP (2007) Reverse engineering of the NK Boolean network and its extension–fuzzy logic network. New Mathematics and Natural Computing Journal, Vol.3, No.1, pp 69–87.
5. Zadeh LA (1965) Fuzzy sets, Information and Control. 8, pp 338–353.

6. Kerre EE and Mordeson JN (2005) A historical overview of fuzzy Mathematics, New Mathematics and Natural Computation, Vol.1, No. 1, pp 1–26.
7. Gavin WJ (1992) William James and the Reinstatement of the Vague, Temple University Press, Philadelphia, pp 70–73.
8. Keefe R and Smith P (1996) Vagueness: A Reader. The MIT Press, Cambridge, Massachusetts.
9. von Neumann J (1966) Theory of Self-Reproducing Automata, edited and completed by Arthur W. Burks, University of Illinois Press, Urbana and London, Library of Congress Catalog Card No.63-7246.
10. Russell B (1916) Introduction to Mathematical Philosophy, Routledge Inc., London and New York, Ninth Impression 1956 of the original publication in 1916. Reprinted 1993.
11. Morris L (1950) "William James,–The Message of a Modern Mind," Charles Scribners Sons, New York and London.
12. Nowak MA (2006) "Evolutionary Dynamics—Exploring the Equation of Life," The Belknap Press of Harvard University Press, Cambridge, Massachusetts and London, England.
13. Lawry J. "Modeling and Reasoning with Vague Concepts". Springer Web site: http://www.springer.com/sgw/cda/frontpage/0,11855,4-175-22-86707466-detailsPage%253Dppmmedia%257Ctoc, 00.html

On Fuzzy Set Theories*

Ana Pradera**, Enric Trillas, Sergio Guadarrama, and Eloy Renedo

Abstract. This paper analyzes some of the main issues involved in the construction of fuzzy set theories. It reviews both standard solutions (based on the well-known t-norms and t-conorms), as well as less conventional proposals that provide alternative views on, e.g. the definition of fuzzy connectives and the study of their properties.

1 Introduction

Forty years have gone by since Zadeh's pioneering paper introducing fuzzy sets and fuzzy logic [94]. During this period, numerous papers have been published on fuzzy topics, the field has experienced an enormous growth, and many of Zadeh's seminal concepts have naturally evolved in different directions. In particular a variety of set theories have been defined such as: L-fuzzy sets [46], flou sets [45], type-2 fuzzy sets [61,95], interval-valued fuzzy sets [75], intuitionistic fuzzy sets [12], twofold fuzzy sets [34], fuzzy rough sets [35], vague sets [44] or loose sets [81].

This paper aims to review the main issues related to the construction of fuzzy set theories from the Zadeh's initial conceptions to some newer proposals. It is an extended and revised version of [72], and is organized as follows. Section 2 presents a general discussion of the construction of fuzzy set theories. Section 3 proposes a general definition for the concept of fuzzy set theory, and Sect. 4 recalls the definition and the main properties of one of the most popular and important theories, so-called standard fuzzy set theories. Finally, Sect. 5 deals with some non-standard aspects in the construction of fuzzy theories, such as the use of non-functionally expressible connectives or the study of properties other than the basic Boolean laws. Most of the examples that are used to illustrate these non-standard views are taken from some of our previous works.

---

* This paper is an extended and revised version of [72]
This work has been partially supported by MEC (Spain) under project TIN2005-08943-C02-01
** Corresponding Author

2 Fuzzy Set Theories: General Discussion

As is well known, given a non-empty set X, the set of all its classical subsets, $\mathcal{P}(X)$, endowed with the set operations for intersection ($\cap$), union ($\cup$) and complementation (^{c}), provides the structure $(\mathcal{P}(X), \cap, \cup, ^c)$, which is a Boolean algebra. Table 1 summarizes the basic properties of Boolean algebras, which are valid for any sets $A, B, C \in \mathcal{P}(X)$ (note that some are redundant since they can be obtained by means of the others).

Remember also that the elements of $\mathcal{P}(X)$ can be identified by their characteristic functions $A : X \longrightarrow \{0,1\}$. The structure $(\mathcal{P}(X), \cap, \cup, ^c)$ is then isomorphic to $(\{0,1\}^X, \mathrm{Min}, \mathrm{Max}, 1 - Id)$, where $\{0,1\}^X$ is the set of all the functions defined on X and taking values in $\{0,1\}$.

Fuzzy sets, as introduced by Zadeh in 1965 [94], are just a generalization of crisp sets in the sense that, instead of associating a value in $\{0,1\}$ to each element of the set X, they accept any value in the closed unit interval $[0,1]$. This means that when dealing with fuzzy sets the characteristic functions of crisp sets are transformed into membership functions $A : X \longrightarrow [0,1]$, where the value $A(x)$ indicates the degree to which the element $x \in X$ belongs to the fuzzy set A. Extending the crisp case, Zadeh endowed the set $\mathcal{F}(X) = \{A; A : X \longrightarrow [0,1]\}$ of all the fuzzy subsets of X with the partial pointwise ordering induced by the usual order of $\mathbb{R}$, i.e. for any $A, B \in \mathcal{F}(X)$, he defined:

$$A \preceq B \quad \text{if and only if} \quad A(x) \leq B(x) \text{ for all } x \text{ in } X$$

Table 1. Basic properties of Boolean algebras

B1. Idempotency	$A \cap A = A$	
	$A \cup A = A$	
B2. Identity	$A \cap X = A$	$A \cap \emptyset = \emptyset$
	$A \cup \emptyset = A$	$A \cup X = X$
B3. Commutativity	$A \cap B = B \cap A$	
	$A \cup B = B \cup A$	
B4. Associativity	$A \cap (B \cap C) = (A \cap B) \cap C$	
	$A \cup (B \cup C) = (A \cup B) \cup C$	
B5. Distributivity	$A \cap (B \cup C) = (A \cap B) \cup (A \cap C)$	
	$A \cup (B \cap C) = (A \cup B) \cap (A \cup C)$	
B6. Absorption	$A \cap (A \cup B) = A$	
	$A \cup (A \cap B) = A$	
B7. Non-contradiction	$A \cap A^c = \emptyset$	
& Excluded-middle	$A \cup A^c = X$	
B8. Involution	$(A^c)^c = A$	
B9. De Morgan laws	$(A \cap B)^c = A^c \cup B^c$	
(duality laws)	$(A \cup B)^c = A^c \cap B^c$	

Note that the above definition entails, by double inclusion, that $A = B$ if and only if $A(x) = B(x)$ for any $x \in X$. In addition, Zadeh proposed to build fuzzy set operations $\cap, \cup : \mathcal{F}(X) \times \mathcal{F}(X) \longrightarrow \mathcal{F}(X)$ and $^c : \mathcal{F}(X) \longrightarrow \mathcal{F}(X)$ from operations defined on $[0,1]$ in a pointwise manner . In particular, Zadeh's seminal paper included the following proposal: for any $A, B \in \mathcal{F}(X)$ and for any $x \in X$,

$$(A \cap B)(x) = \text{Min}(A(x), B(x))$$
$$(A \cup B)(x) = \text{Max}(A(x), B(x))$$
$$A^c(x) = 1 - A(x)$$

Remark 1. Zadeh also proposed other fuzzy set operations, some of which have been accepted later as intersections and unions:

Algebraic product: $(AB)(x) = A(x) \cdot B(x)$
Dual of algebraic product: $(A \oplus B)(x) = A(x) + B(x) - A(x) \cdot B(x)$
Algebraic sum: $(A + B)(x) = \text{Min}(1, A(x) + B(x))$
Absolute difference: $|A - B|(x) = |A(x) - B(x)|$

After forty years of research in fuzzy sets and fuzzy logic, Zadeh's pioneering proposal on how to construct fuzzy sets has naturally been extended in several directions, with just one universally accepted, easy-to-understand and sensible limitation: since crisp sets are particular instances of fuzzy sets, any newly defined fuzzy concept should, when applied to crisp sets, provide the same results as in the classical setting. This limitation, which we will call the *Classical Preservation Principle* (CPP), establishes, for example, that any fuzzy intersection should, when applied to two crisp sets, output the crisp set that would have been obtained if the classical intersection had been used. Some of the aspects related to the construction of fuzzy set theories (FST) that have evolved since they were first introduced are explained below (most will be reviewed in more detail and illustrated with some examples in Sect. 5).

2.1 Membership Degree Scale

The closed unit interval, used to representing the membership degrees of fuzzy sets, can be easily generalized into other spaces. The first ideas on this generalization were presented as early as in 1967 by Goguen in [46] who introduced the L-fuzzy set or L-set, where L can be any partially ordered set, but is usually a complete distributive lattice. For example, Sambuc defined in [75] interval-valued fuzzy sets, where the elements of L are pairs $(\alpha, \beta) \in [0,1] \times [0,1]$ such that $\alpha \leq \beta$ that represent an interval degree of membership. Another proposal by Atanassov in 1983 are intuitionistic fuzzy sets, where $(L, \leq, N)$ is a complemented lattice and the elements of L are pairs $(\mu, \gamma) \in [0,1] \times [0,1]$ such that $\mu \leq N(\gamma)$ representing a membership and a non-membership degree, respectively.

Note that in order to satisfy the CPP, the set L should contain a least element 0_L and a greatest element 1_L used to represent,respectively, an element's crisp non-membership and crisp full-membership of a fuzzy set. For a semantic interpretation of L, see [79].

2.2 Fuzzy Set Ordering

As mentioned earlier, $\mathcal{F}(X)$ is endowed with the pointwise extension of the usual total order of the real line in Zadeh's approach. Even though this is the most common approach, it may be a too restrictive notion in some situations. Different alternatives for ordering fuzzy sets have been explored. For example, DeLuca and Termini [26] proposed the so-called *sharpened* ordering, whose extension to $\mathcal{F}(X)$ provides a different way for comparing fuzzy sets. In general, fuzzy set orderings are defined from the orders established in the underlying set L, even if this is not strictly necessary. In fact, the only constraint required to satisfy the CPP is the preservation of the classical order between crisp sets, i.e. if $A, B \in \mathcal{P}(X)$, then $A \preceq B$ if and only if $A \subseteq B$, where $\subseteq$ denotes the classical inclusion.

2.3 Fuzzy Operation Definition

Contrary to the classical case, which relies on the underlying two-valued logic and where the basic operations on sets are uniquely defined, the fuzzy world allows for a wide variety of alternatives. When dealing with crisp propositions the meaning of the connectives *and, or* and *not* is necessarily unique, whereas these connectives may have many different meanings in the case of fuzzy propositions, depending on the context in which they are used. There are, consequently, different alternatives for modeling the intersection, union and complementation of fuzzy sets.

The most popular way for constructing fuzzy operations is to adopt, as is done in both the classical case and Zadeh's first proposal, a pointwise definition, assuming therefore that the operations on fuzzy sets may be defined from functions on L (when this occurs, we will say that the fuzzy operators are functionally expressible). Within this category, the most common choice is to take $L = [0, 1]$ and use *strong negation functions* [78], continuous *triangular norms* and continuous *triangular conorms* to model, respectively, the complementation, intersection and union of fuzzy sets. This kind of fuzzy set theories, including especially Zadeh's aforementioned theory, are called *standard fuzzy set theories* and will be reviewed in detail in Sect. 4. Triangular norms and conorms were introduced in the framework of probabilistic metric spaces [56, 60, 76] and have been widely researched leading to different characterizations, construction methods and parameterized families now available [4,55]. Nevertheless, as empirically demonstrated in the eighties [96], triangular norms and conorms do not allow for compensation between the input values and, consequently, are not always the best choice. In fact, these functions belong to a wider category of operators defined on $[0, 1]$ known as *aggregation operators* [16, 21, 33], among which there are alternative operators that may be better suited for representing the intersection and/or the union of fuzzy sets (or, more generally, the aggregation of fuzzy sets) in some situations.

When L is a complemented lattice, operations are typically defined as follows:

- $(A \cap B)(x) = \bigwedge(A(x), B(x))$.
- $(A \cup B)(x) = \bigvee(A(x), B(x))$.
- $A^c(x) = N(A(x))$

where $\bigvee$, $\bigwedge$ and N are, respectively, the lattice's meet, join and complement.

Functionally expressible theories (see Sect. 5.2) are easily defined and have the advantage that the properties of the operations $\cap, \cup$ and ^{c} can be directly derived from those of the underlying functions on which they are built. Despite this, it should be pointed out that more general approaches to the construction of FST have also been proposed [81], and there are situations where non-functionally expressible connectives are better suited than functionally expressible ones, see [86, 89].

In sum, the only restrictions that seem to be required when defining connectives for fuzzy sets are the constraints derived from the CPP, meaning that the fuzzy intersection, union and complement chosen should behave in exactly the same way as the classical intersection, union and complement when applied to crisp sets.

2.4 Fuzzy Operation Properties

As mentioned above, there is a series and not just one FST, each one with its own particularities. In general, the choice of a given FST for modeling a specific application heavily depends on the properties exhibited by the selected set operations. Many papers have been published on this topic, most of them dealing with functionally expressible theories, since the connective properties directly translate into functional equations in this case, and can therefore be studied in the light of so-called functional equations theory [1].

Taking into account the Boolean structure of the classical case, the first obvious question is to wonder which Boolean properties are satisfied by the different fuzzy connectives. In one of the first works on this matter, Dubois and Prade [31] proved that, in the framework of functionally expressible fuzzy theories equipped with an involutive complement, idempotency (as well as mutual distributivity) is incompatible with the laws of non-contradiction and the excluded-middle, and, consequently, not all the Boolean laws can hold at the same time. However, depending on which underlying operations are chosen, other kinds of interesting algebraic structures can be obtained (see Sect. 4.2).

Other approaches regarding the study of the properties satisfied by fuzzy operators include the problem of satisfying either non-basic (derived)

Boolean-based properties or properties that do not have a Boolean background, as well as the reinterpretation of some Boolean laws in the fuzzy framework.

2.5 The Fuzzy Set Space

Since the foundation of fuzzy logic, FSTs have almost always been built upon the set $\mathcal{F}_L(X)$ of *all* the possible fuzzy sets with membership degrees in L, i.e. generally *any* function $A : X \longrightarrow L$ has been regarded as a fuzzy set. A recent paper [85] argues that this approach may be too general, and that a restriction to some specific families of fuzzy sets may be necessary in some practical situations. To illustrate this point, the authors show that although the classical power set $\mathcal{P}(X)$ contains a unique self-contradictory element (the empty set), this is no longer the case of its extension $\mathcal{F}_L(X)$, where many self-contradictory fuzzy sets can be found (for further details see [89]). Therefore, if self-contradiction is to be avoided, it is necessary to remove some fuzzy sets from $\mathcal{F}_L(X)$ and to take a subset obtained after excluding self-contradictory fuzzy sets other than the empty set. A number of different FSTs built upon fuzzy set spaces of this kind have been proposed in [86,89], and it is interesting to note that the use of non-functionally expressible connectives is a great aid in their construction.

In general, any subset of $\mathcal{F}_L(X)$ could be considered, subject to the unique CPP derived condition that it contains all the crisp sets. Note nevertheless that the exclusion of some kinds of fuzzy sets has to be analyzed carefully, since some of them may be necessary in practical situations (this may be the case, e.g. of constant fuzzy sets [86, 89]).

3 Fuzzy Set Theories: Classical Preservation Principle

Many authors have written papers relating different *"fuzzy set theories"* and finding equivalences between them. In [17] it was shown that vague sets are intuitionistic fuzzy sets. In [24] it was pointed out that fuzzy rough sets are intuitionistic L-fuzzy sets. In [90] it was proved that intuitionistic L-fuzzy sets [13] and L-fuzzy sets are equivalent, [53] summarizes the links between fuzzy sets and other models such as flou sets, twofold fuzzy sets and L-fuzzy sets, [27] studies the relations between intuitionistic fuzzy sets, L-fuzzy sets, interval-valued fuzzy sets and interval-valued intuitionistic fuzzy sets. Finally, some minimum requirements for extending classical set theory were studied in [81].

There is some confusion in the literature regarding what a fuzzy set theory is and how it should be defined. In this paper we understand FST in a broad sense referring to all kinds of fuzzy theories. Following from the above discussion on the different aspects involved in the construction of a FST the following very general definition can be established, with the Classical Preservation Principle in mind:

Definition 1. *Let* $\mathcal{F}_L(X) = \{A; A : X \to L\}$ *be the set of all fuzzy sets on* X *with membership degrees in the poset* $(L, \leq)$, *endowed with a partial order* $\preceq$, *and three operations* $\cap, \cup : \mathcal{F}_L(X) \times \mathcal{F}_L(X) \longrightarrow \mathcal{F}_L(X)$, ${}^c : \mathcal{F}_L(X) \longrightarrow \mathcal{F}_L(X)$ *that represent, respectively, the intersection* (and), *union* (or) *and complement* (not).

Then, $(\mathcal{F}_L(X), \preceq, \cap, \cup, {}^c)$ *will be called a* fuzzy set theory *whenever the following conditions (expressing the classical preservation principle) are satisfied:*

1. $(L, \leq)$ *is bounded, i.e.* L *contains a least element* 0_L *and a greatest element* 1_L.
2. *For any* $A, B \in \mathcal{P}(X)$, $A \preceq B$ *if and only if* $A \subseteq B$, *where* $\subseteq$ *stands for the classical set inclusion.*
3. *For any* $A, B \in \mathcal{P}(X)$, $A \cap B = A \widehat{\cap} B$, $A \cup B = A \widehat{\cup} B$ *and* $A^c = A^{\widehat{c}}$, *where* $\widehat{\cap}$, $\widehat{\cup}$ *and* $\widehat{c}$ *stand, respectively, for the classical set intersection, union and complement.*

Remark 2. The three conditions that have been established in the above definition are the only ones that are needed to preserve the classical case. Other crisp properties, such as, e.g. the commutativity of the intersection and union operations or the involutive nature of the complement do not seem to be strictly necessary in the fuzzy setting, although they may be needed in many practical situations.

4 Standard Fuzzy Set Theories

As is well-known, Zadeh's FST that was recalled in Sect. 2 is just a particular instance of a general class of theories, known as *standard fuzzy set theories* (SFST). SFSTs are functionally expressible fuzzy set theories where L is the unit interval, fuzzy sets are pointwise ordered by means of the usual order on $[0, 1]$, and $(\cap, \cup, {}^c)$ are defined by means of a continuous triangular norm, a continuous triangular conorm and a strong negation.

Definition 2. *A* standard fuzzy set theory *(SFST) is a fuzzy set theory* $(\mathcal{F}_L(X), \preceq, \cap, \cup, {}^c)$ *with the following properties:*

1. $L = [0, 1]$ *(and, therefore,* $\mathcal{F}_L(X) = \mathcal{F}(X)$*).*
2. *For any* $A, B \in \mathcal{F}(X)$, $A \preceq B$ *if and only if* $A(x) \leq B(x)$ *for any* $x \in X$.
3. *For any* $A, B \in \mathcal{F}(X)$, $(\cap, \cup, {}^c)$ *are defined as:*

$$(A \cap B)(x) = T(A(x), B(x))$$
$$(A \cup B)(x) = S(A(x), B(x))$$
$$A^c(x) = N(A(x))$$

for any $x \in X$, *where* $T : [0, 1] \times [0, 1] \longrightarrow [0, 1]$ *is a continuous triangular norm,* $S : [0, 1] \times [0, 1] \longrightarrow [0, 1]$ *is a continuous triangular conorm and* $N : [0, 1] \longrightarrow [0, 1]$ *is a strong negation.*

In the following, let us recall the definitions and main characterizations related to triangular norms/conorms and strong negations, and then summarize the properties of the theories built using these functions.

4.1 Definitions and Characterizations

Triangular norms (t-norms) are non-decreasing operators $T : [0,1] \times [0,1] \to [0,1]$ that are commutative, associative and have a neutral element 1. Their dual operators, known as *triangular conorms (t-conorms)*, are non-decreasing, commutative and associative operators $S : [0,1] \times [0,1] \to [0,1]$ with a neutral element 0. Since the early eighties (see [10, 11, 30, 32]) both t-norms and t-conorms have been commonly used to model, respectively, the intersection and the union of fuzzy sets. Although there is no universal representation theorem for either t-norms or t-conorms for the time being, some useful characterizations are available for a broad class of these operators, namely the continuous ones, using *automorphisms* of the unit interval. Let us first recall the definition of automorphism of the unit interval and give some examples:

Definition 3. *An* automorphism *of the unit interval is a continuous and strictly increasing function* $\varphi : [0,1] \longrightarrow [0,1]$ *verifying the boundary conditions* $\varphi(0) = 0$ *and* $\varphi(1) = 1$.

There exist many different automorphisms of the unit interval, some of which in particular depend on various parameters. Some examples are given below.

- $\varphi(a) = \frac{2a}{a+1}$
- $\varphi(a) = a^{\lambda}$, $\lambda > 0$
- $\varphi(a) = 1 - (1-a)^{\lambda}$, $\lambda > 0$
- $\varphi(a) = \frac{\lambda^a - 1}{\lambda - 1}$, $\lambda > 0, \lambda \neq 1$
- $\varphi(a) = \frac{a}{\lambda + (1-\lambda)a}$, $\lambda > 0$
- $\varphi(a) = \frac{ln(1+\lambda a^{\alpha})}{ln(1+\lambda)}$, $\lambda > -1, \alpha > 0$

It is easy to verify that, given a t-norm and an automorphism φ, the operator T_{φ}, defined by $T_{\varphi} = \varphi^{-1} \circ T \circ (\varphi \times \varphi)$, is also a t-norm. Similarly, if S is a t-conorm and φ is an automorphism, $S_{\varphi} = \varphi^{-1} \circ S \circ (\varphi \times \varphi)$ is a t-conorm. The following definition can then be established:

Definition 4. *Given a t-norm* T*, the set of all the t-norms* T_{φ}*, where* φ *is any automorphism, is denoted* $\mathbb{F}(T)$ *and is called* the family of T. *Similarly, given a t-conorm* S*, the set of all the t-conorms* S_{φ}*, where* φ *is any automorphism, is denoted* $\mathbb{F}(S)$ *and is called the* family of S.

As is well known, there are three prototypical families of continuous triangular norms:

- The Min family, $\mathbb{F}(\text{Min})$, which contains only the minimum t-norm Min.
- The Product family, $\mathbb{F}(Prod)$, composed of all the t-norms of the form $Prod_\varphi = \varphi^{-1} \circ Prod \circ (\varphi \times \varphi)$, where $Prod(x, y) = x \cdot y$ (also denoted T_P) is the product t-norm and φ is any automorphism.
- The Łukasiewicz family, $\mathbb{F}(W)$, made of all the t-norms of the form $W_\varphi = \varphi^{-1} \circ W \circ (\varphi \times \varphi)$, where $W(x, y) = max(0, x + y - 1)$ (also denoted as T_L) is the Łukasiewicz t-norm and φ is any automorphism.

Continuous t-norms may be characterized as follows [55]:

Proposition 1. *A function* $T : [0,1] \times [0,1] \longrightarrow [0,1]$ *is a continuous t-norm if and only if there exist a uniquely determined (finite or countably infinite) index set* I*, a family of uniquely determined pairwise disjoint open subintervals of the unit interval,* $\{]a_i, b_i[\}_{i \in I}$*, and a family of uniquely determined t-norms* $\{T_i : T_i \in \mathbb{F}(Prod) \cup \mathbb{F}(W)\}_{i \in I}$ *such that for any* $x, y \in [0,1]$*:*

$$T(x,y) = \begin{cases} a_i + (b_i - a_i) T_i \left(\frac{x - a_i}{b_i - a_i}, \frac{y - a_i}{b_i - a_i} \right), & \textit{if } (x,y) \in [a_i, b_i]^2 \\ Min(x,y), & \textit{otherwise} \end{cases}$$

In the above result, the case of the continuous t-norm Min is recovered when the family of subintervals is empty, and the members of the families $\mathbb{F}(W)$ and $\mathbb{F}(Prod)$ appear when the set of subintervals $\{]a_i, b_i[\}$ is reduced to the unique interval $]0,1[$. In the remaining cases, it is said that T is an *ordinal sum* of the family of continuous t-norms $\{T_i\}$. Therefore, it appears that any continuous t-norm T verifies one and only one of the following properties:

1. $T = Min$
2. $T \in \mathbb{F}(Prod)$
3. $T \in \mathbb{F}(W)$
4. T is an ordinal sum of a family of t-norms $\{T_i; T_i \in \mathbb{F}(Prod) \cup \mathbb{F}(W)\}$

When dealing with continuous triangular conorms, the three prototypical families are as follows [55]:

- The Max family, $\mathbb{F}(\text{Max})$, which contains only the maximum t-conorm Max.
- The Probabilistic Sum family, $\mathbb{F}(Prod^*)$, composed of all the t-conorms of the form $Prod^*_\varphi = \varphi^{-1} \circ Prod^* \circ (\varphi \times \varphi)$, where $Prod^*(x, y) = x + y - x \cdot y$ (also denoted S_P) is the probabilistic sum t-conorm and φ is any automorphism.
- The Łukasiewicz family, $\mathbb{F}(W^*)$, composed of all the t-conorms of the form $W_\varphi = \varphi^{-1} \circ W^* \circ (\varphi \times \varphi)$, where $W^*(x, y) = min(1, x + y)$ (also denoted S_L) is the Łukasiewicz t-conorm and φ is any automorphism.

Continuous t-conorms may be characterized in a similar way as continuous t-norms, that is, a continuous t-conorm is either a member of the set $\{Max\} \cup \mathbb{F}(Prod^*) \cup \mathbb{F}(W^*)$ or it is an ordinal sum of t-conorms belonging to this set:

Proposition 2. *A function $S : [0,1]\times[0,1] \longrightarrow [0,1]$ is a continuous t-conorm if and only if there exist a uniquely determined (finite or countably infinite) index set I, a family of uniquely determined pairwise disjoint open subintervals of the unit interval, $\{]a_i, b_i[\}_{i\in I}$, and a family of uniquely determined t-conorms $\{S_i : S_i \in \mathbb{F}(Prod^*) \cup \mathbb{F}(W^*)\}_{i\in I}$ such that for any $x, y \in [0,1]$:*

$$S(x,y) = \begin{cases} a_i + (b_i - a_i)S_i\left(\frac{x-a_i}{b_i-a_i}, \frac{y-a_i}{b_i-a_i}\right), & \text{if } (x,y) \in [a_i, b_i]^2 \\ Max(x,y), & \text{otherwise} \end{cases}$$

We will now deal with *strong negation functions* (non-increasing and involutive functions $N : [0,1] \longrightarrow [0,1]$), which are commonly used to model the complementation of fuzzy sets. By definition, strong negations are continuous and strictly decreasing functions, satisfy the boundary conditions $N(0) = 1$ and $N(1) = 0$, and have a unique fixed point x_N verifying $0 < x_N < 1$ and $N(x_N) = x_N$. The most commonly used strong negation is the so-called *standard negation*, defined by Zadeh as $N(x) = 1 - x$ for all $x \in [0,1]$, whose fixed point is $1/2$. Despite its simplicity, this function plays a fundamental role in the construction of strong negations, since any strong negation can be built from the standard negation using an automorphism of the unit interval [78]:

Proposition 3. *A function $N : [0,1] \longrightarrow [0,1]$ is a strong negation if and only if there exists an automorphism $\varphi : [0,1] \longrightarrow [0,1]$ such that $N = N_\varphi = \varphi^{-1} \circ (1 - Id) \circ \varphi$, i.e. $N(x) = N_\varphi(x) = \varphi^{-1}(1 - \varphi(x))$ for any $x \in [0,1]$.*

Note that the fixed point of a strong negation function N_φ is given by $\varphi^{-1}(1/2)$. Finally, let us recall that strong negation functions can be used to establish a close link between t-norms and t-conorms:

Definition 5. *Let T be a t-norm, S a t-conorm and N a strong negation.*

- *The function $T^* : [0,1] \times [0,1] \longrightarrow [0,1]$, given by $T^*(x,y) = N(T(N(x), N(y)))$ for all $x, y \in [0,1]$, is a t-conorm, which is said to be the* N-dual t-conorm of T.
- *The function $S^* : [0,1] \times [0,1] \longrightarrow [0,1]$, given by $S^*(x,y) = N(S(N(x), N(y)))$ for all $x, y \in [0,1]$, is a t-norm, which is said to be the* N-dual t-norm of S.
- *The triple (T, S, N) is called a* De Morgan triple *if, for all $x, y \in [0,1]$, $T(x,y) = N(S(N(x), N(y)))$ (note that due to the involutive nature of N, this is equivalent to $S(x,y) = N(T(N(x), N(y)))$).*

Some famous De Morgan triples are $(\text{Min}, \text{Max}, N_{\text{id}})$, $(\text{Prod}_\varphi, \text{Prod}^*_\varphi, N_\varphi)$ and $(W_\varphi, W^*_\varphi, N_\varphi)$.

4.2 Structural Properties

Table 2 summarizes the main properties satisfied by SFST, denoted $(\mathcal{F}(X), T, S, N)$ (for details, proofs or references to the original proofs, see e.g. [36]

or [55]). Since SFST are functionally expressible theories endowed with a strong negation function, they can never behave like Boolean algebras (see Sect. 2.4 or [31]). However, depending on the particular triple (T, S, N) that is chosen, it is possible to find theories satisfying different subsets of the whole set of basic Boolean properties.

The following general conclusions regarding the structural properties of SFST can be drawn from Table 2:

1. The theories of the form $(\mathcal{F}(X), \mathrm{Min}, \mathrm{Max}, N)$, where N is any strong negation, are the ones that preserve more Boolean properties.
   Indeed, these theories preserve all the basic Boolean properties given in Table 2 except for the non-contradiction and the excluded-middle laws. They are distributive lattices and Kleene algebras [62], since they obviously verify Kleene's law $A \cap A^c \leq B \cup B^c$, $\mathrm{Min}(x, N(x)) \leq \mathrm{Max}(y, N(y))$ for any $x, y \in [0, 1]$ (note that, actually, any t-norm/t-conorm pair verifies this law). They are the only SFST that have a lattice structure.

Table 2. Satisfaction of basic Boolean properties in standard fuzzy set theories

	T	S	N
B1. Idempotency			
$T(x, x) = x$	Min	–	–
$S(x, x) = x$	–	Max	–
B2. Identity			
$T(x, 1) = x, T(x, 0) = 0$	any	–	–
$S(x, 0) = x, S(x, 1) = 1$	–	any	–
B3. Commutativity			
$T(x, y) = T(y, x)$	any	–	–
$S(x, y) = S(y, x)$	–	any	–
B4. Associativity			
$T(x, T(y, z)) = T(T(x, y), z)$	any	–	–
$S(x, S(y, z)) = S(S(x, y), z)$	–	any	–
B5. Distributivity			
$T(x, S(y, z)) = S(T(x, y), T(x, z))$	any	Max	–
$S(x, T(y, z)) = T(S(x, y), S(x, z))$	Min	any	–
B6. Absorption			
$T(x, S(x, y)) = x$	Min	any	–
$S(x, T(x, y)) = x$	any	Max	–
B7. Non-contradiction & excluded-middle			
$T(x, N(x)) = 0$	W_φ	–	$N \leq N_\varphi$
$S(x, N(x)) = 1$	–	W_φ^*	$N \geq N_\varphi$
B8. Involution			
$N(N(x)) = x$	–	–	any
B9. De Morgan Laws			
$N(T(x, y)) = S(N(x), N(y))$		$T = N \circ S \circ N \times N$	
$N(S(x, y)) = T(N(x), N(y))$		$T = N \circ S \circ N \times N$	

2. There are infinite SFST verifying the non-contradiction and/or the excluded-middle laws.
   The triples (T, S, N) that verify these two Boolean laws can be determined from the results given in Table 2.

   Proposition 4. *A standard fuzzy set theory $(\mathcal{F}(X), T, S, N)$ verifies the non-contradiction and the excluded-middle laws if and only if there exist two automorphisms $\varphi_1, \varphi_2 : [0,1] \longrightarrow [0,1]$ such that:*

   $$T = W_{\varphi_1}, \quad S = W^*_{\varphi_2}, \quad \textit{and} \quad N_{\varphi_2} \leq N \leq N_{\varphi_1}$$

   It is interesting to note that the theories $(\mathcal{F}(X), T, S, N)$ given by the above characterization are not necessarily dual theories. Of course, taking $\varphi_1 = \varphi_2 = \varphi$, the result includes all the De Morgan triples of the form $(W_\varphi, W^*_\varphi, N_\varphi)$, i.e. all the De Morgan triples that are isomorphic to the Łukasiewicz triple (W, W^*, N_{Id}) (that provides a MV-algebra).
   In sum, it is possible to find either dual or non-dual SFST verifying the non-contradiction and the excluded-middle laws.

3. The structures where the intersection and/or the union are represented by t-norms and/or t-conorms in the Product family are the weaker ones as regards to the satisfaction of the basic Boolean properties.
   Indeed, when using t-norms and t-conorms of the form $T = \text{Prod}_\varphi$ or $S = \text{Prod}^*_\varphi$, all the basic Boolean laws of Table 2, except the ones which are trivially satisfied by any t-norm/t-conorm, fail. Note, however, that these operators, when appropriately combined with others, can satisfy some other non-basic Boolean properties (see Sect. 5.5).

5 Some Aspects Concerning the Construction of Fuzzy Set Theories

This section illustrates some of the aspects related to the construction of FST in more detail.

5.1 Pexider-FST and Pexider-SFST

According to a number of authors, Definition 1 could still be insufficient in some particular applications (such as preference modeling [39,40]) or in situations where the connectives *and, or* and *not* are used in different ways and, therefore, deserve FST with several fuzzy connectives [3,5,7]. In such circumstances, theories of the form $(\mathcal{F}_L(X); \cap_1, \ldots, \cap_p;\ \cup_1, \ldots, \cup_q;\ ^{c_1}, \ldots, ^{c_r})$ are needed. In [7] these theories are called *Pexider Fuzzy Set Theories*, since when the fuzzy operations are supposed to be functionally expressible, the study of their properties is equivalent to solving so-called Pexider functional equations [1].

This idea can easily be particularized to the case of SFST, that is, a Pexider-SFST will be a structure $(\mathcal{F}(X); T_1, \ldots, T_p;\ S_1, \ldots, S_q;\ N_1, \ldots, N_r)$ where $T_1, \ldots, T_p$ are continuous t-norms, $S_1, \ldots, S_q$ are continuous t-conorms and $N_1, \ldots, N_r$ are strong negations.

5.2 Functionally Expressible Theories

Definition 6. *A fuzzy set theory (see Definition 1) is said to be a* functionally expressible fuzzy set theory *when there exist functions $F, G : L \times L \longrightarrow L$ and $N : L \longrightarrow L$ such that for any A, B in $\mathcal{F}_L(X)$ and for any $x \in X$:*

$$\begin{aligned} &(A \cap B)(x) = F(A(x), B(x)) \\ &(A \cup B)(x) = G(A(x), B(x)) \\ &A^c(x) = N(A(x)) \end{aligned}$$

This concept can be generalized to a Pexider-FST:

Definition 7. *A Pexider-FST is* functionally expressible *when there exist functions $F_1 \ldots, F_p, G_1, \ldots, G_q : L \times L \to L$ and $N_1, \ldots, N_r : L \to L$ such that for any A, B in $\mathcal{F}_L(X)$ and for any $x \in X$:*

$$\begin{aligned} &(A \cap_1 B)(x) = F_1(A(x), B(x)), \ldots, (A \cap_p B)(x) = F_p(A(x), B(x)) \\ &(A \cup_1 B)(x) = G_1(A(x), B(x)), \ldots, (A \cup_q B)(x) = G_q(A(x), B(x)) \\ &A^{c_1}(x) = N_1(A(x)), \ldots, A^{c_r}(x) = N_r(A(x)) \end{aligned}$$

Functionally expressible theories are very useful for practical purposes for two main reasons. First, it is clearly much easier to build functionally expressible than non-functionally expressible operations. Second, the study of which properties are fulfilled by the operations $\cap, \cup$ and ^{c} is enormously simplified in the first case, since it relies on the properties of the underlying functions F, G and N, which are much easier to determine. For example, the condition (3) required to satisfy the CPP stated in Definition 1 is equivalent, in the case of functionally expressible theories, to the satisfaction of the following boundary conditions for functions F, G and N:

$$\begin{aligned} &F(0_L, 1_L) = F(1_L, 0_L) = F(0_L, 0_L) = 0_L \text{ and } F(1_L, 1_L) = 1_L \\ &G(0_L, 1_L) = G(1_L, 0_L) = G(1_L, 1_L) = 1_L \text{ and } G(0_L, 0_L) = 0_L \\ &N(0_L) = 1_L \text{ and } N(1_L) = 0_L \end{aligned}$$

Functionally Expressible Intersections and Unions

In the case $L = [0, 1]$, triangular norms/conorms are clear examples of functions F/G verifying these boundary conditions. They are by far the most popular choices for building functionally expressible theories, mainly due to their clear conjunctive/disjunctive behavior and the calculation facilities they provide. Nevertheless, many authors have pointed out that these operators are not always the most appropriate and have suggested larger frameworks where associativity and/or commutativity in particular are not necessary [38,42,49,87,88]. Consequently, a vast catalogue of operations able to induce fuzzy intersections and unions is now available. In general, all of them belong to the class of *aggregation operators*, which, in the multi-dimensional case [21], are defined as follows:

Definition 8. *An* aggregation operator *is a function* $M : \bigcup_{n \in \mathbb{N}} [0,1]^n \to [0,1]$ *such that:*

1. $M(x_1, \ldots, x_n) \leq M(y_1, \ldots, y_n)$ *whenever* $x_i \leq y_i$ *for all* $i \in \{1, \ldots, n\}$.
2. $M(x) = x$ *for all* $x \in [0,1]$.
3. $M(0, \ldots, 0) = 0$ *and* $M(1, \ldots, 1) = 1$.

There exists a large collection of distinguished classes of aggregation operators as well as different construction methods, and the satisfaction of various algebraic and analytical properties has been explored (see, e.g. [33], [16] or the recent overview on aggregation theory given in [21]). Any binary aggregation operator M such that $M(0,1) = M(1,0) = 0$ may be used to induce a fuzzy intersection, and, consequently, such operators are sometimes called *conjunctive aggregation operators*. Note that, due to the monotonicity of aggregation operators (first condition of the definition), conjunctive aggregation operators have an annihilator element of 0, i.e. they verify $M(x,0) = M(0,x) = 0$ for any $x \in [0,1]$.

Example 1. Some examples of binary conjunctive aggregation operators are given below:

- Any aggregation operator such that $M \leq \text{Min}$ (i.e. $M(x,y) \leq \text{Min}(x,y)$ for any $x, y \in [0,1]$) is obviously a conjunctive aggregation operator. Apart from the above mentioned t-norms, this class includes the class of *copulas* [55,63] or the so-called *weak t-norms* introduced by Fodor [38].
- It is also possible to find conjunctive operators within the class of *mean* or *averaging* aggregation operators, i.e. operators fulfilling the property $\text{Min} \leq M \leq \text{Max}$ and which are, consequently, idempotent operators (they verify $M(x,x) = x$ for any $x \in [0,1]$). Such operators may be found, for instance, within the well-known class of *quasi-linear means* or *weighted quasi-arithmetic means* [1,21], operators which, in the binary case, are defined as $M_{f,\lambda}(x,y) = f^{-1}[(1-\lambda)f(x)+\lambda f(y)]$ for any $x, y \in [0,1]$, where $\lambda \in]0,1[$ and $f : [0,1] \to [-\infty, \infty]$ is a continuous and strictly monotone (increasing or decreasing) function such that $\{f(0), f(1)\} \neq \{-\infty, +\infty\}$. Quasi-linear means are conjunctive aggregation operators whenever the chosen generating function f is such that $f(0) = \pm\infty$. This is the case, for example, of the well-known *geometric mean*, $M(x,y) = \sqrt{xy}$, generated by $f(x) = log(x)$ with $\lambda = 1/2$.

- Finally, there are also conjunctive aggregation operators that do not belong to either of the above two categories. For example:
  - *Uninorms* [21, 93] are commutative and associative aggregation operators $U : [0,1]^2 \longrightarrow [0,1]$ possessing a neutral element $e \in]0,1[$. They behave like t-norms in $[0,e] \times [0,e]$, like t-conorms in $[e,1] \times [e,1]$ and otherwise verify $\text{Min} \leq U \leq \text{Max}$. They may be divided into two different classes: operators with an annihilator element $a = 0$, known as *conjunctive uninorms* and operators with an annihilator $a = 1$, known as *disjunctive uninorms*. A commonly cited example of conjunctive uninorm (which is a symmetric sum [21,28,33,54,77] when the convention $\frac{0}{0} = 0.5$ is adopted) is the operator given by

  $$U(x,y) = \frac{xy}{xy + (1-x)(1-y)}$$

  for any $x, y \in [0,1]$, with the convention $\frac{0}{0} = 0$.
  - Consider now the family of *quasi-linear T-S operators* [71], i.e. operators which, in the binary case, are of the form $QL_{T,S,\lambda,f}(x,y) = f^{-1}\left[(1-\lambda)f(T(x,y)) + \lambda f(S(x,y))\right]$, where T is a t-norm, S is a t-conorm, $\lambda \in]0,1[$ and $f : [0,1] \to [-\infty,\infty]$ is a continuous and strictly monotone function such that $\{f(0), f(1)\} \neq \{-\infty, +\infty\}$. It is not difficult to prove [71] that such operators have an annihilator element of 0 if and only if $f(0) = \pm\infty$. If $f = log$ is chosen, this class includes the so-called *exponential convex T-S operators* [21], given by $E_{T,S,\lambda}(x,y) = T(x,y)^{1-\lambda} \cdot S(x,y)^{\lambda}$.

Similarly, binary aggregation operators verifying $M(0,1) = M(1,0) = 1$, which can be called *disjunctive aggregation operators*, can induce fuzzy unions (examples of such operators can be easily obtained, by duality, from the conjunctive operators given above).

Many authors have studied the satisfaction of the basic Boolean properties within the aggregation operators world, such as, e.g. distributivity [15,18,22,37,59,74], idempotency [25,29], absorption [19], associativity [58] or non-contradiction [69]. The main results on this topic, as well as appropriate references, can be found in [21].

Functionally Expressible Complements

When $L = [0,1]$, the most usual functionally expressible fuzzy complements $A^c = N \circ A$ are the ones defined choosing N as a strong negation function. Such functions are very easily built (see Sect. 4.1) and have the advantage that they provide involutive and order-reversing fuzzy complements (i.e. for any $A, B \in \mathcal{F}(X)$, $(A^c)^c = A$, and $B^c \preceq A^c$ whenever $A \preceq B$) when $\preceq$ is taken as the pointwise order built from the usual order in $[0,1]$. Strong negation functions are, additionally, continuous and strictly decreasing functions.

In general, any function N verifying the boundary conditions $N(0) = 1$ and $N(1) = 0$ could be used in order to build a functionally expressible fuzzy complement. For example, it is possible to use the so-called *strict negations*, which are functions verifying all the properties of strong negations but involution. A common example of strict negation is given by $N(x) = 1 - x^2$ for any $x \in [0, 1]$. Different kinds of functions can be found if continuity and/or strictness are dropped. This is the case, for example, of the so-called *intuitionist negation* (also known as the Gödel negation), N_i, and the negation defined from the former by duality, N_{di}. They are the limit cases given by [65, 92]:

$$N_i(x) = \begin{cases} 1, & \text{if } x = 0 \\ 0, & \text{otherwise} \end{cases} \qquad N_{di}(x) = \begin{cases} 0, & \text{if } x = 1 \\ 1, & \text{otherwise} \end{cases}$$

Obviously, N_i and N_{di} are neither continuous nor strictly decreasing, and they are not involutive. Note, nevertheless, that N_i provides the only functionally expressible fuzzy complement that verifies the non-contradiction law, $A \cap A^c = \emptyset$, when $\cap$ is functionally defined as $A \cap B = \text{Min} \circ (A \times B)$. Similarly, the functionally expressible fuzzy complement built by means of N_{di} is the only one that verifies the excluded-middle law, $A \cup A^c = X$, when $\cup$ is functionally defined by means of Max.

5.3 Non-Functionally Expressible Theories

Operations on $\mathcal{F}_L(X)$ are not necessarily functionally expressible. According to Definition 1, the CPP-derived requirement, is that, when applied to crisp sets, they must behave like the respective crisp operations. In general, the result of a fuzzy operation, when applied to a given element $x \in X$, may depend on both the concrete element to which it is applied as well as the fuzzy sets that are involved.

Non-functionally Expressible Intersections and Unions

The general form of operations $\cap$ and $\cup$, for any $A, B \in \mathcal{F}_L(X)$ and any $x \in X$, is given by

$$\begin{aligned} (A \cap B)(x) &= F_{A,B,x}(A(x), B(x)) \\ (A \cup B)(x) &= G_{A,B,x}(A(x), B(x)) \end{aligned}$$

where $\{F_{A,B,x} : L \times L \longrightarrow L\}$ and $\{G_{A,B,x} : L \times L \longrightarrow L\}$ are appropriate families of functions verifying the same boundary conditions as the functions F and G in the functionally expressible case. Of course, $F_{A,B,x}$ or $G_{A,B,x}$ may happen not to depend on any of the parameters A, B or x, or the same functions may be taken for some fuzzy sets or for some elements in X, or both. This is the case of functionally expressible intersections and unions, where it is supposed for all triplets (A, B, x) that $F_{A,B,x} = F$ and $G_{A,B,x} = G$. The following example shows that this latter situation is not a requirement:

Example 2. Let $L = [0, 1]$, $\preceq$ be the order induced by the real order of the unit interval, and consider the binary operation $\cup : \mathcal{F}(X) \times \mathcal{F}(X) \longrightarrow \mathcal{F}(X)$ defined, for any $A, B \in \mathcal{F}(X)$ and any $x \in X$, as:

$$(A \cup B)(x) = \begin{cases} \text{Max}(A(x), B(x)), & \text{if } A \text{ or } B \text{ are crisp} \\ \text{Max}(H_A, H_B), & \text{otherwise} \end{cases}$$

with $H_A = \underset{x \in X}{Sup}\{A(x)\}$ and $H_B = \underset{x \in X}{Sup}\{B(x)\}$

It is easy to check that this operation, when applied to crisp sets, coincides with the classical set union and is, therefore, a fuzzy union. It may be expressed as $(A \cup B)(x) = G_{A,B}(A(x), B(x))$ for any $x \in X$, where for any $u, v \in [0, 1]$:

$$G_{A,B}(u, v) = \begin{cases} \text{Max}(u, v), & \text{if } A \text{ or } B \text{ are crisp} \\ \text{Max}(H_A, H_B), & \text{otherwise} \end{cases}$$

But it is not functionally expressible, that is, there is no function $G : [0, 1]^2 \longrightarrow [0, 1]$ such that $(A \cup B)(x) = G(A(x), B(x))$ for any $A, B \in \mathcal{F}(X)$ and any $x \in X$. Indeed, let us suppose that such a function exists and choose $A \in \mathcal{F}(X)$ such that $A(x) = 1/2$ for any $x \in X$. Then:

- Taking the crisp set $B_1 = \emptyset$, $(A \cup B_1)(x) = \text{Max}(1/2, 0) = 1/2$ for any $x \in X$. On the other hand, we have $(A \cup B_1)(x) = G(1/2, 0)$, and hence $G(1/2, 0) = 1/2$.
- If we now choose a fuzzy set $B_2(x) = x$ then we get $(A \cup B_2)(x) = 1$ for any $x \in X$. But also $(A \cup B_2)(0) = G(1/2, 0)$, and then $G(1/2, 0) = 1$, which is a contradiction of the value found above.

Note that this operation is also commutative ($A \cup B = B \cup A$ for any $A, B \in \mathcal{F}(X)$), monotone (if $A \preceq B$, then $A \cup C \preceq B \cup C$ for any $A, B, C \in \mathcal{F}(X)$), it is an upper bound ($A, B \preceq A \cup B$ for any $A, B \in \mathcal{F}(X)$), and verifies the identity laws ($A \cup \emptyset = A$ and $A \cup X = X$ for any $A \in \mathcal{F}(X)$). Nevertheless, it is not idempotent, since whenever A is not crisp, we get $(A \cup A)(x) = H_A$.

Other examples of non-functionally expressible fuzzy intersections and unions will be discussed in Sect. 5.9.

Non-functionally Expressible Complements

Regarding non-functionally expressible fuzzy complements, Lowen [57] provided a general mechanism for building operators of this kind that depend not on the fuzzy set to which they are applied but on the point where they are applied. Let $L = [0, 1]$ and $\preceq$ be the order induced by the real order of the unit interval. Then Lowen fuzzy complements are defined as $A^c(x) = N_x(A(x))$ for any $x \in X$, where $\{N_x : [0, 1] \longrightarrow [0, 1]; x \in X\}$ is a family of strong negation functions.

Example 3. Let us choose $X = (0, +\infty)$, and define the following operation:

$$A^c(x) = \begin{cases} (1 - A(x)^x)^{1/x}, & \text{if } x \text{ is rational} \\ \frac{1-A(x)}{1+xA(x)}, & \text{otherwise} \end{cases}$$

This Lowen's complement is built upon the strong negation functions $N_x(a) = (1 - a^x)^{1/x}$ if x is rational and $N_x(a) = \frac{1-a}{1+xa}$ otherwise. The first functions belong to what is known as Yager's class of strong negations, whose general form is $N(a) = (1 - a^p)^{1/p}$, $p > 0$, and the second functions are members of the so-called Sugeno class, given as $N(a) = \frac{1-a}{1+pa}$, $p > -1$.

It is easy to verify that, thanks to the properties of strong negations, Lowen's fuzzy complements are both order-reversing and involutive. Moreover, Ovchinnikov [65] proved that if L is a complete distributive lattice, Lowen's complements appear to be the only ones that, apart from preserving the classical case, are order-reversing and involutive.

It is, of course, possible to define non-functionally expressible fuzzy complements other than Lowen's, but some properties are lost in this case.

Example 4. Let $X = [0, 1], L = [0, 1]$, $\preceq$ be the order induced by the real order of the unit interval, and let us define the three following operations:

$$A^{c_1}(x) = \begin{cases} 1 - A(x), & \text{if } x \in [0, 0.5] \\ 1 - A(x)^2, & \text{otherwise} \end{cases}$$

$$A^{c_2}(x) = \begin{cases} A^{\widehat{c}}(x), & \text{if } A \in \mathcal{P}(X) \\ 1 - A(1 - x), & \text{otherwise} \end{cases}$$

$$A^{c_3}(x) = 1 - A(1 - x)$$

A^{c_1} and A^{c_2} clearly preserve the classical set complement when applied to crisp sets (remember that $A^{\widehat{c}}$ is used in this chapter to denote the classical complement) and may, therefore, be considered as fuzzy complements. Nevertheless, A^{c_1} is not involutive (if $x > 0.5$, it is $(A^{c_1})^{c_1}(x) \neq A(x)$) and A^{c_2} does not always reverse the ordering between fuzzy sets. In proof of this, let us consider the following example: given an element $x_0 \neq 0.5$, define two fuzzy sets A and B such that:

- $A(x_0) = 1$ and, otherwise, $A(x) = 0$ (A is a crisp set);
- $B(1 - x_0) = 0.5$ and, otherwise, $B(x) = 1$.

Clearly $A \preceq B$, but $B^{c_2} \preceq A^{c_2}$ does not apply since, in particular, it is $B^{c_2}(x_0) = 1 - B(1 - x_0) = 0.5 \npreceq A^{c_2}(x_0) = 0$.

Regarding A^{c_3}, it is clear that it is both involutive and order-reversing. Nevertheless, it does not preserve the classical case. Indeed, take, e.g. any crisp set A such that $A(x_0) = 0$ and $A(1 - x_0) = 1$ for some $x_0 \neq 0.5$; then $A^{c_3}(x_0) = 1 - A(1 - x_0) = 0 \neq 1$, and then $A^{c_3} \neq A^{\widehat{c}}$.

Note finally that the operator A^{c_3} given in the last example is just a particular instance of the class of all the order-reversing and involutive operators on $(\mathcal{F}_L(X), \preceq)$, where $\preceq$ is the pointwise order built from the order of the lattice, which can be defined, when L is a directly indecomposable complete distributive lattice, as [64]:

$$A^c(x) = N_x(A(s(x))) \quad \text{for any } x \in X$$

where $s : X \longrightarrow X$ is a symmetry ($s^2 = Id$) and $\{N_x : L \longrightarrow L; x \in X\}$ is a family of non-increasing functions verifying $N_{s(x)} = N_x^{-1}$ for any $x \in X$. In general, these operators do not preserve the classical case. In fact, they do so if and only if $s = Id$ [65], that is, when they are Lowen's fuzzy complements.

5.4 Other Fuzzy Operators

The linguistic "or" is often better represented by "exclusive or" than by "inclusive or". The "exclusive or" is modeled by the symmetric difference operator.

Definition 9. *A binary operator Δ defined in a fuzzy set theory $(\mathcal{F}_L(X), \preceq, \cap, \cup, ^c)$ will be called a* symmetric difference operator *if the following condition is satisfied:*

> *For any $A, B \in \mathcal{P}(X)$, $A \Delta B = A \widehat{\Delta} B$, where $\widehat{\Delta}$ stands for the classical symmetric difference.*

Similarly to the intersection or union operators, a symmetric difference operator is functionally expressible if there exists a function $D : L \times L \to L$ such that

$$(A \Delta B)(x) = D(A(x), B(x)),$$

for any $A, B \in \mathcal{F}_L(X)$ and any $x \in X$.

To satisfy the CPP, such a function should meet the following conditions:

$$D(0_L, 1_L) = D(1_L, 0_L) = 1_L \text{ and } D(1_L, 1_L) = D(0_L, 0_L) = 0_L$$

In the case of Pexider-SFST it is possible to model a symmetric difference by means of expressions of the form:

$$D_1(x, y) = S(T_1(x, N(y)), T_2(N(x), y))$$
$$D_2(x, y) = T_1(S(x, y), N(T_2(x, y)))$$

that generalize the definition of the classical operator.

5.5 Fuzzy Operations' Laws

As mentioned in the previous sections, the studies on the properties of fuzzy set operations most often deal with basic Boolean laws and have generally been conducted on the functionally expressible case choosing $L = [0, 1]$. It is known that such theories may never be isomorphic to Boolean algebras, but it is interesting to note that this result can be easily extended to a wider class of FST:

Proposition 5. *Let* $(F_L(X), \preceq, \cap, \cup, ^c)$ *be a FST (as given in Definition 1) where* $(L, \leq)$ *is a chain and for any* $A, B \in \mathcal{F}_L(X)$, $A \preceq B$ *if and only if* $A(x) \leq B(x)$ *for any* $x \in X$. *Then* $(F_L(X), \preceq, \cap, \cup, ^c)$ *is not a boolean algebra.*

Proof. $(\mathcal{F}_L(X), \preceq)$ is a (bounded and distributive) lattice when endowed with the functionally expressible operations induced from the meet and the join of L (that we will denote Min and Max). It can therefore only be a Boolean algebra if it is possible to find a complement $^c : \mathcal{F}_L(X) \longrightarrow \mathcal{F}_L(X)$ satisfying both the non-contradiction and the excluded middle laws, but this is impossible. To examine this point, it suffices to choose a non-crisp fuzzy set $A \in \mathcal{F}_L(X)$, for which, due to its fuzziness, there exists an element $x_0 \in X$ such that $A(x_0) \notin \{0_L, 1_L\}$. Then $A \cap A^c = \emptyset$ and $A \cup A^c = X$ imply, respectively, $\text{Min}(A(x_0), A^c(x_0)) = 0_L$ and $\text{Max}(A(x_0), A^c(x_0)) = 1_L$, which means, on the one hand, $A^c(x_0) = 0_L$, and, on the other hand, $A^c(x_0) = 1_L$.

Note, therefore, that a very broad and common class of FST, whether or not they be functionally expressible, may never behave like Boolean algebras and cannot, in particular, be isomorphic to any power set $\mathcal{P}(Y)$ (see [91]).

Next, some additional options for exploring fuzzy operators'properties are reviewed: the satisfaction of non-basic Boolean laws, the verification of non-Boolean properties, and the consideration of non-standard interpretations of some Boolean laws.

5.6 Derived Boolean Laws

As is well known, Boolean algebras verify many other laws apart from the basic laws (such as idempotency, associativity, commutativity, absorption, etc.) which we will call *derived Boolean laws*, and which may be proved from the basic laws by performing substitutions of logically equivalent sub-expressions. For example, Von Neumann's laws establish the following equivalences for any crisp sets A and B:

$$(A \cap B) \cup (A \cap B^c) = A$$
$$(A \cup B) \cap (A \cup B^c) = A$$

These two laws are clearly valid in any Boolean algebra. A straightforward proof of the first law is given below (the second one can, thanks to the duality principle, be proved symmetrically):

$$\begin{aligned} A &= A \cap X && \text{(identity law)} \\ &= A \cap (B \cup B^c) && \text{(excluded-middle)} \\ &= (A \cap B) \cup (A \cap B^c) && \text{(distributivity of } \cap \text{ over } \cup) \end{aligned}$$

Note that this law is valid in Boolean algebras by virtue of the identity, the excluded-middle and one of the distributivity laws. In the case of standard fuzzy set theories, Table 2 shows that it is not possible to find any continuous t-conorm S verifying both the excluded-middle and the distributivity of T over S. Nevertheless, as mentioned later, even though the two Von Neumann's laws can never be simultaneously satisfied in any SFST, there are many SFST verifying either one [2].

From a general perspective, these ideas pose the following problem: given a derived Boolean law, is it possible to find a FST where this law is satisfied? Although, as mentioned in Sect. 4, it is always possible to find a SFST that verifies each basic Boolean law, the problem is not solved for arbitrary derived Boolean laws. Nevertheless, the following results are available (see [7, 9]) for dealing with SFST and Pexider-SFST:

- There are derived Boolean laws that are valid in some SFST.
- There are derived Boolean laws that do not hold in any SFST.
- There are derived Boolean laws that do not hold in any SFST but do hold in some Pexider-SFST (see Sect. 5.1).
- There are derived Boolean laws that do not hold in any Pexider-SFST.

In the following, some examples falling in each of these four categories are provided.

Derived Boolean Laws Valid in Some SFST

An interesting example of derived Boolean laws that are valid in some SFST are the aforementioned Von Neumann's laws. The results obtained are as follows [2]:

Proposition 6. *Let $(\mathcal{F}(X), T, S, N)$ be a standard fuzzy set theory. Then:*

(a) The law $S(T(x,y), T(x, N(y))) = x$ holds for all $x, y \in [0,1]$ if and only if there exists an automorphism $\varphi : [0,1] \longrightarrow [0,1]$ such that

$$T = Prod_{\varphi}, \quad S = W^*_{\varphi}, \quad \textit{and} \quad N = N_{\varphi}$$

(b) The law $T(S(x,y), S(x, N(y))) = x$ holds for all $x, y \in [0,1]$ if and only if there exists an automorphism $\varphi : [0,1] \longrightarrow [0,1]$ such that

$$T = W_{\varphi}, \quad S = Prod^*_{\varphi}, \quad \textit{and} \quad N = N_{\varphi}$$

Remark 3. Von Neumann's equations have also been studied in the two following situations:

- When considering Pexider-SFST, i.e. solving equations of the form $S(T_1(x,y), T_2(x, N(y))) = x$ and $T(S_1(x,y), S_2(x, N(y))) = x$ ([3, 39, 40]).
- When considering functionally expressible theories endowed with uninorms and nullnorms ([20]).

Another example within this category is given by the following two derived dual Boolean laws:

$$(A \cap B^c)^c = B \cup (A^c \cap B^c)$$
$$(A \cup B^c)^c = B \cap (A^c \cup B^c)$$

In a similar way as for Von Neumann's Laws, these two equivalences are valid in Boolean algebras. Indeed, the first one can be proved as follows (the proof of the second one would be similar, substituting the laws used in the proof by their dual laws):

$$\begin{aligned} B \cup (A^c \cap B^c) &= (B \cup A^c) \cap (B \cup B^c) && \text{(distributivity of } \cup \text{ over } \cap) \\ &= (B \cup A^c) \cap X && \text{(excluded-middle)} \\ &= B \cup A^c && \text{(identity law)} \\ &= A^c \cup B && \text{(commutativity of } \cup) \\ &= (A \cap B^c)^c && \text{(De Morgan law + involution)} \end{aligned}$$

Again, Table 2 shows that there are no triples (T, S, N) verifying all the laws used in the above proof. Nevertheless, the following result has been obtained [82, 83]:

Proposition 7. *Let $(\mathcal{F}(X), T, S, N)$ be a standard fuzzy set theory. Then:*

(a) The law $N(T(x, N(y))) = S(y, T(N(x), N(y)))$ holds for all $x, y \in [0,1]$ if and only if there exists an automorphism $\varphi : [0,1] \longrightarrow [0,1]$ such that

$$T = Prod_\varphi, \quad S = W^*_\varphi, \quad \text{and} \quad N = N_\varphi$$

(b) The law $N(S(x, N(y))) = T(y, S(N(x), N(y)))$ holds for all $x, y \in [0,1]$ if and only if there exists an automorphism $\varphi : [0,1] \longrightarrow [0,1]$ such that

$$T = W_\varphi, \quad S = Prod^*_\varphi, \quad \text{and} \quad N = N_\varphi$$

Table 3, which summarizes the two results given above, suggests that there exist two families of non-dual SFST, SFST of the forms $(\mathcal{F}(X), \text{Prod}_\varphi, W^*_\varphi, N_\varphi)$ and $(\mathcal{F}(X), W_\varphi, \text{Prod}^*_\varphi, N_\varphi)$, that appear to be the *only* SFST where the derived Boolean laws B10 and B11 do hold.

Another example of a derived Boolean law that holds in different SFST is

$$(A \cup B) \cup (A \cap B) = A \cup B$$

whose solutions include very large families of non-Archimedean t-norms and t-conorms, as well as ordinal sums [7].

Derived Boolean Laws Not Valid in Any SFST

In [7] the authors explored the satisfaction within SFST of what they call *iterative Boolean-like laws*, i.e. Boolean laws where some variables appear several times because they come from non-simplified Boolean identities. Some of these derived Boolean laws clearly hold in some specific SFST, whilst others never hold in such a framework. A simple example of the first case is the law $A \cap A \cap B = A \cap B$, which, in the case of SFST, is equivalent to the functional equation $T(x, T(x, y)) = T(x, y)$ for all $x, y \in [0, 1]$. This equation is obviously satisfied for $T = \text{Min}$, and it suffices to choose $y = 1$ in order to prove that this is the only solution, since, as mentioned in Sect. 4, Min is the only idempotent t-norm. On the other hand, [7] proves that the iterative Boolean-like law $(A \cup A) \cap (A \cap A)^c = \emptyset$ has no solutions in the framework of SFST.

Derived Boolean Laws Valid in Some Pexider-SFST

As mentioned in Sect. 5.1, some situations require the use of Pexider theories, i.e. theories endowed with several fuzzy intersections, unions and/or complements. Theories of this kind are useful when trying to satisfy some derived Boolean laws which are not satisfied in SFST. For example, the aforementioned Boolean law $(A \cup A) \cap (A \cap A)^c = \emptyset$, for which no SFST exists that makes it true, reduces, in the case of Pexider-like SFST, to solving the Pexider functional equation $T_1(S(x, x), N(T_2(x, x))) = 0$, which has infinite solutions (as, e.g., $T_1 = W$, $S = \text{Max}$, $T_2 = \text{Min}$ and $N = 1 - Id$; for more details see [7]).

Derived Boolean Laws Not Valid in Any Pexider-SFST

There exist some Boolean laws that do not hold in any Pexider-SFST, and obviously not in SFST. In [9] the authors explored the satisfaction of the Boolean law:

Table 3. Satisfaction of some derived Boolean properties in standard fuzzy set theories

	T	S	N
B10. Von Neumann's Laws			
$S(T(x,y), T(x, N(y))) = x$	Prod_φ	W^*_φ	N_φ
$T(S(x,y), S(x, N(y))) = x$	W_φ	Prod^*_φ	N_φ
B11. Other Laws			
$N(T(x, N(y))) = S(y, T(N(x), N(y)))$	Prod_φ	W^*_φ	N_φ
$N(S(x, N(y))) = T(y, S(N(x), N(y)))$	W_φ	Prod^*_φ	N_φ

$$A \cup B = [(A \cup B) \cap (A \cup C)] \cup [(A \cup B) \cap A^c] \tag{1}$$

and concluded that no Pexider-SFST verifies this law. Nevertheless, they did find a non-standard solution, which is restated in the following proposition:

Proposition 8. *Let S_1, S_2, S_3, S_4, S_5 be binary operations in $[0,1]$ which are non-decreasing in each place, 1 is an absorbent element and 0 is a unit. Let T_1, T_2 be two binary operations in $[0,1]$ which are non-decreasing in each place with 1 as a unit and 0 as an absorbent element. Let $N : [0,1] \to [0,1]$ be a function such that $N(0) = 1$ and $N(1) = 0$. Then the general solution of*

$$S_1(a,b) = S_2(T_1(S_3(a,b), S_4(a,c)), T_2(S_5(a,b), N(a))) \tag{2}$$

is given by:

- $S_2 = Max$;
- $S_1 = S_5 \geq S_3$;
- $N(x) = 1$ *for all* x *in* $[0,1)$; *(note that* N *is not a strong negation)*
- S_4, T_1, T_2, *arbitrary.*

That is, there is no Pexider-SFST verifying (2).

Note 1. The law (1) belongs to a collection of postulates for orthomodular lattices [14], and therefore it can be said that neither SFST nor Pexider-SFST verify all the laws of an orthomodular lattice and, obviously, of a Boolean algebra.

5.7 Non-Boolean Properties

In the framework of functionally expressible theories $(\mathcal{F}(X), F, G, N)$ it is also possible to study the satisfaction of properties that do not have a Boolean background.

Perhaps one of the most important non-Boolean properties is *continuity*. Even if the continuity of functions F, G and N is not strictly necessary, it generally appears to be a very desirable property. Informally speaking, continuity ensures that small variations in the input values will not cause important differences in the function's output, and this is an interesting property for functions that are used to represent the intersection, union and complement of fuzzy sets. Indeed, fuzzy sets are usually described by means of continuous functions in order to reflect the flexibility in the use of the fuzzy propositions they are representing. Therefore, if F, G or N are chosen as discontinuous functions, this may entail the discontinuity of the fuzzy sets $A \cap B$, $A \cup B$ or A^c, which will therefore not mirror the flexibility shown by A and B.

Note, however, that non-continuous operators (in particular, non-continuous t-norms and t-conorms) may be useful in different applications, such as in the context of residuated lattices and theories based on left-continuous t-norms [23, 50–52] or in fuzzy preference modeling [40].

Another important and well-known non-Boolean property is Frank's equation, given by $F(x,y) + G(x,y) = x + y$ for any $x, y \in [0,1]$, which has been solved both for the case of t-norms/t-conorms [43] as well as for aggregation operators belonging to the classes of uninorms and nullnorms [20].

Other examples of functional properties of this kind are, for any binary aggregation operator $M : [0,1]^2 \longrightarrow [0,1]$, as follows [1, 8, 21, 33, 41, 55, 66–68]:

- *Bisymmetry.* For any $x_1, x_2, y_1, y_2 \in [0,1]$,

$$M(M(x_1,x_2), M(y_1,y_2)) = M(M(x_1,y_1), M(x_2,y_2))$$

- *Shift-invariance.* For any $b \in]0,1[$ and for any $x_1, x_2 \in [0, 1-b]$,

$$M(x_1 + b, x_2 + b) = M(x_1, x_2) + b$$

- *Homogeneity.* For any $b \in]0,1[$ and for any $x_1, x_2 \in [0,1]$,

$$M(bx_1, bx_2) = bM(x_1, x_2)$$

- *Additivity.* For any $x_1, x_2, y_1, y_2 \in [0,1]$ such that $x_1 + y_1, x_2 + y_2 \in [0,1]$,

$$M(x_1 + y_1, x_2 + y_2) = M(x_1, x_2) + M(y_1, y_2)$$

- *Stability.* For any $x_1, x_2 \in [0,1]$, and any automorphism $\varphi : [0,1] \longrightarrow [0,1]$,

$$M(\varphi(x_1), \varphi(x_2)) = \varphi(M(x_1, x_2))$$

- *φ-comparability.* For any $x_1, x_2, y_1, y_2 \in [0,1]$, and any automorphism $\varphi : [0,1] \longrightarrow [0,1]$,

$$\text{If } M(x_1,x_2) < M(y_1,y_2) \text{ then } M(\varphi(x_1), \varphi(x_2)) < M(\varphi(y_1), \varphi(y_2))$$

- *Principle of incremental coherence.* For any $x_1, x_2, y \in [0,1]$ such that $x_1 + y \in [0,1]$,

$$M(x_1 + y, x_2) - M(x_1, x_2) = M(y, x_2) - M(0, x_2)$$

- *Principle of incremental linearity.* For any $x_1, x_2, y \in [0,1]$ such that $x_1 + y \in [0,1]$,

$$M(x_1 + y, x_2) - M(x_1, x_2) = \alpha y + \beta$$

  where $\alpha = \alpha(x_1, x_2)$ and $\beta = \beta(x_1, x_2)$ are arbitrary functions on $[0,1]$.

5.8 Reinterpretation of Boolean Properties

Some Boolean laws do not have a unique interpretation. This is the case, for example, of the well-known *Non-Contradiction* (NC) law, which, in its ancient Aristotelian formulation, can be stated as follows: for any statement p, the statements p and *not* p cannot be at the same time, i.e. $p \wedge \neg p$ is *impossible*, where the binary operation $\wedge$ represents the *and* connective and the unary operation $\neg$ models the negation. This formulation may be interpreted in two (at least) different ways, depending on how the term *impossible* is understood [84]:

- Taking the approach that is common in modern logic, the term *impossible* may be thought of as *false*, and then the NC principle may be expressed in a structure with the minimum element $\mathbf{0}$ as $p \wedge \neg p = \mathbf{0}$ for any statement p.
- Another possibility, which can be considered closer to ancient logic, is to interpret *impossible* as *self-contradictory* in the understanding that an object is self-contradictory whenever it entails its negation. In this case, the NC principle may be written as $p \wedge \neg p \models \neg(p \wedge \neg p)$ for any statement p, where $\models$ represents an entailment relation.

Notice that the latter treatment is syntactic rather than semantic as is the approach taken in modern logic, where the NC law is only considered after the introduction of truth values.

In the setting of orthocomplemented lattices (and therefore in both quantum and classical logics) the modern and ancient interpretations coincide, since the minimum element is the only self-contradictory object. Nevertheless, it can be proved [84] that this is not the case for more general structures, where the first approach is clearly stronger than the second. Indeed, if $p \wedge \neg p = \mathbf{0}$, then obviously $\mathbf{0} = p \wedge \neg p \models \neg(p \wedge \neg p)$ is also verified, since $\mathbf{0}$ is the minimum element, but the opposite is not always true.

One of the structures where the two interpretations differ is the lattice $([0,1], \leq)$, the framework associated with binary aggregation operators acting on $[0,1]$, which, as mentioned in Sect. 5.2, may be used to build functionally expressible FST. Then, if $F : [0,1]^2 \longrightarrow [0,1]$ is an aggregation operator that is used to generate a fuzzy intersection $\cap$, and a strong negation $N : [0,1] \longrightarrow [0,1]$ is used to generate a fuzzy complement, the NC law $A \cap A^c = \emptyset$ can be interpreted in the following two ways:

- $\forall x \in [0,1], F(x, N(x)) = 0$ (NC in modern logic)
- $\forall x \in [0,1], F(x, N(x)) \leq N(F(x, N(x)))$ (NC in ancient logic)

Similar arguments may be applied to the excluded-middle (EM) law. Indeed, this law may be formulated as follows: for any statement p, either p or *not* p are true, or, equivalently, $\neg(p \vee \neg p)$ is *impossible*. This allows again for two different interpretations: $\neg(p \vee \neg p) = \mathbf{0}$ (modern logic) and $\neg(p \vee \neg p) \models \neg(\neg(p \vee \neg p))$ (ancient logic). Then, in the case of a functionally expressible FST whose fuzzy union operator $\cup$ is generated by an aggregation operator $G : [0,1]^2 \longrightarrow [0,1]$ and, again, a strong negation function

$N : [0,1] \longrightarrow [0,1]$ is used to induce a fuzzy complement, the EM law $A \cup A^c = X$ can be interpreted in either of the following ways:

- $\forall x \in [0,1], G(x, N(x)) = 1$ (EM in modern logic)
- $\forall x \in [0,1], N(G(x, N(x))) \leq G(x, N(x))$ (EM in ancient logic)

The results obtained when studying the satisfaction of the NC law within the aggregation operators world have been studied in [69] (modern logic interpretation) and in [70] (ancient logic interpretation). The latter interpretation provides a new criterion for studying and comparing the behavior of aggregation operators.

5.9 The Fuzzy Set Space

As already mentioned in Sect. 2.5, there are situations where it is interesting to restrict the fuzzy set space to a subset of the whole set $\mathcal{F}_L(X)$. This happens, e.g. if one wants to avoid self-contradiction in fuzzy set theory [85]. Indeed, the logical concept of *contradiction* is, generally, defined by means of a negation $\neg$ and an entailment relation $\leq$ in such a way that "p and q are contradictory" if and only if $p \leq \neg q$. This means that "p is self-contradictory" if and only if $p \leq \neg p$. In Boolean algebras, the latter is equivalent to $p \wedge \neg p = 0$, and therefore, the only self-contradictory object is the minimum element 0 (the empty set when considering the algebra of classical sets). But this is not the case in more general settings as in FST, where there are many self-contradictory fuzzy sets.

Definition 10. *Let* $(\mathcal{F}_L(X), \preceq, \cap, \cup, ^c)$ *be a FST. Then a fuzzy set* $A \in \mathcal{F}_L(X)$ *is said to be* self-contradictory *when* $A \preceq A^c$.

If $L = [0,1]$ and the fuzzy complements are considered to be built by means of strong negation functions, the following definitions can be established [85].

Definition 11. *Let us consider the set* $(\mathcal{F}(X), \preceq)$, *where* $\preceq$ *is the pointwise order built from the usual real order. Then for any* $A \in \mathcal{F}(X)$:

- *A is said to be* strongly self-contradictory *when it is* $A \preceq N \circ A$ *for any strong negation* N.
- *A is said to be* weakly self-contradictory *when it is* $A \preceq N \circ A$ *for some strong negation* N.

Therefore, the only strongly self-contradictory fuzzy set is the empty set, and the necessary and sufficient condition for a fuzzy set A not being weakly self-contradictory is $SupA = 1$, that is, A has to be a normalized fuzzy set. This means, also taking into account that $A \preceq N \circ A$ is equivalent to $A(x) \leq x_N$ for all $x \in [0,1]$, where x_N is the negation's fixed point, that there are two possible ways of avoiding self-contradictory fuzzy sets (as well as fuzzy sets whose complements are self-contradictory):

1. If one wants to avoid any weakly self-contradictory fuzzy set (except the empty set), it is necessary to consider the fuzzy set space defined as [86]:

$$\mathcal{F}_{01}(X) = \{A : X \longrightarrow [0,1]; \mathrm{Sup}A = 1; \mathrm{Inf}A = 0\} \cup \{\emptyset, X\}$$

2. If, given a strong negation N with a fixed point x_N, one wants to avoid fuzzy sets that are self-contradictory with respect to N (except the empty set), then the fuzzy set space to be considered is [89]:

$$\mathcal{F}_{nc}(X) = \{A : X \longrightarrow [0,1]; \mathrm{Sup}A > x_N; \mathrm{Inf}A < x_N\} \cup \{\emptyset, X\}$$

Note that both $\mathcal{F}_{01}(X)$ and $\mathcal{F}_{nc}(X)$ contain all the crisp sets and prevent self-contradiction. Nevertheless, the latter property is obtained at quite a high price, since constant fuzzy sets (which may be needed in many applications) are to be discarded.

Notice also that building fuzzy intersections and unions for these kind of fuzzy set spaces is not so straightforward as for the general case. For example, it is no longer possible to consider functionally expressible operators induced by t-norms and t-conorms, since the fuzzy sets obtained in this way clearly do not necessarily belong to the fuzzy set spaces $\mathcal{F}_{01}(X)$ or $\mathcal{F}_{nc}(X)$. When studying these fuzzy set spaces, the authors proposed to use the following non-functionally expressible connectives (where T is a t-norm and S is a t-conorm) [86,89]:

1. In $\mathcal{F}_{01}(X)$:
$$(A \cap B)(x) = \begin{cases} T(A(x), B(x)), & \text{if } \mathrm{Sup}(T \circ (A \times B)) = 1 \\ 0, & \text{otherwise} \end{cases}$$
$$(A \cup B)(x) = \begin{cases} S(A(x), B(x)), & \text{if } \mathrm{Inf}(S \circ (A \times B)) = 0 \\ 1, & \text{otherwise} \end{cases}$$
2. In $\mathcal{F}_{nc}(X)$:
$$(A \cap B)(x) = \begin{cases} T(A(x), B(x)), & \text{if } \mathrm{Sup}(T \circ (A \times B)) > x_N \\ 0, & \text{otherwise} \end{cases}$$
$$(A \cup B)(x) = \begin{cases} S(A(x), B(x)), & \text{if } \mathrm{Inf}(S \circ (A \times B)) < x_N \\ 1, & \text{otherwise} \end{cases}$$

6 Conclusions

This paper intends just to draw attention to the actual theories of fuzzy sets viewed not only through the properties of functions taking values in the unit interval, a technique of great importance when the theories are supposed to be functionally expressible. In a sense, the paper tries to go back to the origins of the mathematical study of fuzzy sets, where what really mattered was the semantic contents and not just the obtention of mathematical results. The interest of the mathematical study of fuzzy sets mainly lies in its fertility in the field of both mathematics itself and in its applications, once its productiveness has been recognized by the leading experts in the respective fields.

From the very beginning, fuzzy sets had an impaction in the important problem of *meaning*, which, as is well known, is considered a crucial issue for the future of Computational Intelligence. In fact, a fuzzy set is more than a function only when it represents something in a language. This is the way in which fuzzy sets can be used to generalize Cantor's axiom of specificity to imprecise properties of elements in a universe of discourse. What really counts for fuzzy set theories is the world of imprecision and its related uncertainties, and, perhaps, a new axiom of specificity could be informally stated by saying that, for any family of predicates on X, there is an associate fuzzy sets theory $(\mathcal{F}_L(X), \preceq, \cap, \cup, ^c)$ that is sound with respect to the respective uses of the linguistic terms *and* (see [47]), *or* (see [48]) and *not* between the predicates in the family. This is why the mathematical study of fuzzy sets, like that of classical sets, acquires relevance outside mathematics and, namely, in knowledge representation.

In the above spirit, there is a lot of room for mathematics in fuzzy set theories. On the subject of this paper, the study and usefulness of non-functionally expressible theories, the genesis of fuzzy sets coming from the aggregation of precise information and the respective theories, or the study of families of predicates whose degrees cannot be directly given in $[0, 1]$ but in a non-numerical poset L, e.g. are just some instances of a large collection of open problems. In our view, the future of fuzzy sets will partly hinge on an experimental science of fuzziness where mathematics is called upon not only to supply models fitting the experimental facts, but suggesting new ideas and experiments.

References

1. J. Aczél. *Lectures on Functional Equations and their Applications* (Academic Press, New York, 1966).
2. C. Alsina. On a family of connectives for fuzzy sets. *Fuzzy Sets and Systems* **16** (1985) 231–235.
3. C. Alsina. On connectives in fuzzy logic satisfying $S(T_1(x, y), T_2(x, N(y))) = x$, in *Proc. FUZZ'IEEE-97*, Barcelona, (1999), 149–153.
4. C. Alsina, M.J. Frank and B. Scheiwzer. *Associative functions. Triangular Norms and Copulas* (World Scientific, Singapure, 2006).
5. C. Alsina and E. Trillas. On the functional equation $S_1(x, y) = S_2(x, T(N(x), y))$, in *Functional equations results and advances*, eds. Z. Daróczy and Z. Páles (Kluwer, Dordrecht, 2002), 323–334.
6. C. Alsina and E. Trillas. On the symmetric difference of fuzzy sets. *Fuzzy Sets and System* **153** (2005) 181–194.

7. C. Alsina and E. Trillas. On iterative Boolean-like laws of fuzzy sets, in *Proceedings of the 4th International Conference in Fuzzy Logic and Technology* (Barcelona, Spain, 2005) 389–394.
8. C. Alsina, E. Trillas and C. Moraga. Combining degrees of impairment: the case of the index of Balthazard. *Mathware & Softcomputing* **X**(1) (2003) 23–41.
9. C. Alsina, E. Trillas and E. Renedo. On two classical laws with fuzzy sets, in *Proc. ESTYLF'06*, (Ciudad Real, Spain, 2006) 19–21.
10. C. Alsina, E. Trillas and L. Valverde. On non-distributive logical connectives for fuzzy set theory. *BUSEFAL* **3** (1980) 18–29.
11. C. Alsina, E. Trillas and L. Valverde. On some logical connectives for fuzzy set theory. *J. Math. Anal. Appl.* **93** (1983) 15–26.
12. K.T. Atanassov. Intuitionistic fuzzy sets. *VII ITKR's Session* Sofia June 1983.
13. K.T. Atanassov and S.Stoeva. Intuitionistic L-fuzzy sets. R. Trapple (Ed.), *Cybernetics and Systems Research* **2**, Elsevier, Amsterdam (1984) 539–540.
14. L. Beran, *Orthomodular Lattices*, D. Reidel Pubs., 1985.
15. C. Bertoluzza and V. Doldi. On the distributivity between t-norms and t-conorms. *Fuzzy Sets and Systems* **142** (2004) 85–104.
16. B. Bouchon-Meunier, editor. *Aggregation and Fusion of Imperfect Information* (Studies in Fuzziness and Soft Computing, volume 12, Physica Verlag, Springer, 1998).
17. H. Bustince and P. Burillo. Vague sets are intuitionistic fuzzy sets. *Fuzzy Sets and Systems*, **79** (3) (1996) 403–405.
18. T. Calvo. On some solutions of the distributivity equation. *Fuzzy Sets and Systems* **104** (1999) 85–96.
19. T. Calvo and B. De Baets. On a generalization of the absorption equation. *J. Fuzzy Math.* **8** (2000) 141–149.
20. T. Calvo, B. De Baets and J. Fodor. The functional equations of Frank and Alsina for uninorms and nullnorms. *Fuzzy Sets and Systems* **120**(3) (2001) 385–394.
21. T. Calvo, G. Mayor and R. Mesiar, editors. *Aggregation Operators: New Trends and Applications* (Studies in Fuzziness and Soft Computing, volume 97, Physica Verlag, Springer, 2002).
22. M. Carbonell, M. Mas, J. Suñer and J. Torrens. On distributivity and modularity in De Morgan triplets. *Internat.J.Uncertainty Fuzz. Knowledge-Based Syst.* **4** (1996) 351–368.
23. R. Cignoli, F. Esteva, L. Godo and F. Montagna. On a class of left-continuous t-norms. *Fuzzy Sets and Systems* **131**(3) (2002) 283–296.
24. D. Coker. Fuzzy rough sets are intuitionistic L-fuzzy sets. *Fuzzy Sets and Systems* **96** (1998) 381–383.
25. B. De Baets. Idempotent uninorms. *European J.Oper.Res.* **118** (1999) 631–642.
26. A. DeLuca and S. Termini. Entropy of L-fuzzy sets. *Infor. & Control* **24** (1974) 55–73.
27. G. Deschrijver and E. Kerre. On the relationship between some extensions of fuzzy set theory. *Fuzzy Sets and Systems* **133** (2003) 227–235.
28. J. Dombi. Basic concepts for a theory of evaluation: The aggregative operator. *European J. Oper. Res.* **10** (1982) 282–293.
29. P. Drigaś. A characterization of idempotent nullnorms. *Fuzzy Sets and Systems* **145** (2004) 455–461.
30. D. Dubois. Triangular norms for fuzzy sets, in *Proceedings Second International Seminar on Fuzzy Set Theory, Linz*, 1980, 39–68.

31. D. Dubois and H. Prade. New results about properties and semantics of fuzzy-set-theoretic operators, in *Fuzzy Sets: Theory and Applications to Policy Analysis and Information Systems*, eds. P.P. Wang and S.K. Chang (Plenum Publ. 1980), 59–75.
32. D. Dubois and H. Prade. *Fuzzy Sets and Systems. Theory and Applications* (Academic Press, New York, 1980).
33. D. Dubois and H. Prade. A review of fuzzy set aggregation connectives. *Information Sciences* **36** (1985) 85–121.
34. D. Dubois and H. Prade. Twofold fuzzy sets and rough sets — Some issues in knowledge representation. *Fuzzy Sets and Systems* **23** (1987), 3–18.
35. D. Dubois and H. Prade. Rough fuzzy sets and fuzzy rough sets. International *Journal of General Systems*, **17**(1990) 191–209.
36. D. Dubois and H. Prade (editors). *Fundamentals of Fuzzy Sets* (Kluwer, Upper Saddle River, NJ, 2000).
37. Q. Feng and Z. Bin. The distributive equations for idempotent uninorms and nullnorms. *Fuzzy Sets and Systems* **155**(3) (2005) 446–458.
38. J. Fodor. Strict preference relations based on weak t-norms. *Fuzzy Sets and Systems* **43** (1991) 327–336.
39. J. Fodor and M. Roubens. Valued preference structures. *European J. Oper. Res.* **97** (1994) 277–286.
40. J. Fodor and M. Roubens. *Fuzzy Preference Modelling and Multicriteria Decision Support* (Kluwer Academic Publishers, Dordrecht, 1994).
41. J. Fodor and M. Roubens. On meaningfulness of means. *Journal of Computational and Applied Mathematics* (1995) 103–115.
42. J. Fodor and T. Keresztfalvi. Nonstandard Conjunctions and Implications in Fuzzy Logic. *International Journal of Approximate Reasoning* **12** (1995) 69–84.
43. M.J. Frank. On the simultaneous associativity of $F(x, y)$ and $x + y - F(x, y)$. *Aequationes Math.* **19** (1979) 194–226.
44. W.L. Gau and D.J. Buehrer. Vague sets. *IEEE Trans. Systems Man Cybernet.* **23** (2) (1993) 610–614.
45. Y. Gentilhomme. Les ensembles flous en linguistique. *Cahiers de linguistique théorique et appliquée* **5** (1968) 47–65.
46. J.A. Goguen. L-fuzzy sets. *J. Math. Anal. Appl.* **18** (1967) 145–174.
47. S. Guadarrama and E. Renedo. A reflection on the use of *And. Int. Jour. of Applied Mathematics and Computer Science* (in press).
48. S. Guadarrama, E. Renedo and E. Trillas. Some fuzzy counterparts of the language uses of *And* and *Or*. *Proc. of 9th Fuzzy Days'06*, (Dortmund 2006) (in press).
49. J. Harmse. Continuous Fuzzy Conjunctions and Disjunctions. *IEEE Transactions on Fuzzy Systems* **4**(3) (1996) 295–314.
50. S. Jenei. New family of triangular norms via contrapositive symmetrization of residuated implications. *Fuzzy Sets and Systems* **110** (1999) 157–174.
51. S. Jenei. Structure of Girard Monoids on [0,1], in *Topological and Algebraic Structures in Fuzzy Sets, A Handbook of Recent Developments in the Mathematics of Fuzzy Sets*, eds. E.P. Klement, S.E. Rodabaugh (Trends in Logic, volume 20, Kluwer Academic Publishers, Dordrecht, 2003), 277–308.
52. S. Jenei. How to construct left-continuous triangular norms-state of the art. *Fuzzy Sets and Systems* **143** (2004) 27–45.

53. E. Kerre. A First view on the alternatives of fuzzy set theory, in B. Reusch, H.-H. Temme (Eds.), *Computational Intelligence in Theory and Practice*, Physica-Verlag, Heidelberg (2001), 55–72.
54. E. Klement, R. Mesiar and E. Pap. On the relationship of associative compensatory operators to triangular norms and conorms. *Int. Journal of Uncertainty, Fuzziness and Knowledge-Based Systems.* **4**(2) (1996) 129–144.
55. E.P. Klement, R. Mesiar and E. Pap. *Triangular Norms* (Trends in Logic, volume 8, Kluwer Academic Publishers, Dordrecht, 2000).
56. C.H. Ling. Representation of Associative Functions. *Publ. Math. Debrecen* **12** (1965) 189–212.
57. R. Lowen. On fuzzy complements. *Inform. Sci.* **14** (1978) 107–113.
58. J.L. Marichal. On the associativity functional equation. *Fuzzy Sets and Systems* **114** (2000) 381–389.
59. M. Mas, G. Mayor and J. Torrens. The distributivity condition for uninorms and t-operators. *Fuzzy Sets and Systems* **128** (2002) 209–225.
60. K. Menger. Statistical Metric Spaces. *Proc. Nat. Acad. Sci. USA* **28** (1942) 535–537.
61. M. Mizumoto and K. Tanaka. Some properties of fuzzy sets of type-2. *Information and Control*, **31** (1976) 312–340.
62. M. Mukaidono. A set of independent and complete axioms for a fuzzy algebra (Kleene algebra). In *Proceedings of the 11th International Symposium of Multiple-Valued Logic* (IEEE, 1981)
63. R.B. Nelsen. *An Introduction to Copulas* (Lecture Notes in Statistics, volume 139, Springer, New York, 1999).
64. S.V. Ovchinnikov. Involutions in Fuzzy Set Theory. *Stochastica* **IV** (1980) 227–231.
65. S.V. Ovchinnikov. General negations in Fuzzy Set Theory. *Journal of Mathematical Analysis and Applications* **92**(1) (1983) 234–239.
66. S.V. Ovchinnikov. Means on ordered sets. *Mathematical Social Sciences* **32** (1996) 39–56.
67. S.V. Ovchinnikov. An Analytic Characterization of some Aggregation Operators. *International Journal of Intelligent Systems.* **13** (1998) 59–68.
68. S.V. Ovchinnikov. On robust aggregation procedures. In [16] 3–10.
69. A. Pradera and E. Trillas. Aggregation and Non-Contradiction, in *Proceedings of the 4th International Conference in Fuzzy Logic and Technology* (Barcelona, Spain, 2005) 165–170.
70. A. Pradera and E. Trillas. Aggregation Operators from the ancient NC and EM point of view. *Kybernetika* **42** (3) (2006) 243–260.
71. A. Pradera, E. Trillas and T. Calvo. A general class of triangular norm-based aggregation operators: quasi-linear T-S operators. *International Journal of Approximate Reasoning* **30** (2002) 57–72.
72. A. Pradera, E. Trillas and E. Renedo. An overview on the construction of fuzzy set theories. *New Mathematics and Natural Computation* **1** (3), (2005), 329–358.
73. E. Renedo, E. Trillas and C. Alsina. A Note on the Symmetric Difference in Lattices. *Mathware & Soft Computing* **12** (2005) 75–81.
74. D. Ruiz and J. Torrens. Distributive idempotent uninorms. *Internat. J.Uncertainty Fuzz.Knowledge-Based Syst.* **11** (2003) 413–428.
75. R. Sambuc. Fonctions Φ-floues. Application à l'aide au diagnostic en pathologie thyroidienne, Ph.D. Thesis, Univ. Marseille, France, 1975.

76. B. Schweizer and A. Sklar. *Probabilistic Metric Spaces.* (North-Holland, 1983).
77. W. Silvert. Symmetric summation: A class of operations on fuzzy sets. *IEEE Trans. Systems, Man Cybernet.* **9** (1979) 659–667.
78. E. Trillas. Sobre funciones de negación en la teoría de los subconjuntos difusos. *Stochastica, III-1* (1979) 47-59 (in Spanish). Reprinted (English version) in *Advances of Fuzzy Logic* S. Barro et al. (eds), Universidad de Santiago de Compostela 31–43, 1998.
79. E. Trillas. On words and fuzzy sets. *Information Science* **176**, 11, (2006), 1463–1487.
80. E. Trillas and C. Alsina. On Sheffer Stroke, in *Proc. XIth International Symposium on Multiple-Valued Logic,* (1981) 244–245.
81. E. Trillas and C. Alsina. Fuzzy Sets from a Mathematical-Naïve Point of View, in *Technologies for Constructing Intelligent Systems,* **2**, Studies in Fuzziness and Soft Computing, Springer Verlag (2002), 381–391.
82. E. Trillas and C. Alsina. Elkan's theoretical argument, reconsidered. *International Journal of Approximate Reasoning* **26** (2001) 145–152.
83. E. Trillas and C. Alsina. Standard Theories of Fuzzy Sets with the law $(\mu \wedge \sigma')' = \sigma \vee (\mu' \wedge \sigma')$. *International Journal of Approximate Reasoning,* **37**(2) (2004) 87–92.
84. E. Trillas, C. Alsina and A. Pradera. Searching for the roots of non-contradiction and excluded-middle. *International Journal of General Systems* **31** (2002) 499–513.
85. E. Trillas, C. Alsina and J. Jacas. On contradiction in fuzzy logic. *Soft Computing* **3** (1999) 197–199.
86. E. Trillas, C. Alsina and J. Jacas. On Logical Connectives for a Fuzzy Set Theory with or without Nonempty Self-Contradictions. *International Journal of Intelligent Systems* **15** (2000) 155–164.
87. E. Trillas, S. Cubillo and J.L. Castro. Conjunction and disjunction on ([0,1], $\leq$). *Fuzzy Sets and Systems* **72** (1995) 155–165.
88. E. Trillas, A. Pradera, and S. Cubillo. A mathematical model for fuzzy connectives and its application to operators behavioural study. In B. Bouchon-Meunier, R.R. Yager, and L.A. Zadeh, editors, *Information, Uncertainty and Fusion,* The Kluwer International Series in Engineering and Computer Sciences, pages 307–318. Kluwer Academic Publishers, 1999.
89. E. Trillas, E. Renedo and S. Guadarrama. On a new theory of fuzzy sets with just one self-contradiction. *Proceedings of the 10th IEEE International Conference on Fuzzy Systems,* Melbourne, Australia, **2**, (2001), 105–108.
90. G. Wang and Y. He. Intuitionistic fuzzy sets and L-fuzzy sets. *Fuzzy Sets and Systems,* **110** (2), (2000), 271–274.
91. A. Weinberger. Reducing Fuzzy Algebra to Classical Algebra. *New Mathematics and Natural Computation* **1**(1) (2005) 27–64.
92. R.R. Yager. On the measure of fuzziness and negation II. Lattices. *Inform. and Control* **44** (1980) 236–260.
93. R.R. Yager and A. Rybalov. Uninorm aggregation operators. *Fuzzy Sets and Systems* **80** (1996) 111–120.
94. L.A. Zadeh. Fuzzy Sets. *Information and Control* **8** (1965) 338–353.
95. L.A. Zadeh. Quantitative fuzzy semantics. *Information Sciences* **3** (1971) 159–176.
96. H.J. Zimmermann and P. Zysno. Latent connectives in human decision-making. *Fuzzy Sets and Systems* **4** (1980) 37–51.

Uninorm Basics

János Fodor* and Bernard De Baets

Abstract. In this paper, we give an overview of some basic results concerning uninorms. Uninorms generalize both t-norms and t-conorms as they allow for a neutral element anywhere in the unit interval. We revise their structure and the main classes characterized so far. We also look at related operators such as residual implicators and nullnorms. Finally, we mention some interesting results on functional equations involving uninorms.

1 Introduction

The appropriate definition of connectives (conjunction, disjunction, negation, implication, etc.) is one of the most basic problems in fuzzy logic and its applications. By now, triangular norms, triangular conorms, strong negations, and related implications have become standard tools both in theoretical investigations and practical applications. These classes of operations are mathematically sound and contain a wide variety of particular members.

Uninorms were introduced by Yager and Rybalov [21] as a generalization of t-norms and t-conorms. For uninorms, the neutral element is not forced to be either 0 or 1, but can be any value e in the unit interval. Apart from their theoretical interest there are some practical reasons behind uninorms. Consider for instance the setting of multicriteria decision making, where aggregation is one of the key issues. Suppose that some alternatives are evaluated from several points of view, and each evaluation is a number from the unit interval. Let us choose a level of satisfaction $e \in [0,1]$. Two semantical ways of expressing aggregation of the obtained numbers are as follows [21]:

(i) If **all** criteria are satisfied to at least the extent e, then we are satisfied with any of them; else we want all of them satisfied.
(ii) If **any** of the evaluations is above e, then we are satisfied with any of them; else we want them all satisfied.

* This research has been supported in part by OTKA T046762

Such situations can be modelled perfectly by uninorms and this leads to the particular classes introduced in [21], and also to the general forms studied in [9].

By no means, this paper gives a complete overview of what is known about uninorms. Instead, we present some carefully selected basic pieces that fit together and provide literature pointers to further work. We have organized this paper as follows. In Sect. 2 we revise the basic notions related to uninorms and emphasize their block-wise structure. The main classes of uninorms are discussed in Sect. 3 with ample attention given to idempotent uninorms and representable uninorms. De Morgan triplets based on uninorms are touched upon in Sect. 4. In Sect. 5 it is explained how an implicator can be obtained from a uninorm through the application of the residuation technique. The case of representable uninorms is discussed in detail drawing the analogy with nilpotent t-norms. Nullnorms, a class of operations related to uninorms, are introduced in Sect. 6, and the interaction between uninorms and nullnorms in the context of functional equations is treated in the final section.

2 Fundamentals of Uninorms

2.1 Basic Notions

Throughout this paper, a $[0,1] \longrightarrow [0,1]$ mapping is called a unary operator, and a $[0,1]^2 \longrightarrow [0,1]$ mapping a binary operator.

Definition 1. *[21]* A *uninorm* U is a commutative, associative and increasing binary operator with a neutral element $e \in [0,1]$, i.e. for all $x \in [0,1]$ we have $U(x,e) = x$.

This definition is a rather general one and includes the following special cases. If $e = 1$ then Definition 1 gives back the notion of *t-norms*, which are well-accepted models for **AND** in fuzzy logic [7]. If $e = 0$ then the class of *t-conorms* is obtained. These operations model logical **OR** in fuzzy logic [7]. Therefore, a uninorm U with a neutral element $e \in]0,1[$ is called *proper.*

Example 1. [21] Consider the binary operator U_c defined by

$$U_c(x,y) = \begin{cases} \max(x,y) & , \text{ if } (x,y) \in [e,1]^2 , \\ \min(x,y) & , \text{ elsewhere.} \end{cases}$$

This is a uninorm with a neutral element e, satisfying also $U_c(0,1) = U_c(1,0) = 0$. The binary operation U_d defined by

$$U_d(x,y) = \begin{cases} \min(x,y) & , \text{ if } (x,y) \in [0,e]^2 , \\ \max(x,y) & , \text{ elsewhere} \end{cases}$$

is also a uninorm with a neutral element e and satisfies $U_d(0,1) = U_d(1,0) = 1$.

Proposition 1. *[9] For any uninorm U, the common value $U(0,1) = U(1,0)$ is an* annihilator *of U, i.e., for any $x \in [0,1]$ we have*

$$U(0,1) = U(U(0,1), x).$$

In addition, one of the following two cases always holds:

(i) $U(0,1) = U(1,0) = 0$;
(ii) $U(0,1) = U(1,0) = 1$.

As a consequence, there are only two consistent ways of defining $U(0,1)$. If $U(0,1) = 0$ then we call the uninorm U *conjunctive*, since it satisfies the classical boundary conditions of **AND**. If $U(0,1) = 1$ then U is called *disjunctive*, for a similar reason.

Another interesting particular case is given by the so-called *aggregative operators* introduced by Dombi [6]. He investigated associative, strictly increasing binary operators that are continuous except in the points $(0,1)$ and $(1,0)$, and these are *self-dual* with respect to an involutive negator N:

$$N(U(x,y)) = U(N(x), N(y)), \quad \text{for all } x, y \in [0,1] .$$

Recall that a negator N is a decreasing unary operator that exchanges 0 and 1. It is called strict when it is strictly decreasing, and involutive when $N(N(x)) = x$ for all $x \in [0,1]$. If $e \in]0,1[$ is the unique fixpoint of N (i.e. a value for which $N(e) = e$) then e is the neutral element of these aggregative operators. Therefore, this class consists of uninorms.

2.2 The Structure of Uninorms

The structure of uninorms has been studied by Fodor et al. [9].

Proposition 2. *[9] Consider a proper uninorm U with a neutral element $e \in]0,1[$.*

(i) The binary operator T_U defined by

$$T_U(x,y) = \frac{U(ex, ey)}{e}, \quad \textit{for all } (x,y) \in [0,1]^2$$

is a t-norm.

(ii) The binary operator S_U defined by

$$S_U(x,y) = \frac{U(e + (1-e)x, e + (1-e)y) - e}{1-e}, \quad \textit{for all } (x,y) \in [0,1]^2$$

is a t-conorm.

The structure of a uninorm with the neutral element e on the squares $[0,e]^2$ and $[e,1]^2$ is therefore closely related to t-norms and t-conorms. For $e \in]0,1[$, we denote by ϕ_e and ψ_e the linear transformations defined by $\phi_e(x) = x/e$ and $\psi_e(x) = (x-e)/(1-e)$. According to Proposition 2, to any uninorm U with a neutral element $e \in]0,1[$, there corresponds a t-norm T and a t-conorm S such that:

(i) $U(x,y) = \phi_e^{-1}(T(\phi_e(x), \phi_e(y)))$ for $(x,y) \in [0,e]^2$;
(ii) $U(x,y) = \psi_e^{-1}(S(\psi_e(x), \psi_e(y)))$ for $(x,y) \in [e,1]^2$.

Concerning the other parts of the unit square, we have the following proposition.

Proposition 3. *[21] Consider a proper uninorm U with the neutral element e. For any (x,y) in $[0,e] \times [e,1] \cup [e,1] \times [0,e]$ it holds that*

$$\min(x,y) \leq U(x,y) \leq \max(x,y)\,.$$

Therefore, on the rectangular parts $[0,e] \times [e,1]$ and $[e,1] \times [0,e]$ a uninorm is *compensatory* (e.g. like means): $U(x,y)$ is always between the smallest and the largest of the two values x, y.

3 Some Particular Classes of Uninorms

3.1 The Classes $\mathcal{U}_{\min}$ and $\mathcal{U}_{\max}$

The following two propositions investigate the extreme cases in Proposition 3, where U acts as the minimum operator or as the maximum operator on $[0,e[\times]e,1] \cup]e,1] \times [0,e[$.

Theorem 1. *A binary operator U is a conjunctive uninorm with a neutral element $e \in]0,1[$ such that $U(\cdot,1)$ is continuous on $[0,e[$ if and only if there exists a t-norm T and a t-conorm S such that*

$$U(x,y) = \begin{cases} \phi_e^{-1}(T(\phi_e(x), \phi_e(y))) & , \text{ if } (x,y) \in [0,e]^2\,, \\ \psi_e^{-1}(S(\psi_e(x), \psi_e(y))) & , \text{ if } (x,y) \in [e,1]^2\,, \\ \min(x,y) & , \text{ elsewhere.} \end{cases}$$

For obvious reasons, the class of uninorms characterized by Theorem 1 is denoted $\mathcal{U}_{\min}$. A similar class of disjunctive uninorms, denoted $\mathcal{U}_{\max}$, is characterized in the next theorem. The implications from left to right in Theorems 1 and 2 were shown in [9]; the implications from right to left are a matter of direct verification.

Theorem 2. *A binary operator U is a disjunctive uninorm with a neutral element $e \in]0,1[$ such that $U(\cdot,0)$ is continuous on $]e,1]$ if and only if there exists a t-norm T and a t-conorm S such that*

$$U(x,y) = \begin{cases} \phi_e^{-1}(T(\phi_e(x),\phi_e(y))) & , \textit{if } (x,y) \in [0,e]^2 , \\ \psi_e^{-1}(S(\psi_e(x),\psi_e(y))) & , \textit{if } (x,y) \in [e,1]^2 , \\ \max(x,y) & , \textit{elsewhere.} \end{cases}$$

Given a t-norm T and t-conorm S, Theorems 1 and 2 show how to construct a conjunctive and disjunctive uninorm that both have T and S as underlying t-norm and t-conorm. In case the underlying t-norm and t-conorm are min and max, we obtain the first uninorms considered by Yager and Rybalov [21]: the conjunctive right-continuous uninorm U_c and the disjunctive left-continuous uninorm U_d listed in Example 1. Note that left- or right-continuity of a binary operator is shorthand for the corresponding type of continuity of all of its partial mappings.

The following statement gives the weakest and the strongest uninorms with a given neutral element e.

Proposition 4. *[9] Consider a uninorm U with the neutral element e. It then holds that*

$$\underline{U}_e \leq U \leq \overline{U}_e ,$$

with $\underline{U}_e \in \mathcal{U}_{\min}$ and $\overline{U}_e \in \mathcal{U}_{\max}$ defined by

$$\underline{U}_e(x,y) = \begin{cases} 0 & , \textit{if } (x,y) \in [0,e[^2 , \\ \max(x,y) & , \textit{if } (x,y) \in [e,1]^2 , \\ \min(x,y) & , \textit{elsewhere.} \end{cases}$$

$$\overline{U}_e(x,y) = \begin{cases} \min(x,y) & , \textit{if } (x,y) \in [0,e]^2 , \\ 1 & , \textit{if } (x,y) \in]e,1]^2 , \\ \max(x,y) & , \textit{elsewhere.} \end{cases}$$

3.2 Idempotent Uninorms

A uninorm U with a neutral element $e \in]0,1[$ and min and max as underlying t-norm and t-conorm is clearly idempotent: $U(x,x) = x$ for all $x \in [0,1]$. For instance, the uninorms U_c and U_d mentioned above are idempotent. The class of *idempotent* uninorms is implicitly described by Czogała and Drewniak [3], as well as the fact that there does not exist a continuous idempotent uninorm with a neutral element $e \in]0,1[$. All idempotent uninorms are, of course, *non-compensatory*. Explicit characterizations can be given for the class of left-continuous idempotent uninorms and the class of right-continuous idempotent uninorms.

Theorem 3. *[4] A binary operator U is a left-continuous idempotent uninorm with a neutral element $e \in]0,1]$ if and only if there exists a decreasing unary operator g with a fixpoint e satisfying*

(i) $g(g(x)) \geq x$ for all $x \in [0, g(0)]$, and
(ii) $g(x) = 0$ for all $x \in]g(0), 1]$

such that U is given by

$$U(x,y) = \begin{cases} \min(x,y) & , \text{ if } y \leq g(x) \text{ and } x \leq g(0)\,, \\ \max(x,y) & , \text{ elsewhere.} \end{cases}$$

Theorem 4. *[4] A binary operator U is a right-continuous idempotent uninorm with a neutral element $e \in [0,1[$ if and only if there exists a decreasing unary operator g with a fixpoint e satisfying*

(i) $g(g(x)) \leq x$ for all $x \in [g(1), 1]$, and
(ii) $g(x) = 1$ for all $x \in [0, g(1)[$

such that U is given by

$$U(x,y) = \begin{cases} \max(x,y) & , \text{ if } y \geq g(x) \text{ and } x \geq g(1)\,, \\ \min(x,y) & , \text{ elsewhere.} \end{cases}$$

In the class of left-continuous idempotent uninorms, the conjunctive ones correspond to the case $g(0) = 1$, and, hence, the unary operator g is *super-involutive*: $g(g(x)) \geq x$ for all $x \in [0,1]$; the disjunctive ones correspond to the case $g(0) < 1$. Similarly, in the class of right-continuous idempotent uninorms, the disjunctive ones correspond to the case $g(1) = 0$, and, hence, the unary operator g is *sub-involutive*: $g(g(x)) \leq x$ for all $x \in [0,1]$; the conjunctive ones correspond to the case $g(1) > 0$. The unary operator g_d corresponding to the uninorm U_d is given by

$$g_d(x) = \begin{cases} e & , \text{ if } x \leq e\,, \\ 0 & , \text{ if } x > e\,. \end{cases}$$

Modifying g_d into g'_d in the following way

$$g'_d(x) = \begin{cases} 1 & , \text{ if } x = 0\,, \\ g_d(x) & , \text{ elsewhere,} \end{cases}$$

we obtain a *conjunctive* left-continuous idempotent uninorm U'_d given by

$$U'_d(x,y) = \begin{cases} \min(x,y) & , \text{ if } y \leq g'_d(x)\,, \\ \max(x,y) & , \text{ elsewhere.} \end{cases}$$

Also, starting from an involutive negator N, we can construct both a conjunctive left-continuous idempotent uninorm U_c^N and a disjunctive right-continuous idempotent uninorm U_d^N as follows:

$$U_c^N(x,y) = \begin{cases} \min(x,y) & \text{, if } y \le N(x)\,, \\ \max(x,y) & \text{, elsewhere,} \end{cases}$$

$$U_d^N(x,y) = \begin{cases} \max(x,y) & \text{, if } y \ge N(x)\,, \\ \min(x,y) & \text{, elsewhere.} \end{cases}$$

In a recent paper [15] the general characterization and representation of idempotent uninorms is presented, in terms of locally internal operations. A binary operator F is called *locally internal* if $F(x,y) \in \{x,y\}$ for all $x,y \in [0,1]$. Obviously, any locally internal operator is idempotent, but the converse is not true. It is also interesting to know that if a locally internal, monotonic operation is commutative, then it is associative [15].

The following result can be viewed as a characterization of locally internal, associative and increasing binary operators with a neutral element, and also as an improvement of Czogała–Drewniak's theorem [3] including also the converse statement.

Theorem 5. *[15] A binary operator F is increasing, associative, idempotent and has a neutral element $e \in [0,1]$ if and only if there exists a decreasing unary operator g with $g(e) = e$, $g(x) = 0$ for all $x > g(0)$, $g(x) = 1$ for all $x < g(1)$ satisfying*

$$\inf\{y \mid g(y) = g(x)\} \le g^2(x) \le \sup\{y \mid g(y) = g(x)\} \quad \text{, for all } x \in [0,1]\,,$$

and such that

$$F(x,y) = \begin{cases} \min(x,y) & \text{, if } y < g(x) \text{, or } y = g(x) \text{ and } x < g^2(x)\,, \\ \max(x,y) & \text{, if } y > g(x) \text{, or } y = g(x) \text{ and } x > g^2(x)\,, \\ \min(x,y) \text{ or } \max(x,y) & \text{, if } y = g(x) \text{ and } x = g^2(x)\,. \end{cases}$$

Moreover, in this case F must be commutative except perhaps in points (x,y) such that $y = g(x)$ with $x = g^2(x)$.

Let us point out that Theorem 5 provides a characterization of all idempotent uninorms, requiring only commutativity in the points (x,y) such that $y = g(x)$ and $x = g^2(x)$. In particular, this characterization includes those given in Theorems 3 and 4 for left- and right-continuous idempotent uninorms. In fact, if the binary operator F is left-continuous then it must coincide with the minimum in all points (x,y) such that $y = g(x)$, and thus the unary operator g must verify $g^2(x) \ge x$ for all $x \in [0,1]$. A similar argumentation can be applied for right-continuity.

3.3 Representable Uninorms

In analogy to the representation of continuous Archimedean t-norms and t-conorms, Fodor et al. [9] have investigated the existence of uninorms with a similar representation in terms of a single variable function. This search turns out to lead back to the class of aggregative operators introduced by Dombi in [6]. This work is also closely related to that of Klement *et al.* on associative compensatory operators [13].

Proposition 5. *[9] Consider $e \in]0,1[$ and a strictly increasing continuous $[0,1] \longrightarrow \overline{\mathbb{R}}$ mapping h with $h(0) = -\infty$, $h(e) = 0$ and $h(1) = +\infty$. The binary operator U defined by*

$$U(x,y) = h^{-1}(h(x) + h(y)), \quad \text{for all } (x,y) \in [0,1]^2 \setminus \{(0,1),(1,0)\},$$

and either $U(0,1) = U(1,0) = 0$ or $U(1,0) = U(0,1) = 1$, is a uninorm with a neutral element e.

The mapping h in the foregoing proposition is called an *additive generator* of the uninorm U.

Theorem 6. *[9] Consider a proper uninorm U with a neutral element $e \in]0,1[$. There exists a strictly increasing continuous $[0,1] \longrightarrow \overline{\mathbb{R}}$ mapping h with $h(0) = -\infty$, $h(e) = 0$ and $h(1) = +\infty$ such that*

$$U(x,y) = h^{-1}(h(x) + h(y)) \quad , \text{for all } (x,y) \in [0,1]^2 \setminus \{(0,1),(1,0)\}$$

if and only if

(i) U is strictly increasing and continuous on $]0,1[^2$;
(ii) there exists an involutive negator N with a fixpoint e such that for all $(x,y) \in [0,1]^2 \setminus \{(0,1),(1,0)\}$ we have

$$U(x,y) = N(U(N(x),N(y)))\,.$$

The uninorms characterized by Theorem 6 are called *representable uninorms*. Note that any representable uninorm comes in a conjunctive and a disjunctive version, i.e. there always exist two representable uninorms that only differ in the points $(0,1)$ and $(1,0)$. The disjunctive version of a conjunctive representable uninorm U will be denoted U^*. Note that the partial mapping $U(\cdot,1)$ is not continuous at 0 and that the partial mapping $U^*(\cdot,0)$ is not continuous at 1. It is immediately clear that the classes $\mathcal{U}_{\min}$ and $\mathcal{U}_{\max}$ do not contain representable uninorms. Note that the additive generator of a representable uninorm is unique up to a positive multiplicative constant. The involutive negator N_U corresponding to a representable uninorm U with the additive generator h, as mentioned in Theorem 6, is given by $N_U(x) = h^{-1}(-h(x))$.

Example 2. Consider $h(x) = \log \frac{x}{1-x}$ and the corresponding conjunctive representable uninorm U:

$$U(x, y) = \begin{cases} 0 & , \text{ if } (x, y) \in \{(1, 0), (0, 1)\}, \\ \dfrac{xy}{(1-x)(1-y)+xy} & , \text{ otherwise.} \end{cases}$$

Its neutral element is $\frac{1}{2}$, and $N_U(x) = 1 - x$. This uninorm is often called the "3Π" operator, since the product (Π) appears three times in its formula.

Any representable uninorm U with the neutral element e satisfies

$$U(x, x) < x \text{ for all } x \in]0, e[, \quad \text{and} \quad U(x, x) > x \text{ for all } x \in]e, 1[,$$

an obvious generalization of the Archimedean property of continuous t-norms and t-conorms. Note that Theorem 6 implies that the t-norm and t-conorm underlying a representable uninorm are both strict.

On the other hand, we will show next that representable uninorms are in some sense also generalizations of nilpotent t-norms and nilpotent t-conorms. We first recall the following characterizations. A continuous t-norm T is nilpotent if and only if there exists an involutive negator N such that

$$T(x, N(x)) = 0, \quad \text{for all } x \in]0, 1[.$$

Similarly, a continuous t-conorm S is nilpotent if and only if there exists an involutive negator N such that

$$S(x, N(x)) = 1, \quad \text{for all } x \in]0, 1[.$$

Proposition 6. *Consider a representable uninorm U with the neutral element e. Then there exists an involutive negator N such that*

$$U(x, N(x)) = e, \quad \textit{for all } x \in]0, 1[.$$

In order to prove this proposition, it suffices to consider the involutive negator N_U.

3.4 Uninorms Continuous on the Open Unit Square

It is clear from [9] that a proper uninorm cannot be continuous on $[0, 1]^2$. Therefore, Hu and Li [12] studied uninorms that are continuous on the *open* unit square $]0, 1[^2$. Their results can be reinterpreted as follows.

Theorem 7. *[12] A proper uninorm with a neutral element $e \in]0, 1[$ is continuous on $]0, 1[^2$ if and only if one of the following two conditions is satisfied:*

(a) There exists $a \in [0, e[$ such that

$$U(x,y) = \begin{cases} U^*(x,y) & , \text{if } (x,y) \in [a,1]^2 , \\ \min(x,y) & , \text{elsewhere,} \end{cases}$$

where U^* is a representable uninorm with the neutral element $a+(1-a)\cdot e$.

(b) There exists $b \in]e, 1]$ such that

$$U(x,y) = \begin{cases} U^*(x,y) & , \text{if } (x,y) \in [0,b]^2 , \\ \max(x,y) & , \text{elsewhere,} \end{cases}$$

where U^* is a representable uninorm with the neutral element $b \cdot e$.

4 De Morgan Triplets Based on Uninorms

Suppose that U_1 and U_2 are two uninorms and let N be an involutive negator.

Definition 2. *[9] The triplet (U_1, U_2, N) is called a* De Morgan triplet *if it holds that*

$$N(U_1(x,y)) = U_2[N(x), N(y)] \quad , \text{for all } x, y \in [0,1] .$$

We recall some necessary conditions on (U_1, U_2, N) to be a De Morgan triplet.

Proposition 7. *[9] Consider two proper uninorms U_1 and U_2 with the neutral element $e_1 \in]0,1[$ and $e_2 \in]0,1[$, resp., and an involutive negator N. Let T_i and S_i be the underlying t-norm and t-conorm of U_i $i = 1,2$). If (U_1, U_2, N) is a De Morgan triplet, then*

(a) $e_2 = N(e_1)$ and $U_2(0,1) = N(U_1(0,1))$;

(b) there exist two strict negators n_1 and n_2 such that both (T_1, S_2, n_2) and (T_2, S_1, n_1) are De Morgan triplets and

$$n_2(x) = \frac{1-e_1}{1-e_2} n_1\left(\frac{e_1}{e_2}x\right) + \frac{e_1-e_2}{1-e_2}, \quad \text{for all } x \in [0,1]$$

holds if $e_1 \leq e_2$. (If $e_1 \geq e_2$ then a similar equality holds with reversed subscripts.)

Fortunately, these necessary conditions are sufficient if we consider the classes $\mathcal{U}_{\min}$ and $\mathcal{U}_{\max}$.

Theorem 8. *[9] Consider a proper uninorm $U_1 \in \mathcal{U}_{\min}$ with the neutral element $e_1 \in]0,1[$ and a proper uninorm $U_2 \in \mathcal{U}_{\max}$ with the neutral element $e_2 \in]0,1[$. Let T_i and S_i $(i = 1,2)$ be the underlying t-norms and t-conorms of U_1 and U_2. Then there exists an involutive negator N such that (U_1, U_2, N)*

is a De Morgan triplet if and only if there exist two strict negators n_1 and n_2 such that both (T_1, S_2, n_2) and (T_2, S_1, n_1) are De Morgan triplets and

$$n_2(x) = \frac{1-e_1}{1-e_2} n_1\left(\frac{e_1}{e_2}x\right) + \frac{e_1 - e_2}{1-e_2}, \quad \text{for all } x \in [0,1]$$

if $e_1 \leq e_2$. (If $e_1 \geq e_2$ then the same equality holds with reversed subscripts.)

5 Residual Implicators of Uninorms

5.1 Definitions

The residuation technique is well known in fuzzy set theory for constructing new logical operators from given ones. For instance, for any t-norm T, the binary operator I_T defined by

$$I_T(x,y) = \sup\{z \in [0,1] \mid T(x,z) \leq y\}$$

is an implicator, i.e. is hybrid monotonous (decreasing first and increasing second partial mappings) and satisfies $I_T(1,0) = 0$ and $I_T(1,1) = I_T(0,0) = 1$. For the minimum operator $T_\mathbf{M}$ for instance, the residual implicator $I_\mathbf{M}$, also known as the Gödel implicator, is given by

$$I_\mathbf{M}(x,y) = \begin{cases} 1, & \text{if } x \leq y\,, \\ y, & \text{elsewhere.} \end{cases}$$

Similarly, for a t-conorm S we can consider the binary operator R_S defined by

$$R_S(x,y) = \sup\{z \in [0,1] \mid S(x,z) \leq y\},$$

but apart from its obvious hybrid monotonicity, this operator has no particular logical interpretation.

We can now consider the construction of a residual operator I_U from a uninorm U. For a comprehensive study see [5], where all the proofs of the following results can also be found.

Proposition 8. *[5] Consider a proper uninorm U with a neutral element $e \in]0,1[$. The binary operator I_U defined by*

$$I_U(x,y) = \sup\{z \in [0,1] \mid U(x,z) \leq y\}$$

is an implicator if and only if

$$U(0,z) = 0 \quad , \text{for all } z \in]e,1[\,.$$

Corollary 1. *[5] For any conjunctive uninorm U or representable uninorm U (conjunctive or disjunctive), the residual operator I_U is an implicator.*

Apart from representable uninorms, no other disjunctive uninorms U are known for which the residual operator I_U yields an implicator. Note that the residual operator I_U of a uninorm U with a neutral element e satisfies

$$I_U(e, y) = y \quad , \text{for all } y \in [0, 1] ,$$

a generalization of the neutrality principle which holds for the residual implicator of a t-norm.

5.2 Residual Implicators of Members of the Class $\mathcal{U}_{\min}$

The first question that comes to mind when studying residual implicators of uninorms is whether they also show some kind of block structure (or ordinal sum structure). The following theorem investigates this structure for members of the class $\mathcal{U}_{\min}$.

Theorem 9. *[5] Consider a proper uninorm $U \in \mathcal{U}_{\min}$ with the neutral element $e \in]0, 1[$ and underlying t-norm T and t-conorm S. The residual implicator I_U is given by*

$$I_U(x, y) = \begin{cases} \phi_e^{-1}(I_T(\phi_e(x), \phi_e(y))) & , \textit{if } (x, y) \in [0, e]^2 \textit{ and } y < x , \\ \psi_e^{-1}(R_S(\psi_e(x), \psi_e(y))) & , \textit{if } (x, y) \in [e, 1]^2 , \\ I_{\mathbf{M}}(x, y) & , \textit{elsewhere.} \end{cases}$$

One might have expected that $I_U(x, y) = \phi_e^{-1}(I_T(\phi_e(x), \phi_e(y)))$, for any $(x, y) \in [0, e]^2$. However, in case $x \leq y$, this would yield $I_U(x, y) = e$, whereas Theorem 9 yields $I_U(x, y) = 1$.

5.3 Residual Implicators of Representable Uninorms

For a continuous Archimedean t-norm T with an additive generator f, the residual implicator I_T is given by $I_T(x, y) = f^{(-1)}(f(y) - f(x))$ [14]. The following theorem brings a similar result for representable uninorms.

Theorem 10. *[5] Consider a representable uninorm U with an additive generator h, then its residual implicator I_U is given by*

$$I_U(x, y) = \begin{cases} h^{-1}(h(y) - h(x)) & , \textit{if } (x, y) \in [0, 1]^2 \setminus \{(0, 0), (1, 1)\} , \\ 1 & , \textit{elsewhere.} \end{cases}$$

Proposition 9. *[5] Consider a representable uninorm U, then I_U is contrapositive w.r.t. N_U, i.e. for all $(x, y) \in [0, 1]^2$ we have*

$$I_U(x, y) = I_U(N_U(y), N_U(x)) .$$

Note that the involutive negator N_U corresponding to a representable uninorm U with a neutral element e can be written as $N_U(x) = I_U(x, e)$.

For any disjunctive uninorm U and negator N, the binary operator $I_{U,N}$ defined by

$$I_{U,N}(x, y) = U(N(x), y)$$

clearly is an implicator.

Proposition 10. *[5] Consider a conjunctive representable uninorm U. Then the following equality holds:*

$$I_{U^*,N_U} = I_U .$$

Propositions 9 and 10 are generalizations of results that are characteristic for nilpotent t-norms [8]. Recall, for instance, that for a nilpotent t-norm T, the residual implicator I_T is contrapositive w.r.t. the involutive negator N_T defined by $N_T(x) = I_T(x, 0)$. Moreover, it holds that $I_T(x, y) = S(N_T(x), y)$, with S the N_T-dual t-conorm of T.

6 Nullnorms

In [2] we studied two functional equations for uninorms. One of those required to introduce a new family of associative binary operations on $[0, 1]$ as follows.

Definition 3. *[2] A* nullnorm *V is a commutative, associative and increasing binary operator with an absorbing element $a \in [0, 1]$ (i.e. $V(x, a) = a$ for all $x \in [0, 1]$), and that satisfies*

(i) $V(x, 0) = x$ for all $x \in [0, a]$,
(ii) $V(y, 1) = y$ for all $y \in [a, 1]$.

On the other hand, in [1] the following notion was also defined, with the superfluous requirement of continuity [16].

Definition 4. *[1, 16] A* t-operator *F is a commutative, associative and increasing binary operator such that*

(i) $F(0, 0) = 0$ and $F(1, 1) = 1$;
(ii) the partial mappings $F(\cdot, 0)$ and $F(\cdot, 1)$ are continuous.

It turns out that Definitions 3 and 4 yield exactly the same operations; that is, t-operators coincide with nullnorms. When $a = 1$ we obtain t-conorms, while $a = 0$ gives back t-norms. The basic structure of nullnorms is similar to that of uninorms, as we state in the following theorem.

Theorem 11. *[2] A binary operator V is a nullnorm with absorbing element $a \in]0,1[$ if and only if there exists a t-norm T_V and a t-conorm S_V such that*

$$U(x,y) = \begin{cases} \phi_a^{-1}(S_V(\phi_a(x), \phi_a(y))) & , \text{ if } x, y \in [0,a]\,, \\ \psi_a^{-1}(T_V(\psi_a(x), \psi_a(y))) & , \text{ if } x, y \in [a,1]\,, \\ a & , \text{ elsewhere.} \end{cases}$$

Thus, nullnorms are generalizations of the well-known *median* operator. More important properties of nullnorms (such as duality and self-duality, classification) can be found in [16].

7 Some Functional Equations for Uninorms and Nullnorms

First, it is interesting to notice that uninorms U with a neutral element in $]0,1[$ are just those binary operators that turn the structures $([0,1], \sup, U)$ and $([0,1], \inf, U)$ distributive semirings in the sense of Golan [11].

Subclasses of uninorms and nullnorms satisfying some algebraic properties (such as modularity, distributivity, reversibility) as well as their structure on finite totally ordered scales were studied in [17–20]. These properties are usually formulated as functional equations.

For instance, in [18] the authors studied the *modularity* condition given by the functional equation

$$F(x, G(y,z)) = G(F(x,y), z), \quad \text{for all } z \leq x\,,$$

where the unknown functions F and G are uninorms and/or nullnorms.

Another algebraic property is *distributivity.* It has been studied in [19], formulated as the functional equation

$$F(x, G(y,z)) = G(F(x,y), F(x,z)), \quad \text{for all } x, y, z \in [0,1]\,.$$

The authors characterized all the solutions for both equations in the four possible cases: (i) when F and G are nullnorms, (ii) when F is a nullnorm and G a uninorm, (iii) when F is a uninorm and G a nullnorm and, finally, (iv) when F and G are uninorms.

The famous functional equation of Frank [10] (together with its solutions) is well known for t-norms and t-conorms. The task is to find all pairs (T,S) of t-norms and t-conorms which are solutions of the functional equation

$$T(x,y) + S(x,y) = x + y, \quad \text{for all } x, y \in [0,1]\,. \tag{1}$$

Since uninorms are generalizations of t-norms and t-conorms, it is quite natural to study the Frank equation (1) in case one of the unknown binary operators is a uninorm U with a given neutral element $e \in [0,1]$. This has been done in [2]. In view of the importance of the Frank t-norms in preference modelling, we are particularly interested in the case of a conjunctive uninorm. The very first observations are formulated in the following proposition, and leads to the notion of nullnorms in a natural way.

Proposition 11. *[2] Consider a uninorm U with a neutral element $e \in [0,1]$ and an associative and increasing binary operator V. If U and V satisfy the functional equation*

$$U(x,y) + V(x,y) = x + y \quad , \text{for all } (x,y) \in [0,1]^2 \,, \tag{2}$$

then V is commutative, has as an absorbing element e and satisfies $V(x,0) = x$ for all $x \in [0,e]$, and $V(x,1) = x$ for all $x \in [e,1]$.

These properties of V determine nullnorms, see Definition 3 above.

Somewhat disappointingly, we proved in [2] that all considerations lead back to the already known t-norm and t-conorm solutions of (2), as is obvious from the following theorems.

Theorem 12. *[2] Consider a proper uninorm U with a neutral element $e \in]0,1[$, then there exists no nullnorm V with an absorbing element e such that (U,V) is a solution of (2).*

Theorem 13. *[2] Consider a nullnorm V with absorbing element $a \in]0,1[$, then there exists no uninorm U with a neutral element a such that (U,V) is a solution of (2).*

8 Conclusion

In this chapter we summarized our present knowledge on basics of uninorms. This included their fundamental properties, their structure based on t-norms and t-conorms, some known particular classes like idempotent and representable uninorms. De Morgan triplets of two uninorms and an involutive negator were also studied. We investigated residual implicators of uninorms playing a key role in inference processes. We explained the natural entry of another interesting class of operations called nullnorms. Finally we summarized results concerning functional equations involving uninorms and nullnorms.

References

1. Calvo T, Fraile A, Mayor G (1986) Algunes consideracions sobre connectius generalitzats. In: Actes del VI Congrés Català de Lògica. Barcelona pp. 45–46
2. Calvo T, De Baets B, Fodor J (2001) The functional equations of Frank and Alsina for uninorms and nullnorms. Fuzzy Sets and Systems 120:385–394

3. Czogała E, Drewniak J (1984) Associative monotonic operations in fuzzy set theory. Fuzzy Sets and Systems 12:249–269
4. De Baets B (1999) Idempotent uninorms. European J. Oper. Res. 118:631–642
5. De Baets B, Fodor J (1999) Residual operators of uninorms. Soft Computing 3:89–100
6. Dombi J (1982) Basic concepts for the theory of evaluation: The aggregative operator. Europ. J. Oper. Res. 10:282–293
7. Dubois D, Prade H (1985) A review of fuzzy sets aggregation connectives. Information Sciences 36:85–121
8. Fodor J (1995) Contrapositive symmetry of fuzzy implications. Fuzzy Sets and Systems 69:141–156
9. Fodor J, Yager R, Rybalov A (1997) Structure of uninorms. Internat. J. Uncertain. Fuzziness Knowledge-Based Systems 5:411–427
10. Frank M (1979) On the simultaneous associativity of $F(x, y)$ and $x+y-F(x, y)$. Aeq. Math. 19:194–226
11. Golan J (1992) The theory of semirings with applications in mathematics and theoretical computer science. Pitman Monographs and Surveys in Pure and Applied Mathematics, Vol. 54. Longman Scientific and Technical
12. Hu SK, Li ZF (2001) The structure of continuous uni-norms. Fuzzy Sets and Systems 124:43–52
13. Klement EP, Mesiar R, Pap E (1996) On the relationship of associative compensatory operators to triangular norms and conorms. Int. J. Uncertainty, Fuzziness and Knowledge-Based Systems 4:129–144
14. Klement EP, Mesiar R, Pap E (2000) Triangular Norms. Kluwer, Boston London Dordrecht
15. Martín J, Mayor G, Torrens J (2003) On locally internal monotonic operations. Fuzzy Sets and Systems 137:27–42
16. Mas M, Mayor G, Torrens J (1999) T-operators. Internat. J. Uncertainty Fuzziness Knowledge-Based Systems 7:31–50
17. Mas M, Mayor G, Torrens J (1999) T-operators and uninorms in a finite totally ordered set. Internat. J. Intell. Systems 14:909–922
18. Mas M, Mayor G, Torrens J (2002) The modularity condition for uninorms and t-operators. Fuzzy Sets and Systems 126:207–218
19. Mas M, Mayor G, Torrens J (2002) The distributivity condition for uninorms and t-operators. Fuzzy Sets and Systems 128:209–225
20. Monserrat M, Torrens J (2002) On the reversibility of uninorms and t-operators. Fuzzy Sets and Systems 131:303–314
21. Yager R, Rybalov A (1996) Uninorm aggregation operators. Fuzzy Sets and Systems 80:111–120

Structural Interpolation and Approximation with Fuzzy Relations: A Study in Knowledge Reuse

Witold Pedrycz

Abstract. In this study, we are concerned with a problem of structural interpolation and approximation realized in the framework of fuzzy relations. Fuzzy relations are fundamental concepts which represent, reveal and process key dependencies between systems' variables, factors and concepts thus capturing an underlying knowledge about systems. Different relations could deliver various insights into the nature of the system. In particular, such relations could be constructed at different levels of granularity and in this sense may provide a variety of perspectives (views) at the same system. Here we focus attention on the issue of structural knowledge reconstruction and demonstrate how for a given collection of fuzzy relations, one can effectively reconstruct fuzzy relations at some required level of granularity. We also cast this problem in the context of knowledge re-use by showing how the existing fuzzy relations could be effectively employed (re-used) in this setting. The ensuing optimization task is formed and a detailed gradient-based learning scheme is presented. Some numeric illustration is also included.

1 Introductory Notes

Fuzzy relations are important formal structures that represent and quantify dependencies between variables defined in the same space or expressed in different spaces, cf. [5], [7], [8], [13]. The degree of membership $R(x_i, y_j) = r_{ij}$ where $r_{ij} \in [0,1]$ denotes an extent to which the two variables ("i" and "j") are *related* (associated). The higher are the values of the entry r_{ij}, the stronger the relationship (association) between the corresponding variables (factors). There is a wealth of conceptual developments, algorithmic constructs and resulting applications in fuzzy relations and relational calculus. With this regard, the reader may refer to the recent conceptual and application-oriented advancements reported in [1], [2], [3], [4], [6], [9], [11], [12]. In the context of granular computing and the underlying concept of information granules ([14]; [17] [18]), fuzzy relations have started to play a pivotal role by offering some new views and algorithmic pursuits into the development of granular models.

In essence, one could stress that a wealth of studies in fuzzy sets is inherently associated with the operations on fuzzy relations, their formation as mechanisms of knowledge representation ([6]; [7]) and various mappings between them as directly linked with various ways of knowledge processing. The most visible areas of applications concern fuzzy modeling (and relational fuzzy modeling, in particular), fuzzy associative memories, relational networks, etc. Interestingly, as fuzzy relations generalize relations and relational calculus, the results developed there could be of significant interest and relevance to the Boolean relations and their computing.

In this study, we are concerned with a new class of reconstruction problems whose essence could be posed as follows. For a given collection of fuzzy relations defined in finite spaces; we are interested in a reconstruction of fuzzy relations defined in some associated universes of discourse. The formulation of this nature, whose details will be revealed in Sect. 2, brings forward an interesting concept of structural interpolation and extrapolation as well as raises an important issue of knowledge re-use being captured in the form of the already available fuzzy relations.

The material of this study is organized in the following manner. We start with a problem formulation in Sect. 2. In Sect. 3, we look into the structural optimization and link the reconstruction task to the problem of structural interpolation and extrapolation. This is followed by the detailed optimization scheme (Sect. 4). Numeric experiments are reported in Sect. 5. We include them to help reveal the most essential features of the optimization algorithm. Concluding comments are covered in Sect. 6.

Throughout this study, we adhere to the standard notation encountered in the theory of fuzzy sets. Capital letters will be used to denote fuzzy relations and fuzzy sets. We assume that all of these constructs are defined in finite universes of discourse or Cartesian products of such finite universes of discourse. Logic operators (*and*- and *or*-type) are realized with the use of triangular norms and conorms (t-norms and t-conorms) ([13]). The composition operator applied to fuzzy relations (typically involving a certain category of s-t composition) is denoted here by $\circ$.

2 Problem Formulation

The knowledge reconstruction and knowledge re-use as being realized on a basis of a finite collection of fuzzy relations can be presented in the following manner. Given are two collections of fuzzy relations A[ii]: $\mathbf{X}_{\mathrm{ii}} \times \mathbf{Y} \rightarrow [0,1]$, $\mathrm{ii} = 1, 2, \ldots, \mathrm{p}$; $\mathrm{B[jj]} : \mathbf{Y}_{\mathrm{jj}} \times \mathbf{X} \rightarrow [0,1], \mathrm{jj} = 1, 2, \ldots, \mathrm{r}$. The cardinalities of the universes of discourse are finite and are denoted as $\mathrm{n[ii]} = \mathrm{card}(\mathbf{X}_{\mathrm{ii}}), \mathrm{m[jj]} = \mathrm{card}(\mathbf{Y}_{\mathrm{jj}})$, $\mathrm{n} = \mathrm{card}(\mathbf{X})$, $\mathrm{m} = \mathrm{card}(\mathbf{Y})$. These relations can be treated as a means to capture knowledge about the system(s) and dependencies between the variables of the system. The indexes (ii and jj) being used in conjunction with the fuzzy relations A[ii] and B[jj] emphasize the fact that we are provided with a family of relationships being reflective of various sources of knowledge. Depending upon the cardinality of the universes of discourse, we could encounter a very detailed (specific) knowledge about the system when the cardinality of the corresponding spaces (say $\mathbf{X}_{ii}$ or $\mathbf{Y}_{jj}$) is high meaning that the corresponding fuzzy relation is very much specific by capturing a lot of detailed relationships between the variables. If, on the other hand, we encounter only a very limited space (viz. the one of low dimensionality) $\mathbf{X}_{ii}$ or $\mathbf{Y}_{jj}$ (so only very few variables are taken into account) then we may allude to the corresponding fuzzy relation as being quite general and therefore far less specific. The collection of all fuzzy relations available so far, $\{A[ii], B[jj]\}$ constitutes a knowledge about the system. Given this family of fuzzy relations, we are interested in "reconstructing" a fuzzy relation R: $\mathbf{X} \times \mathbf{Y} \rightarrow [0, 1]$. As displayed graphically in Fig. 1, such reconstruction becomes possible because of the formulation of the "unknown" relation is realized in the spaces being "shared" by the already available fuzzy relations A[ii] and B[jj]. More specifically, the co-domains of all A[ii]s and B[jj]s constitute the Cartesian product of the space over which R is being formulated.

The formulation of the problem requires a careful attention. As R is defined in the Cartesian product of $\mathbf{X}$ and $\mathbf{Y}$, one has to be aware that before moving forward with any optimization we have to operate on objects–fuzzy relations whose dimensionality is compatible with the dimensionality exhibited by R. Given this, the main step of the construct is then to work with some transformed fuzzy relations so that we can reach a required structural compatibility between the available fuzzy relations and the fuzzy relation we are interested to determine. As A[ii] is defined in $\mathbf{X}_{ii} \times \mathbf{Y}$, we can form its s-t composition $A[ii]^T \circ A[ii]$ which gives rise to the fuzzy relation defined in $\mathbf{Y} \times \mathbf{Y}$. Likewise when computing $R^T \circ R$ we end up with a fuzzy relation formed in the same Cartesian product as formed before for the composition of A[ii], that is $\mathbf{Y} \times \mathbf{Y}$

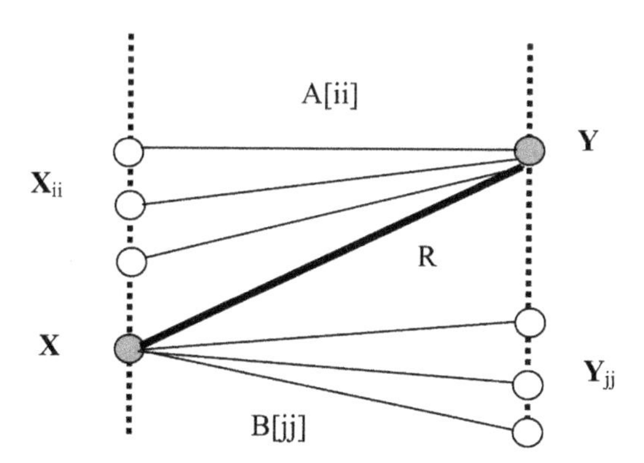

Fig. 1. The schematic view at the reconstruction problem: a collection of fuzzy relations, A[ii], B[jj] and a fuzzy relation (R) to be reconstructed. Each fuzzy relation is schematically visualized in a form of a link between corresponding spaces

so as these two constructs are consistent as far as dimensions are concerned. Because of that we could compare them by computing some distance $||.||$ of the form $||A^T[ii] \circ A[ii] - R^T \circ R||^2$. Likewise the distance computed over "p" fuzzy relations reads then as $\sum_{ii=1}^{p} ||A^T[ii] \circ A[ii] - R^T \circ R||^2$. The same scheme of dimensionality reconciliation applies to the fuzzy relations B[jj] leading to the computations of the following sum $\sum_{jj=1}^{r} ||B[jj] \circ B^T[jj] - R \circ R^T||^2$. The overall performance index Q(R) is then formed in such a way that the two components become involved,

$$Q(R) = \sum_{ii=1}^{p} ||A^T[ii] \circ A[ii] - R^T \circ R||^2 + \sum_{jj=1}^{r} ||B[jj] \circ B^T[jj] - R \circ R^T||^2 \quad (1)$$

We have stressed the dependency of Q on the unknown fuzzy relation R by using the corresponding notation. In particular, one may consider $||.||$ to be implemented as the Euclidean distance. The minimization of Q realized with respect to R, $\min_R Q$, leads to the fuzzy relation that is formed in a way so that it dwells upon the structural dependencies captured by the individual fuzzy relations A[ii] and B[jj]. Before arriving at the detailed solution to the problem, let us focus on the nature of the overall optimization process to show the essence of the reconstruction problem posed in this manner.

3 A Concept of Structural Reconstruction of Fuzzy Relations

As indicated in the formulation of the problem, the unknown fuzzy relation is determined on a basis of a collection of fuzzy relations (A[ii] and B[jj]). We showed that the optimization (1) is in fact realized by the minimization of distances between the corresponding results of the s-t compositions of the corresponding fuzzy relations. Furthermore we have stressed that the composition plays a pivotal role in the formation of the constructs that are compatible with regard to their dimensionality (and hence the optimization could be realized). The available fuzzy relations capture various facets of domain knowledge about the system. Given the variable dimensionality of the spaces in which they are defined, we clearly see that such domain knowledge comes at different levels of specificity. The reconstruction of R could concern the relation at the level of specificity that could be quite different from the specificity of individual fuzzy relations. Given that and alluding to the nature of the fuzzy relations, we may view the process of constructing of R as knowledge reconstruction and knowledge re-use. In essence, we are building the new fuzzy relation (R) by explicitly exploiting the knowledge conveyed by the available fuzzy relations. There is another interesting facet of this optimization task that relates to the dimensionality of spaces in which the fuzzy relations are available and the dimensionality of the spaces over which R is now being formed. This gives rise to an interesting problem of structural approximation, interpolation and extrapolation. Each of them may occur in the considered setting depending upon the dimensionality of the individual spaces. Some typical situations are envisioned in Fig. 2.

In contrast to what becomes well known in approximation theory while alluding to the notions of interpolation and extrapolation, those tasks are primarily concerned with some parametric character of the problem. Here we are faced with the optimization whose nature pertains to the inherently *structural* aspects of the problem. The situation illustrated in Fig. 2(a) concerns the approximation aspects of the determination of R. The situation illustrated in Fig. 2(b) shows the extrapolation character of the problem – we are concerned with the availability of fuzzy relations of lower or higher dimensionality in comparison to the one that needs to be reconstructed. As indicated, the extrapolation could involve the relations available at the lower or higher level of abstraction (or specificity). The interpolation character of the reconstruction is visualized in Fig. 2(b) –(c). We thus may refer to the extrapolation that occurs at the reduced level of granularity or leads to the construct formed at the increased level of granularity.

A convincing example of the reconstruction problem of this character emerges in the framework of fuzzy modeling. Let us elaborate on this matter in more detail. A simple example can be used to illustrate the underlying concept.

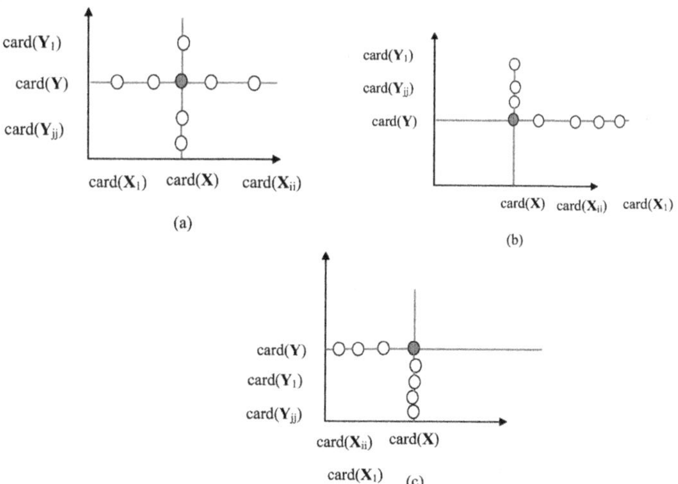

Fig. 2. Examples of structural reconstruction of R: (a) interpolation, (b) extrapolation – reduced level of granularity, and (c) extrapolation – increased level of granularity

Consider a single input – single output system. As commonly encountered in fuzzy modeling, we define (construct) a family of referential fuzzy sets (information granules) in the corresponding input and output space. Those usually come with a well-defined semantics such as e.g. *small, medium, large*, etc. Given their membership functions, any numeric experimental data could be expressed in terms of these fuzzy sets by computing the membership grades of numeric data in the respective fuzzy sets. The relationship between the information granules in the input and output spaces is captured in the form of some fuzzy relation defined in the Cartesian product of the finite number of fuzzy sets in the input and output spaces. More specifically, a collection of fuzzy sets defined in the input space gives rise to a finite space of fuzzy sets (membership grades) denoted by **X** [1] while for the output we form the space of membership **Y** being the result of existence of some fuzzy sets in the output. The construction of the fuzzy relation expressing relationships between fuzzy sets in the spaces is optimized based on the available experimental data. The result is the fuzzy relation A [1]. Considering the same experimental data, we establish a different conceptual view at the system by defining some other families of fuzzy sets in the input and output space. In particular, we use a new family of fuzzy sets in the input space (which gives rise to a finite space **X**) and output space (which leads to the family denoted here as **Y** [1]). Following the same development scheme as before, we construct a fuzzy relation B [1] defined in the Cartesian product of these spaces. An illustration of the obtained constructs is presented in Fig. 3.

Once the fuzzy relations have been constructed, the experimental data are not available any longer. Now, consider that we would like to form another view at the system by discovering relationships expressed in the Cartesian product of **X** and **Y**. At this point the reconstruction mechanism comes into play. As the data are not available any longer, this means that if we want

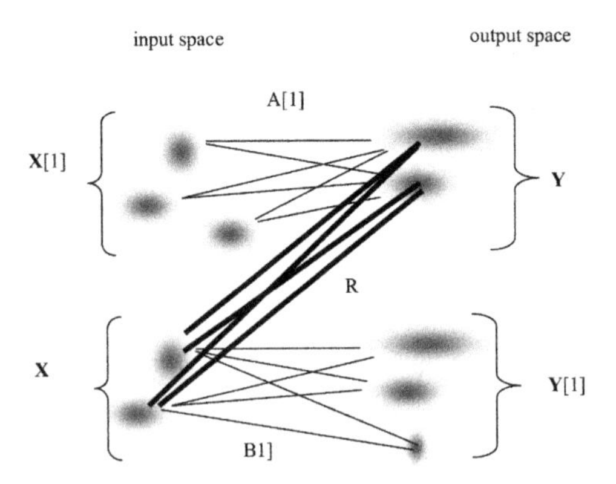

Fig. 3. Fuzzy model of relationships for a single input – single output system. The dependencies are expressed on Cartesian products of fuzzy sets predefined in the corresponding spaces

to form a relational model for some other input and output spaces, there is no other alternative but to re-use the existing fuzzy relations of the existing models.

4 The Optimization Process and Details of the Learning

The optimization problem described by (1), $\mathrm{Min}_R Q$, could be handled in many different ways including gradient-based techniques (gradient based learning), methods of evolutionary optimization (such as genetic algorithms and evolutionary strategies) and particle swarm optimization. Here, we proceed with the gradient-based learning as such schemes are encountered quite commonly in the optimization of relational structures. In what follows, we elaborate on the computational details. The point of departure is the general formula governing the updates of the entries of the fuzzy relation R realized through the gradient of Q,

$$R(iter+1) = R(iter) - \alpha \nabla_{R(iter)} Q \tag{2}$$

where *iter* denotes an index of consecutive iterations (*iter, iter*+1, etc) of the optimization scheme. Given the individual entries of the fuzzy relation R, the above expression reads in the form

$$r_{st}(iter+1) = r_{st}(iter) - \alpha \frac{\partial Q}{\partial r_{st}(iter)} \tag{3}$$

The calculations of the gradient following (3) need to be further quantified. We rewrite the performance index by expressing it in an explicit manner and showing the individual indexes of the fuzzy relations. We obtain

$$\frac{\partial Q}{\partial r_{st}} = -2\sum_{ii=1}^{p} ||A^T[ii] \circ A[ii] - R^T \circ R|| \\ \frac{\partial}{\partial r_{st}}(R^T \circ R) - 2\sum_{ii}^{p} ||B[ii] \circ B^T[ii] - R \circ R^T|| \frac{\partial}{\partial r_{st}}(R \circ R^T) \tag{4}$$

$s = 1, 2, \ldots, n; t = 1, 2, \ldots, m$. When moving down to the detailed computing, we need to specify the form of the t-norm and t-conorm occurring in the s-t composition. For illustrative purposes, we consider the product (t-norm) and the probabilistic sum (t-conorm). Those implementations of t-norms and t-conorms are differentiable and have been used quite frequently in the literature. Note that the composition of R^T and $R, R^T \circ R$ gives rise to the square matrix (relation) of the form $\mathop{S}_{i=1}^{n}(r_{ij} t r_{ik}), j, k = 1, 2, \ldots, m$. In the sequel, the derivative of this expression $\frac{\partial}{\partial r_{st}} \underset{i=1}{\overset{n}{S}} (r_{ij} t r_{ik})$ involves four separate conditions to be considered

$$\frac{\partial}{\partial r_{st}} \underset{i=1}{\overset{n}{S}} (r_{ij} t r_{ik}) = \begin{cases} r_{sk}(1 - AA(s,k,t)) & \text{if } j = t \text{ and } k \neq t \\ r_{sj}(1 - BB(s,j,t)) & \text{if } j \neq t \text{ and } k = t \\ 0 & \text{if } j \neq t, k \neq t \\ 2r_{st}(1 - CC(s,t)) & \text{if } j = k = t \end{cases} \tag{5}$$

where we define the following expressions

$$AA(s,k,t) = \underset{\substack{i=1 \\ i \neq s}}{\overset{n}{S}} (r_{it} t r_{sk})$$

$$BB(s,j,t) = \underset{\substack{i=1 \\ i \neq s}}{\overset{n}{S}} (r_{ij} t r_{it})$$

$$CC(s,t) = \underset{\substack{i=1 \\ i \neq s}}{\overset{n}{S}} (r_{it} t r_{it}) \tag{6}$$

For the second component of the derivative, we have to compute the expression $\frac{\partial}{\partial r_{st}}(R \circ R^T)$. Here the detailed formula reads as follows

$$\frac{\partial}{\partial r_{st}} \underset{i=1}{\overset{m}{S}} (r_{ji} t r_{ki}) = \begin{cases} r_{kt}(1\text{-}DD(s,t,k) & \text{if } j = s \text{ and } k \neq s \\ r_{jt}(1\text{-}EE(j,t,s)) & \text{if } j \neq s \text{ and } k = s \\ 0 & \text{if } j \neq s,\ k \neq s \\ 2r_{st}(1\text{-}FF(s,t)) & \text{if } j = k = s \end{cases} \tag{7}$$

In the sequel, the corresponding components of the above expression of the derivative are defined in the form

$$DD(s,t,k) = \underset{\substack{i=1 \\ i \neq t}}{\overset{n}{S}} (r_{si} t r_{ki})$$

$$EE(j,t,s) = \underset{\substack{i=1 \\ i \neq t}}{\overset{n}{S}} (r_{jt} t r_{si})$$

$$FF(s,t) = \underset{\substack{i=1 \\ i \neq t}}{\overset{n}{S}} (r_{si} t r_{si}) \tag{8}$$

5 Numeric Experimentation

In this section, we present several numeric examples to illustrate the performance of the optimization of the reconstruction process and interpret the resulting fuzzy relations. The learning was run for 1,000 epochs with the learning rate α set up to 0.05. The numeric values of these two parameters were chosen experimentally. The value of the learning rate provided a smooth process of learning (we avoided eventual oscillations when minimizing the performance index). The selected number of iterations was more than enough; even before reaching this limit, most of the learning has already been accomplished and no further improvement (reduction) of the performance index has been observed afterwards. As we are considering only single fuzzy relations A[ii] and B[ii], their corresponding indexes could be dropped leading to the shorthand notation of A and B, respectively. The t-norm is specified as the product while the probabilistic sum ($a + b - ab$, $a, b \in [0, 1]$) is used to realize the t-conorm.

Example 1. Consider the fuzzy relations A and B defined in the spaces $\mathbf{X}_1 \times \mathbf{Y}$ and $\mathbf{X} \times \mathbf{Y}_1$, respectively where $\dim(\mathbf{X}_1) = 4$, $\dim(\mathbf{Y}_1) = 3$ and $\dim(\mathbf{X}) = \dim(\mathbf{Y}) = 5$. Their entries are equal to

$$A = \begin{bmatrix} 0.7 & 0.1 & 0.3 & 0.1 & 1.0 \\ 0.3 & 0.5 & 0.0 & 0.5 & 0.0 \\ 0.3 & 1.0 & 0.4 & 1.0 & 0.3 \\ 0.4 & 0.0 & 0.9 & 0.2 & 1.0 \end{bmatrix} \quad B = \begin{bmatrix} 0.7 & 0.1 & 0.3 \\ 0.3 & 0.5 & 0.0 \\ 0.3 & 1.0 & 0.4 \\ 0.9 & 0.3 & 1.0 \\ 0.4 & 0.0 & 0.9 \end{bmatrix}$$

The value of the performance index produced after 1,000 learning epochs reaches 0.31 while the resulting fuzzy relation R defined in $\mathbf{X} \times \mathbf{Y}$ has the following entries

$$R = \begin{bmatrix} 0.258 & 0.161 & 0.000 & 0.629 & 0.329 \\ 0.129 & 0.451 & 0.000 & 0.260 & 0.000 \\ 0.429 & 0.971 & 0.000 & 0.503 & 0.000 \\ 0.058 & 0.578 & 0.678 & 0.950 & 0.573 \\ 0.750 & 0.051 & 0.867 & 0.000 & 1.000 \end{bmatrix}$$

As clearly seen, the strongest connections (dependencies) occur for the pairs (x_5, y_5), (x_3, y_2), and (x_4, y_4).

Example 2. Here the fuzzy relations A and B are given as the following matrices

$$A = \begin{bmatrix} 0.3 & 0.1 & 0.5 \\ 0.2 & 0.2 & 0.7 \\ 0.9 & 0.7 & 0.3 \\ 0.5 & 0.3 & 0.0 \\ 1.0 & 0.0 & 0.8 \end{bmatrix} \quad B = \begin{bmatrix} 0.6 & 0.0 & 0.1 & 0.9 & 1.0 & 0.4 \\ 0.4 & 1.0 & 0.3 & 0.0 & 0.2 & 0.9 \\ 0.9 & 0.3 & 0.2 & 0.1 & 0.0 & 0.6 \end{bmatrix}$$

Here the dimensions of $\mathbf{X}$ and $\mathbf{Y}$ ("n" and "m") are lower than n_1 and m_1. Thus the fuzzy relation to be optimized is of lower dimensionality than the relations contributing to its structural reconstruction, $\dim(\mathbf{X}_1) = 6$, $\dim(\mathbf{Y}_1) = 6$, $\dim(\mathbf{X}) = \dim(\mathbf{Y}) = 3$. The optimization leads to a far lower value of the performance index than reported in Example 1 (Q = 0.049). This could have been somewhat anticipated considering that in the reconstructed fuzzy relation R we have to deal with a lower number of constraints.

Example 3. In this case the dimensionality of the spaces over which fuzzy relations are built are the following $\dim(\mathbf{X}) = 3$, $\dim(\mathbf{Y}) = 5$, $\dim(\mathbf{X}_1) = 7$, $\dim(\mathbf{Y}_1) = 2$ meaning that the structural reconstruction is positioned somewhere in-between as illustrated in Fig. 2. The corresponding fuzzy relations are equal to

$$A = \begin{bmatrix} 0.7\ 0.1\ 0.6\ 0.9\ 0.8 \\ 0.4\ 1.0\ 0.2\ 0.3\ 0.4 \\ 0.0\ 0.7\ 0.5\ 0.1\ 0.8 \\ 0.8\ 0.2\ 0.7\ 0.9\ 0.2 \\ 0.3\ 0.1\ 0.5\ 0.9\ 0.1 \\ 0.1\ 0.3\ 0.5\ 0.7\ 0.9 \\ 1.0\ 0.8\ 0.6\ 0.4\ 0.2 \end{bmatrix} \qquad B = \begin{bmatrix} 0.8\ 0.3 \\ 0.1\ 0.9 \\ 0.5\ 0.6 \end{bmatrix}$$

The resulting fuzzy relation R (with the associated value of the performance index Q equal to 0.455) comes with the following entries

$$R = \begin{bmatrix} 0.963\ 0.792\ 0.954\ 0.997\ 0.953 \\ 0.000\ 0.761\ 0.000\ 0.000\ 0.000 \\ 0.141\ 0.685\ 0.000\ 0.000\ 0.000 \end{bmatrix}$$

The strongest associations are reported between x_1 and almost all y_j's, x_2 and y_2. The variables x_3 associates with y_2 but to a far lesser extent than the previous associations.

Example 4. Fuzzy relations can be conveniently interpreted as images with entries representing levels of brightness of the individual pixels. Let us consider a fuzzy relation which represents a 5 by 5 image with the entries

$$\begin{bmatrix} 0.7\ 0.9\ 0.5\ 0.2\ 0.0 \\ 0.2\ 1.0\ 0.6\ 0.0\ 0.0 \\ 0.0\ 0.6\ 0.0\ 0.1\ 0.1 \\ 0.1\ 0.3\ 1.0\ 0.2\ 0.1 \\ 0.0\ 0.4\ 0.5\ 0.0\ 0.0 \end{bmatrix}$$

Given this image, we consider its projections along the two coordinates giving rise to the fuzzy relations A and B with the entries $A = [0.7\ 1.0\ 1.0\ 0.2\ 0.1]$, $B = [0.9\ 1.0\ 0.6\ 1.0\ 0.5]$. Note that the projection has been completed by taking the maximal values of the entries of the fuzzy relation in the corresponding rows and columns. The reconstructed fuzzy relation R comes with the value of the performance index Q equal to 0.001689 and has the following entries

$$R = \begin{bmatrix} \mathbf{0.534} & \mathbf{0.687} & \mathbf{0.696} & \mathbf{0.186} & 0.026 \\ 0.405 & \mathbf{1.000} & \mathbf{0.803} & \mathbf{0.083} & \mathbf{0.037} \\ 0.294 & 0.269 & 0.474 & 0.020 & \mathbf{0.107} \\ 0.091 & 0.995 & \mathbf{0.937} & 0.000 & \mathbf{0.000} \\ 0.242 & 0.167 & \mathbf{0.416} & \mathbf{0.005} & \mathbf{0.043} \end{bmatrix}$$

When compared with the original fuzzy relation (image), we note that several regions of the image (relation) have been reconstructed quite well (the corresponding entries of the relation are shown in boldface).

To further explore the role of different projections and their impact on the reconstruction, we consider sub-projections involving several overlapping rows and columns of the relation and being afterwards organized in the fuzzy relations A and B,

$$A = \begin{bmatrix} 0.7 & 1.0 & 0.6 & 0.2 & 0.0 \\ 0.2 & 1.0 & 0.6 & 0.1 & 0.1 \\ 0.1 & 0.6 & 1.0 & 0.2 & 0.1 \\ 0.1 & 0.4 & 1.0 & 0.2 & 0.1 \end{bmatrix} \quad B = \begin{bmatrix} 0.9 & 0.9 & 0.5 & 0.2 \\ 1.0 & 1.0 & 0.7 & 0.0 \\ 0.6 & 0.6 & 0.1 & 0.1 \\ 0.3 & 1.0 & 1.0 & 0.2 \\ 0.4 & 0.5 & 0.0 & 0.0 \end{bmatrix}$$

In this case the reconstructed fuzzy relation R (with the value of the performance index Q = 0.0198) comes with the following entries

$$R = \begin{bmatrix} \mathbf{0.686} & \mathbf{0.957} & \mathbf{0.672} & \mathbf{0.186} & \mathbf{0.000} \\ \mathbf{0.182} & \mathbf{1.000} & 1.000 & 0.341 & \mathbf{0.183} \\ 0.112 & \mathbf{0.702} & 0.473 & \mathbf{0.000} & \mathbf{0.059} \\ \mathbf{0.036} & 1.000 & 0.000 & \mathbf{0.000} & \mathbf{0.000} \\ 0.207 & \mathbf{0.570} & 0.191 & \mathbf{0.000} & \mathbf{0.000} \end{bmatrix}$$

Noticeably, these more detailed representations of the original relation have led to a better reconstruction in several regions; here the entries of R get closer to the ones shown in the original fuzzy relation. Those entries where the coincidence between the original fuzzy relation and its reconstruction is higher are again marked in boldface.

Example 5. Now we consider two fuzzy relations A and B defined in $\mathbf{X}_1 \times \mathbf{Y}$ and $\mathbf{X} \times \mathbf{Y}_1$ $\dim(\mathbf{X}_1) = 5$, $\dim(\mathbf{Y}_1) = 3$, $\dim(\mathbf{X}) = 4$, $\dim(\mathbf{Y}) = 3$ with the following entries

$$A = \begin{bmatrix} 1.0 & 0.7 & 0.4 \\ 0.2 & 0.5 & 0.9 \\ 0.9 & 0.0 & 0.6 \\ 0.4 & 0.6 & 0.9 \\ 0.5 & 0.1 & 0.7 \end{bmatrix} \quad B = \begin{bmatrix} 0.2 & 0.4 & 0.8 \\ 0.7 & 1.0 & 0.1 \\ 0.9 & 0.0 & 0.5 \\ 1.0 & 0.0 & 0.9 \end{bmatrix}$$

The fuzzy relation R defined in $\mathbf{X} \times \mathbf{Y}$ being "reconstructed" on the basis of A and B leads to the value of the performance index equal to 0.052. The corresponding entries of R are now equal to

$$R = \begin{bmatrix} 0.82 & 0.35 & 0.04 \\ 0.37 & 0.90 & 0.43 \\ 0.47 & 0.42 & 0.88 \\ 0.89 & 0.46 & 0.95 \end{bmatrix}$$

hence the most dominant relationships come in the pairs (x_1, y_1), (x_2, y_2), (x_3, y_3), and (x_4, y_3).

6 Conclusions

Fuzzy relations capture dependencies between system variables. Different views at the same system usually lead to different spaces of variables and result in different fuzzy relations of dependencies. In this study, we formulated and solved the problem of reconstruction and re-use of fuzzy relations defined in pertinent spaces. We have shown the conditions under which such reconstruction processes are made possible by discussing the character of the individual spaces (**X**, **Y**, etc.) and showing a series of explicit linkages that must be satisfied. We also brought a concept of structural interpolation and extrapolation by discussing the relationships between the spaces in which the individual fuzzy relations are formed. The detailed gradient-based optimization scheme has been developed in order to construct the entries of the fuzzy relations under search. A series of numeric experiments was used to demonstrate the effectiveness of the learning procedure.

While the study was focused on the fuzzy relations defined in the Cartesian products of two spaces as our primary intent was to offer a simplest possible version of the problem, the framework of reconstruction and re-use of fuzzy relations could be instantaneously extended to fuzzy relations defined in multidimensional Cartesian products involving a higher number of system variables.

Acknowledgments

Support from the Canada Research Chair (CRC) program and the Natural Sciences and Engineering Research Council (NSERC) is gratefully acknowledged.

References

1. Bui, L.D. and Kim, Y.G. (2006), An obstacle-avoidance technique for autonomous underwater vehicles based on BK-products of fuzzy relation, *Fuzzy Sets and Systems*, 157, 4, 560–577.

2. Chen, Y., Zhou, M. and Wang, H. (1994), Qualitative modeling based on fuzzy relation and its application, *Proceedings of the 3rd IEEE Conference on Fuzzy Systems, 1994. IEEE World Congress on Computational Intelligence*, 26–29 June 1994, vol. 3, 852–1856.
3. Dawyndt, P., De Meyer, H. and De Baets, B. (2004), On the min-transitive approximation of symmetric fuzzy relations, *Proc. 2004 IEEE International Conference on Fuzzy Systems*, 25–29 July 2004, Vol. 1, 67–171.
4. De Meyer, H., Naessens, H. and De Baets B.(2004), Algorithms for computing the min-transitive closure and associated partition tree of a symmetric fuzzy relation, *European Journal of Operational Research*, 155, 1, 226–238.
5. Fernández, M. J. and Gil, P. (2004), Some specific types of fuzzy relation equations, *Information Sciences*, 164, 1–4, 189–195.
6. Iliadis, L.S. and Spartalis, S.I.(2005), Fundamental fuzzy relation concepts of a D.S.S. for the estimation of natural disasters' risk (The case of a trapezoidal membership function), *Mathematical and Computer Modelling*, 42, 7–8, 747–758.
7. Li, H.X., Miao Z.H, Han, S.C and Wang J.Y.(2005), A new kind of fuzzy relation equations based on inner transformation, *Computers & Mathematics with Applications*, 50, 3–4, 623–636.
8. Liu, D.R., Wu, I.C. and Yang, K.S. (2005), Task-based -Support system: disseminating and sharing task-relevant knowledge, *Expert Systems with Applications*, 29, 2, 408–423.
9. Loia, V. and Sessa, S. (2005), Fuzzy relation equations for coding/decoding processes of images and videos, *Information Sciences*, 171, 1–3, 145–172.
10. Luo Y. and Li, Y. (2004), Decomposition and resolution of min-implication fuzzy relation equations based on S-implications, *Fuzzy Sets and Systems*, 148, 2, 305–317.
11. Luoh; L., Wang, W.J. and Liaw, Y.K. (2003), Matrix-pattern-based computer algorithm for solving fuzzy relation equations, *IEEE Transactions on Fuzzy Systems*, 11, 1, 100–108.
12. Oura, K., Kitazawa, R., Akizuki, K. and Miyazaki, M. (1999), Construction of a fuzzy relation with reduced dimension for multivariable systems by using genetic algorithm, *Proc. IEEE Conf. Systems, Man, and Cybernetics, SMC '99*, vol. 5, 12–15 Oct. 1999, 320–325.
13. Pedrycz, W., Hirota, K. and Sessa, S. (2001), A decomposition of fuzzy relations, *IEEE Transactions on Systems, Man and Cybernetics, Part B*, 31, 4, 657–663.
14. Pedrycz, W. (2005), *Knowledge-Based Fuzzy Clustering*, J. Wiley, Hoboken, NJ.
15. Raghuvanshi, P.S. and Kumar, S. (1999), On the structuring of systems with fuzzy relations *IEEE Transactions on Systems, Man and Cybernetics, Part B*, 29, 4, 547–553.
16. Stamou, G.B. and Tzafestas, S.G. (1999), Fuzzy relation equations and fuzzy inference systems: an inside approach, *IEEE Transactions on Systems, Man and Cybernetics, Part B*, 29, 6, 694–702.
17. Zadeh, L.A. (1999), From computing with numbers to computing with words-from manipulation of measurements to manipulation of perceptions, *IEEE Trans. on Circuits and Systems*, Vol. 45, 105–119.
18. Zadeh, L.A. (2005), Toward a generalized theory of uncertainty (GTU) – an outline, *Information Sciences*, 172, 1–2,1–40.

On Fuzzy Logic and Chaos Theory
— *From an Engineering Perspective*

Zhong Li and Xu Zhang

Abstract. A review to the current studies on the interactions between fuzzy logic and chaos theory will be presented in this article with focus on fuzzy modeling of chaotic systems using Takagi-Sugeno (TS) models, linguistic descriptions of chaotic systems, fuzzy control of chaos, complex fuzzy systems, and a combination of fuzzy control technology and chaos theory for an engineering practice. The aim is to provide some heuristic research achievements and insightful ideas to attract more attention on the topic, interactions or relationship between fuzzy logic and chaos theory, which are related at least within the context of human reasoning and information processing.

1 Introduction

The naissance of fuzzy logic and the prosperity of scientific research on chaos theory occurred almost at the same time in the 1960s, a decade full of confusion, when scientists faced difficulties in dealing with imprecise information and complex dynamics. A set theory and then an infinite-valued logic of Lotfi A. Zadeh were so confusing that they were called fuzzy set theory and fuzzy logic in 1965; a deterministic system found by Edward N. Lorenz in 1963 to have random behaviors was so unusual that it was lately named a chaotic system. Just like irrational and imaginary numbers, negative energy, anti-matter, ..., fuzzy logic and chaos were gradually and eventually accepted by many, if not all, scientists and engineers as fundamental concepts, theories, as well as technologies.

Since then, fuzzy logic (or say, fuzzy set theory) and chaos theory have independently developed along their own ways and matured as sciences (although still evolving). They have provided many insights into previously intractable and inherently imprecise or complex nonlinear natural phenomena.

In particular, fuzzy systems technology has achieved its maturity with widespread applications in many industrial, commercial and technical fields, ranging from control, automation, and artificial intelligence to image/signal processing, pattern recognition, and electronic commerce. Chaos, on the other hand, was considered one of the three monumental discoveries of the twentieth century together with the theory of relativity and quantum mechanics. As a very special nonlinear dynamical phenomenon, chaos has reached its current outstanding status from being merely a scientific curiosity in the mid-1960s to an applicable technology in the late 1990s.

Why we bring together the two seemingly unrelated concepts to study their interactions and relationships? On the one hand, finding the intrinsic relationship between fuzzy logic and chaos theory is certainly of significant interest and of potential importance. The past twenty years have indeed witnessed some serious explorations of the interactions between fuzzy logic and chaos theory, leading to such research topics as fuzzy modeling of chaotic systems using Takagi-Sugeno models, linguistic descriptions of chaotic systems, fuzzy control of chaos, and a combination of fuzzy control technology and chaos theory for various engineering practices. On the other hand, the reason to study the interactions between fuzzy logic and chaos theory lies in that they are related at least within the context of human reasoning and information processing. In fact, fuzzy logic resembles human approximate reasoning using imprecise and incomplete information with inaccurate and even self-conflicting data to generate reasonable decisions under such uncertain environments, while chaotic dynamics play a key role in human brains for processing massive amounts of information instantly. It is believed that the capability of humans in controlling chaotic dynamics in their brains is more than just an accidental by-product of the brain's complexity, but rather, it could be the chief property that makes the human brain different from any artificial-intelligence machines. It is also believed that to understand the complex information processing within the human brain, fuzzy data and fuzzy logical inference are essential, since precise mathematical descriptions of such models and processes are clearly out of question with today's scientific knowledge.

What is necessary to mention is that Lotfi A. Zadeh has integrated fuzzy logic and chaos theory into the concept of *soft computing* (SC), where he states that SC consists of fuzzy logic (FL), neural network theory (NN) and probabilistic reasoning (PR), with the latter subsuming parts of belief networks, genetic algorithms, chaos theory and learning theory. It is noted that SC is not a melange of FL, NN and PR. Rather, it is an integration in which each of the partners contributes a distinct methodology for addressing problems in their common domain. In this perspective, the principal contributions of FL, NN and PR are complementary rather than competitive.

This chapter aims to provide some heuristic research achievements and insightful ideas to attract more attention on the topic, through reviewing the current studies on the interactions between fuzzy logic and chaos theory, including fuzzy modeling of chaotic systems using Takagi-Sugeno (TS) models, linguistic descriptions of chaotic systems, fuzzy control of chaos, complex fuzzy systems, and a combination of fuzzy control technology and chaos theory for an engineering practice.

2 Fuzzy Definitions of Chaos

The term *chaos* associated to an interval map was first formally introduced into mathematics by Li and Yorke in 1975 [1], where they established a simple criterion for the existence of chaos in one-dimensional difference equations, i.e. "period three implies chaos."

However, a definitive, universally accepted, and completely rigorous mathematical definition of chaos is not yet available in the scientific literature to provide a fundamental basis for studying such exotic phenomena. Instead, various alternative, but closely related definitions of chaos have been proposed, along with mechanisms giving rise to such behavior. Among those, the original definition of Li and Yorke and its fine tuning by Devaney seem to be the most popular. These definitions of chaos have been generated to difference equations in $\mathcal{R}^n$, Banach spaces and complete metric spaces. In particular, an application of the definition in complete metric spaces to chaotic dynamics on the metric space $(\mathcal{E}^n, D)$ of fuzzy sets on the base space $\mathcal{R}^n$ may lay a foundation for further study on the interactions between fuzzy logic and chaos theory [2].

It is shown by the result of Li and Yorke that such chaotic behavior could arise in quite simple systems and could be generated by quite simple mechanisms. In particular, the "period three implies chaos" result in scalar difference equations involves *noninvertible, continuous* maps rather than more demanding diffeomorphisms. Consequently, one may ask to what extent the one-dimensional result of Li and Yorke carries over to a higher dimensional difference equation

$$x_{k+1} = f(x_k), \quad k = 0, 1, 2, \ldots, \tag{1}$$

with a continuous map $f : X \to X$, where X is a closed subset of $\mathcal{R}^n$ for $n \geq 2$.

A counter example shows that the result does not carry over to higher dimensions without some suitable restriction on the class of maps f [2,3]. To determine a suitable class of maps for which the result of the Li and Yorke might hold in higher dimensions, we found that the one-dimensional maps for which the difference equation (1) have cycles of period three all have graphs with a hump, i.e. which fold over on themselves, namely, they are not one-to-one maps. This suggests that attention might profitably be restricted to maps that are not one-to-one. This was done by Marotto [4] who showed that difference equations on $\mathcal{R}^n$ defined in terms of continuously differentiable maps with *snap-back repellers*, so consequently not one-to-one, behave chaotically in the sense of Li and Yorke. His proof used the inverse function theorem for one-to-one local restrictions of the maps and the Brouwer fixed point theorem, but otherwise paralleled the proof of Li and Yorke for one-dimensional maps. This was extended in 1981 to maps with a saddle point by Kloeden [5] and Shiraiwa and Kurata [6], where the maps in the difference equations are assumed to be continuous rather than continuously differentiable, and the Brouwer fixed point theorem is used on a homeomorphism of an l-ball for some $1 \leq l \leq n$ rather than on a homeomorphism of an n-ball as in Marotto's proof. Further, the result of Kloeden for the case of saddle points in a finite dimensional Euclidean space $\mathcal{R}^n$ can be easily extended by using the Schauder fixed point theorem to a Banach space. The Schauder fixed point theorem states that a compact map f from a closed bounded convex set K in a Banach space X into itself has a fixed point. Even more generally, some criteria for chaos of difference equations in general complete metric spaces have been given in [10, 11].

What we concern here is to generalize the Li-Yorke and Marotto definitions to be applicable to maps from a space of fuzzy sets into itself, namely the metric space $(\mathcal{E}^n, D)$ of fuzzy sets on the base space $\mathcal{R}^n$. The result to be given below is essentially an adaptation of the result in a Banach space, which is possible because the metric spaces of fuzzy sets under consideration can be embedded as a cone in a certain Banach space.

Here, we simply give the Kaleva fixed point theorem without giving some basic terminologies [7, 8].

Theorem 1. (Kaleva)
Let $f : \mathcal{E}^n \to \mathcal{E}^n$ be continuous and let $\mathcal{X}$ be a non-empty compact convex subset of $\mathcal{E}^n$ such that $f(\mathcal{X}) \subseteq \mathcal{X}$. Then f has a fixed point $\bar{u} = f(\bar{u}) \in \mathcal{X}$.

Consider an iterative scheme of fuzzy sets

$$u_{k+1} = f(u_k), \quad k = 1, 2, \ldots, \tag{2}$$

where f is a continuous map from the space of fuzzy sets $\mathcal{E}^n$ into itself. Using the Kaleva fixed point theorem, sufficient conditions will be given below for a map on fuzzy sets to be chaotic.

Theorem 2. (Kloeden [9])
Let $f : \mathcal{E}^n \to \mathcal{E}^n$ be continuous and suppose that there exist non-empty compact subsets $\mathcal{A}$ and $\mathcal{B}$ of $\mathcal{E}^n$, and integers $n_1, n_2 \geq 1$ such that

(i) $\mathcal{A}$ is homeomorphic to a convex subset of $\mathcal{E}^n$,
(ii) $\mathcal{A} \subseteq f(\mathcal{A})$,
(iii) f is expanding on $\mathcal{A}$, that there exists a constant $\lambda > 1$ such that

$$\lambda D(u, v) \leq D(f(u), f(v))$$

for all $u, v \in \mathcal{A}$,
(iv) $\mathcal{B} \subset \mathcal{A}$,
(v) $f^{n_1}(\mathcal{B}) \cap \mathcal{A} = \emptyset$,
(vi) $\mathcal{A} \subseteq f^{n_1+n_2}(\mathcal{B})$,
(vii) $f^{n_1+n_2}$ is one-to-one on $\mathcal{B}$.

Then the map f is chaotic.

It is noted that difference equations generated by Poincaré section maps provide a link between the dynamics of discrete-time dynamical systems and continuous-time dynamical systems. However, it has been much more difficult to give a mathematically rigorous proof of the existence of chaos in a continuous-time nonlinear autonomous systems. Even one of the classic icons of modern nonlinear dynamics, the Lorenz attractor, now known for 40 years, was not proven rigorously to be chaotic until 1999 by Warwick Tucker of the University of Uppsala in his Ph.D. dissertation [12, 13]. A commonly agreed analytic criterion for proving the existence of chaos in continuous-time systems is based on the fundamental work of Shil'nikov, known as the Shil'nikov method or Shil'nikov criterion [14], whose role is in some sense equivalent to that of the Li-Yorke definition in the discrete setting. The Shil'nikov criterion guarantees that complex dynamics will occur near homoclinicity or heteroclinicity when an inequality (Shil'nikov inequality) is satisfied between the eigenvalues of the linearized flow around the saddle point(s), i.e., if the real eigenvalue is larger in modulus than the real part of the complex eigenvalue. Complex behavior always occurs when the saddle set is a limit cycle.

3 From Chaotic Systems to Fuzzy Systems

This section discusses fuzzy modeling of chaotic systems, meaning to transform chaotic systems into fuzzy formulations.

Fuzzy system models basically fall into two categories, which differ fundamentally in their abilities to represent different types of information. The first category includes linguistic models, which have been referred to so far as Mamdani fuzzy models. They are based on collections of IF-THEN rules with vague predicates and use fuzzy reasoning [15, 16]. In these models, fuzzy quantities are associated with linguistic labels, and a fuzzy model is essentially a qualitative expression of the underlying system. Models of this type form a basis for qualitative modeling that describes the system behavior by using natural language [17]. A corresponding fuzzy logic controller is a prototypical example of such a linguistic model, in which its rules give a linguistic expression of the control strategy in a common sense.

The second category of fuzzy models is based on the Takagi-Sugeno (TS) method of reasoning [18–20]. These models are formed by logical rules that have a fuzzy antecedent part and a functional consequent. They are combinations of fuzzy and nonfuzzy models. Fuzzy models based on the TS method of reasoning integrate the ability of linguistic models for qualitative knowledge representation with great potential for expressing quantitative information.

A general Mamdani fuzzy rule, for either fuzzy control or fuzzy modeling, can be expressed as,

$$R_M^l : \text{IF } x_1 \text{ is } F_1^l \text{ and } \cdots \text{ and } x_n \text{ is } F_n^l, \text{ THEN } y \text{ is } G^l, \tag{3}$$

where F_i^l and G^l are fuzzy sets, $x = (x_1, \ldots, x_n)^T \in U$ and $y \in V$ are input and output linguistic variables, respectively, and $l = 1, 2, \ldots, q$. This kind of fuzzy IF-THEN rules provides a convenient framework to incorporate human experts' knowledge.

The reasoning method uses fuzzy relations and their composition. A fuzzy IF-THEN rule (3) is interpreted as a fuzzy implication $F_1^l \times \cdots \times F_n^l \to G^l$ on $U \times V$. Let a fuzzy set A' on U be the input to a fuzzy inference engine. Then each fuzzy IF-THEN rule (3) determines a fuzzy set B' on V using the so-called sup-star composition, which writes

$$\mu_{B'}(y) = \sup_{x \in U}\{\mu_{F_1^l \times \cdots \times F_n^l \to G^l}(x, y) \star \mu_{A'}(x)\}. \tag{4}$$

Instead of using fuzzy sets in the consequence part of (3), TS fuzzy rules adopt linear functions, which represent input-output relations. A TS fuzzy rule is described as,

$$\begin{aligned} R_T^l : \text{IF } x_1 \text{ is } F_1^l \text{ and } \cdots \text{ and } x_n \text{ is } F_n^l, \\ \text{THEN } y^l = c_0^l + c_1^l x_1 + \cdots + c_n^l x_n, \end{aligned} \tag{5}$$

where F_i^l are fuzzy sets, c_i are real-valued parameters, y^l is the system output, and $l = 1, 2, \ldots, q$. That is, in the TS fuzzy rules, the IF part (premise) is fuzzy but the THEN part (consequence) is crisp – the output is a linear combination of input variables.

For the TS fuzzy model (5), fuzzy reasoning is carried out by the weighted mean,

$$y(x) = \frac{\sum_{l=1}^{M} \omega^l y^l}{\sum_{l=1}^{M} \omega^l}, \tag{6}$$

where the weight ω^l is the overall truth value of the premise of rule R_T^l for the input, and is given by

$$\omega^l = \prod_{i=1}^{n} \mu_{F_i^l}(x_i), \tag{7}$$

where $\mu_{F_i^l}(x_i)$ is the membership degree of the fuzzy set F_i^l.

3.1 Fuzzy Modeling of Chaotic Systems Based on Mamdani Model

Fuzzy logic allows to model processes in a linguistic manner. The basic configuration of a fuzzy logic system is composed of a fuzzyfier, a fuzzy rule-base, a fuzzy inference engine, and a defuzzyfier, where the fuzzy rule-base consists of a collection of fuzzy IF-THEN rules, and the fuzzy inference engine uses these fuzzy IF-THEN rules to determine a map from fuzzy inputs to fuzzy outputs based on fuzzy composition rules.

A systematic approach for modeling chaotic systems using Mamdani model is never available, Baglio et al. [21,22] have, however, subtly derived Mamdani fuzzy models of some typical chaotic systems. To do so, a good description of chaotic systems is required. The definition of chaotic behaviors involves the three fundamental concepts of transitivity, density of periodic orbits, and sensitivity to initial conditions [23]. Furthermore, from a qualitative point of view, chaos can be defined by monitoring the time evolution of trajectories emanating from nearby points on the attractor. In a chaotic system, points that are close to each other repel themselves so that the flow stretches. Then, a folding action must take place for the chaotic behavior to combine with the boundedness of the attractor. The stretching and folding features of the flow are responsible for the sensitivity to initial conditions, and characterize the chaotic behavior.

To develop a fuzzy model of the evolution of a chaotic signal x, two variables can be considered as inputs, i.e. the center value $x(k)$, which is the nominal value of the state x at the instant k, and the uncertainty $d(k)$ on the center value. In terms of fuzzy description, this means that the model contains four linguistic variables, i.e. $x(k)$, $x(k+1)$, $d(k)$ and $d(k+1)$. The whole set of rules has to determine the values $x(k+1)$ and $d(k+1)$ from the values $x(k)$ and $d(k)$.

Take the logistic map $x(k+1) = \mu x(k)(1 - x(k))$, $\mu = 4$, which shows a single-scroll attractor, as an example to illustrate the modeling procedure. In this single scroll system, x tends to move out from the trivial equilibrium point $x_1^* = 0$ until x begins to oscillate around the nontrivial equilibrium point $x_2^* = 3/4$. The increasing amplitude of the oscillations forces the trajectory to enter again the neighborhood of $x_1^* = 0$, where, due to its instability, the above process repeats. The linguistic variables of the system, $(x(k), x(k+1), d(k), d(k+1))$, take five linguistic values: zero (Z), small (S), medium (M), large (L) and very large (VL). The fuzzy sets associated to these linguistic values are shown in Fig. 1. They are constructed in such a way that the equilibrium point $x_2^* = 3/4$ is between the fuzzy set M and the fuzzy set L.

In other words, when x is smaller than the nontrivial equilibrium point $x_2^* = 3/4$, it tends to increase and, when x is very large, it tends to decrease, which can be summarized in the following rules.

$$R_1: \quad \text{IF } x(k) \text{ is } S \text{ THEN } x(k+1) \text{ is } M;$$
$$R_2: \quad \text{IF } x(k) \text{ is } VL \text{ THEN } x(k+1) \text{ is } Z.$$

In this way, a complete set of fuzzy rules to generate a single-scroll chaotic system is summarized in Table 1.

The trajectory of the fuzzy system in the phase space is shown in Fig. 2, which is obviously similar to that of the original logistic map.

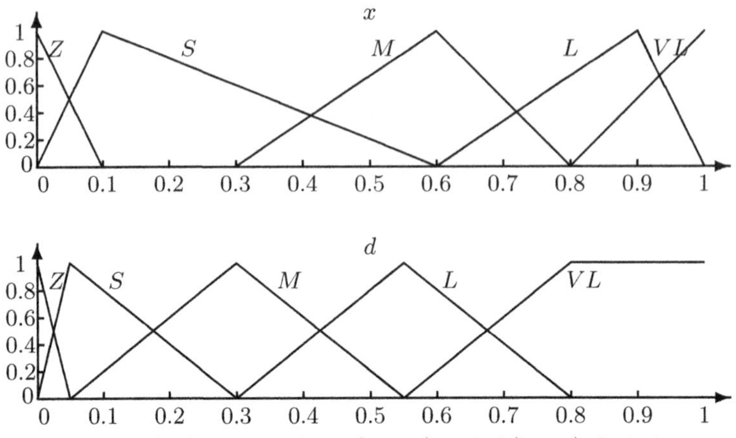

Fig. 1. The fuzzy sets for x (upper) and d (lower): logistic map

3.2 Fuzzy Modeling of Chaotic Systems Based on TS Model

As mentioned above, the TS fuzzy model adopts linear functions rather than fuzzy sets in the consequence part, thus, linearization methods are often used in modeling, where the following theorem can be used to convert the nonlinear terms in the nonlinear systems to weighted linear sums of some linear functions [24].

Theorem 3. Consider the following nonlinear term:

$$f_n = x_1 x_2 \cdots x_n \,, \tag{8}$$

where $x_i \in \left[M_1^i, M_2^i\right]$. Formula (8) can exactly be represented by a linear weighted sum of the form

$$f_n = \left(\sum_{i_2, i_3, \ldots, i_n = 1}^{2} \mu_{i_2 i_3 \cdots i_n} \cdot g_{i_2 i_3 \cdots i_n} \right) x_1, \tag{9}$$

Table 1. Fuzzy rules implementing a single scroll chaotic system

$x(k)/d(k)$	Z	S	M	L	VL
Z	Z/Z	Z/M	Z/M	S/VL	L/L
S	M/Z	M/M	M/M	M/VL	L/S
M	L/Z	L/M	L/M	L/VL	VL/S
L	M/Z	M/M	M/M	M/VL	Z/S
VL	Z/Z	Z/M	Z/M	Z/VL	Z/L

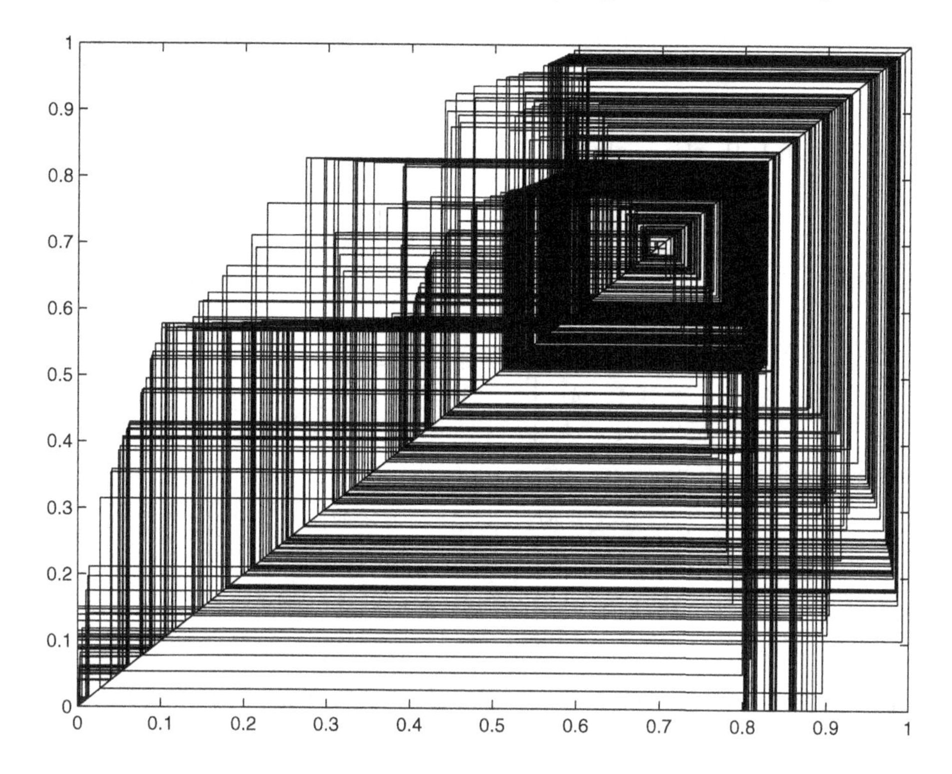

Fig. 2. The logistic map generated by the fuzzy model

where

$$g_{i_2 i_3 \cdots i_n} = \prod_{j=2}^{n} M_{i_j}^j , \quad \mu_{i_2 i_3 \cdots i_n} = \prod_{j=2}^{n} \Gamma_{i_j}^j ,$$

in which $\Gamma_{i_j}^j$ is positive semi-definite for all $x_j \in [M_1 , M_2]$, defined as follows:

$$\Gamma_1^j = \frac{-x_j + M_2^j}{M_2^j - M_1^j} , \quad \Gamma_2^j = \frac{x_j - M_1^j}{M_2^j - M_1^j} .$$

Herewith, for most of chaotic systems, their exact TS fuzzy models can be easily derived to be with only two fuzzy rules. Here, the word "exact" means that the defuzzified output of the TS fuzzy models are mathematically identical to that of the original nonlinear systems.

For instance, the Lorenz equations,

$$\frac{d}{dt} \begin{bmatrix} x \\ y \\ z \end{bmatrix} = \begin{bmatrix} -\sigma x + \sigma y \\ rx - y - xz \\ xy - bz \end{bmatrix} , \tag{10}$$

where $\sigma, r, b > 0$ are parameters, can be expressed as the following TS fuzzy model:

$$\text{Rule 1: IF } x(t) \text{ is about } M_1 \text{ THEN } \frac{d}{dt}\begin{bmatrix} x(t) \\ y(t) \\ z(t) \end{bmatrix} = A_1 \begin{bmatrix} x(t) \\ y(t) \\ z(t) \end{bmatrix}$$

$$\text{Rule 2: IF } x(t) \text{ is about } M_2 \text{ THEN } \frac{d}{dt}\begin{bmatrix} x(t) \\ y(t) \\ z(t) \end{bmatrix} = A_2 \begin{bmatrix} x(t) \\ y(t) \\ z(t) \end{bmatrix}$$

where

$$A_1 = \begin{bmatrix} -\sigma & \sigma & 0 \\ r & -1 & -M_1 \\ 0 & M_1 & -b \end{bmatrix}, \quad A_2 = \begin{bmatrix} -\sigma & \sigma & 0 \\ r & -1 & -M_2 \\ 0 & M_2 & -b \end{bmatrix},$$

and the membership functions are

$$\Gamma_1 = \frac{-x + M_2}{M_2 - M_1}, \quad \Gamma_2 = \frac{x - M_1}{M_2 - M_1},$$

where $\Gamma_i, i = 1, 2$, are positive semi-definite for all $x \in [M_1, \quad M_2]$.

4 From Fuzzy Systems to Chaotic Systems

It is clear to see that fuzzy modeling of chaotic systems implies that fuzzy systems can be also chaotic. In contrast to Sect. 3, "from fuzzy systems to chaotic system" means to make originally stable or non-chaotic fuzzy systems chaotic. This is of practical significance, since chaos can actually be useful under certain circumstances, and there is growing interest in utilizing the very nature of chaos [25–29].

One simple, yet mathematically rigorous control method from the engineering feedback control approach was developed [30–32], where a linear state-feedback controller with an uniformly bounded control-gain sequence can be designed to make all Lyapunov exponents of the controlled system strictly positive and arbitrarily assigned. Moreover, such a controller can be designed for an arbitrarily given, n-dimensional dynamical system that could originally be nonchaotic or even asymptotically stable. The goal of chaotification is finally achieved with a simple modulus operation or a sawtooth (or even a sine) function. The design criterion is to use the definition of chaos given by Devaney [33] or Li-Yorke [1], while for the n-dimensional case the Marotto theorem [4] was used for a proof. For the continuous-time case, a general approach to make an arbitrarily given autonomous system chaotic has also been proposed recently [34–37]. Here, the main tool to use is time-delay feedback perturbation on a system parameter or as an exogenous input [35].

These chaotification techniques can be applied to TS fuzzy systems, where the so-called parallel distributed compensation (PDC) technique is employed to determine the structure of a fuzzy controller [38–43].

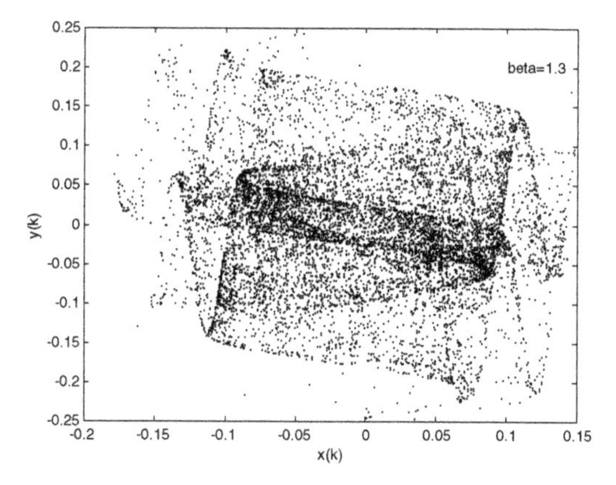

Fig. 3. Phase portrait at $\beta = 1.3$

For example, a simple controller employing a sinusoidal function is designed to chaotify the discrete-time TS fuzzy model of stable Lorenz system, where the controller can be with an arbitrarily small magnitude and of the form $u(k) = \Phi(\beta x(k)) \equiv [\varphi(\beta x_1(k)), \varphi(\beta x_2(k)), \ldots, \varphi(\beta x_n(k))]^T$, in which $x(k) = [x_1(k), x_2(k), \ldots, x_n(k)]^T$, β is a constant, and $\varphi : \Re \to \Re$ is a continuous sinusoidal function defined by $\varphi(x) = \sigma \sin\left(\frac{\pi}{\sigma}x\right)$. Assume $\sigma = 0.1$ and let β be the control parameter. The simulation result is shown in Fig. 3.

It is remarked that modeling of chaotic systems based on Mamdani models discussed above provides a means to generate chaos from Mamdani fuzzy systems in a linguistic manner.

5 Control of Fuzzy Chaotic Systems

For many years, the main feature of chaos, i.e. the extreme sensitivity to initial conditions, made chaos undesirable, and most experimentalists consider such characteristics as something to be strongly avoided [45, 46]. In addition to this feature, chaotic systems have two other important ones. First, there are infinite many unstable periodic orbits embedded in the underlying chaotic attractor, and second, the dynamics in the chaotic attractor is ergodic, which implies that during its temporal evolution the system ergodically visits any small neighborhood of every point in each of the unstable periodic orbits embedded within the chaotic attractor.

Owing to these properties, a fuzzy chaotic system, the fuzzy formulation of a chaotic system, can be seen as shadowing some periodic behavior at a given time, and erratically jumping from one to another periodic orbit. Thus, when a trajectory approaches ergodically a desired periodic orbit embedded in the chaotic attractor, one can apply small perturbations to stabilize such an orbit. Therefore, we can say that the extreme sensitivity of a chaotic system to changes in its initial conditions may be very desirable in practical experimental situations [46]. It suffices to note that, due to chaos, using the same chaotic system one is able to produce infinite many desired dynamical behaviors (either periodic or not periodic) only with properly chosen tiny perturbations. This property is not shared by non-chaotic systems, because the perturbations needed therein for producing a desired behavior must, in general, be of the same order of magnitude as the unperturbed dynamical variables.

Generally, chaos control approaches can be divided into two broad categories: feedback and nonfeedback (or say, open-loop) control approaches. Feedback control methods do not change the controlled systems and stabilize unstable periodic orbits embedded in chaotic attractors, while nonfeedback control methods slightly change the controlled system, mainly by a small tuning of control parameter, changing the system behavior from chaotic attractor to periodic orbit, which is close to the initial attractor.

It is known that the nonfeedback approach is much less flexible, and requires more prior knowledge of motion. To apply such an approach, one does not have to follow the trajectory. The control can be activated at any time, and one can switch from one periodic orbit to another without returning into the chaotic behavior. This approach can be very useful in mechanical systems, where the feedback control systems are often very large (sometimes larger than the system controlled). The extremely simple, easily implementable, low-cost, and reliable nonfeedback approaches are widely applied in many physical experiments and industrial processes today, particularly for nonlinear dynamical systems, as a unified feedback control approach has not been fully established for general nonlinear dynamical systems. Roughly speaking, the nonfeedback approaches include the entrainment and migration control method [47–52], control through external forcing [45, 53–58], while the feedback approaches include the Ott-Grebogi-Yorke (OGY) method, engineering feedback control method, and Pyragas's time-delayed feedback control method [59–61]. In addition, chaos control approaches include conventional linear and nonlinear control, adaptive control, neural networks-based control, fuzzy contro, and another very important topic, synchronization of chaos.

Roughly speaking, synchronization refers to the tendency of two or more appropriately coupled fuzzy chaotic systems to undergo resembling evolution in time, due to coupling or forcing. This ranges from a complete agreement of trajectories to locking of phases [62, 63]. The basic idea in the seminal paper by Pecora and Carroll [64] was to take two identical three-dimensional dynamical systems described as $\dot{x} = f(x)$ with $x = (x, y, z)^T$ and $f(x)$ being a vector field, and to use one of them as the so-called drive system $\dot{x}_d = f(x_d)$ to unidirectionally drive the second one, called response system $\dot{x}_r = f(x_r)$, by a suitable replacement of the dynamical variables in the response system. If $e(t) =: x_d - x_r \to 0$ as $t \to 0$, the dynamics of the response system approaches the time evolution of the drive system, and synchronization of these two systems is achieved.

Synchronization of fuzzy chaotic systems has been extensively investigated [38], where a fuzzy feedback law was proposed for synchronization, which is realized via exact linearization techniques and by solving LMI problems. Although the exact linearization techniques ensure stability, the scheme is no longer suitable to chaotic communications due to the effects of signal masking and modulation. As synchronization issues are closely related to observer/controller design, generalized fuzzy response systems were proposed to solve the synchronization problem from the observer and controller points of view. This leads to a different concept for the design of LMI-based fuzzy chaotic synchronization and communication, which relaxes the EL condition [65–68].

6 Complex Fuzzy Systems

Many real-world systems of interest, like ecosystems, power grids or the Internet, can be modeled by complex dynamical networks. The overall dynamics of these models have two sources of complexity, the dynamics of their components and their interactions. Considering simple regular structures, the overall dynamics can be studied in terms of the dynamical behavior of their components, a scenario that has been investigated, and for which interesting results have been obtained. Simple architectures allow us to focus on the complexity caused by the nonlinear dynamics of the nodes, without considering additional complexity in the network structure itself. Recently, the focus has shifted to consider the complexity arising from the network structure. Along this line significant advances have been reported over the last few years, most significantly the small-world and scale-free complex network models. These models capture essential aspects common to most complex real-world systems. Furthermore, it can be shown that its structural characteristics determine the overall dynamics of a network. Another particularly interesting phenomenon of a network's overall behavior is synchronization. One way to break up synchronization by changing a deterministic protocol in, for instance, Internet traffic is likely to generate another synchrony. Thus, it suggests that a more efficient solution requires a better understanding of the nature of synchronization behavior in complex networks.

The artificial reproduction of some exotic phenomena such as spatiotemporal chaos and synchronization provides a means to study and understand collective dynamics that are commonly observed in nature [69]. In particular, the synchronization of oscillators in a complex system is a classical topic related to different research fields: from circuit theory to biological and social systems, which has attracted ever increasing attention in the fascinating field of complex systems and networks.

The behavior of arrays of large numbers of oscillators by considering both the effects due to the dynamics of the single units and those related to network topology so that the global system achieves synchronization has been studied in [70, 71]. More interestingly, a class of complex systems, arrays of coupled fuzzy chaotic oscillators, has been proposed and analyzed, and the synchronization of these systems has been investigated, both in a qualitative and quantitative way [72, 73].

Similar to the Mamdani fuzzy modeling discussed above, the fundamental element of this network, a discrete chaotic fuzzy oscillator, is modeled, and its chaotic behavior is characterized by varying the Lyapunov exponent in a suitable range. Then, the macro-system is built up by connecting a large number of identical chaotic fuzzy oscillators (C) in a regular configuration. The study of the global dynamics of this fuzzy system starts from the characterization of the spatiotemporal patterns through visual inspection and frequency analysis. The equations of the array's single element are rewritten according to the number of oscillators (N), and the diffusion coefficient (D) that weights the information exchange, as described as follows.

$$\begin{cases} x_i(k+1) = \Psi\left(x_i(k) + D\left(-2Cx_i(k) + \sum\limits_{j=-C, j\neq 0}^{C} x_{i+j}(k)\right), d_i(k)\right) \\ d_i(k+1) = \Phi(x_i(k), d_i(k)), \qquad i = 1, \ldots, N, \end{cases} \tag{11}$$

where Ψ and Φ are the fuzzy inference functions described through a set of fuzzy rules for each variable.

The state value of each fuzzy oscillator depends on its own state value at the previous sample time instant, and on the contributions coming from the state values of the other fuzzy oscillators connected through a bi-directional information exchange. In this way, the one-dimensional array can be designed by coupling $N = 200$ discrete fuzzy chaotic oscillators with $l = 0.1$ and a constant diffusion coefficient $D = 0.005$, and the single unit starts its evolution from random initial conditions.

By varying the number of connections for each unit ($2 \cdot C$) in the fuzzy chain, four different global dynamics can be observed. The pattern formations related to these four different collective behaviors are displayed in Fig. 4. Each spatiotemporal map is built for a particular value of C, where the index i associated to each cell is plotted versus sampling time, meanwhile the state value x_i of each cell is represented through a 64-color scale, with blue for low value and red for high values.

It is illustrated in Fig. 4 that four behaviors can be observed by increasing the numbers of connections (C), i.e. from the initial spatiotemporal chaos at $C = 4$ (Fig. 4a), then a regular synchronized behavior at $C = 16$ (Fig. 4b), through a transition phase at $C = 25$ (Fig. 4c), to finally the chaotic synchronized behavior at $C = 46$ (Fig. 4d).

To characterize various spatiotemporal patterns, a mathematical indicator, i.e. a synchronization index, was introduced [72, 73]. Using this indicator, the

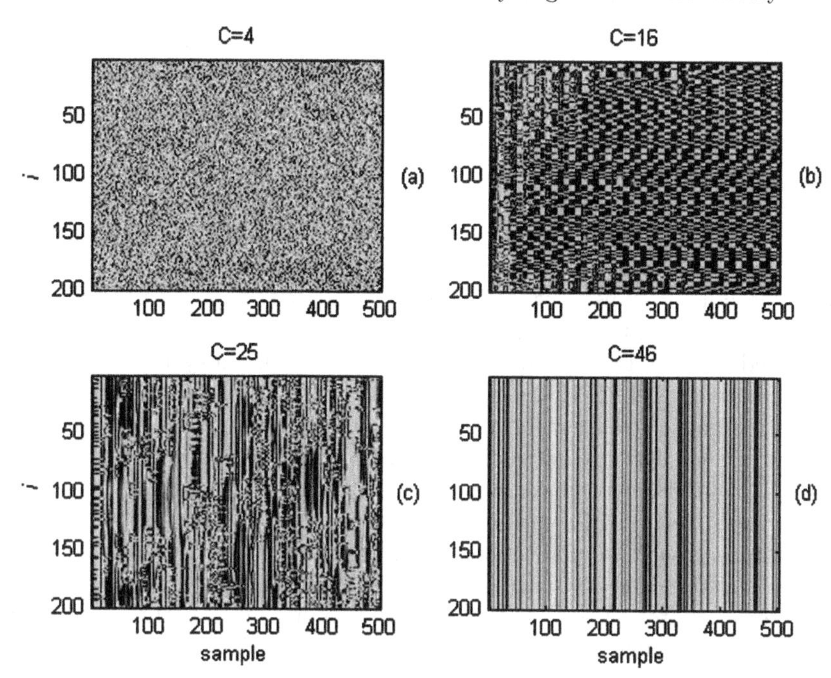

Fig. 4. Spatiotemporal patterns of the fuzzy chains ($l = 0.1$): (a) $C = 4$, (b) $C = 16$, (c) $C = 25$, (d) C=46

analysis of spatiotemporal chaos analysis as carried out above can be extended and detailed by tuning the Lyapunov exponent and the number of connections in suitable ranges. Furthermore, this evaluation index eases comparison and identification of complex network features versus adopted system parameters and, thus, avoids tedious computations in inspecting all spatiotemporal maps versus parameter variations.

7 Fuzzy-Chaos-based Applications: An Example

Combination of fuzzy logic and chaos theory may provide a new means in engineering practice. An example is a fuzzy-model-based chaotic cryptosystem introduced in [65–68].

Cryptography concerns the ways in which communications and data can be encoded to prevent disclosure of their contents through eavesdropping or message interception, using codes, ciphers, or other methods, so that only certain people can see the real messages. So far, varieties of cryptographic methods have been proposed to secure Internet communication. For instance, the Data Encryption Standard (DES) is adopted as a U.S. Federal Information Processing Standard for encrypting unclassified information. Others include IDEA

(International Data Encryption Algorithm), and RSA (developed by Rivest, Shamir and Adleman). These encryption algorithms are based on number theory. However, none of them is absolutely secure. Therefore, some emerging theories, such as chaos theory, are always desirable to be adopted to strengthen existing cryptography. The reason of applying chaos theory in cryptography lies in its intrinsic essential properties, such as sensitivity to initial conditions (or control parameters) and ergodicity, which meet Shannon's requirements of confusion and diffusion for cryptography. In addition, chaotic signals are typically broadband, noise-like, and difficult to predict. Therefore, they can be used in various context for masking information-bearing waveforms. They can also be used as modulating waveforms in spread-spectrum systems. The idea of chaotic masking [74,75] is to directly add the message in a noise-like chaotic signal at the end of the transmitter, while chaotic modulation [76–79] is by injecting the message into a chaotic system as a spread-spectrum transmission. Later, at the receiver, a coherent detector with some signal processing is employed to recover the message. But the signal masking or parameter modulation approach to chaotic communication only provides a lower level of security as stated in [80]. Using basic cryptosystem theory, a fuzzy-model-based chaotic cryptosystem has been proposed to provide a methodology with a higher level of security [81–83]. There, Luré type discrete-time chaotic systems are first exactly represented by TS fuzzy models. Then, a superincreasing sequence is generated by using a chaotic signal, which can flexibly be used as an output of the TS fuzzy chaotic drive system, or any state in which the synchronization error approaches zero. In terms of a cryptosystem, the plaintext (message) is encrypted using the superincreasing sequence at the drive system side, which results in the ciphertext. The ciphertext may be added to the output or state of the drive system using the methodologies proposed in [65,66,84]. Further, this the ciphertext embedding scalar signal is sent to the response system end. Following the design of a response system, the chaotic synchronization between a drive and a response system is achieved by solving LMIs. By the synchronization, one can regenerate the same superincreasing sequence and recover the ciphertext at the response system end. Finally, using

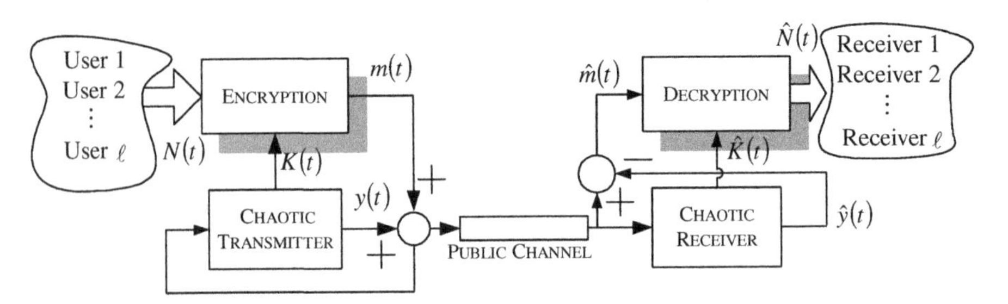

Fig. 5. Block diagram of chaotic encryption methodology

the regenerated superincreasing sequence, the ciphertext is decrypted into the plaintext. The block diagram of the whole cryptosystem is shown in Fig. 5.

8 Conclusions

This chapter has reviewed the current studies on the interactions between fuzzy logic and chaos theory, from the following aspects: fuzzy modeling of chaotic systems using Takagi-Sugeno models, linguistic descriptions of chaotic systems, fuzzy control of chaos, complex fuzzy systems, and a combination of fuzzy control technology and chaos theory for various engineering practices. What needs to emphasize is that in spite of the efforts on exploring the interactions between fuzzy logic and chaos theory, it is still far away from fully understanding their mutual relationships. Although this chapter may not give insight into their relations or may raise more questions than it can provide answers, we hope that it nevertheless contains seeds for future brooming research.

References

1. Li TY, Yorke JA (1975) Amer. Math. Monthly 82:481–485
2. Kloeden P, Li Z (2006) J. of Difference Equations and Applications 12(3/4): 247–279
3. Kloeden PE (1984) Proc. 9th Int. Conf. Nonlinear Oscillations Mitropolsky YA(Ed.) 2:184–187
4. Marotto FR (1978) J. of Math. Anal. Appl. 63:199–223
5. Kloeden PE (1981) J. Austral. Math. Soc. (Series A) 31:217–225
6. Shiraiwa K, Kurata M (1981) Nagoya Math J. 82:83–97
7. Kaleva O (1985) Fuzzy Sets and Systems 17:53–65
8. Diamond P, Kloeden PE (1994) Metric Spaces of Fuzzy Sets: Theory and Application. World Scientific, Singapore
9. Kloeden PE (1991) Fuzzy Sets and Systems 42(1):37–42
10. Shi YM, Chen G (2004) Chaos, Solitons and Fractals 22:555–571
11. Du BS (2005) Journal of Difference Equations and Applications 11:823–828
12. Stewart I (2000) Nature 406:948–949
13. Tucker W (1999) C.R. Acad. Sci. Paris 328:1197–1202
14. Silva CP (1993) IEEE Trans. Circuits Syst.-I 40:675–682
15. Tong RM (1979) The construction and evaluation of fuzzy models. In: Gupta MM, Ragade RK, Yager RR (eds) Advances in Fuzzy Sets Theory and Applications. North-Holland, Amsterdam
16. Pedrycz W (1989) Fuzzy Control and Fuzzy Systems Wiley: New York
17. Sugeno M, Yasukawa T (1993) IEEE Trans. Fuzzy Systems 1:7–31
18. Takagi T, Sugeno M (1983) Proc. IFAC Symposium on Fuzzy Information Marseille 55–60
19. Takagi T, Sugeno M (1985) IEEE Trans. Syst., Man, and Cyber. 15:116–132
20. Sugeno M, Kang GT (1986) Fuzzy Sets and Systems 18:329–346

21. Baglio S, Fortuna L, Manganaro G (1996) Electronics Letters 32:292–293
22. Porto M, Amato P (2000) Proc. 9th IEEE Intl. Conf. on Fuzzy Systems 1:435–440
23. Peitgen HO, Jürgens H,Saupe D (2004) Chaos and fractals – New frontiers of science. Springer, New York
24. Lee HJ (2000) Robust Fuzzy-Model-Based Controller Design for Nonlinear Systems with Parametric Uncertainties. MA Thesis, Yonsei University, Seoul, Korea
25. Chen G, Dong X (1998) From Chaos to Order: Methodologies, Perspectives and Applications. World Scientific: Singapore
26. Chen G (1998) IEEE Circuits Syst. Soc. Newsletter 9:1–5
27. Schiff SJ, Jerger K, Duong DH, Chang T, Spano ML, Ditto WL (1994) Nature 370:615–62
28. Yang W, Ding M, Mandell AJ, Ott E (1995) Phys. Rev. E 51:102–110
29. Triandaf I, Schwartz IB (2000) Phys. Rev. E 62:3529–2534
30. Chen G, Lai D (1996) Int. J. of Bifurcation and Chaos 6:1341–1349
31. Chen G, Lai D (1997) Proc. IEEE Conf. Decision and Control. San Diego, CA 367–372
32. Chen G, Lai D (1998) Int. J. of Bifurcation and Chaos 8:1585–1590
33. Devaney RL (1987) An Introduction to chaotic Dynamical Systems. Addison-Wesley, New York
34. Wang XF, Chen G (2000) IEEE Trans. Circuits and Systems-I 47(3):410–415
35. Wang XF, Chen G, Yu X (2000) Chaos 10(4):1–9
36. Zhou TS, Chen G, Yang QG (2004) Chaos 14:662–668
37. Zhang HG, Wang ZL, Liu DR (2004) Intl. J. of Bifurcation and Chaos 14(10):3505–3517
38. Tanaka K, Ikeda T, Wang HO (1998) IEEE Trans. Circ. Syst. – I 45(10): 1021–1040
39. Wang H, Tanaka K, Griffin M (1995) Proc. FUZZ-IEEE'95 531–538
40. Wang H, Tanaka K, Griffin M (1996) IEEE Trans. Fuzzy Syst. 4:14–23
41. Li Z, Park JB, Chen G, Joo YH, Choi YH (2002) Int. J. Bifur. Chaos 12(10):2283–2291
42. Li Z, Park JB, Joo YH, Chen G, Choi YH (2002) IEEE Trans. Circ. Syst.-I 49(2):249–253
43. Li Z, Park JB, Joo YH (2001) IEEE Trans. Circ. Syst.-I 48(10):1237–1243
44. Li Z, Halang W, Chen G, L.F. Tian (2003) Dynamic of Disc, Cont., Impul. Syst. 10(6):813–832
45. Kapitaniak T (1996) Controlling chaos – Theoretical and practical methods in nonlinear dynamics. Academic Press
46. Boccaletti S, Grebogi C, Lai YC, Mancini H, Maza D (2000) Physics reports 39:103–197.
47. Jackson EA (1990) Phys. Lett. A 151:478–484
48. Jackson EA, Hübler A (1990) Physica D 44:407–420
49. Jackson EA (1991) Physica D 50:341–366
50. Jackson EA (1991) Perspectives of nonlinear dynamics. Cambridge University Press, New York
51. Jackson EA (1991) Phys. Rev. A 44:4839–4853
52. Jackson EA (1995) Physica D 85:1–9
53. Braiman Y, Goldhirsch I (1991) Phys. Rev. Lett. 66:2545–2548
54. Chacón R (1996) Phys. Rev. Lett. 77:482–485

55. Kapitaniak T (1992) Chaos, Solitons and fractals 2:519–530
56. Kapitaniak T, Kocarev L, Chua LO (1993) Int. J. Bifurcat. Chaos 3:459–468
57. Pettini M (1989) Controlling chaos through parametric excitations. In: Lima R, Streit L, Mendes RV (eds) Dynamics and stochastic processes. Springer-Verlag, New York
58. Steeb WH, Louw JA, Kapitaniak T (1986) J. Phys. Soc. Japan 55:3279–3281
59. Pyragas K (1992) Phys. Lett. A 170:421–428
60. Pyragas K (2001) Phys. Rev. Lett. 86:2265–2268
61. Socolar JES, Sukow DW, Gauthier DJ (1994) Phys. Rev. E 50:3245–3248
62. Boccaletti S, Kurths J, Osipov G, Valladares DL, Zhou CS (2002) Physics Reports 366:1–101
63. Callenbach L, Linz S, Hänggi P (2001) Phys. Lett. A 287:90–98
64. Pecora LM, Carroll TL (1990) Phys. Rev. Lett. 64:821–824
65. Lian KY, Liu P, Chiu CS (2003) Int. J. Bifurc. Chaos 13(1):215–225
66. Lian KY, Chiu CS, Chiang TS, Liu P (2001) IEEE Trans. Fuzzy Systems 9: 539–553
67. Lian KY, Chiang TS, Liu P, Chiu CS (20019 Int. J. Bifur. Chaos 11.1397–1410
68. LianKY, Chiang TS, Chiu CS; Liu P (2001) IEEE Trans. Syst., Man, and Cybern. – Part B: Cybernetics 31:66–83
69. Bar-Yam Y (1997) Dynamics of Complex Systems. Addison-Wesley
70. Strogatz SH (2001) Nature 410:268–276
71. Watts DJ (1999) Small Worlds. Princeton University Press, Princeton, NJ
72. Bucolo M, Fortuna L, La Rosa M (2004) IEEE Trans. Fuzzy Systems 12(3): 289–295
73. Fortuna L, La Rosa M, Nicolosi D, Sicurella G. (2004) Proc. 12th Intl. IEEE Workshop on Nonlinear Dynamics of Electronic Syst. Evora, Portugal
74. Cuomo KM, Oppenheim AV, Strogatz SH (1993) IEEE Trans Circuits Syst – II 40:626–633
75. Kocarev LJ, Halle KD, Eckert K, Chua LO, Parlitz U (1992) Int. J. Bifurcation Chaos 2:709–713
76. LiaoTL, Huang NS (1999) IEEE Trans Circuits Syst I 46:1144–1150
77. Wu CW, Chua LO (1993) Int. J. Bifurcation Chaos 3:1619–1627
78. Halle KS, Wu CW, Itoh M, Chua LO (1993) Int. J. Bifurcation Chaos 3:469–477
79. Lian KY, Chiang TS, Liu P (2000) Int. J. Bifurcation Chaos 10:2193–2206
80. Short K (1994) Int. J. Bifurcation Chaos 4:959–977
81. Brucoli M, Cafagna D, Carnimeo L, Grassi G (1999) Int. J. Bifur. Chaos 9: 2027–2037
82. Grassi G, Mascolo S (1999) IEEE Trans Circuits Syst–II 46:478–483
83. Yang T, Wu CW, Chua LO (1997) IEEE Trans. Circuits Syst–I 44:469–472
84. Lian KY, Liu P, Chiu CS, Chiang TS (2002) Int. J. Bifur. Chaos 12:835–846

Upper and Lower Values for the Level of Fuzziness in FCM

Ibrahim Ozkan and I.B. Turksen

Abstract. The level of fuzziness is a parameter in fuzzy system modeling which is a source of uncertainty. In order to explore the effect of this uncertainty, one needs to investigate and identify effective upper and lower boundaries of the level of fuzziness. For this purpose, Fuzzy c-means (FCM) clustering methodology is investigated to determine the effective upper and lower boundaries of the level of fuzziness in order to capture the uncertainty generated by this parameter. In this regard, we propose to expand the membership function around important information points of FCM. These important information points are, cluster centers and the mass-center. At these points, it is known that, the level of fuzziness has no effect on the membership values. In this way, we identify the counter intuitive behaviour of membership function near these particular information points. It will be shown that the upper and lower values of the level of fuzziness can be identified. Hence the uncertainty generated by this parameter can be encapsulated.

1 Introduction

Perfect knowledge or information is not achievable for most, perhaps all, cases of data analyses. After all, if perfect knowledge is available then the data analysis is a mechanical process and there is no need to analyze the uncertainty together with errors in estimations. Naturally pattern recognition under imperfect information may lead to imperfect patterns.

Sources of uncertainty in data analysis context may be classified as:

(1) Model Uncertainty: Uncertainty related with the techniques used to find the data structure and its relations. Examples of this type of sources of uncertainty in Fuzzy System Modeling (FSM) are: similarity measures, inference techniques, clustering techniques, input selection techniques, etc.
(2) Parametric Uncertainty: Uncertainty in parameters. For example, in FCM clustering method, level of fuzziness and the number of clusters can be classified under parameter uncertainty.
(3) Data Uncertainty: Imprecision, incompleteness, vagueness, contradictory examples, etc.

Fuzziness is a specific type of uncertainty. Type-2 Fuzzy Sets represent the uncertainty associated with Type-1 fuzzy sets. Type-2 fuzzy sets were first defined by [1]. [2] investigated the formal representation of Type 2 Fuzzy sets. They are fuzzy sets with fuzzy membership values. [3–6] investigated interval-valued Type-2 uncertainty generated by fuzzy (linguistic) connectives used in combinations of Type-1 fuzzy sets and logic algebra. [7] pointed out that all the parameters of fuzzy system modeling are natural candidates as sources of uncertainties in Type-2 fuzzy sets.

Naturally the parameters used in FCM clustering contribute to the generation of the uncertainty in predictions. In general, in the calculation of membership values with FCM method, an analyst assumes that some parameters are given. The parameters used in FCM are number of clusters, cluster centers, level of fuzziness and similarity measure, i.e. distance measure. In this chapter, we only try to get an understanding of the behavior of FCM methodology for the case when the cluster center values and the distance measure are given but the value of the level of fuzziness is uncertain. Hence we try to answer the question: "Is it possible to find the effective upper and the lower bounds of the level of fuzziness that capture the associated uncertainty?" These effective upper and lower boundaries are determined such that the levels of fuzziness higher than or lower than these values do not bring any significant information about the membership values distribution.

[8] suggested that the value of the level of fuzziness should be between 1.5 and 2.5 based on their analysis on the performance of cluster validity indices. It will be shown that the values of the level of fuzziness we find are close to those found by Pal and Bezdek. However, our main goal is to capture the uncertainty of the level of fuzziness. Furthermore, our approach for this analysis is completely different. [9] suggested that the proper value of the level of fuzziness depends on the data itself. They generated two rules to select the proper value based on the local optimality test. Whereas, our aim is not to find the proper value of the level of fuzziness, but a range of values of the level of fuzziness that capture the uncertainty generated by the level of fuzziness. In particular, we propose to identify the effective upper and lower boundary values of the levels of fuzziness in order to capture this uncertainty. For this purpose we analyze the particular information points which are basically cluster centers and the mass center that are generated with FCM clustering. We show that, in fact, these effective upper and lower boundary values of the levels of fuzziness are a function of the number of clusters. In other words, these boundaries depend on the structure of the data.

For this purpose, we present FCM briefly in Sect. 2. In Sects. 2.1 and 2.2, the behavior of the FCM methodology is analyzed near particular information points where the membership values do not depend on the level of fuzziness. In particular, these information points are: (i) the mass-center where the cluster centers have equal distance to this information point. Mass center is a unique information point defined in n-dimensional data space where n>1. It is identified by cluster centers and continuous membership function such that it has equal distance from all the cluster centers. In addition, when the level of fuzziness goes to infinity, cluster centers collapse to this information point; (ii) all the cluster center values. As it is known, the membership value is one at every cluster center and (1/nc) at the mass center where nc is the number of clusters. As well, it is known that at such points, the level of fuzziness does not have any effect on membership values. In this chapter, we analyze the behavior of FCM membership function by small perturbations around these information points, i.e. the mass center and cluster center values. We suspect these perturbations could at times lead to counter-intuitive results. Thus, we investigate the behavior of the response of the membership values to these information points. Finally we state our conclusions.

2 Fuzzy c-means Membership Function

Reference [10] showed that by minimizing the objective function;

$$J_m(U,V:X)=\sum_{k=1}^{nd}\sum_{c=1}^{nc}\mu_{c,k}^{m}\left\|x_k-v_c\right\|_A^2,$$

where, $\mu_{c,k}$: Membership value of kth vector in c^{th} cluster such that $\mu_{c,k}\in[0,1]$, nd is the number of vectors, nc is the number of clusters, $\|.\|_A$ is norm function and m is the level of fuzziness, FCM membership function is calculated as:

$$\mu_{i,k}=\left[\sum_{c=1}^{nc}\left(\frac{\|x_k-v_i\|_A}{\|x_k-v_c\|_A}\right)^{\frac{2}{m-1}}\right]^{-1}$$

where, $\sum_{c=1}^{nc}\mu_{k,c}=1$ for a given m > 1. This means that the sum of the degrees of membership values of any data is one, or in other words, any data should be a member of at least one of the clusters with a membership value greater then zero, i.e. $\mu_{k,c}\in(0,1]$.

There are a number of information points which we have specified above, such that two points that this membership function does not depend on the level of fuzziness. One of the points is the mass center that has equal distance to all cluster centers and thus has a membership value $\frac{1}{nc}$ to all cluster centers. Hence this value clearly does not depend on the level of fuzziness. The other points are the cluster center values which have a membership value 1 in its cluster and 0 to all others. Hence these values also do not depend on the level of fuzziness.

2.1 Upper Bound Identification

In order to find the behavior of the membership function based on the level of fuzziness, one can expand this function around the mass center by using

Taylor series expansion. Recall that the Taylor series is an expansion of a function about a point. One-dimensional Taylor series of a real function $f(x)$ about a point $x = x_0$ is given by,

$$f(x) = f(x_0) + f'(x_0)(x - x_0) + \frac{1}{2!} f''(x_0)(x - x_0)^2 + \dots + \frac{1}{n!} f^{(n)}(x_0)(x - x_0)^n + \dots ; f(x) = f(x_0) + f'(x_0)(x - x_0) + R$$

where R is the remainder. Let d^* denote the distance measure to all cluster centers from the mass center, d_i denote the distance to ith cluster center of a point very close to the mass center. Taylor series of the membership function around the mass center then can be written as:

$$\mu_i = \mu_i|_{d_i^*} + \frac{\partial}{\partial d_i} \mu_i|_{d_i^*} (d_i - d_i^*) + R$$

Hence the membership value of the mass center is $\mu_i|_{d_i^*} = \frac{1}{nc}$ then,

$$\mu_i = \frac{1}{nc} + \frac{\partial}{\partial d_i} \mu_i|_{d^*} (d_i - d^*) + R \quad (1)$$

The derivative of the membership function with respect to the distance to the ith cluster center is:

$$\frac{\partial \mu_i}{\partial d_i} = -\frac{\left(\frac{1}{m-1}\left(\frac{d_i}{d_1}\right)^{\frac{2-m}{m}}\right) + \dots + \left(\frac{1}{m-1}\left(\frac{d_i}{d_{i-1}}\right)^{\frac{2-m}{m}}\right) + \left(\frac{1}{m-1}\left(\frac{d_i}{d_{i+1}}\right)^{\frac{2-m}{m}}\right) + \dots + \left(\frac{1}{m-1}\left(\frac{d_i}{d_{nc}}\right)^{\frac{2-m}{m}}\right)}{\left(\sum_{c=1}^{nc}\left(\frac{d_i}{d_c}\right)^{\frac{1}{m-1}}\right)^2} \quad (2)$$

Since this derivative should be evaluated at the mass center where $d_j = d^*$, for j = 1, .., nc, we obtain:

$$\frac{\partial}{\partial d_i} \mu_i|_{d^*} = -\frac{\left(\frac{1}{m-1}\right) + \left(\frac{1}{m-1}\right) + \dots + \left(\frac{1}{m-1}\right) + \left(\frac{1}{m-1}\right) + \dots + \left(\frac{1}{m-1}\right)}{\left(\sum_{c=1}^{nc} 1\right)^2} \quad (3)$$

$$= -\frac{(nc-1)\left(\frac{1}{m-1}\right)}{nc^2}$$

then (1) becomes when the remainder R is assumed to be negligible:

$$\mu_i \cong \frac{1}{nc} - \frac{(nc-1)\left(\frac{1}{m-1}\right)}{nc^2}(d_i - d^*) \quad (4)$$

This approximation holds for the points close to the mass center and hence $(d_i - d^*)$ is small. We can write the above equation as;

$$\mu_i \cong \frac{1}{nc} + \frac{(nc-1)\left(\frac{1}{m-1}\right)}{nc^2}(\mp\Delta) \tag{5}$$

Given that nc is a constant for a particular structure of data, this function depends only on the level of fuzziness for small perturbations $\Delta \equiv d_i - d^*$. As the level of fuzziness, m, goes to infinity, membership values goes to 1/nc, and as m gets closer to one this value tends to change rapidly. For example, Fig. 1 shows the approximated membership values around the mass center for the number of clusters 4 and 20. This figure has two vertical axes. Left axis shows the membership value for number of clusters is equal to 20 and the right axis shows the membership value for the number of clusters is equal to 4 (In Figs. 1 and 2, we used these two axes representation in order to show the effect of the level of fuzziness with respect to different values of the nc parameter more explicitly.). Thus Fig. 1 suggests that the big change occur for the values of the level of fuzziness less than 2.6. On the other hand, setting the level of fuzziness very close to 1 in value clearly results in an undesirable effect for Δ=0.1.

Figure 2 shows the effect of the level of fuzziness value around the mass center when Δ is set to a smaller value, such as 0.01. Again this figure has two vertical axes. Left axis shows the membership value for number of clusters is equal to 20 and the right axis shows the membership value for the number

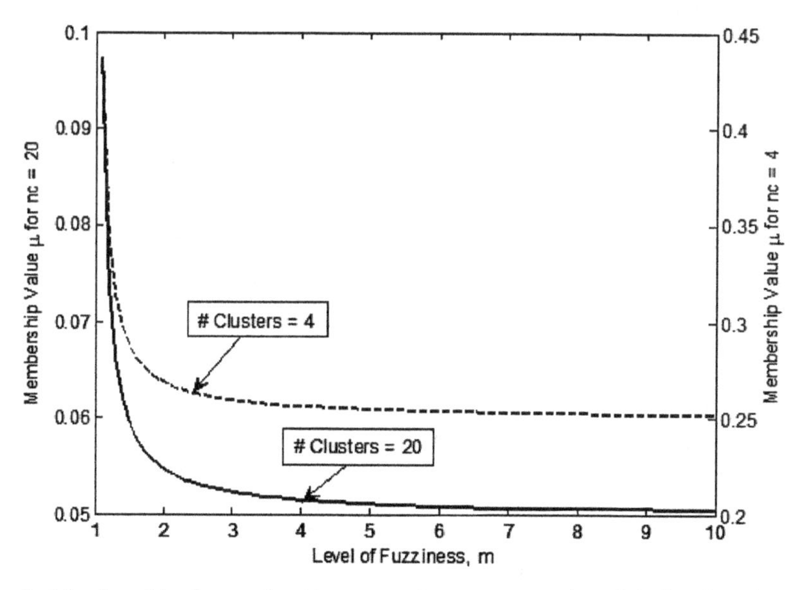

Fig. 1. Membership Approximation around mass center $\Delta = 0.1$ showing the effect of the level of fuzziness with respect to the nc parameter

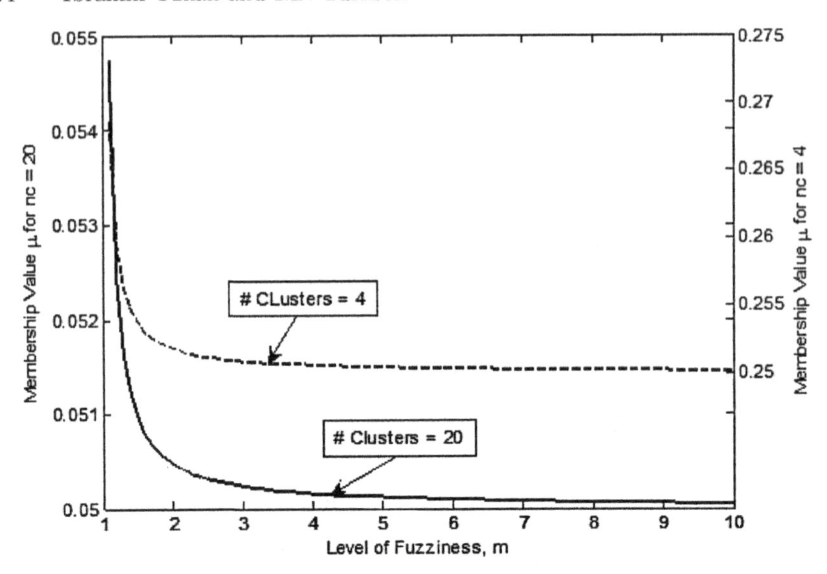

Fig. 2. Membership Approximation around mass center, Δ=0.01 showing the effect of the level of fuzziness with respect to the nc parameter

of clusters is equal to 4. It is seen thus from the Fig. 2 that even for such small value of Δ, the level of fuzziness do effect the degree of membership in a counter-intuitive manner when one sets the value of the level of fuzziness, m, close to one. As well, we observe that the change in the membership value in response to the change in the level of fuzziness almost approaches to zero for the values higher than 2.6. Therefore, there should be an approximate upper limit to be used in FCM clustering methodology in practice.

The changes in membership values depending on the changes in the level of fuzziness become smaller for the higher values of the level of fuzziness. For example, changing the level of fuzziness from 2.6 to 3.6 results in;

$$\begin{aligned} \mu_{i,k}|_{m=2.6} - \mu_{i,k}|_{m=3.6} &\cong \frac{(nc-1)}{1.6 \cdot nc^2}(\mp\Delta) - \frac{(nc-1)}{2.6 \cdot nc^2}(\mp\Delta) \\ &= \frac{(nc-1)}{nc^2}(\mp\Delta)\left[\frac{1}{1.6} - \frac{1}{2.6}\right] \end{aligned}$$

As it can be seen from Fig. 3, the change in the membership values when the value of the level of fuzziness changes from 2.6 to 3.6 for $\Delta = 0.05$ is very small (0.003 for the particular case of the number of clusters equals to 2 and smaller than this value for the higher number of clusters). It can be observed from Fig. 4 that for various values such as $\{m_1 = 2.6,\ m_2 = 3.6\}$, $\{m_1 = 2.6,\ m_2 = 4.6\}$, $\{m_1 = 2.6,\ m_2 = 5.6\}$, $\{m_1 = 2.6,\ m_2 = 6.6\}$, the change in membership values are also negligible.

Thus as they are shown, higher values of the level of fuzziness do not change the membership values significantly. Therefore, this mass center helps

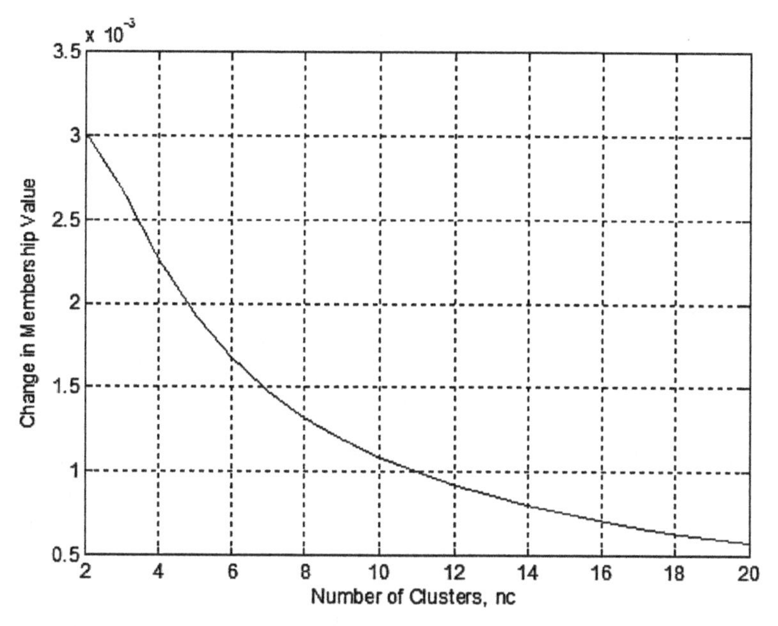

Fig. 3. Change in Membership Value around mass center for, m_1=2.6, m_2=3.6, Δ=0.05

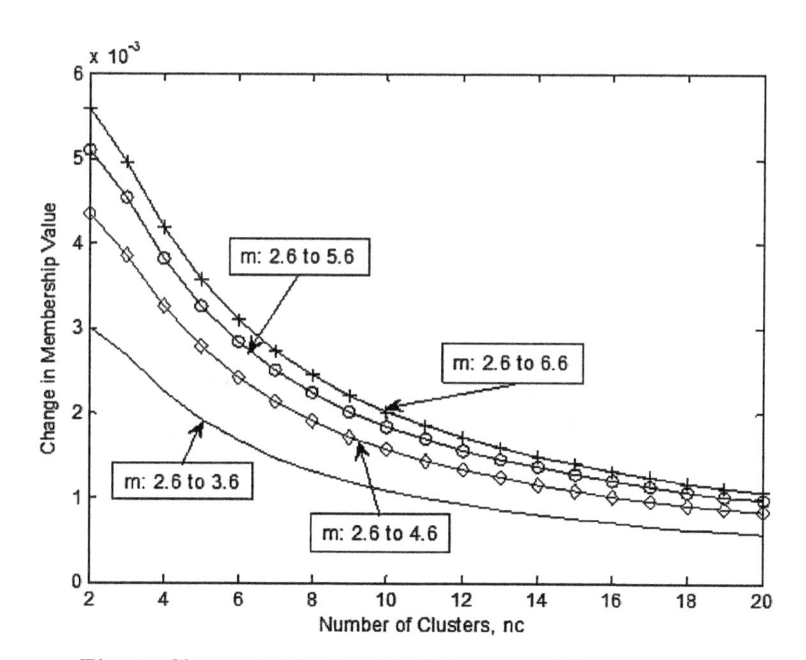

Fig. 4. Change in Membership Value around the mass center

us to approximately determine the upper boundary value. Thus, changes in the level of fuzziness simply do not bring any additional information.

Three-D plots of m, Δ and μ in Fig. 5 and Fig. 6 show the relation between these three parameters for the number of clusters 4 and 20. The influence of the level of fuzziness to the degree of membership clearly states that the higher values of m do not bring extra information. As it can be observed, the influence appears below m=2.6

2.2 Lower Bound Identification

To identify the lower bound, we analyze the behaviour of the membership function for small perturbations around cluster centers. For the behavior of the membership function very close to cluster centers, we will use the membership function directly. Let d_i denote the distance to the ith cluster center from a point which is very close to this cluster center. Thus, d_i is very small in value compared to the distance to all the other cluster centers from this point. For example when there are 3 clusters, let d_1 be the distance to 1st cluster from the perturbation point and d_2, d_3 be the distance to those two other cluster centers from this perturbation point. One can observe that, $d_1 \ll d_2$ and $d_1 \ll d_3$ then $\frac{d_1}{d_2} \approx \frac{d_1}{d_3}$. Thus in general, we can write, $\frac{d_i}{d_1} \approx \frac{d_i}{d_2} \approx .. \approx \frac{d_i}{d_{i-1}} \approx \frac{d_i}{d_{i+1}} \ldots \approx \frac{d_i}{d_{nc}} \cong d$. Thus the membership function can now be written as follows for the perturbation point around a cluster center i,

$$\mu_i = \frac{1}{1 + (nc - 1)d^{\frac{1}{m-1}}}$$

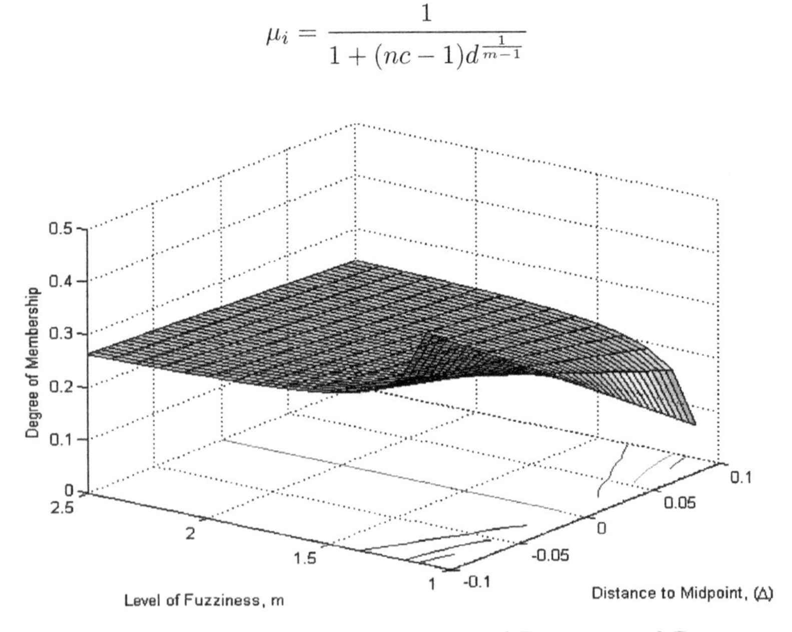

Fig. 5. 3D plot of Degree of membership, Level of Fuzziness and Distance to the mass center (Δ), number of cluster is set to 4

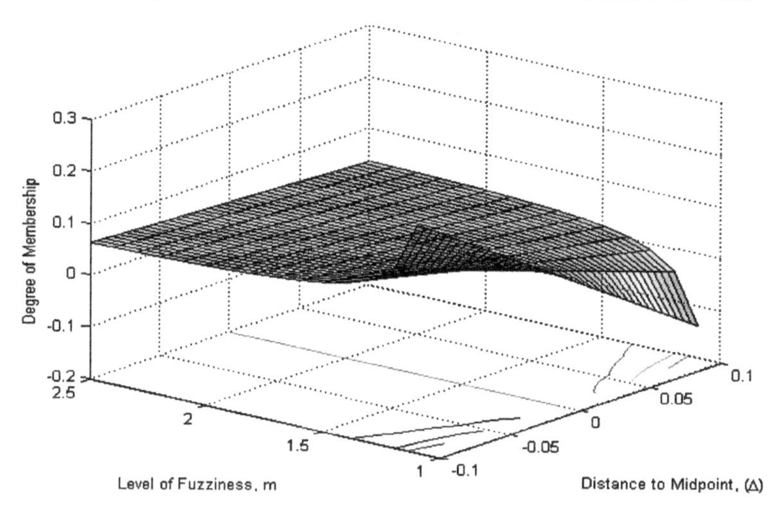

Fig. 6. 3D plot of Degree of membership, Level of Fuzziness and Distance to the mass center (Δ), number of cluster is set to 20

We observe in Fig. 7 that no significant change occurs in membership values around each cluster center value when the level of fuzziness is approximately less than 1.4. On the other hand, very high values of the level fuzziness make very significant change which is a counter intuitive result. Since d represents the ratio of the distance to the nearest cluster center and the distance to the all the other cluster centers and d is assumed to be very small (in this example d=0.05), assigning a very low membership value to this point becomes unacceptable.

We observe in Fig. 8, the effect of the level of fuzziness for the values up to 2 and d=0.05. The membership values do change for the values of the level of fuzziness higher than 1.33 (in case of the number of clusters, nc is 20). But this change is very small in value. We observe that to keep the value of the level of fuzziness approximately lower than 1.4 assigns the value of membership 1 to all perturbation points near to cluster centers and this result is also undesirable. Therefore, this counter-intuitive result suggests that there should be a lower limit for the level of fuzziness to be used in FCM clustering in practice. The values of the level of fuzziness less than 1.4 result in almost no change or very small changes even if the ratio, d, is set to a bit higher value such that 0.1, as shown in Fig. 9.

3D plots of m, d and μ shown in Fig. 10 and Fig. 11 demonstrate the relation between these three parameters. For very low values of d, the membership values do not change significantly if the values of the level of fuzziness are set to values less than 1.4. As well the number of clusters also plays an important role. The effect of the level of fuzziness is increasing with the increasing number of clusters as shown in Fig. 11.

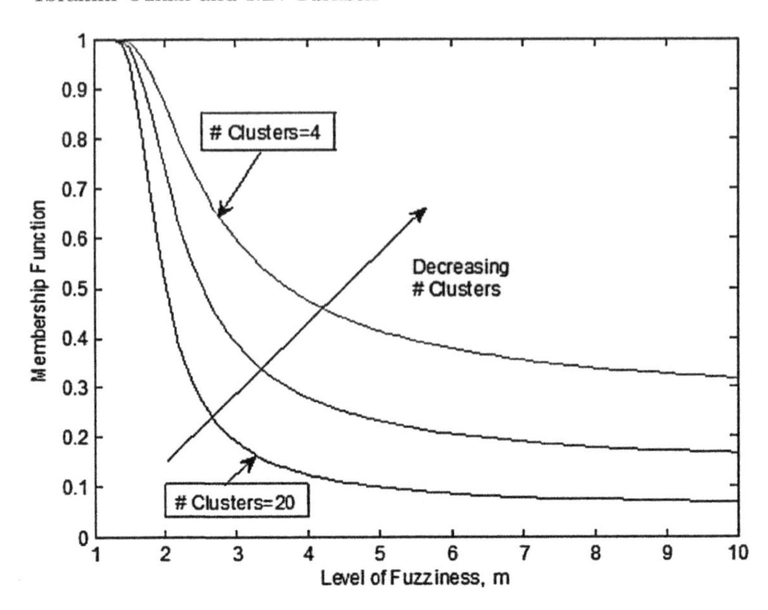

Fig. 7. Approximated membership function around the cluster center, d = 0.05

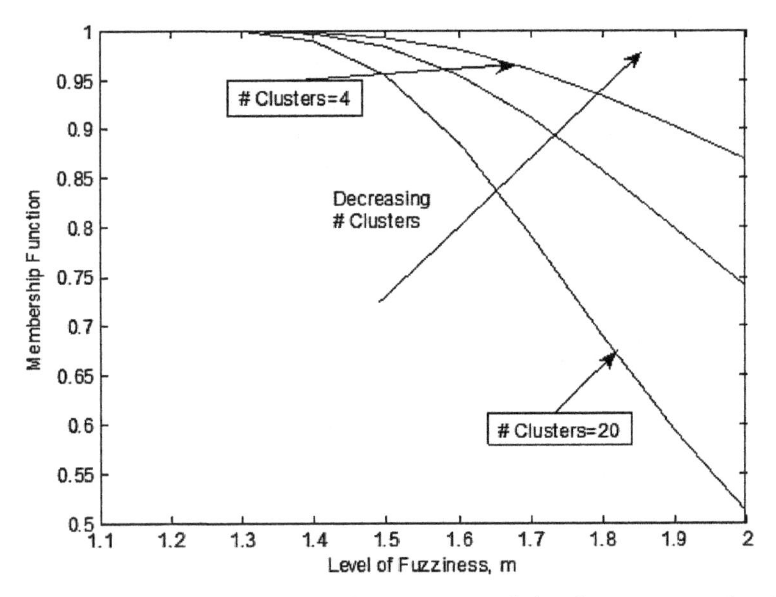

Fig. 8. Approximated membership function around the cluster center, d = 0.05

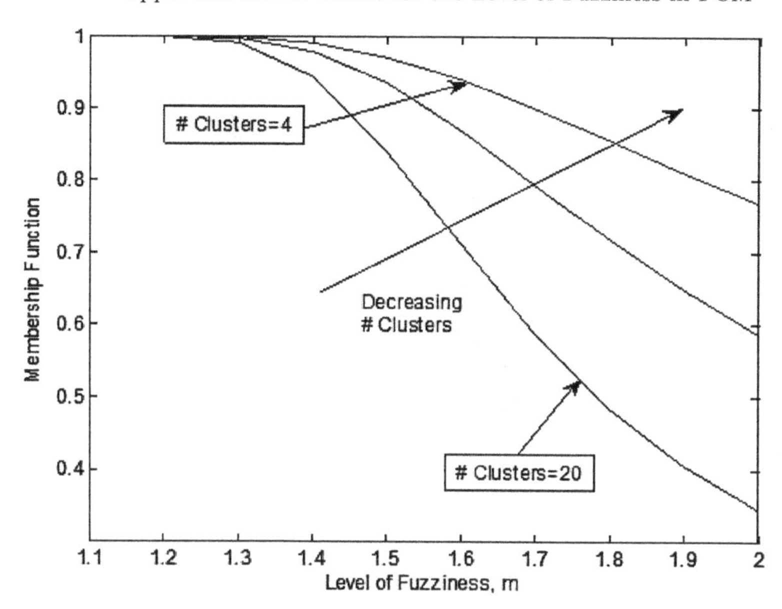

Fig. 9. Approximated membership function around the cluster center, d = 0.1

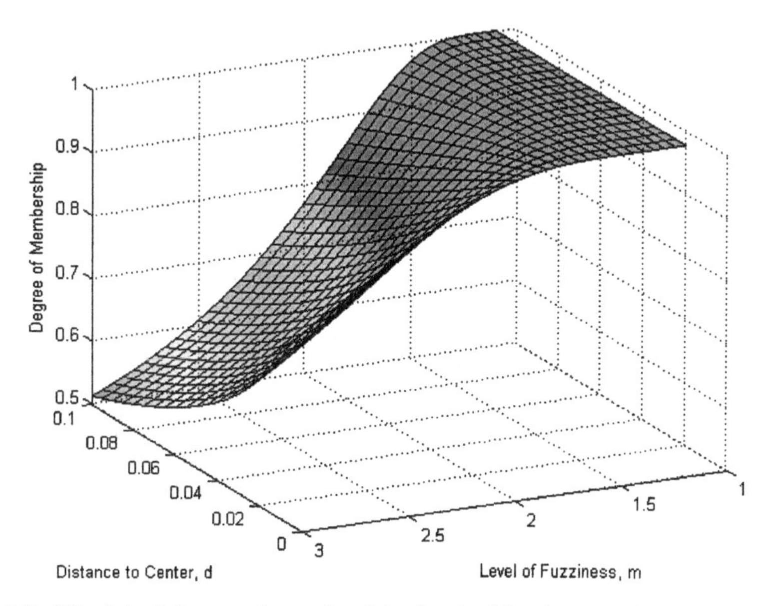

Fig. 10. 3D plot of degree of membership, level of fuzziness and ratio of distance to the cluster center, number of cluster is set to 4

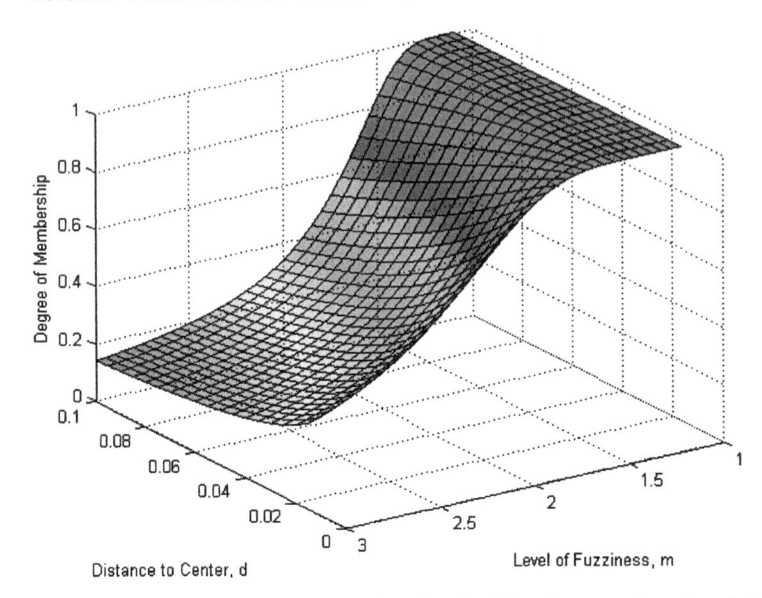

Fig. 11. 3D plot of degree of membership, level of fuzziness and ratio of distance to the cluster center, number of cluster is set to 4

3 Conclusions

In this chapter, we examine the behavior of FCM membership function at distinct information points. There exist such information points in the n-dimensional data space, n>1, where the membership function does not depend on the level of fuzziness. These are: (i) the mass center, where the cluster centers have equal distance to this information point. Mass center is a unique information point defined in n dimensional data space where n>1. It is identified by cluster centers and continuous membership function such that it has equal distance from all the cluster centers. In addition, when the level of fuzziness goes to infinity, cluster centers collapse to this information point; (ii) all cluster centers where membership value is 1. The information obtained from this perturbation analysis of the behavior of FCM methodology around these distinct information points helps us to identify the effective upper and lower bounds of the level of fuzziness to be used in practice. The result of this analysis lowers the search space and hence the search time in system development analysis.

It is found that increases in the level of fuzziness beyond the upper boundary value identified in this analysis do not provide additional information since the membership values do change in a counter-intuitive manner. In particular, around the mass center, higher values of the level of fuzziness simply do not affect the membership value beyond approximately m=2.6. It just makes the slope of membership function flatter. Very small values of the level of fuzziness make very steep slope of the membership function and do set membership values to zero or one around the mass center which is a counter-intuitive result.

It is also found that the small values of the level of fuzziness do not affect the membership function around cluster centers. We observe membership values start to change when the values of the level of fuzziness are approximately greater than m=1.4.

In summary, it is found that in applications significant changes in FCM membership values do occur for the values of the level of fuzziness that are approximately between 1.4 and 2.6. These values create effective boundaries for the level of fuzziness. Therefore, we suggest based on this analysis that the upper boundary value of the level of fuzziness should be approximately 2.6 and the lower boundary value of the level of fuzziness should be approximately 1.4 for FCM methodology in system development practices. For these reasons it is recommended that an analyst should not be concerned about the changes of the membership values outside of these boundaries.

Furthermore, these two effective boundary values of the level of fuzziness encapsulate the uncertainty associated with the level of fuzziness parameter. In addition, we hypothesize based on this analysis that interval valued and/or full type 2 fuzzy system models with FCM can be explored in more depth. This investigation is left for future analysis.

Acknowledgments

This research is partially supported by Knowledge/Intelligence System Laboratory (KIS) of the department of Mechanical and Industrial Engineering at the University of Toronto and the Natural Sciences and Engineering Research Council of Canada (NSERC) Discovery Grant.

References

1. Zadeh, L.A. (1975) The Concept of a Linguistic Variable and its Application to Approximate Reasoning, Information Sciences, vol. 8, pp.199–249, 1975
2. Mizumoto, M. and Tanaka, K. (1976), Some properties of Fuzzy Sets of Type 2", Information and Control, vol. 31, pp. 312–340.
3. Turksen, I.B. (1986), Interval Valued Fuzzy Sets Based on Normal Forms, Fuzzy Sets and Systems, vol. 20, pp. 191–210.
4. Turksen, I.B. (1992), Interval-valued fuzzy sets and 'compensatory AND', Fuzzy Sets and Systems, vol. 51, pp. 295–307.
5. Turksen, I.B. (1994), Interval valued fuzzy sets and fuzzy measures, Proc. of First Int. Conf. of NAFIPS, pp. 317–321.
6. Turksen, I.B. (1999), Type I and Type II fuzzy system modeling, Fuzzy Sets and Systems, vol. 106, pp. 11–34.
7. Mendel, J. M. (2001), Uncertain Rule-Based Fuzzy Logic Systems: introduction and new directions, Prentice Hall, Upper Saddle River.

8. Pal N.R. and Bezdek, J.C. (1995), On cluster validity for the fuzzy c-means model, IEEE Trans. Fuzzy Syst., vol. 3, pp. 370–379.
9. Yu, J., Cheng, Q., and Huang, H. (2004), Analysis of the Weighting Exponent in the FCM, IEEE Trans. System, Man and Cybernetics-B, vol. 34, no. 1, pp. 634–639.
10. Bezdek, J.C. (1973), Fuzzy Mathematics in Pattern Recognition, Ph.D. Thesis, Applied Mathematics Center, Cornell University, Ithaca.

Mathematical Modeling of Natural Phenomena: A Fuzzy Logic Approach

Michael Margaliot

Abstract. In many fields of science human observers have provided verbal descriptions and explanations of various systems. A formal mathematical model is indispensable when we wish to rigorously analyze these systems. In this chapter, we survey some recent results on transforming verbal descriptions into mathematical models using fuzzy modeling. This is a simple and direct approach that offers a unique advantage: the close relationship between the verbal description and the mathematical model can be used to verify the validity of the verbal explanation suggested by the observer. We review two applications of this approach from the field of ethology: fuzzy modeling of the territorial behavior of fish, and of the orientation to light of a flat worm. The fuzzy modeling approach demonstrated here may supply a suitable framework for biomimicry, that is, the design of artificial systems based on mimicking a natural behavior observed in nature.

Keywords: Linguistic modeling, animal and human behavior, territorial behavior, mathematical modeling in biology, hybrid systems, discrete-event systems, biomimicry.

1 Introduction

In many of the "soft sciences" (e.g. psychology, sociology, ethology) scientists provide *verbal* descriptions and explanations of various phenomena based on field observations. It is obvious that obtaining a suitable *mathematical* model, describing the observed system or behavior, can greatly enhance our ability to understand and study it in a scientific manner. Indeed, mathematical models are very useful in summarizing and interpreting empirical data. Furthermore, once derived, such models allow us to analyze the system both qualitatively and quantitatively using mathematical tools.

Fuzzy logic theory provides the most suitable tool for transforming verbal descriptions into mathematical models. Indeed, the real power of fuzzy logic is in its ability to handle and manipulate verbally-stated information based on perceptions rather than equations (see, e.g. [1–4]). Fuzzy modeling is routinely used to transform the knowledge of an expert, be it a physician or a process operator, into a computer algorithm [5,6]. Yet, not enough attention has been given to its possible use as a tool to assist human observers in transforming their verbal descriptions into mathematical models.

A fuzzy model represents the real system in a form that corresponds closely to the way humans perceive it. Thus, the model is understandable, even by non-professionals, and each parameter has a readily perceivable meaning. The model can be easily altered to incorporate new phenomena, and if its behavior is different than expected, it is usually possible to find which rule/term should be modified and how. In addition, fuzzy modeling offers a unique advantage: the close relationship between the verbal description and the resulting mathematical model can be used to verify the validity of the verbal explanations suggested by the observer. Thus, the derived mathematical model can be used to prove or refute the modeler's ideas as to how the natural system behaves and why.

In this chapter, we review this approach and its application to two examples from the field of animal behavior [7, 8]. There are several reasons why our work focuses on models from ethology. The first reason is that for many animal (and human) actions, the all-or-none law does not hold – the behavior itself is "fuzzy". For example, the response to a (low intensity) stimulus might be what HEINROTH called *intention movements*, that is, a slight indication of what the animal is tending to do. TINBERGEN [9, Chap. IV] states: "As a rule, no sharp distinction is possible between intention movements and more complete responses; they form a continuum."[1] Hence, fuzzy modeling seems the most appropriate tool for studying such behaviors.

The second reason is that studies of animal behavior often provide a *verbal* description of both field observations and interpretations. For example, Fraenkel and Gunn [11, p. 23] describe the behavior of a cockroach, that becomes stationary when a large part of its body surface is in contact with a solid object, as: "A high degree of contact causes low activity ...".

Another reason, discussed in detail in Sect. 6 below, is that a great deal of research is being conducted in the field of *biomimicry*–the development of artificial products or machines that mimic biological phenomena. In many cases, verbal descriptions of the natural behavior already exist in the literature, so fuzzy modeling may be very suitable for addressing biomimicry in a systematic manner.

This chapter is organized as follows. Sect. 2 reviews several known approaches for transforming verbal descriptions of a system or process into a well-defined mathematical model. Sect. 3 describes the fuzzy modeling approach. The approach is demonstrated using two case studies from the field of ethology in Sects. 4 and 5. For each case study, we provide the original verbal

[1] It is interesting to recall that Zadeh [10] defined a fuzzy set as "a class of objects with a continuum of grades of membership"

description and apply fuzzy modeling to derive a well-defined mathematical model. We use simulations and rigorous analysis to analyze the mathematical models and compare their behavior to the behavior actually observed in nature. Sect. 6 describes how the fuzzy modeling approach may be utilized in the field of biomimicry. The final section concludes.

2 From Verbal Descriptions to a Mathematical Model

During the 1950s, Forrester and his colleagues developed *system dynamics*[2] as a method for modeling the dynamic behavior of complex systems. The basic idea is to represent the causation structure of the system using *elementary structures* that include positive, negative, or combined positive and negative feedback loops. These are depicted graphically, and then transformed into a set of differential equations. The method was applied successfully to numerous real-world applications in social, economic, and industrial sciences. However, the inherent fuzziness and vagueness of the linguistic description are ignored and exact terms and phrases (e.g. a temperature of 30° Celsius) are modeled in precisely the same way as fuzzy terms (e.g. warm weather).

A more systematic approach to modeling physical systems is *qualitative reasoning* (QR) [13] which transforms qualitative descriptions of a system into *qualitative differential equations.* These are generalizations of differential equations that include two main components: (1) functional relationships between variables can be represented by functions that are either monotonically increasing or decreasing, but do not have to be *completely* specified; and (2) the values of the variables are described using a set of *landmark values* rather than exact numerical values. QR was applied to simulate and analyze many real world systems. However, its applications seem to indicate that it is suitable when modeling with accurate, yet incomplete, knowledge rather than with a verbal description of the system.

Zadeh laid down the foundations of fuzzy sets and fuzzy logic and linked them to human linguistics [10,14]. Mamdani designed the first fuzzy controller based on a verbal control protocol [15]. His work demonstrated the applicability of fuzzy expert systems and led the way to numerous applications (see, e.g. [16–20].) However, most of these applications are based not on modeling real world phenomena, but rather on transforming the knowledge of a human expert into a fuzzy expert system, that can replace the human expert.

The pioneering work of WENSTOP [21, 22] was aimed at building *verbal models* capable of representing and processing *linguistic* information. The basic components of this model include (1) *generative grammar*–used for defining the semantics of the linguistic statements; (2) *fuzzy logic based inferencing*–used in the deductive process; and (3) *linguistic approximation*–used for attaching linguistic labels to the outputs. These components were

[2] To date, more than 30 books on *system dynamics* have appeared. A very readable one is [12]. The journal *Systems Dynamics Review* contains up-to-date papers.

implemented using the APL computer language so the entire process was automated. WENSTOP and KICKERT developed *verbal models* of several interesting systems from the social sciences [21, 23] [24, Chap. 7]. However, *verbal models* are not standard mathematical models, as their input and output are linguistic values rather than numerical values. As such, *verbal models* suffer from two drawbacks. First, there are no methods for analyzing the behavior of such a model so it can only be used as a simulation tool. Second, when modeling dynamic systems the degree of fuzziness (or uncertainty) increases with every iteration. This implies that the system can be simulated effectively only over relatively short time spans.

KOSKO [25] suggested *fuzzy cognitive maps* (FCMs) as a tool for the representation of causal relationships between various linguistic concepts. FCMs were used to model several interesting real world phenomena (see [26] and the references therein). However, the inferencing process used in FCMs yields a discrete-time *linear* system in the form $\mathbf{x}(k+1) = A\mathbf{x}(k)$, which is clearly too simplified to faithfully depict many real world systems.

3 The Fuzzy Modeling Approach

The starting point in the fuzzy modeling approach is a complete *verbal* description of the system. This should include the following:

1. The "agents". For example, in a model from the field of ethology describing territorial behavior, the agents might be two animals of the same species.
2. The "environment", that is, the surrounding factors that influence the agents' behavior and interaction. For example, in a model of humans reacting to a fire alarm this should include the size of the room they are in and the location and size of the exit.
3. The behavior of each agent, that is, its reaction to the other agents and the environment.
4. The overall patterns observed in the natural system as a result of each agent's behavior and the interaction between the agents and their environment. For example, in a model describing foraging ants, an observed pattern might be that eventually all the ants follow a trail from their nest to the *nearest* food source. This information is vital because it allows us to validate the mathematical model, once derived.

The verbal information is transformed into a mathematical model using the following steps (see the detailed examples in the following sections):

1. Identify the variables. For example, if the model describes an animal that moves in a 3D world, then three variables will be needed to describe the animal's position.
2. Transform the verbal description into fuzzy rules stating the relations between the variables.

3. Define the fuzzy terms (logical operators) in the rules using suitable fuzzy membership functions (mathematical operators).
4. Transform the fuzzy rule base into a mathematical model using fuzzy inferencing.

At this point we can analyze and simulate the mathematical model. Its suitability is determined, among other factors, by how well it mimics the patterns that were actually observed in the natural system.

When creating a mathematical model from a verbal description there are always numerous degrees of freedom. In the fuzzy modeling approach, this is manifested in the freedom in choosing the components of the fuzzy model: the type of membership-functions, logical operators, inferencing method, and the values of the different parameters.

The following guidelines may be helpful in selecting the different components of the fuzzy model (see also [27] for details on how the various elements in the fuzzy model influence its behavior). First, it is important that the resulting mathematical model have a simple (as possible) form, in order to be amenable to analysis. Thus, for example, a Takagi-Sugeno model with singleton consequents might be more suitable than a model based on Zadeh's compositional rule of inference.

Second, when modeling real-worlds systems, the variables are physical quantities with dimensions (e.g. length, time). *Dimensional analysis* [28, 29], that is, the process of introducing dimensionless variables, can often simplify the resulting equations and decrease the number of parameters.

Third, sometimes the verbal description of the system is accompanied by quantitative data such as measurements of various quantities in the system. In this case, methods such as fuzzy clustering, neural learning, or least squares approximation can be used to fine-tune the model using the discrepancy between the measurements and the model's output (see, e.g. [30–32] and the references therein).

4 Example 1: Territorial Behavior in the Stickleback

Territory has a major role in social animal behavior [33] and results in a rich set of phenomena, but how is the territory created? Nobel Laureate Konrad Lorenz describes a specific example [34, p. 47]:

> "... a real stickleback fight can be seen only when two males are kept together in a large tank where they are both building their nests. The fighting inclinations of a stickleback, at any given moment, are in direct proportion to his proximity to his nest. ... The vanquished fish invariably flees homeward and the victor ... chases the other furiously, far into its domain. The farther the victor goes from home, the more his courage ebbs, while that of the vanquished rises in proportion.

Arrived in the precincts of his nest, the fugitive gains new strength, turns right about and dashes with gathering fury at his pursuer ..."

Note that Lorenz provided us with a complete verbal description including the agents (the fish), the relevant factors in their environment (their nests), and their behavior and interaction. Furthermore, he also described the resulting patterns as observed in nature:

"The pursuit is repeated a few times in alternating directions, swinging back and forth like a pendulum which at last reaches a state of equilibrium at a certain point."

4.1 Fuzzy Modeling

We apply the fuzzy modeling approach to derive a mathematical model of this system. In the first step, we identify the state-variables: fish i, $i = 1, 2$, is located at $\mathbf{x}^i(t) \in \mathbb{R}^n$, and has a *fighting inclination* $w_i(t) \in \mathbb{R}$, and a nest located at $\mathbf{c}^i \in \mathbb{R}^n$.

Next, we transform Lorenz's description of the change in fighting inclination into the following fuzzy rules:

- if $near_i(\mathbf{x}^i, \mathbf{c}^i)$ then $\dot{w}_i = +1$
- if $far_i(\mathbf{x}^i, \mathbf{c}^i)$ then $\dot{w}_i = -1$

that is, the *fighting inclination* increases (decreases) when the fish is near (far) its nest.

Similarly, the description of the movement of fish i is transformed into:

- if $near_i(\mathbf{x}^i, \mathbf{x}^j)$ and $high_i(w_i)$ then $\dot{\mathbf{x}}^i = \mathbf{x}^j - \mathbf{x}^i$
- if $near_i(\mathbf{x}^i, \mathbf{x}^j)$ and $low_i(w_i)$ then $\dot{\mathbf{x}}^i = \mathbf{c}^i - \mathbf{x}^i$

where $\mathbf{x}^j$ is the location of the other fish. That is, when the other fish is near me, and my fighting inclination is high (low), I move in the direction of the other fish (my nest).

The third step is to determine the membership functions for the fuzzy terms. We define $near_i(\mathbf{x}, \mathbf{y})$ using the membership function $n_i(\mathbf{x}, \mathbf{y}) = \exp(-\frac{||\mathbf{x}-\mathbf{y}||^2}{k_i^2})$ (where $k_i > 0$ determines the spread of the Gaussian function). The term $high_i$ is defined using $h_i(w) = \frac{1}{2}(1 + \tanh(\frac{w}{a_i}))$ (where $a_i > 0$ determines the slope of h_i). Note that $\lim_{w \to -\infty} h_i(w) = 0$ and $\lim_{w \to +\infty} h_i(w) = 1$. The opposite terms far_i and low_i are defined using $f_i(x, y) = 1 - n_i(x, y)$, and $l_i(w) = 1 - h_i(w)$, respectively.

The last step is to calculate the output of the rules using inferencing. We use multiplication for the "and" operator, and center of gravity defuzzification, so the rules yield

$$\dot{w}_i = \frac{n_i(\mathbf{x}^i, \mathbf{c}^i) - f_i(\mathbf{x}^i, \mathbf{c}^i)}{n_i(\mathbf{x}^i, \mathbf{c}^i) + f_i(\mathbf{x}^i, \mathbf{c}^i)},$$
$$\dot{\mathbf{x}}^i = \frac{(\mathbf{x}^j - \mathbf{x}^i)h_i(w_i)n_i(\mathbf{x}^i, \mathbf{x}^j) + (\mathbf{c}^i - \mathbf{x}^i)l_i(w_i)n_i(\mathbf{x}^i, \mathbf{x}^j)}{h_i(w_i)n_i(\mathbf{x}^i, \mathbf{x}^j) + l_i(w_i)n_i(\mathbf{x}^i, \mathbf{x}^j)}$$

for $i = 1, 2$. Substituting the membership functions and simplifying yields:

$$\dot{w}_i = 2\exp(-\frac{||\mathbf{x}^i - \mathbf{c}^i||^2}{k_i^2}) - 1,$$
$$\dot{\mathbf{x}}^i = \mathbf{c}^i - \mathbf{x}^i + h_i(w_i)(\mathbf{x}^j - \mathbf{c}^i), \tag{1}$$

for $i = 1, 2$. Summarizing, the fuzzy modeling approach allowed us to transform the verbal description into the well-defined mathematical model (1). This model can now be used to simulate and analyze the system.

4.2 Simulations

We first simulated the one-dimensional case ($n = 1$) using the parameters:

$$c^1 = -1,\ c^2 = 1,\ a_1 = a_2 = k_1 = k_2 = 1, \tag{2}$$

and initial values $x^1(0) = -0.4$, $x^2(0) = 0.8$, $w_1(0) = w_2(0) = 1$.

Figure 1 depicts $x^1(t)$ and $x^2(t)$ as a function of t. It may be seen that the fish follow an oscillatory movement with one fish advancing, the other retreating until a point is reached where they switch roles. Finally, they converge to a steady state point at $\overline{x}^1 = -0.1674$ and $\overline{x}^2 = 0.1674$.

We also simulated the three-dimensional case ($n = 3$), this being the one actually found in nature. Figure 2 depicts the behavior of the model with

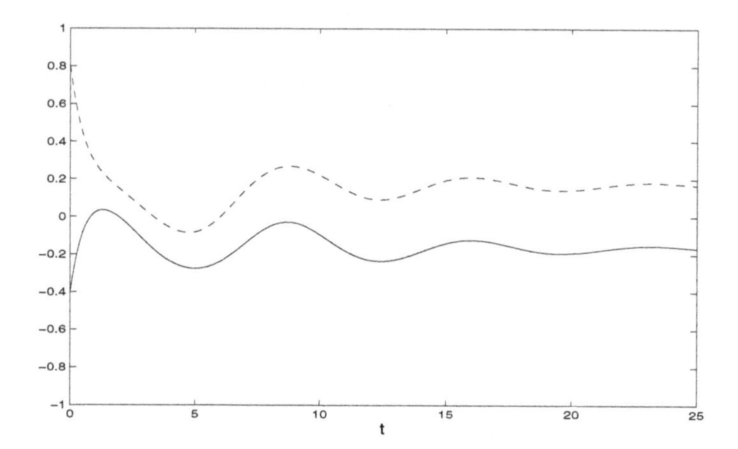

Fig. 1. Trajectories $x^1(t)$ (solid line) and $x^2(t)$ (dashed line) for initial positions $x^1(0) = -0.4$ and $x^2(0) = 0.8$. The nests are located at $c^1 = -1$ and $c^2 = 1$

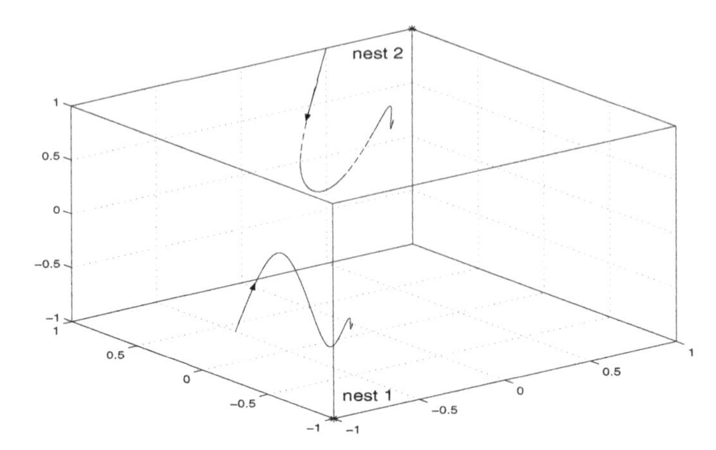

Fig. 2. Trajectories $\mathbf{x}^1(t)$ (solid line) and $\mathbf{x}^2(t)$ (dashed line)

the parameters: $\mathbf{c}^1 = (-1,-1,-1)^T$, $\mathbf{c}^2 = (1,1,1)^T$, $a_1 = a_2 = k_1 = k_2 = 1$, initial values $\mathbf{x}^1(0) = (-0.5, 0.4, -1)^T$, $\mathbf{x}^2(0) = (0.5, 1, 1)^T$, and $w_1(0) = w_2(0) = 1$. It may be seen that again, we have an oscillatory movement converging to a steady state point $\mathbf{x}^i(t) \to \overline{\mathbf{x}}^i$, $i = 1, 2$, with $\overline{\mathbf{x}}^i$ on the line connecting the two nests.

Note that the oscillatory movement of the fish in our fuzzy model and the behavior observed in nature, as described by LORENZ, are congruent.

4.3 Analysis

We analyzed the case $n = 1$, that is, the fish live in a 1D world, and the system is described by four state-variables: x^1, x^2, w_1, and w_2. We assume, without loss of generality, that $c^2 = c$ and $c^1 = -c$, for some $c > 0$. In this case, it is easy to verify that (1) admits an equilibrium point:

$$\begin{aligned} \overline{x}^1 &= -c + k_1\sqrt{\log 2}, & \overline{w}_1 &= a_1 \tanh^{-1}\left(2\tfrac{\overline{x}^1+c}{\overline{x}^2+c} - 1\right), \\ \overline{x}^2 &= c - k_2\sqrt{\log 2}, & \overline{w}_2 &= a_2 \tanh^{-1}\left(2\tfrac{\overline{x}^2-c}{\overline{x}^1-c} - 1\right), \end{aligned} \quad (3)$$

where $\tanh^{-1}$ denotes the inverse hyperbolic tangent function. We assume that the parameters are chosen such that $-c < \overline{x}^1 < \overline{x}^2 < c$ (that is, the equilibrium point is between the two nests, with each fish closer to its nest than the other fish), which implies that the $\overline{w}_i$'s are well-defined.

The next result shows that the arrangement (nest1,fish1,fish2,nest2) is an invariant of the mathematical model.

Proposition 1 *[7] If $-c < x^1(0) < x^2(0) < c$, then*

$$-c \leq x^1(t) \leq x^2(t) \leq c \textit{ for all } t \geq 0.$$

Recall that an equilibrium point $\mathbf{x}_e$ of the differential equation $\dot{\mathbf{x}}(t) = \mathbf{f}(\mathbf{x}(t))$ is said to be *locally asymptotically stable* (see, e.g. [35]) if the following properties hold: (1) for any $\epsilon > 0$ there exists a $\delta > 0$ such that $||\mathbf{x}(0) - \mathbf{x}_e|| \leq \delta$ implies $||\mathbf{x}(t) - \mathbf{x}_e|| \leq \epsilon$ for all $t \geq 0$; and (2) there exists a $c > 0$ such that $||\mathbf{x}(0) - \mathbf{x}_e|| \leq c$ implies that $\lim_{t \longrightarrow \infty} \mathbf{x}(t) = \mathbf{x}_e$. In other words, for sufficiently small initial perturbations the trajectories both remain in the vicinity of the equilibrium point, and converge to the equilibrium point.

Proposition 2 *[7] The equilibrium point* (3) *of* (1) *is locally asymptotically stable.*

4.4 Parameter Influence

A mathematical model can be used to study the effect of different parameters on the system's behavior.

Consider for example the case $n = 1$ and the parameter k_i which determines the spread of the Gaussian function defining the term $near_i$. As k_i decreases, the Gaussian function becomes narrower so the rule:

- if $near_i(\mathbf{x}^i, \mathbf{c}^i)$ then $\dot{w}_i = +1$

will fire only when $\mathbf{x}^i$ is very close to $\mathbf{c}^i$. In other words, the fish will have a somewhat lower "fighting potential" since its fighting inclination will begin to increase only in a very small neighborhood around the nest.

Figure 3 depicts $x^1(t)$ and $x^2(t)$ for the same values as in (2) except that we decreased k_2 from 1.0 to 0.5. It may be seen that in this case the convergence to the steady state is faster and that $x_1(t) \rightarrow -0.1674$ (as before), but $x^2(t) \rightarrow 0.5387$, that is, the equilibrium point is no longer symmetrical. Indeed, it follows from (3) that when k_2 decreases, $\overline{x}^2$ increases. If we define the territory of fish 2 as $[\overline{x}^2, c^2]$, then decreasing k_2 yields a decrease in the size of the territory.

In fact, this too is congruent with the behavior observed in nature as described in [34]: "...the relative fighting potential of the individual is shown by the size of the territory which he keeps clear of rivals."

4.5 Prediction

We can also use the mathematical model to analyze and simulate new scenarios that were not described by the observer. The results can be regarded as *predictions* of the behavior of the real system.

For example, suppose that before the fish are placed in the aquarium, they are "irritated" in a form that increases their fighting inclinations. How will this affect their mutual behavior? We can simulate this scenario in our model by increasing the values $w_i(0)$, $i = 1, 2$.

Figure 4 depicts $x^1(t)$ and $x^2(t)$ for the same parameters as in Sect. 4.2, but with $w_1(0) = w_2(0) = 3$ instead of $w_1(0) = w_2(0) = 1$. Comparing this

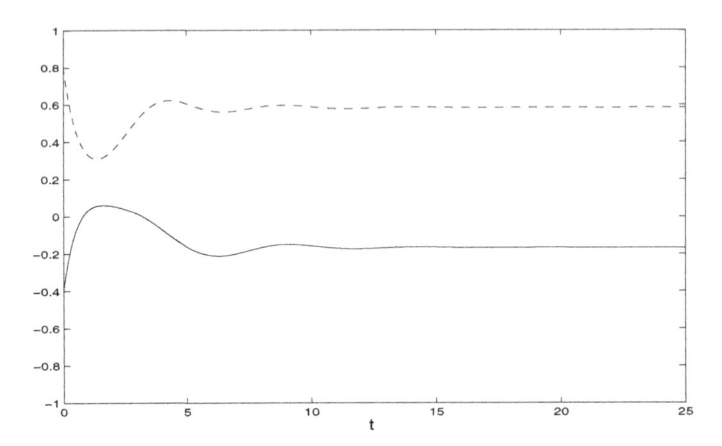

Fig. 3. Trajectories $x^1(t)$ (solid line) and $x^2(t)$ (dashed line) for $k_1 = 1$, $k_2 = 0.5$. The nests are located at $c^1 = -1$ and $c^2 = 1$

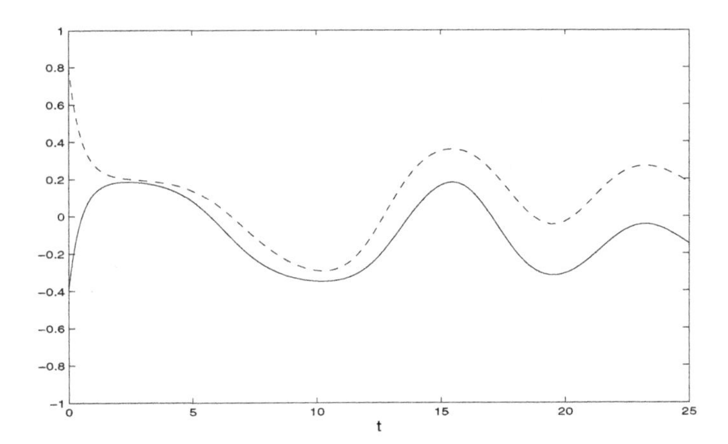

Fig. 4. Trajectories $x^1(t)$ (solid line) and $x^2(t)$ (dashed line) for $w_1(0) = w_2(0) = 3$

with Fig. 1, we can see that now the fish come closer together before they eventually retreat, and that the amplitude of their oscillations is increased. Thus, the model provides a prediction of the outcome of new experiments, as well as stimulus for further study.

5 Example 2: Klino-Kinesis in the *Dendrocoleum lacteum*

An animal's life depends on oriented movements. Such movements guide the animal into its normal habitat or into other situations which are of importance to it. For example, predators must move towards potential prey, and away from possible danger. Various external factors–light, smell, currents, humidity, heat, and so on–activate the living mechanisms and lead to orientation.

A movement whose direction is correlated with the direction of the stimulus is called *taxis* (see, e.g. [36]). For example, positive (negative) photo-taxis is the directed movement towards (away from) a source of light. Taxes require sensory organs that can accurately detect the direction of the stimulus, and a brain sophisticated enough to process the sensory data and determine the appropriate direction of movement.

Simple organisms do not necessarily have the physiological equipment needed to perform taxes. Their eyes, for example, may do little more than indicate the general intensity of light, but not its direction. Hence, in such organisms the locomotory action is affected by the intensity of the stimulus, but *not* by the direction of the stimulus. This type of response is referred to as *kinesis*.[3] In particular, *Klino-kinesis*[4] is defined as a movement where the rate of turning, but *not* the direction of turning, depends on the intensity of the stimulus. This type of movement appears in many flat-worms: in regions with higher light intensity their rate of turning increases. As a result, the animals eventually aggregate in shadier parts of the available habitat.

In a seminal paper, Philip Ullyott studied this type of behavior in the *Dendrocoleum lacteum* [37]. He analyzed the effects of light stimulus on the animal's behavior, and found that the stimulus did not affect the animal's linear velocity. Instead, increased light intensity yielded an increase in the *rate of change of direction* (r.c.d.) in which the animal moved. Ullyott defined the r.c.d. as the sum of all the deviations in the animal's path during one time unit, summing up both right-hand and left-hand deflections as positive changes, and expressing the result in units of angular degrees per minute. As the light was switched on, the r.c.d. immediately increased. With time, an adaptation process caused it to fall off, converging to a constant level, designated the *basal r.c.d.*

Ullyott summarized his findings as follows [37, p. 274]: "(1) An increase in stimulating intensity produces an increase in r.c.d.; (2) This initial increase in r.c.d. falls off under constant stimulation owing to adaptation; (3) There is a basal r.c.d., which is an expression of the fact that turning movements occur even in absolute darkness or at complete adaptation."

Ullyott also provided a verbal description of the overall pattern observed in his experiments [37, p. 277]: "This alternate stimulation and adaptation has an effect on the r.c.d. of such a kind that the animal is led automatically to the place of minimal intensity."

Fraenkel and Gunn [11, Chap. V] reviewed and refined Ullyott's work. They developed a simplified deterministic model for the "averaged" animal's movement. The most important simplification is the assumption that the

---

[3] from the Greek *kinesis*, movement

[4] from the Greek *klino*, incline

animal always turns to the right and always through exactly 90°. As the r.c.d. increases, the time between these right-hand turns decreases.

Fraenkel and Gunn provided a heuristic explanation of how this behavior drives the animal to the darker regions. Suppose that the animal is placed in a plane described by two coordinates (x, y), and the light intensity becomes stronger as we move along the positive direction of the x-axis (see Fig. 5). Beginning at a point A (and assuming that the animal is fully adapted to the light at A), the animal continues in the positive x direction until making a right-hand turn at point B, and so on. Along the segment AB, the light intensity increases and the adaptation level lags behind, so the r.c.d. increases. Along the BC segment the light intensity is constant and the r.c.d. decreases back to its basal level. Along the CD segment the light intensity decreases and the r.c.d. remains constant (note that r.c.d. is affected only by an *increase* in the light's intensity). Finally, the behavior along the DE segment is as in the BC segment. Since the r.c.d. increases along the AB segment and, due to the adaptation process, decreases along the other segments (until it converges to the basal r.c.d.), the segment AB will be *shorter* than the segment CD. Hence, after a set of consecutive turns the animal ends up at a point E, which is closer to the darker part of the plane than the initial point A.

Note that the adaptation process plays an essential role in this explanation since it allows the animal to (indirectly) compare the current light intensity with previous intensities.

Both Ullyott and Fraenkel and Gunn provided only a verbal description of the animal's behavior. Patlak [38] studied the behavior of particles under the following assumptions: (1) the particles move in a random way, but with persistence of direction; and (2) their movement is also influenced by an external force. He derived a suitable Fokker-Planck-type equation characterizing the particles' movement and used this stochastic model to analyze klinokinesis [39] (see also [40]). However, his model ignores the adaptation process which plays a vital role in the description given by Ullyott.

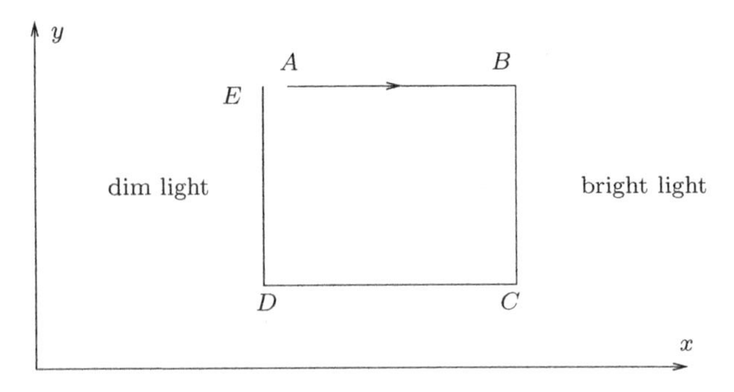

Fig. 5. The path followed by the "average" animal. Adapted from [11, p. 49]

5.1 Fuzzy Modeling

We now apply the fuzzy modeling approach to derive a mathematical model of this system. In the first step, we identify the variables in the model: the animal's location in the plane $(x(t), y(t))$; the light intensity at every point in the plane $l(x, y)$; the light intensity that the animal is currently adapted to $l_a(t)$; the r.c.d. $r(t)$; the direction of movement $\theta(t)$; and the discrete set of times: $t_1, t_2, \ldots$, in which the animal performs a turn. The model also includes two constants: the animal's linear velocity v, and the basal r.c.d. r_b.

The next stage is to transform the description given above into a set of fuzzy rules. The adaptation process changes $l_a(t)$, the level of adaptation to light, in accordance with the level of light $l(t)$. We state this as two fuzzy rules

- If $l - l_a$ is *positive*, then $\dot{l}_a = c_1$
- If $l - l_a$ is *negative*, then $\dot{l}_a = -c_1$

where $c_1 > 0$ is a constant.

The r.c.d. decreases when it is above the basal r.c.d., and increases when the light stimulus is above the adaptation level. We state this as

- If $r - r_b$ is *large*, then $\dot{r} = -c_2$
- If $l - l_a$ is *high*, then $\dot{r} = c_3$

where $c_2, c_3 > 0$ are constants.

Finally, following Fraenkel and Gunn, we add a crisp rule

- If $\int_{t_l}^{t} r(\tau)d\tau = q$, then $\theta \leftarrow \theta - \pi/2$

Here, $t_l < t$ is the time when the last turn took place, and $q > 0$ is a constant. In other words, the r.c.d. is accumulated and whenever it reaches the threshold q, the animal makes a right-hand turn.

The third step is to define the various fuzzy membership functions, operators, and the fuzzy inferencing method used. For our first set of rules, we model the fuzzy sets using the membership functions

$$\mu_{positive}(x) = \frac{e^{k_1 x}}{e^{k_1 x} + e^{-k_1 x}}, \qquad \mu_{negative}(x) = \frac{e^{-k_1 x}}{e^{k_1 x} + e^{-k_1 x}}, \tag{4}$$

where $k_1 > 0$ is a constant.

For the second set of rules, we use $\mu_{large}(x) = S_{k_2}(x)$, $\mu_{high}(x) = S_{k_3}(x)$, where $k_2, k_3 > 0$ are constants, and S_k is the piecewise-linear function

$$S_k(z) := \begin{cases} 0, & \text{if } z \leq 0 \\ z/k, & \text{if } 0 < z < k \\ 1, & \text{if } z \geq k. \end{cases} \tag{5}$$

We can now inference the fuzzy rules to obtain a set of differential equations. Applying simple additive inferencing (see, e.g. [41]), the first set of rules yields

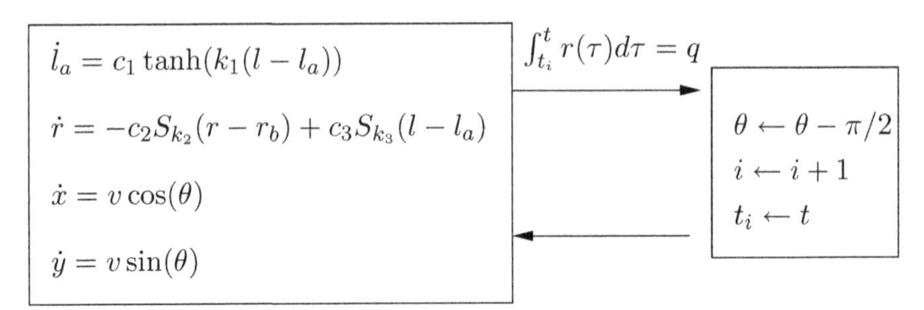

$\dot{l}_a = c_1 \tanh(k_1(l - l_a))$
$\dot{r} = -c_2 S_{k_2}(r - r_b) + c_3 S_{k_3}(l - l_a)$
$\dot{x} = v\cos(\theta)$
$\dot{y} = v\sin(\theta)$
$\int_{t_i}^t r(\tau)d\tau = q$
$\theta \leftarrow \theta - \pi/2$
$i \leftarrow i + 1$
$t_i \leftarrow t$

Fig. 6. The hybrid model

$$\begin{aligned} \dot{l}_a &= c_1 \mu_{positive}(l - l_a) - c_1 \mu_{negative}(l - l_a) \\ &= c_1 \tanh(k_1(l - l_a)). \end{aligned} \tag{6}$$

Similarly, the second set of rules yields

$$\dot{r} = -c_2 S_{k_2}(r - r_b) + c_3 S_{k_3}(l - l_a). \tag{7}$$

The actual movement of the animal is given by

$$\dot{x} = v\cos(\theta), \quad \dot{y} = v\sin(\theta). \tag{8}$$

The crisp rule implies that the value θ "jumps" at the discrete times $t_1, t_2, \ldots$ satisfying $\int_{t_i}^{t_{i+1}} r(\tau)d\tau = q$, $i = 0, 1, \ldots$ (with t_0 defined as zero). Thus, the system combines continuous-time dynamics and discrete events and is, therefore, a *hybrid system* [42].

Figure 6 summarizes the mathematical model. The upper arrow in this figure corresponds to the conditional transition described by the crisp rule. When the condition holds, the value of θ "jumps", and the evolution in time proceeds using the continuous-time dynamics.

In the next two sections, we study the behavior of this mathematical model using simulations and rigorous analysis.

5.2 Simulations

Our simulations were motivated by the experiments performed by Ullyott [37].

Response to a Sudden Increase in Light Intensity

In one of his experiments, Ullyott first determined the basal r.c.d. by measuring the r.c.d. after the animals were placed in total darkness for three hours (to make them completely dark adapted). He then exposed the animals to a sudden increase in light intensity, and measured the r.c.d. at different times. Ullyott repeated each experiment several times and averaged the resulting measurements. Figure 7 depicts the averaged r.c.d. as a function of time.

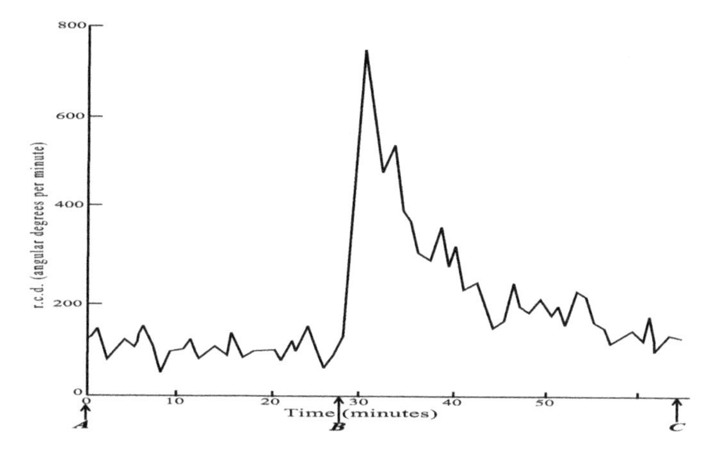

Fig. 7. The relationship between rate of change of direction and duration of stimulus. AB, r.c.d. of the animal in darkness (basal r.c.d.). At B a light intensity of 2500 *regs./cm.*^{2}*/sec.* was switched on. BC, adaptation to the stimulus. Each point on the curve represents the average of fourteen experiments. (Reproduced from [37] with permission from The Company of Biologists Ltd.)

We simulated a similar scenario in our model using (6)–(7) with

$$c_1 = 5, c_2 = 1, c_3 = 2, k_1 = 1, k_2 = 1, k_3 = 2, r(0) = r_b = 2, l_a(0) = 0, \quad (9)$$

and the light intensity $l(t) = 0$ for $t \in [0, 1)$, and $l(t) = 1$ for $t \geq 1$.

Figure 8 depicts $r(t)$ as a function of time. Note the rapid increase of $r(t)$ near the switching time $t = 1$, and then the convergence back to the basal r.c.d. It is quite interesting to compare this figure with the results actually measured by Ullyott as depicted in Fig. 7. The qualitative resemblance is obvious.

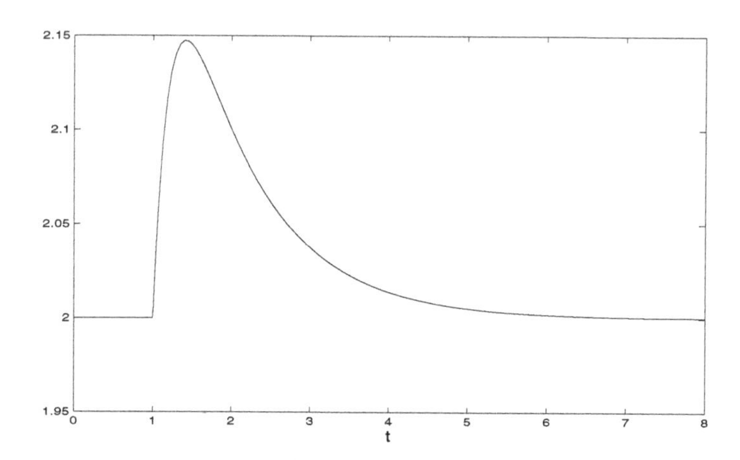

Fig. 8. $r(t)$ as a function of time. $r(0) = r_b = 2$, and the light is switched on at $t = 1$

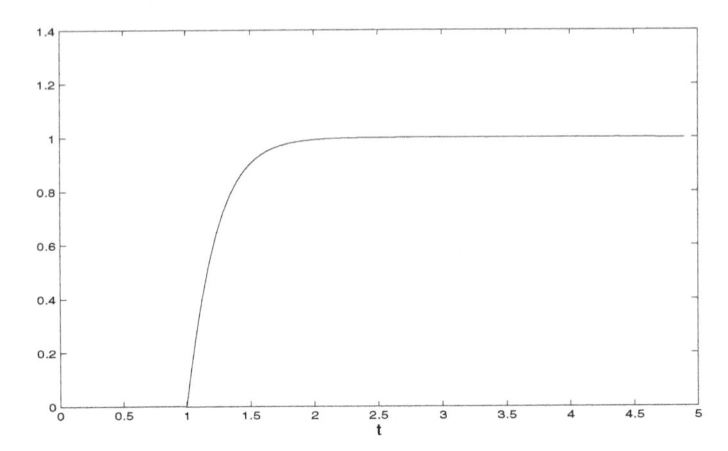

Fig. 9. $l_a(t)$ as a function of time. The light is switched on at $t = 1$

Figure 9 depicts $l_a(t)$ as a function of time. It may be seen that the adaptation level increases when the light is switched on, and then converges to $l_a(t) = 1(= l(t))$.

Response in a Plane with Different Light Intensities

Our second simulation was motivated by the explanation provided by Fraenkel and Gunn (see Fig. 5). We simulated our hybrid model with: $c_1 = c_2 = c_3 = k_1 = k_2 = v = 1, \quad r_b = 2, \quad k_3 = 3, \quad q = 10$, initial conditions: $r(0) = r_b$, $x(0) = y(0) = l_a(0) = 0$, and light intensity $l(x, y) = x$, that is, the light intensity increases linearly along the x-axis.

Figure 10 depicts the trajectory $(x(t), y(t))$. Denoting the times of the right-hand turns by $t_1, t_2, \ldots$, (with t_0 defined as zero), it may be seen that

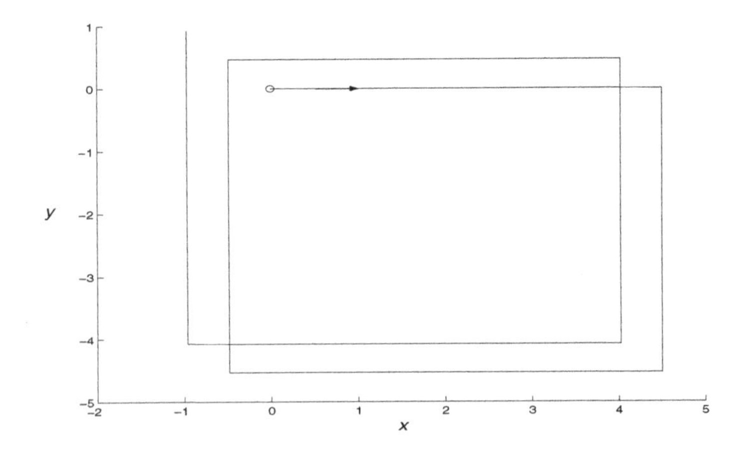

Fig. 10. The trajectory $(x(t), y(t))$ for $x(0) = y(0) = 0$ and $l(x, y) = x$

$$x(t_0) > x(t_4) > x(t_8) > \dots.$$

Thus, the trajectory indeed moves toward the darker region of the plane. This is in agreement with Ullyott's observations [37, p. 272]: "The shift of the position of the animal towards the end of the gradient was rather a gradual process, but in each case with the steep gradient, the animal was to be found at the darker end within 2 hours from the beginning of the experiment."

5.3 Analysis

We analyzed the behavior of the hybrid model for the scenarios described in the simulations above.

Response to a Sudden Increase in Light Intensity

The next result describes the effect of a "jump" in the light intensity on the r.c.d. Let $1(t)$ denote the step function.[5]

Proposition 3 *[8] Consider the hybrid model with initial conditions* $l_a(0) = 0$, $r(0) = r_0$, *and light intensity* $l(t) = 1(t)$. *Denote* $w := c_1k_1k_2 - c_2$, $p_1 := \sinh(k_1)$, *and assume that the model's parameters satisfy:*

$$k_3 \geq 1, \quad w > 0, \quad \text{and } k_2 \geq r_0 - r_b + \frac{p_1 k_2 c_3}{k_1 k_3 w}. \tag{10}$$

Then

$$\begin{aligned} r(t) =& r_b + \exp(-\frac{c_2}{k_2}t)(r_0 - r_b) \\ &+ \frac{c_3}{k_1 k_3}\int_0^t \exp(\frac{c_2}{k_2}(\tau - t))\sinh^{-1}(p_1\exp(-c_1k_1\tau))d\tau, \end{aligned} \tag{11}$$

and

$$l_a(t) = 1 - \frac{1}{k_1}\sinh^{-1}(p_1\exp(-c_1k_1t)), \tag{12}$$

where $\sinh^{-1}$ *denotes the inverse hyperbolic sine function.*

It is easy to verify that the values (9) used in the simulations satisfy (10). Hence, (11) and (12) actually provide an analytical description of the responses depicted in Figs. 8 and 9, respectively.

The proof in [8] implies that the response of $r(t)$ to a step function in the light intensity decays exponentially. This agrees with the observations of Ullyott, who states [37, p. 270]: "...it is possible to see that the falling off of r.c.d. with time is exponential ...".

---

[5] i.e. $1(t) = 0$ for $t < 0$, and $1(t) = 1$ for $t \geq 0$

Response in a Plane with Different Light Intensities

The next result describes the trajectory of the hybrid model when exposed to a light with a directional gradient. Specifically, assume that $l(x, y) = x$, that is, the light intensity increases as we move along the positive x direction.

Proposition 4 *[8] Consider the hybrid model with:*

$$c_1 = c_2 = c_3 = k_1 = k_2 = v = 1, \quad r_b = 2, \quad k_3 = 3, \quad q = 10, \quad l(x, y) = x,$$

and initial conditions: $r(0) = r_b$, $x(0) = y(0) = l_a(0) = 0$. *Then, there exist times* $0 = t_0 < t_1 < t_2 \ldots$ *such that* $\int_{t_i}^{t_{i+1}} r(\tau)d\tau = q$, *and*

$$x(t_4) < x(t_0).$$

In other words, after one set of turns, the animal ends up with $x(t_4) < x(t_0)$, so that it has indeed moved towards the darker region of the plane.

6 Fuzzy Modeling and Biomimicry

Many natural beings possess the capability to reason, learn, evolve, adapt, and heal. Modern computers and machines still lack many of these desirable properties. Furthermore, over the course of evolution, living systems developed efficient and robust solutions to various problems. Some of these problems are also encountered in engineering applications. For example, plants had to develop efficient mechanisms for absorbing and utilizing solar energy. Engineers that design solar cells, face a very similar challenge. It is thus natural that considerable research is recently devoted to the development of artificial products or machines that mimic (or are inspired by) biological phenomena [43, 44].

Examples from this emerging field of science, known as *Biomimicry* or *Bio-inspired systems*, include the following: optimization algorithms that are inspired by Darwin's theory of evolution [45, 46]; the design of structures that mimic shape laws in nature and, in particular, the structural properties of trees [47]; autonomous robots that mimic the behavior of various animals [48, 49]; the design of computer security systems inspired by the natural immune system [50] (see also [51]); and the design of optimization, search, and clustering algorithms based on the behavior of social insects [52, 53].

An important component in biomimicry is the ability to perform reverse engineering of an animal's functioning and then implement this behavior in an artificial system. In many cases, verbal descriptions of the natural behavior already exist in the literature. For example, as noted in [54], it is customary to describe the behavior of simple organisms (e.g. ants) using simple rules of thumb [55].

The fuzzy modeling approach may be very suitable for addressing biomimicry in a systematic manner. Namely, start with a verbal description of an animal's behavior (e.g. foraging in ants) and, using fuzzy logic theory, obtain a mathematical model of this behavior which can be implemented by artificial systems (e.g. autonomous robots).

7 Concluding Remarks

Verbal descriptions and explanations of various phenomena appear throughout many of the "soft sciences". Constructing a suitable mathematical model, based on this information, can greatly enhance our understanding of these phenomena. Fuzzy modeling seems to be the natural tool for this purpose. We demonstrated this using two examples from ethology: the territorial behavior in the stickleback and the orientation to light of the planarian *Dendrocoleum lacteum.* The fuzzy modeling approach transforms the verbal descriptions of the behavior into well-defined mathematical models, that can be simulated on a computer and analyzed using mathematical tools.

We believe that fuzzy modeling can and should be utilized in many fields of science, including biology, economics, psychology, sociology, history [56] and more. In such fields, many verbal models exist in the research literature, and they can be directly transformed into mathematical models using the method described herein.

Recently, the field of biomimicry, that is, the design of artificial algorithms and machines that imitate biological behavior, is attracting considerable research interest. In many cases, there exist verbal descriptions of the natural behavior we aim to mimic. Hence, we believe that the paradigm presented here, namely, using fuzzy modeling to transform a behavior described in words into a mathematical model, is particularly suitable for biomimicry. Work in this direction is now under way.

Acknowledgments

I thank The Company of Biologists Ltd for granting me permission to reproduce some of the material from [37]. I am grateful to David Eilam for many illuminating discussions.

References

1. D. Dubois, H. T. Nguyen, H. Prade, and M. Sugeno, "Introduction: the real contribution of fuzzy systems," in *Fuzzy Systems: Modeling and Control*, H. T. Nguyen and M. Sugeno, Eds. Kluwer, 1998, pp. 1–17.
2. M. Margaliot and G. Langholz, "Fuzzy Lyapunov based approach to the design of fuzzy controllers," *Fuzzy Sets Systems*, vol. 106, pp. 49–59, 1999.
3. ——, *New Approaches to Fuzzy Modeling and Control - Design and Analysis.* World Scientific, 2000.

4. L. A. Zadeh, "Fuzzy logic = computing with words," *IEEE Trans. Fuzzy Systems*, vol. 4, pp. 103–111, 1996.
5. W. Siler and J. J. Buckley, *Fuzzy Expert Systems and Fuzzy Reasoning*. Wiley-Interscience, 2004.
6. A. Kandel, Ed., *Fuzzy Expert Systems*. CRC Press, 1992.
7. E. Tron and M. Margaliot, "Mathematical modeling of observed natural behavior: a fuzzy logic approach," *Fuzzy Sets Systems*, vol. 146, pp. 437–450, 2004.
8. ——, "How does the Dendrocoleum lacteum orient to light? a fuzzy modeling approach," *Fuzzy Sets Systems*, vol. 155, pp. 236–251, 2005.
9. N. Tinbergen, *The Study of Instinct*. Oxford University Press, 1969.
10. L. A. Zadeh, "Fuzzy sets," *Information and Control*, vol. 8, pp. 338–353, 1965.
11. G. S. Fraenkel and D. L. Gunn, *The Orientation of Animals: Kineses, Taxes, and Compass Reactions*. Dover Publications, 1961.
12. G. P. Richardson and A. L. Pugh, *Introduction to System Dynamics Modeling with Dynamo*. MIT Press, 1981.
13. B. Kuipers, *Qualitative Reasoning: Modeling and Simulation with Incomplete Knowledge*. The MIT Press, 1994.
14. L. A. Zadeh, "Outline of a new approach to the analysis of complex systems and decision processes," *IEEE Trans. Systems, Man, Cybernetics*, vol. 3, pp. 28–44, 1973.
15. E. H. Mamdani, "Applications of fuzzy algorithms for simple dynamic plant," *Proc. IEE*, vol. 121, pp. 1585–1588, 1974.
16. D. Dubois, H. Prade, and R. R. Yager, Eds., *Fuzzy Information Engineering*. Wiley, 1997.
17. K. Tanaka and M. Sugeno, "Introduction to fuzzy modeling," in *Fuzzy Systems: Modeling and Control*, H. T. Nguyen and M. Sugeno, Eds. Kluwer, 1998, pp. 63–89.
18. T. Terano, K. Asai, and M. Sugeno, *Applied Fuzzy Systems*. AP Professional, 1994.
19. R. R. Yager and D. P. Filev, *Essentials of Fuzzy Modeling and Control*. John Wiley & Sons, 1994.
20. B. A. Sproule, C. A. Naranjo, and I. B. Türksen, "Fuzzy pharmacology: theory and applications," *TRENDS in Pharmacological Science*, vol. 23, pp. 412–417, 2002.
21. F. Wenstop, "Deductive verbal models of organizations," in *Fuzzy Reasoning and its Applications*, E. H. Mamdani and B. R. Gaines, Eds. Academic Press, 1981, pp. 149–167.
22. ——, "Quantitative analysis with linguistic rules," *Fuzzy Sets Systems*, vol. 4, pp. 99–115, 1980.
23. W. J. M. Kickert, "An example of linguistic modeling: The case of Mulder's theory of power," in *Advances in Fuzzy Set Theory and Applications*, M. M. Gupta, R. K. Ragade, and R. R. Yager, Eds. North-Holland, 1979, pp. 519–540.
24. ——, *Fuzzy Theories on Decision-Making*. Martinus Nijhoff, 1978.
25. B. Kosko, *Neural Networks and Fuzzy Systems: A Dynamical Systems Approach to Machine Intelligence*. Prentice-Hall, 1992.
26. J. Aguilar, "Adaptive random fuzzy cognitive maps," in *Lecture Notes in Artificial Intelligence 2527*, F. J. Garijo, J. C. Riquelme, and M. Toro, Eds. Springer-Verlag, 2002, pp. 402–410.
27. J. M. C. Sousa and U. Kaymak, *Fuzzy Decision Making in Modeling and Control*. World Scientific, 2002.

28. G. W. Bluman and S. C. Anco, *Symmetry and Integration Methods for Differential Equations*. Springer-Verlag, 2002.
29. L. A. Segel, "Simplification and scaling," *SIAM Review*, vol. 14, pp. 547–571, 1972.
30. S. Guillaume, "Designing fuzzy inference systems from data: an interpretability-oriented review," *IEEE Trans. Fuzzy Systems*, vol. 9, pp. 426–443, 2001.
31. G. Bontempi, H. Bersini, and M. Birattari, "The local paradigm for modeling and control: from neuro-fuzzy to lazy learning," *Fuzzy Sets Systems*, vol. 121, pp. 59–72, 2001.
32. J. S. R. Jang, C. T. Sun, and E. Mizutani, *Neuro-Fuzzy and Soft Computing: A Computational Approach to Learning and Machine Intelligence*. Prentice-Hall, 1997.
33. E. O. Wilson, *Sociobiology: The New Synthesis*. Harvard University Press, 1975.
34. K. Z. Lorenz, *King Solomon's Ring: New Light on Animal Ways*. Methuen & Co., 1957.
35. M. Vidyasagar, *Nonlinear Systems Analysis*. Prentice Hall, 1993.
36. G. Goodenough, B. McGuire, and R. A. Wallace, *Perspectives on Animal Behavior*. Second edition, John Wiley & Sons, 2001.
37. P. Ullyott, "The behavior of Dendrocoelum lacteum II: responses in non-directional gradients," *J. Experimental Biology*, vol. 13, pp. 253–278, 1936.
38. C. S. Patlak, "Random walk with persistence and external bias," *Bulletin of Mathematical Biophysics*, vol. 15, pp. 311–338, 1953.
39. ——, "A mathematical contribution to the study of orientation of organisms," *Bulletin of Mathematical Biophysics*, vol. 15, pp. 431–476, 1953.
40. P. Turchin, *Quantitative Analysis of Movement: Measuring and Modeling Population Redistribution in Animals and Plants*. Sinauer Associates, 1998.
41. J. M. Benitez, J. L. Castro, and I. Requena, "Are artificial neural networks black boxes?" *IEEE Trans. Neural Networks*, vol. 8, pp. 1156–1164, 1997.
42. A. van der Schaft and H. Schumacher, *An Introduction to Hybrid Dynamical Systems*. LNCIS 251, Springer, 2000.
43. M. Sipper, *Machine Nature: The Coming Age of Bio-Inspired Computing*. McGraw-Hill, 2002.
44. K. M. Passino, *Biomimicry for Optimization, Control, and Automation*. Springer, 2004.
45. J. R. Koza, *Genetic Programming: on the Programming of Computers by Means of Natural Selection*. MIT Press, 1992.
46. D. E. Goldberg, *Genetic Algorithms in Search, Optimization, and Machine Learning*. Addison-Wesley Professional, 1989.
47. C. Mattheck, *Design in Nature: Learning from Trees*. Springer-Verlag, 1998.
48. Y. Bar-Cohen and C. Breazeal, Eds., *Biologically Inspired Intelligent Robots*. SPIE Press, 2003.
49. C. Chang and P. Gaudiano, Eds., *Robotics and Autonomous Systems, Special Issue on Biomimetic Robotics*, vol. 30, 2000.
50. S. Forrest, S. A. Hofmeyr, and A. Somayaji, "Computer immunology," *Communications of the ACM*, vol. 40, pp. 88–96, 1997.
51. E. Hart and J. Timmis, "Application areas of AIS: the past, the present and the future," in *Artificial Immune Systems*, ser. Lecture Notes in Computer Science, C. Jacob, M. L. Pilat, P. J. Bentley, and J. Timmis, Eds. Springer-Verlag, 2005, vol. 3627, pp. 483–497.
52. E. Bonabeau, M. Dorigo, and G. Theraulaz, *Swarm Intelligence: from Natural to Artificial Systems*. Oxford University Press, 1999.

53. M. Dorigo and T. Stutzle, *Ant Colony Optimization.* MIT Press, 2004.
54. S. Schockaert, M. De Cock, C. Cornelis, and E. E. Kerre, "Fuzzy ant based clustering," in *Ant Colony, Optimization and Swarm Intelligence*, ser. Lecture Notes in Computer Science, M. Dorigo, M. Birattari, C. Blum, L. M. Gambardella, F. Mondada, and T. Stutzle, Eds. Springer-Verlag, 2004, vol. 3172, pp. 342–349.
55. B. Holldobler and E. O. Wilson, *The Ants.* Belknap Press, 1990.
56. P. Turchin, *Historical Dynamics: Why States Rise and Fall.* Princeton University Press, 2003.

Mathematical Fuzzy Logic in Modeling of Natural Language Semantics*

Vilém Novák

Abstract. This chapter contains an overview of some achievements in the modeling of semantics of some parts of natural languages using the calculus of mathematical fuzzy logic, namely the higher order one. In the second part, we briefly describe selected applications in geology, fuzzy control and decision-making.

1 Introduction

One of the goals that science has struggled with has been to develop and use new, precise languages that would not suffer from vagueness and that could be used to find precise and optimal solutions of various complicated problems. Work in this area has led to several calculi, among which the leading role was taken by set theory and classical mathematical logic. These have replaced natural languages in most parts of science (sometimes even in humanistic ones).

Fuzzy logic, however, posed a question whether such a result is not too radical since there are many tasks in which natural language still plays a non-substitutable role. Thus, solutions provided by pure mathematical means are sometimes too rough and non-satisfactory, or even impossible. Namely, natural languages are too tied with the human way of seeing the world, making it much more fruitful to utilize them than to suppress them. Of course, this requires a proper understanding of their structure and to master their semantics. For technical use, it is also very important to develop a kind formalization of natural languages that could have applications. This raised the idea of *precisiated natural language* (L. A. Zadeh, [37]), that there should be a specific formal system capturing semantics of natural languages and giving it a certain precision to make various kinds of technical applications possible. At the very least, the outcome is twofold: first, information that cannot be

* The chapter has been supported by projects 1M0572 and MSM 6198898701 of MŠMT ČR

formalized using classical means can be utilized, allowing new solutions to be obtained. Second, the resulting systems behave in a way that is close to the behavior of people.

This chapter has two parts. First, we briefly present mathematical fuzzy logic in a narrow sense (FLn) as a specific mathematical theory providing tools for modeling the vagueness phenomenon. This is further extended to fuzzy logic in the broader sense (FLb), whose paradigm is to provide a model of natural human reasoning in which a crucial role is played by natural language. Constituents of FLb include all means of the basic logical calculus, the theory of evaluating linguistic expressions, the logical theory of fuzzy IF-THEN rules and perception-based logical deduction. In the second part, we demonstrate the power of this model on some selected applications.

The chapter is organized as follows. In the next section, we briefly introduce mathematical fuzzy logic and focus on fuzzy type theory that is a fuzzy logic of higher order. In Sect. 3 we present the theory of evaluating linguistic descriptions and outline basic principles of the mathematical model of their semantics. Section 4 is devoted to the theory of fuzzy IF-THEN. Unlike the generally accepted interpretation, we understand them as special sentences of natural language. In this section we also explain the main ideas of the *perception-based* logical deduction as the main tool for reasoning in FLb. Section 5 consists of several applications. The first is one specific application in geology where we show that it is possible to take a vague geologist's description as an algorithm that provides practically the same results as that of a skillful geologist. In the second application, we show that the linguistic approach can also work very well in fuzzy control. We thus accomplish the original idea of fuzzy control — to directly transfer description of the control provided by human operators. The third application is in complex managerial decision-making, which combines objective and subjective information. This application, in our opinion, demonstrates that mathematical methods, when properly joined with qualitative expert knowledge, can be an effective decision-making tool for managers.

The scope of possible applications of our theory is by no means finished. For example, we have also developed a model of complex detective reasoning [9,10] when a murderer is to be convicted on the basis of found evidence and his testimony. This is a sophisticated application demonstrating the power of FLb. It is also notable that when modeling human reasoning, it was not only necessary to deal with natural language but also to apply principles of nonmonotonic reasoning.

Another application involves discovering linguistic associations [30] from numerical data that may be further taken as fuzzy IF-THEN rules, providing us with some special knowledge. This list of potential applications, of course, is by no means finished. In particular, the theory of evaluating linguistic expressions presented below turns to be an extremely powerful theory, whose applications can be found in many, and quite often, very unexpected places.

2 Mathematical Fuzzy Logic in the Narrow and Broader Sense

2.1 General Overview

In the past 17 years, mathematical fuzzy logic has undergone intense development. However, the first, mathematically deep and advanced formal system was already published in 1979 by J. Pavelka in [32]. In fact, he established a limit generalization of logic by allowing evaluation of formulas in syntax. After the seminal monograph of P. Hájek [12], we now distinguish two fundamental approaches to the theory of mathematical fuzzy logic: that with *traditional syntax* and that with *evaluated syntax*. The former encompasses *basic fuzzy logic* BL, *involutive monoidal t-norm-based logic* IMTL, *monoidal t-norm-based logic* MTL ([11]), *Łukasiewicz logic* Ł, *Gödel logic*, *product logic* Π, ŁΠ and, possibly, some other systems of fuzzy logics. For more details, together with various references, see [26]. Here, we can also rank the *fuzzy type theory* (FTT), that is a higher-order fuzzy logic serving as a basis for the development of FLb. The aim of all these logics is to provide formal means that can be used to develop a working mathematical model of the vagueness phenomenon. The principal idea is based on introduction of degrees (of truth) taken from some scale that is further endowed by a specific algebraic structure.

The program of fuzzy logic in the broader sense (FLb) was announced in 1995 in [19]. FLb has been specified as an extension of FLn, whose main goal is to develop a model of natural human reasoning which is characteristic by its use of natural language. Therefore, a necessary constituent of FLb is a mathematical model of the meaning of specific expressions of natural language which should cover the so called evaluating linguistic expressions and predications, and generalized quantifiers. Another constituent of FLb are special inference rules enabling one to deal with these expressions. Below, we will demonstrate some of the achievements in this theory.

In this chapter, we will use the standard fuzzy set and fuzzy logic terminology and technique. Recall that by a fuzzy set A in the universe U we understand a function $A : U \longrightarrow [0,1]$ and often write $A \underset{\sim}{\subset} U$. The set of all fuzzy sets in the universe U is denoted by $\mathcal{F}(U)$. Let $\Box : [0,1] \times [0,1] \longrightarrow [0,1]$ be a binary operation on $[0,1]$ and $A \underset{\sim}{\subset} U$, $B \underset{\sim}{\subset} V$ be fuzzy sets. Then we can extend the operation $\Box$ to the operation $\Box : \mathcal{F}(U) \times \mathcal{F}(V) \longrightarrow \mathcal{F}(U \times V)$ on fuzzy sets pointwise by the formula

$$(A \Box B)(x,y) = A(x) \Box B(y), \qquad x \in U, y \in V. \tag{1}$$

2.2 Relevant Fuzzy Logics

As mentioned in the previous subsection, there are many kinds of fuzzy logics. This naturally raises the question: which fuzzy logic should we consider in our reasoning and applications. The answer is not easy, since it depends on the goal we want to reach. A general scheme presenting several distinguished fuzzy logics has been presented in [26]. With respect to the agenda of fuzzy logic that includes modeling of the vagueness phenomenon, parts of natural language semantics and human reasoning, we present a modified scheme in Fig. 1. The scheme has two parts: fuzzy logic in the narrow sense (FLn) and that in the broader sense (FLb). FLn is the ground for the preparation of necessary tools and it includes MTL as the initial logic. This raises IMTL, which preserves the contraposition (double negation) property essential for linguistic considerations and BL, the logic of continuous t-norms (continuity is essential for mathematical model of vagueness). Ł (Łukasiewicz logic) has properties of both previous logics and is very suitable for most kinds of deliberation about vagueness. All fuzzy logics are put together in ŁΠ whose only disadvantage is its great complexity.

All the mentioned logics have traditional syntax, which means that the syntax of classical logic is extended by some additional connectives but the concept of the inference rule and provability remain the same. $\mathrm{Ev}_{\text{Ł}}$ (fuzzy logic with evaluated syntax) stands aside, as it also introduces grades into syntax so that axioms that are not fully convincing can also be considered.

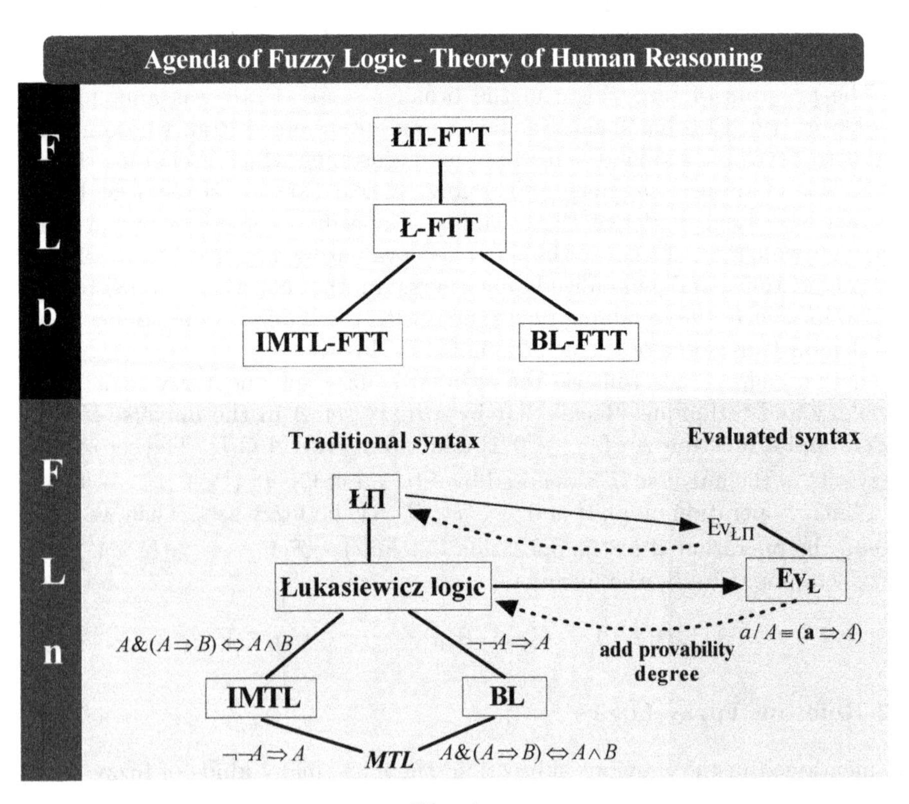

Fig. 1.

This logic extends the power of Ł (e.g. it enables the introduction of the concept of intension into first-order fuzzy logic — cf. [12], Chap. 6). The shift from traditional to evaluated syntax is possible only for Ł, and in a sense, also for ŁΠ (this requires a special infinitary rule). The dotted arrows back express representability in the corresponding logic without evaluated syntax (with some additional conditions).

As mentioned in the introduction, the paradigm of FLb is to provide a model of natural human reasoning in which natural language plays a crucial role. This is also the essential constituent of the agenda of fuzzy logic. For this purpose, extension of first-order fuzzy logic to higher order (fuzzy type theory) is necessary. The scheme demonstrates that this has been done for four of the above mentioned logics. Thus, it is conceivable to develop FLb on the basis of the corresponding fuzzy type theory.

2.3 Fuzzy Type Theory

A fundamental formal frame for FLb and the applications described below is fuzzy type theory. This is a higher order fuzzy logic extensively presented in [24] (see also [21,23]). It is a generalization of classical type theory described in details in [1]. In this chapter, we only briefly touch on some of the main points of Łukasiewicz style fuzzy type theory (denoted by Ł-FTT).

Truth Values

For developing fuzzy type theory, we must first clarify the structure of truth values to be used. In general, this can be one of the following: IMTL$_\Delta$-algebra, Łukasiewicz$_\Delta$-algebra, BL$_\Delta$-algebra or ŁΠ-algebra.

Since this chapter focuses on FLb and its ability to model (a part of) natural language semantics, we will only consider Łukasiewicz$_\Delta$-algebra, which is the algebraic structure

$$\mathcal{L} = \langle [0,1], \vee, \wedge, \otimes, \oplus, \Delta, \rightarrow, 0, 1\rangle \tag{2}$$

where

$$\begin{aligned} \wedge &= \text{minimum}, & \vee &= \text{maximum}, \\ a \otimes b &= 0 \vee (a + b - 1), & a \rightarrow b &= 1 \wedge (1 - a + b), \\ \neg a &= a \rightarrow 0 = 1 - a, & a \oplus b &= 1 \wedge |a - b|, \end{aligned}$$

$$\Delta(a) = \begin{cases} \mathbf{1} & \text{if } a = \mathbf{1}, \\ \mathbf{0} & \text{otherwise.} \end{cases}$$

Syntax

Each formula in Ł-FTT has a certain type which, in general, represents objects of special kind. Types are defined recurrently as follows.

Let ϵ, o be distinct objects. The set of types is the smallest set *Types* satisfying:

(i) $\epsilon, o \in \mathit{Types}$,
(ii) If $\alpha, \beta \in \mathit{Types}$ then $(\alpha\beta) \in \mathit{Types}$.

The type ϵ represents *elements* and o *truth values.* More complex types $\beta\alpha$ represent functions from a set of type α to a set of type β (note the reverse order!).

A set of formulas of type $\alpha \in \mathit{Types}$, denoted by Form_α, is the smallest set satisfying:

(i) Variables $x_\alpha \in \mathit{Form}_\alpha$ and constants $c_\alpha \in \mathit{Form}_\alpha$,
(ii) if $B \in \mathit{Form}_{\beta\alpha}$ and $A \in \mathit{Form}_\alpha$ then $(BA) \in \mathit{Form}_\beta$,
(iii) if $A \in \mathit{Form}_\beta$ then $\lambda x_\alpha\, A \in \mathit{Form}_{\beta\alpha}$,

If $A \in \mathit{Form}_\alpha$ is a formula of the type $\alpha \in \mathit{Types}$ then we write A_α. The symbol λ is a lambda abstractor, which is used for construction of new functions.

Special constants of Ł-FTT are $\mathbf{E}_{(o\alpha)\alpha}$ (*equality*), $\mathbf{C}_{(oo)o}$ (*conjunction*) and $\mathbf{D}_{oo}$ (*Baaz delta*) and $\iota_{\epsilon(o\epsilon)}, \iota_{o(oo)}$ (*description operators*).

Let us remark that just as in classical type theory, an essential connective in Ł-FTT is the fuzzy equality $\equiv$. Its interpretation depends on the given type. For example, if the type is o (truth values) then $\equiv$ becomes equivalence.

The following are basic definitions of symbols of Ł-FTT:

(a) *Fuzzy equivalence/fuzzy equality* $\equiv := \lambda x_\alpha\, (\lambda y_\alpha\, \mathbf{E}_{(o\alpha)\alpha}\, y_\alpha) x_\alpha$.
(b) *Conjunction* $\wedge := \lambda x_o\, (\lambda y_o\, \mathbf{C}_{(oo)o}\, y_o) x_o$.
(c) *Baaz delta* $\boldsymbol{\Delta} := \lambda x_o\, \mathbf{D}_{oo}\, x_o$.
(d) *Representation of truth* $\top := (\lambda x_o\, x_o \equiv \lambda x_o\, x_o)$
(e) *Representation of falsity* $\bot := (\lambda x_o\, x_o \equiv \lambda x_o\, \top)$.
(f) *Negation* $\neg := \lambda x_o\, (\bot \equiv x_o)$.
(g) *Implication* $\Rightarrow\ := \lambda x_o\, (\lambda y_o\, ((x_o \wedge y_o) \equiv x_o))$.
(h) *Special connectives*

$$\begin{aligned} \vee &:= \lambda x_o(\lambda y_o(x_o \Rightarrow y_o) \Rightarrow y_o), && \textit{(disjunction)} \\ \mathbf{\&} &:= \lambda x_o(\lambda y_o(\neg(x_o \Rightarrow \neg y_o))), && \textit{(strong conjunction)} \\ \nabla &:= \lambda x_o(\lambda y_o(\neg(\neg A_o \,\&\, \neg B_o))). && \textit{(strong disjunction)} \end{aligned}$$

(i) *General quantifier* $(\forall x_\alpha) A_o := (\lambda x_\alpha\, A_o \equiv \lambda x_\alpha\, \top)$.
(j) *Existential quantifier* $(\exists x_\alpha) A_o := \neg(\forall x_\alpha)\neg A_o$.

The syntax of Ł-FTT consists of definitions of fundamental formulas, axioms and inference rules. The Ł-FTT has 17 axioms that may be divided into the following subsets: fundamental equality axioms, truth structure axioms, quantifier axioms and axioms of descriptions.

The following is the list of all axioms:

(FT1) $\boldsymbol{\Delta}(x_\alpha \equiv y_\alpha) \Rightarrow (f_{\beta\alpha}\, x_\alpha \equiv f_{\beta\alpha}\, y_\alpha)$

(FT2$_1$) $(\forall x_\alpha)(f_{\beta\alpha}\, x_\alpha \equiv g_{\beta\alpha}\, x_\alpha) \Rightarrow (f_{\beta\alpha} \equiv g_{\beta\alpha})$

(FT2$_2$) $(f_{\beta\alpha} \equiv g_{\beta\alpha}) \Rightarrow (f_{\beta\alpha}\, x_\alpha \equiv g_{\beta\alpha}\, x_\alpha)$

(FT3) $(\lambda x_\alpha B_\beta)A_\alpha \equiv C_\beta$ where C_β is obtained from B_β by replacing all free occurrences of x_α in it by A_α, provided that A_α is substitutable to B_β for x_α (*lambda conversion*).

(FT4) $(x_\epsilon \equiv y_\epsilon) \Rightarrow ((y_\epsilon \equiv z_\epsilon) \Rightarrow (x_\epsilon \equiv z_\epsilon))$

(FT5) $(A_o \equiv \top) \equiv A_o$

(FT6) $A_o \Rightarrow (B_o \Rightarrow A_o)$

(FT7) $(A_o \Rightarrow B_o) \Rightarrow ((B_o \Rightarrow C_o) \Rightarrow (A_o \Rightarrow C_o))$

(FT8) $(\neg B_o \Rightarrow \neg A_o) \equiv (A_o \Rightarrow B_o)$

(FT9) $(A_o \vee B_o) \equiv (B_o \vee A_o)$

(FT10) $A_o \wedge B_o \equiv B_o \wedge A_o$

(FT11) $A_o \wedge B_o \Rightarrow A_o$

(FT12) $(A_o \wedge B_o) \wedge C_o \equiv A_o \wedge (B_o \wedge C_o)$

(FT13) $(g_{oo}(\boldsymbol{\Delta} x_o) \wedge g_{oo}(\neg\boldsymbol{\Delta} x_o)) \equiv (\forall y_o) g_{oo}(\boldsymbol{\Delta} y_o)$

(FT14) $\boldsymbol{\Delta}(A_o \wedge B_o) \equiv \boldsymbol{\Delta} A_o \wedge \boldsymbol{\Delta} B_o$

(FT15) $(\forall x_\alpha)(A_o \Rightarrow B_o) \Rightarrow (A_o \Rightarrow (\forall x_\alpha)B_o)$ where x_α is not free in A_o.

(FT16) $\iota_{\alpha(o\alpha)}(\mathbf{E}_{(o\alpha)\alpha}\, y_\alpha) \equiv y_\alpha, \qquad \alpha = o, \epsilon,$
where $\mathbf{E}_{(o\alpha)\alpha}$ is a special constant representing fuzzy equality.

There are two inference rules in Ł-FTT, namely

(R) *Let $A_\alpha \equiv A'_\alpha$ and $B \in Form_o$. Then, infer B' where B' comes from B by replacing one occurrence of A_α, which is not preceded by λ, by A'_α.*

(N) *Let $A_o \in Form_o$. Then, from A_o infer $\boldsymbol{\Delta} A_o$.*

A theory T is a set of formulas of type o (determined by a subset of special axioms, as usual). Provability is defined as usual. Inference rules of *modus ponens* and *generalization* are derived rules of Ł-FTT. It can be proved that all formulas provable in first-order Łukasiewicz logic are also provable in Ł-FTT.

Semantics

A further step in defining semantics of Ł-FTT is to consider a *basic frame*. This is a system of sets

$$(M_\alpha)_{\alpha \in Types} \tag{3}$$

assigned to each type $\alpha \in$ *Types* such that $M_o = L$ is a set of truth values, M_ϵ is an arbitrary set of objects and further, if $\gamma = \beta\alpha$ then $M_\gamma \subseteq M_\beta^{M_\alpha}$, i.e. the corresponding set $M_{\beta\alpha}$ contains (not necessarily all) functions $f : M_\alpha \longrightarrow M_\beta$.

To define interpretation of formulas of Ł-FTT, we need the concept of *fuzzy equality* (see, e.g. [2, 6, 14] and many other authors) which interprets the connective $\equiv$. In general, it is a a fuzzy relation $=_\alpha$: $M_\alpha \times M_\alpha \longrightarrow L$ considered for each type $\alpha \in$ *Types*, which has the following properties:

$$[x =_\alpha x] = 1, \qquad \text{(reflexivity)}$$

$$[x =_\alpha y] = [y =_\alpha x], \qquad \text{(symmetry)}$$

$$[x =_\alpha y] \otimes [y =_\alpha z] \leq [x =_\alpha z] \qquad \text{(transitivity)}$$

where $[x =_\alpha y] \in L$ denotes a truth value of the formula $x =_\alpha y$, i.e. an equality between the objects $x, y \in M_\alpha$.

There are numerous examples of fuzzy equalities w.r.t. specific structures of truth values. A typical fuzzy equality in $M_\epsilon = \mathbb{R}$ for Łukasiewicz algebra is

$$[x =_\epsilon y] = 1 - (1 \wedge |x - y|), \qquad x, y \in \mathbb{R}.$$

A *frame* is a tuple

$$\mathcal{M} = \langle (M_\alpha, =_\alpha)_{\alpha \in \mathit{Types}}, \mathcal{L}_\Delta \rangle \tag{4}$$

such that the following holds:

(i) The $\mathcal{L}_\Delta$ is a structure of truth values (in our case, the Łukasiewicz$_\Delta$ algebra),

(ii) $=_\alpha$ is a fuzzy equality on M_α and $=_\alpha \in M_{(o\alpha)\alpha}$ for every $\alpha \in$ *Types*.

A precise definition of interpretation of formulas of Ł-FTT is described in [24]. We will only recall that if A_α is a formula of type α then its interpretation in the frame $\mathcal{M}$ is an element of the set M_α, i.e.

$$\mathcal{I}^{\mathcal{M}}(A_\alpha) \in M_\alpha. \tag{5}$$

This means that, if $\alpha = o$ (truth value) then $\mathcal{I}^{\mathcal{M}}(A_o) \in [0, 1]$ is a truth value. If $\alpha = \epsilon$ then $\mathcal{I}^M(A_\epsilon) \in M_\epsilon$. Finally, if $\alpha = \beta\gamma$ then interpretation of $A_{\beta\gamma}$ is a function $\mathcal{I}^M(A_{\beta\gamma}) \in M_\beta^{M_\gamma}$. Moreover, each such function must comply with the corresponding fuzzy equalities $=_\beta, =_\gamma$ determined in the frame $\mathcal{M}$, so that axioms (FT1) and (FT2) are fulfilled (for the details — see [24]).

A *general model* is a frame such that (5) is assured in it, i.e. for each formula A_α there is an element in the corresponding set M_α of the frame that interprets it. If T is a theory of Ł-FTT then $T \models A_o$ means that $\mathcal{I}^{\mathcal{M}}(A_o) = 1$ in each general model $\mathcal{M}$ of T.

The proof of the following theorem can be found in [24].

Theorem 1 (Completeness).

(a) A theory T of fuzzy type theory is consistent iff it has a general model $\mathcal{M}$.
(b) For every theory T and a formula A_o,

$$T \vdash A_o \quad \textit{iff} \quad T \models A_o.$$

3 Theory of Evaluating Linguistic Expressions

3.1 Importance of Evaluating Expressions

Evaluating linguistic expressions (or, simply, evaluating expressions) are special expressions of natural language such as *small, medium, big, about twenty five, roughly one hundred, very short, more or less deep, not very tall, roughly warm or medium hot, quite roughly strong, roughly medium size*, and many others. They form a small but very important part of natural language and are present in its everyday use any time. The reason is that people very often need to evaluate phenomena around them. Furthermore, they often make important decisions in situations described using evaluating expressions or use these expressions, e.g. when learning to drive or to control a specific complex process, and in many other activities. Therefore, evaluating expressions constitute a very important subject for our study.

The role of evaluating expressions in natural language and possibilities for their use in technical applications were first pointed out by L. A. Zadeh (see [36]). His proposal to model the meaning of evaluating expressions can be ranked among the most important contributions of fuzzy logic for the practice. Among applications of their theory are fuzzy control, decision making, classification and various industrial applications. Most of them have been obtained using fuzzy IF-THEN rules as special expressions of natural language consisting of evaluating ones.

3.2 Grammatical Structure of Evaluating Expressions

The grammatical structure of evaluating expressions is fairly simple and can be characterized as follows:

(i) *Simple evaluating expressions*:
 (a) ⟨pure evaluating expression⟩ :=
  ⟨linguistic hedge⟩⟨atomic evaluating expression⟩
 (b) ⟨fuzzy quantity⟩ := ⟨linguistic hedge⟩⟨numeral⟩
 ⟨numeral⟩ – a name of an element from the scale
 ⟨linguistic hedge⟩ := ⟨empty⟩ | ⟨narrowing adverb⟩ |
  ⟨widening adverb⟩ | ⟨specification adverb⟩

(ii) *Negative evaluating expressions*

$$\mathsf{not}\ \langle\text{pure evaluating expression}\rangle$$

(iii) *Compound evaluating expressions*
  (a) ⟨pure evaluating expression⟩ or ⟨pure evaluating expression⟩
  (b) ⟨pure evaluating expression⟩ and/but⟨negative evaluating expression⟩

Example 1. Examples of atomic evaluating expressions are, e.g. *small, medium, big short, medium long, long, light, medium heavy, heavy,* etc. Hedges can be classified as those with *narrowing effect,* for example *very, significantly, extremely,* etc., *widening effect,* for example *roughly, more or less, quite roughly,* etc. and *mixed,* for example *rather, approximately,* etc.

Examples of simple evaluating expressions are, e.g. *very short, more or less strong, more or less medium, roughly big, about twenty five,* etc. Fuzzy numbers are *twenty five, roughly one hundred.* Examples of negative evaluating expressions are, e.g. *not big, not [very short], not [extremely strong],* etc. Examples of compound evaluating expressions are, e.g. *roughly small or medium, small but not very (small),* and others.

Evaluating expressions form *antonyms*

$$\langle\text{nominal adjective}\rangle \longleftrightarrow \langle\text{antonym}\rangle,$$

for example "young — old"; "ugly — nice"; "stupid — clever", etc. These expressions together with the middle term then form a so called *fundamental evaluating trichotomy*

$$\langle\text{nominal adjective}\rangle \text{ — } \langle\text{middle member}\rangle \text{ — } \langle\text{antonym}\rangle,$$

for example, "young — medium age — old"; "ugly — normal — nice"; "stupid — medium intelligent — clever", etc.

Evaluating expressions occur in *evaluating predications* that are expressions of the form:

(i) '⟨noun⟩ is $\mathcal{A}$ ' where $\mathcal{A}$ is an evaluating expression.
(ii) *Compound evaluating predications:*
  (a) IF$\langle\text{noun}\rangle_1$ is $\mathcal{A}$ THEN $\langle\text{noun}\rangle_2$ is $\mathcal{B}$,
  (b) $\langle\text{noun}\rangle_1$ is $\mathcal{A}$ AND $\langle\text{noun}\rangle_2$ is $\mathcal{B}$,
  (c) $\langle\text{noun}\rangle_1$ is $\mathcal{A}$ OR $\langle\text{noun}\rangle_2$ is $\mathcal{B}$.

Example 2. Examples of evaluating predications are *temperature is very high, size is roughly medium, sea is rather deep,* etc.

Compound evaluating predications are, for example, *IF distance is very long THEN fuel consumption is more or less high, low temperature AND roughly medium pressure, nice book OR good movie,* and many others.

A noun (or a noun phrase) is usually a name of some feature (characteristics) of objects, or just an object, evaluated by the given evaluating expression. For various kinds of applications, however, the concrete feature is unimportant. Therefore, '⟨noun⟩' is quite often replaced by a variable X.

A special case of linguistic expressions are *fuzzy IF-THEN rules.* These are compound evaluating predications characterizing some kind of dependence between characteristics. Here, "dependence" refers to the phenomenon when values of one characteristic influence values of the other one. Since the "OR" connective cannot express mutual dependence, we can think of a a fuzzy IF-THEN rule as one of the following expressions:

$$\mathcal{R}^I := \mathsf{IF}\ X \text{ is } \mathcal{A}\ \mathsf{THEN}\ Y \text{ is } \mathcal{B}, \tag{6}$$

$$\mathcal{R}^A := X \text{ is } \mathcal{A}\ \mathsf{AND}\ Y \text{ is } \mathcal{B}. \tag{7}$$

Expression (7) characterizes the "positive part" of the dependence — objects Y having the property named by $\mathcal{B}$ depend on objects X having the property named by $\mathcal{A}$. Thus, (7) can be understood as a special part of (6) whose meaning is more general because it also takes into consideration situations when objects X *do not have* the property $\mathcal{A}$*.

3.3 Formalization of the Meaning of Evaluating Expressions

In this subsection, we will briefly describe a theory T^{Ev} which is a formal theory of Ł-FTT and which provides a model of the meaning of evaluating linguistic expressions. Since a detailed description of the formal theory would go out of the scope of this chapter, we will focus mainly on the semantics. All the details and proofs of theorems can be found in [27].

Basic Principles

The formal model of the meaning of evaluating expressions and predications is inspired by concepts developed during past centuries by philosophers and logicians W. Leibniz, G. Frege, R. Carnap, R. Quine and others. The following concepts (cf. [5, 16]) are fundamental for our exposition.

A *possible world* is a state of the world at a given time moment and place (particular context in which the linguistic expression is used). *Intension* of a linguistic expression, of a sentence, or of a concept, is identified with the property denoted by it. It leads to different truth values in various contexts but is invariant with respect to them. It is important to note that expressions $\mathcal{A}$ of natural language are, in general, names of intensions.

* Let us remark that quite often, people use the surface form (6) but in fact mean (7). This is the reason why the so called Mamdani-Assilian rules use the min connective as an operation joining membership degrees of the corresponding fuzzy sets

Extension is determined by an intension and it is a class of elements which fall into the meaning of a linguistic expression in the given context. Thus, extension depends on the particular context of use and changes whenever the context is changed. For example, the expression "high" is a name of an intension being a property of some feature of objects, i.e. of their height. Its meaning can be, for example, 30 cm when a beetle needs to climb a straw, 30 m for an electrical pylon, but 4 km or more for a mountain.

An extremely important feature of the meaning of most natural language expressions, and namely, of evaluating ones, is vagueness of their meaning. This is a consequence of the indiscernibility between objects that can be reasonably modeled in fuzzy logic (see [13]). Our goal is to develop a formal theory of the meaning of a part of natural language that could be used for various applications.

The theory presented below can be classified as part of the methodology introduced by L. A. Zadeh in his paper [37] and called *Precisiated Natural Language* (PNL). It is an attempt to develop a unified formalism for various tasks involving natural language propositions. A simple application of this methodology to economy has been published in [7]. Another, more sophisticated application is logical analysis of a detective story presented in [9, 10].

The theory of the meaning of evaluating expressions is built as a formal theory of *fuzzy type theory*. All the properties are formulated syntactically and can thus have various interpretations depending on the chosen model.

General Characteristics

The following are global characteristics of the meaning of evaluating expressions:

(A) Extensions of evaluating expressions are classes of elements taken from nonempty, linearly ordered and bounded scales. Each specific scale represents a *context* in which the given evaluating expression can be used. On each scale, three distinguished limit points can be determined: *left bound* v_L, *right bound* v_R, and a *central point* v_S.

(B) Each of the above limit points is a starting point of a horizon running towards the next limit point in the sense of the ordering and vanishing beyond. Thus, three horizons can be distinguished on each scale (see Fig. 3), namely *left horizon* running from v_L towards v_S, *right horizon* running from v_R toward v_S and *middle horizon* running from v_S to both sides. Each horizon is delineated by a reasoning analogous to the reasoning leading in classical logic to the Sorites paradox. Each position before the horizon is characterized by a truth value.

(C) Extension of any evaluating expression is delineated by a specific horizon resulting from a shift of the basic horizon determined in item (B) by modifying the truth values. The shift corresponds to a linguistic hedge and the modification is "small for big truth values" and "big for small ones".

(D) Each scale (context) is vaguely partitioned by the fundamental evaluating trichotomy consisting of a pair of antonyms, and a middle member (typically, "small, medium, big"). Any element of the scale is contained in extensions of at most two neighboring expressions.

The general situation for each context is depicted in Fig. 2. One can see that extension of the atomic evaluating expression modified using a hedge is changed (in basic cases, it is made smaller or larger).

As mentioned above, the theory of evaluating expressions is a formal theory of Ł-FTT. In this chapter, however, we have no means to describe it more precisely. Thus, we will demonstrate its features on the semantics.

Remark 1. Figure 2 also demonstrates the widespread mistake to model extensions of evaluating expressions using triangle membership functions forming a partition of some sort. Let us compare *extremely small* and *very small* and consider for a while that their extensions form triangles depicted in the lower part of Fig. 2. Then, e.g. 1m is surely an extremely small value in the given context, and of course, it is also very small as well as small. But when accepting the depicted triangles then it would mean that it is almost surely extremely small, but in *very little degree* very small and *not at all* small. But this is strictly counterintuitive and in contradiction with the way how people understand these expressions. The same also holds for big values. We conclude

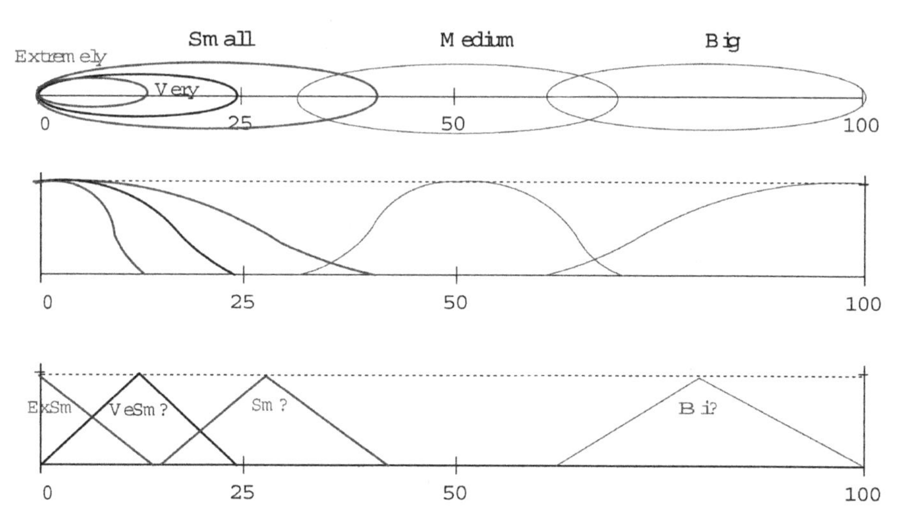

Fig. 2. Scheme characterizing extensions of evaluating expressions in the given context determined by the points $v_L = 0, v_S = 50, v_R = 100$. It also demonstrates incorrectness of triangle partition

that *partition of the context by triangles cannot be considered as a model of extensions of pure evaluating expressions*!

Mathematical Principles

Context

In our theory, the context is interpreted as a surjective non-decreasing function $w : [0,1] \longrightarrow M_\alpha$. This idea enables us to transfer certain properties from the algebra of truth values to the set M_α and thus, to make it a specific scale*. More explicitly, we suppose existence of an ordering $\preceq$ induced in the set $w([0,1]) \subseteq M_\alpha$ by in the implication: if $a \leq b$ then $w(a) \preceq w(b)$, $a, b \in [0,1]$ so that

$$w(0) = v_L, \qquad \text{(left bound)}$$
$$w(0.5) = v_S, \qquad \text{(central point)}$$
$$w(1) = v_R, \qquad \text{(right bound)}$$

where v_L and v_R are minimal and maximal elements w.r.t. $\preceq$. Consequently, $w([0,1]) = [v_L, v_R]$. In the sequel, we will write the contexts as triples $w = \langle v_L, v_S, v_R \rangle$. The set of all contexts will be denoted by W.

Horizon

An essential concept in our theory is that of *horizon*. Its formal properties are analogous to Sorites paradox (the detailed analysis can be found in see [12,13]). In the theory of evaluating expressions, horizon is modeled using a specific fuzzy equality $\sim$ on L, which induces a fuzzy equality $\approx_w$ in the given context $w \in W$ as follows:

$$[x \approx_w y] = [a \sim b]$$

where $a, b \in [0,1]$ are (unique) elements such that $x = w(a)$ and $y = w(b)$.

For example, in standard Łukasiewicz MV-algebra of truth values we can define

$$[x \sim y] = \frac{0 \vee (0.5 - |x - y|)}{0.5},$$

where $x, y \in [0,1]$.

The following are left, middle and right horizons defined in $[0,1]$ using the fuzzy equality $\sim$ and the corresponding horizons in the context $w \in W$ introduced using the induced fuzzy equality $\approx_w$:

---

* Given $x \in w([0,1])$, one must be careful with uniqueness of $a = w(x)$ to be assured that the mentioned transfer of properties works. This is achieved by using the so called description operator ι. For the details, see [27].

$$
\begin{aligned}
LH(a) &= [0 \sim a], & LH_w(x) &= [v_L \approx_w x], & x &= w(a), && \text{(left horizon)}\\
MH(a) &= [0.5 \sim a], & MH_w(x) &= [v_S \approx_w x], & x &= w(a), && \text{(middle horizon)}\\
RH(a) &= [1 \sim a], & RH_w(x) &= [v_R \approx_w x]. & x &= w(a) && \text{(right horizon)}
\end{aligned}
$$

for all $a \in [0, 1]$. The horizons are depicted in Fig. 3.

We will demonstrate our theory on the example of the original form of the Sorites paradox:

One grain does not form a heap. Adding one grain to what is not yet a heap does not make a heap. Consequently, there are no heaps.

Let $\mathbb{N}$ be a set of natural numbers. A heap is formed by a big number of stones. This means that any number of stones that is not big cannot form a heap. Since the concept of a heap is vague, the number of stones that cannot form a heap belongs to a fuzzy set $\mathbb{FN} \underset{\sim}{\subset} \mathbb{N}$. Note that we may also call $\mathbb{FN}$ a fuzzy set of *finite numbers* since the number of stones forming a heap is non-transparent and thus, a sort of infinite (P. Vopěnka speaks about *natural infinity*, cf. [34]). There is also a number $p \in \mathbb{N}$ (it lays somewhere beyond the horizon of finite numbers) such that any number $n \geq p$ of stones already forms a heap. This means that we may consider a context w_N in which $w_N(0)=0$, $w_N(0.5) = p$ and $w_N(1) = q$ where q is a number significantly bigger than $p^{\dagger}$.

The core of the resolution of Sorites paradox lays in the assumption that the implication $\mathbb{FN}(n) \Rightarrow \mathbb{FN}(n+1)$ is not fully true but only "almost true" since the shape of the given grouping of stones has imperceptibly changed and became a bit closer to a "heap". Let us put $\mathbb{FN}(n) = [0 \approx_{w_N} n]$. This can be interpreted as a truth value of the statement "n does not form a heap". The theorem below demonstrates that $\mathbb{FN}$ behaves according to the intuition. It is important to stress that its original formulation is syntactical and its (syntactical) proof can be found in [27].

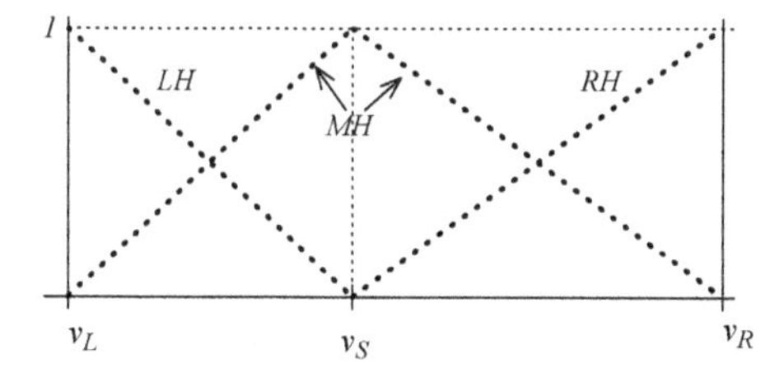

Fig. 3. Three horizons determining the meaning of pure evaluating expressions

† This may be some arbitrary number that is out of any specific considerations — billion, trillion or whatever. Clearly, this depends on the concrete situation. For example, when speaking about bricks, we may put $p = 1000$ and $q = 10^5$, for sand stones $p = 10^6$ and $q = 10^9$, etc

Theorem 2.

(a) $\mathbb{FN}(0) = 1$,
(b) $p \leq n \Rightarrow \mathbb{FN}(n) = 0$,
(c) $\mathbb{FN}(n+1) \leq \mathbb{FN}(n)$,
(d) There is no n *such that* $\mathbb{FN}(n) = 1$ *and* $\mathbb{FN}(n+1) = 1$,
(e) For all n, $\mathbb{FN}n \Rightarrow ((n \approx_{w_N} n+1) \Rightarrow \mathbb{FN}(n+1))$.

Let us comment this theorem in words: (a) expresses that 0 surely does not form a heap (it is surely a finite number); (b) says that each n behind the horizon of finite numbers already forms a heap; (c) says that the truth value of the statement "n does not form a heap" diminishes with the increase of n; (d) expresses the intuitive requirement that there is no abrupt change in the horizon, i.e. there is no specific n by which the horizon ends. In other words, there is no specific number n such that n stones do not form a heap but $n+1$ stones already form it. Finally, (e) explicitly expresses what "almost true" means. Namely, the formula $\mathbb{FN}n \Rightarrow ((n \approx_{w_N} n+1) \Rightarrow \mathbb{FN}(n+1))$ can be interpreted as "*it is almost true* that $\mathbb{FN}n$ implies $\mathbb{FN}(n+1)$". Note that this is in full accordance with the analysis contained in [13]. The *change of grouping of stones* after adding one stone to it is characterized by the formula $n \approx_{w_N} n+1$.

Hedges

The theory of linguistic hedges was first proposed by L. A. Zadeh [36] and then elaborated by several authors, e.g. [3, 15, 18, 35]. In our theory, we interpret hedges as horizon shifts (modifications). They are assigned special functions $\nu : [0,1] \longrightarrow [0,1]$ fulfilling the following conditions:

(a) There are $a, c \in [0,1]$ such that $a < c$, $\nu(a) = 0$ and $\nu(c) = 1$.
(b) For every $t, z \in [0,1]$, $t \leq z$ implies $\nu(t) \leq \nu(z)$.
(c) There is $b \in [0,1]$ such that $t \leq b$ implies $\nu(t) \leq t$, and $b \leq t$ implies $t \leq \nu(t)$.
(d) There is $b \in [0,1]$ such that $t \leq b$ implies $\nu(t) \leq t$, and $b \leq t$ implies $t \leq \nu(t)$.

Typical shapes of the function ν are depicted in Fig. 4. In each context, these functions are applied to the above introduced horizons to obtain extensions of the corresponding evaluating expression.

Fundamental Evaluating Trichotomy

The form and properties of the fundamental evaluating trichotomy in each context are clear from Figs. 2 and 5. Note that extensions of the expressions *small* — *medium* — *big* (which form the fundamental evaluating trichotomy)

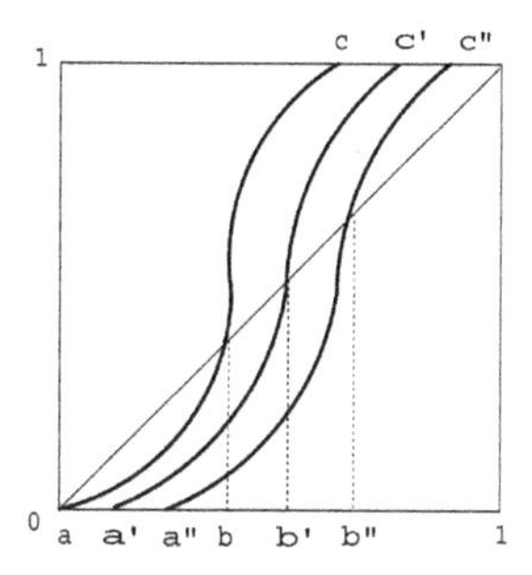

Fig. 4. Typical shapes of functions interpreting hedges

are pairwise overlapping, i.e. in each context there are elements that are both small (big) as well as medium, but no elements that are both small as well as big.

Construction of extensions of evaluating expressions is depicted in more detail in Fig 5. The hedge functions ν are depicted on the left side of the y-axis and turned 90 degrees counterclockwise.

Sharpness of Evaluating Expressions

An important property of evaluating expressions is a possibility to define partial ordering $\lessdot$ in the set of all of them which expresses sharpness of their meaning. This ordering is based on the natural (partial) ordering of linguistic hedges. Namely, hedges with a narrowing effect are sharper than those with a widening effect and the corresponding property holds for the pure evaluating expressions. We will not give a precise definition in this chapter and refer to Figs. 3 and 4 instead. For a precise definition, see [25, 27, 29].

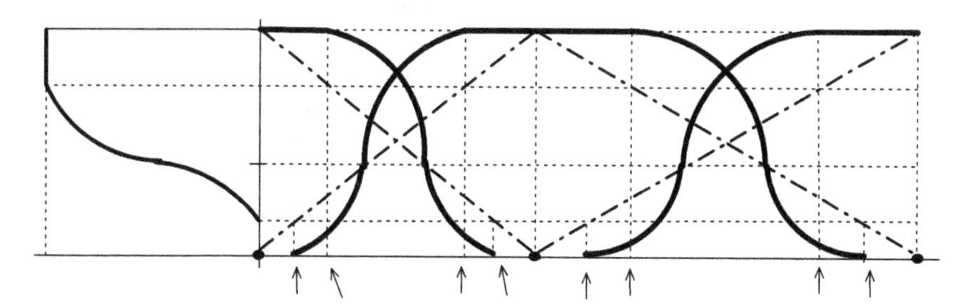

Fig. 5. Scheme of the construction of extension of evaluating linguistic expressions

Intension of Linguistic Expressions

In accordance with R. Carnap's [5] proposal, we will model intensions of linguistic expressions as functions from the set W of contexts into the set of fuzzy sets over a certain universe.

Intension of evaluating predications

Let 'X is $\mathcal{A}$' be an evaluating predication. Let us stress that this expression assigns a property represented by the evaluating expression $\mathcal{A}$ to elements of some set M_α. Therefore, we define its intension as a function

$$\mathrm{Int}(X \text{ is } \mathcal{A}) : \mathrm{W} \longrightarrow \mathcal{F}(\mathrm{M}_\alpha) \tag{8}$$

where W is the set of contexts (possible worlds) introduced above and M_α is a set of elements of a specific type. This means that, given a context $w \in W$, the extension of $\mathcal{A}$ in the context w is a fuzzy set

$$\mathrm{Ext}_w(X \text{ is } \mathcal{A}) = \mathrm{Int}(X \text{ is } \mathcal{A})(w) \underset{\sim}{\subset} w([0,1]). \tag{9}$$

Let $\langle$linguistic hedge$\rangle$ be assigned a function $\nu : [0,1] \longrightarrow [0,1]$ introduced above. Then we define intension of pure evaluating predications as functions

(i) $\mathrm{Int}(X$ is $\langle$linguistic hedge$\rangle$ small$) : w \mapsto \nu(LH_w)$ for all $w \in W$,
(ii) $\mathrm{Int}(X$ is $\langle$linguistic hedge$\rangle$ medium$) : w \mapsto \nu(MH_w)$ for all $w \in W$,
(iii) $\mathrm{Int}(X$ is $\langle$linguistic hedge$\rangle$ big$) : w \mapsto \nu(RH_w)$ for all $w \in W$.

It is clear from (9) that extensions of pure evaluating expressions in the context $w \in W$ are just the fuzzy sets $\nu(LH_w), \nu(MH_w), \nu(RH_w)$, respectively.

Example 3. Among the most often used linguistic hedges we can rank the following: *extremely (Ex), significantly (Si), very (Ve), more or less (ML), roughly (Ro), quite roughly (QR), very roughly (VR).* They can be used in each context for computing extensions of the following evaluating predications (*Sm* means *small*, *Me* means *medium* and *Bi* means *big*):

X is $Ex\,Sm$		X is $Ex\,Bi$
X is $Si\,Sm$		X is $Si\,Bi$
X is $Ve\,Sm$		X is $Ve\,Bi$
X is Sm	X is Me	X is Bi
X is $ML\,Sm$	X is $ML\,Me$	X is $ML\,Bi$
X is $Ro\,Sm$	X is $Ro\,Me$	X is $Ro\,Bi$
X is $QR\,Sm$	X is $QR\,Me$	X is $QR\,Bi$
X is $VR\,Sm$	X is $VR\,Me$	X is $VR\,Bi$

Possible concrete formulas for computing extensions of these expressions can be found in [22]. Shapes of all these extensions are depicted in Fig 6.

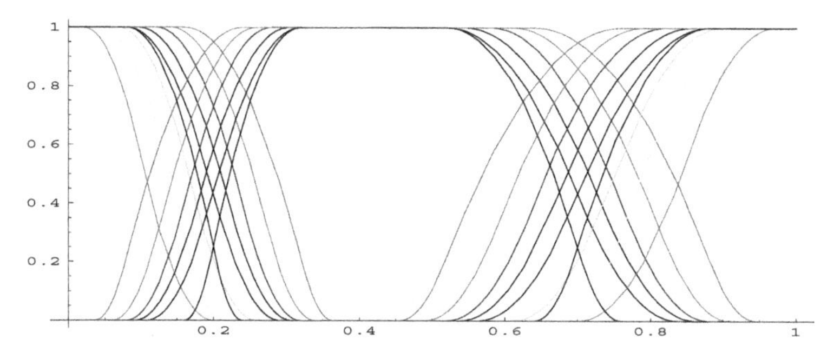

Fig. 6. Experimentally estimated extensions of selected evaluating predications in the context $\langle 0, 0.4, 1\rangle$

Intensions of pure evaluating expressions

A question that arises is, what is the semantical difference between pure evaluating expressions and predications? The clue is the elements. Namely, in the predication 'X is $\mathcal{A}$' we explicitly point the elements that have the given property in question while the pure expression $\mathcal{A}$ characterizes this property on universal unnamed elements. A proper set of such elements in our theory is the set of truth values $[0, 1]$. Consequently, our construction suggests the idea to take intension $\mathrm{Int}(\mathcal{A})$ as a function assigning a fuzzy set in the universe $[0, 1]$ to the context $\langle 0, 0.5, 1\rangle$. Since there is only one such context, we obtain $\mathrm{Int}(\mathcal{A}) = \mathrm{Ext}(\mathcal{A})$ and consequently

(i) $\mathrm{Int}(\langle\text{linguistic hedge}\rangle\ \mathsf{small}) = \nu(LH)$,
(ii) $\mathrm{Int}(\langle\text{linguistic hedge}\rangle\ \mathsf{medium}) = \nu(MH)$,
(iii) $\mathrm{Int}(\langle\text{linguistic hedge}\rangle\ \mathsf{big}) = \nu(RH)$.

Using the means outlined above it is possible to construct a model of the formal theory of evaluating expressions. By completeness of the fuzzy type theory, we immediately obtain the following theorem.

Theorem 3. *The formal theory of evaluating linguistic expressions is consistent.*

4 Fuzzy IF-THEN Rules as Expressions of Natural Language

4.1 Linguistic Description

One of the reasons why we take fuzzy IF-THEN rules (6) or (7) as genuine expressions of natural language is the goal of developing human-like robots, since it must be possible to communicate with them in natural language, i.e., to give them instructions, to teach them, etc. Hence, we must be able to capture linguistic meaning of these expressions. Of course, to describe a more complicated situation, one fuzzy IF-THEN rule is not sufficient. This leads us to the concept of linguistic description.

A *linguistic description* is a set $LD = \{\mathcal{R}_1, \ldots, \mathcal{R}_m\}$ of fuzzy IF-THEN rules

$$\begin{aligned}
\mathcal{R}_1 &= \mathsf{IF}\ X \text{ is } \mathcal{A}_1\ \mathsf{THEN}\ Y \text{ is } \mathcal{B}_1, \\
\mathcal{R}_2 &= \mathsf{IF}\ X \text{ is } \mathcal{A}_2\ \mathsf{THEN}\ Y \text{ is } \mathcal{B}_2, \\
&\ldots\ldots\ldots\ldots\ldots\ldots\ldots\ldots \\
\mathcal{R}_m &= \mathsf{IF}\ X \text{ is } \mathcal{A}_m\ \mathsf{THEN}\ Y \text{ is } \mathcal{B}_m.
\end{aligned}$$

From the linguistic point of view, a linguistic description can be understood as a specific kind of a (structured) text that we use to describe various situations or processes. On this basis and using new additional information we may realize control, decision-making, classification and many other applications. Some of them are described below.

An important linguistic phenomenon which must also be considered in our theory is topic-focus articulation. Namely, in natural language, each sentence is divided into two parts: topic and focus. *Topic* is a part of a sentence which presents a theme (what is spoken about). *Focus* is a part of a sentence, which conveys new information.

For example, let us consider a simple sentence "John is very clever man". This sentence has several readings depending on the topic-focus articulation, namely "JOHN is very clever" which means that very clever is (topic) just John (focus) and not anybody else. Similarly, we can have "John is VERY CLEVER man", i.e. a man John is (topic) very clever (focus) and not, e.g. very tall, etc.

In case of fuzzy IF-THEN rules, the situation is simpler since we can clearly distinguish between topic and focus. For example, "IF temperature is small THEN change of position of control cock is very big". The topic is *temperature is small* since this is the given situation — it precedes the action determined by the consequent *change of position of control cock is very big*[*].

We conclude that the topic of a fuzzy IF-THEN rule is its antecedent, i.e. the linguistic predication 'X is $\mathcal{A}$', while the focus is its consequent 'Y is $\mathcal{B}$'. It follows that we may also speak about topic and focus of the linguistic description LD. Explicitly we put

$$\begin{aligned}
\mathit{Topic}_{LD} &= \{\mathrm{Int}(X \text{ is } \mathcal{A}_i) \mid i = 1, \ldots, m\}, \\
\mathit{Focus}_{LD} &= \{\mathrm{Int}(Y \text{ is } \mathcal{B}_i) \mid i = 1, \ldots, m\}.
\end{aligned}$$

[*] Of course, in the living language we would say something like "we must turn the control cock very much to the right" but this is unnecessary dissecting of the problem since fuzzy IF-THEN rules are a sort of first formalization of such detailed sentences

4.2 Intension of a Fuzzy IF-THEN Rule

The intension of a fuzzy IF-THEN rule can be constructed from intensions of pure evaluating predications occurring inside it.

Let '$\mathcal{R} = \mathsf{IF}\ X$ is $\mathcal{A}$ $\mathsf{THEN}\ Y$ is $\mathcal{B}$' be a fuzzy IF-THEN rule. Furthermore, let $\text{Int}(X \text{ is } \mathcal{A}) : W \longrightarrow \mathcal{F}(M_\alpha)$ and $\text{Int}(Y \text{ is } \mathcal{B}) : W \longrightarrow \mathcal{F}(M_\beta)$ be intensions of the evaluating predications occurring in $\mathcal{R}$. Then intension of $\mathcal{R}$ is a function

$$\text{Int}(\mathcal{R}) := W \times W \longrightarrow \mathcal{F}(M_\alpha) \times \mathcal{F}(M_\beta) \tag{10}$$

which assigns to each couple $w, w' \in W$ a fuzzy relation

$$\langle w, w' \rangle \mapsto \text{Int}(\mathcal{A})(w) \,\Box\, \text{Int}(\mathcal{B})(w') \tag{11}$$

where $\Box$ is one of the operations $\rightarrow, \otimes, \wedge$ between fuzzy sets defined pointwise (cf. (1)). We will schematically write intension (11) in the form

$$\text{Int}(\mathcal{R}) = \text{Int}(X \text{ is } \mathcal{A}) \,\Box\, \text{Int}(Y \text{ is } \mathcal{B}) \tag{12}$$

where $\Box \in \{\Rightarrow, \&, \wedge\}$ is a suitable logical connective.

On the basis of this definition, it makes good sense to call $w \in W$ *local contexts* and the couples $\langle w, w' \rangle$, $w, w' \in W$ *general contexts*. This approach enables us to extend the concept of a general context to more complicated structures over W. Consequently, we may define intensions of fuzzy IF-THEN rules as *functions from the set of general contexts into a set of fuzzy fuzzy relations*. This definition is in good accordance with the original Carnap's concept of intension.

4.3 Perception-Based Logical Deduction

This is the principal method for derivation of a conclusion on the basis of a linguistic description. We will explain the main idea using the following example.

Example 4. Given a linguistic description

$$\mathcal{R}_1 := \mathsf{IF}\ X \text{ is } \textit{small}\ \mathsf{THEN}\ Y \text{ is } \textit{very big},$$
$$\mathcal{R}_2 := \mathsf{IF}\ X \text{ is } \textit{very big}\ \mathsf{THEN}\ Y \text{ is } \textit{small}.$$

Each rule provides us with a certain knowledge (related to the concrete application). We are able to *distinguish* between the rules despite the fact that their meaning is vague.

Let us now consider specific linguistic contexts: $w \in W$ for values of X, say, $w = \langle 150, 330, 600 \rangle$ (for example, temperature in some oven) and $w' \in W$ for values of Y, $w' = \langle 0, 36, 90 \rangle$ (for example, a turncock position in degrees). Then "small X" are values of X around 180–210 (and smaller) and "very big X" are values around, at least, 550 or higher. Similarly, "small Y" are values of Y around 8 (and smaller) and "very big Y" values around, at least, 80–85 (and higher).

Let a value $X = 180$ be given as an observation of X (this may be, e.g., a result of some measurement). To derive a conclusion on the basis of the given linguistic description, we must first test whether this value falls in its topic. Since the value of 180 is in the context $w = \langle 150, 330, 600 \rangle$ apparently *small*, we may apply knowledge determined by LD and because of rule $\mathcal{R}_1$, we expect a *very big* value of Y (a value around 85). Similarly, $X = 560$ which is *very big* leads to a value of Y around 8 due to the rule $\mathcal{R}_2$.

Perception and evaluation

By perception we will understand an evaluating expression assigned to the given value in the given context. The choice of perception, however, is not arbitrary and it also depends on the topic of the specified linguistic expression.

Let us return to Example 4. The topic of the given linguistic description is

$$Topic_{LD} = \{\mathrm{Int}(X \text{ is } small), \mathrm{Int}(X \text{ is } very\ big)\}.$$

Then the evaluating expression *small* that has been assigned to $X = 180$ in the given context is a *perception* of 180. Analogously, the evaluating expression *very big* is assigned to $X = 560$ and it is a *perception* of 560 (again in the given context).

More formally, we define a special function of *local perception*

$$LPerc^{LD} : M_\alpha \times W \longrightarrow Topic_{LD} \tag{13}$$

assigning to each value $x \in w$ intension

$$LPerc^{LD}(x, w) = \mathrm{Int}(X \text{ is } \mathcal{A}) \tag{14}$$

where $\mathrm{Int}(X \text{ is } \mathcal{A})$ is an intension of the *sharpest* evaluating predication 'X is $\mathcal{A}$' w.r.t. the ordering $\lessdot$ so that $x \in w$ is the *most typical* element for the extension $\mathrm{Ext}_w(X \text{ is } \mathcal{A})$. To be *typical* means that the membership degree $\mathrm{Ext}_w(X \text{ is } \mathcal{A})(x)$ is non-zero and greater than some reasonable threshold a^0 (we usually put $a^0 = 0.9$ or even $a^0 = 1$).

The definition of local perception (14) can be justified by the empirical finding that in the given context, each value can be classified by some evaluating expression. Since the expressions are more, or less specific, the most specific one gives the most precise information. If there is no evaluating expression being most specific and typical then (14) gives nothing. This concept of perception is depicted in Fig. 7. The observer stands at point v_L and observes a person approaching him/her. At each moment, he/she describes a distance of the person using an evaluating expression proper in the given context. The upper part of the figure depicts the extension of the corresponding evaluating expression.

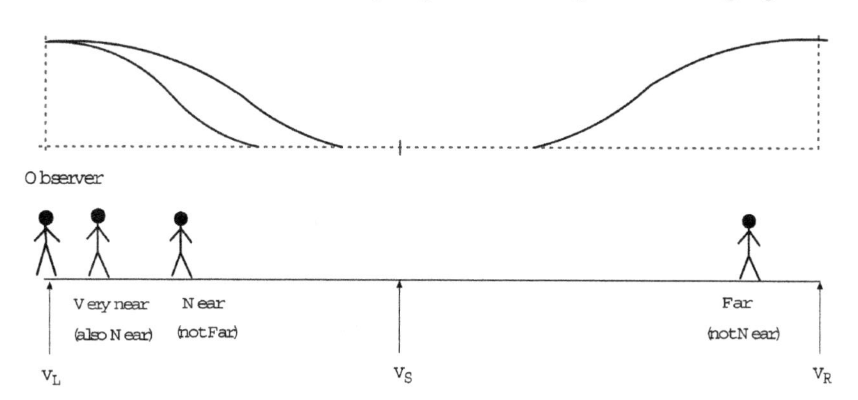

Fig. 7. Schematic characterization of perception

Note that the nearest position can be characterized by two different perceptions: *near* as well as *very near*. This means that we can distinguish between them but the second, sharper perception is included in the first one. Which of these possibilities will actually be used by the observer depends on the topic of his/her description. Note that our theory well captures this double feature of evaluating expressions.

The concept of perception is connected with the concept of evaluation. We say that an element $x \in w$ is *evaluated* by an expression $\mathcal{A}$, if there is a truth value $a \neq 0$ such that $a \rightarrow \mathrm{Ext}_w(X \text{ is } \mathcal{A})(x) = 1^\star$. If this is the case then we will formally write

$$Eval(x, w, \mathcal{A}). \tag{15}$$

Defuzzification of Evaluating Expressions

To deal with evaluating expressions, we must also be able to defuzzify them correspondingly. This means we must assign to extension $\mathrm{Ext}_w(X \text{ is } \mathcal{A})$ a proper element representing its meaning in the best way. . Therefore, a special method called Defuzzification of Evaluating Expressions (DEE) must be used. Its idea is to classify the given fuzzy set first and then find the most proper element. The use of DEE is depicted in Fig. 8. There also exists an improvement of this method, the so called *smooth DEE method*, which was described in [29].

Formalization of Perception-based Logical Deduction

The procedure of perception-based logical deduction can be formally characterized as follows (for the detailed explanation using FTT — see [25]).

Let a linguistic description $LD = \{\mathrm{Int}(\mathcal{R}_1), \ldots, \mathrm{Int}(\mathcal{R}_m)\}$ be given by intensions

* The $\rightarrow$ is the residuation operation in the algebra of truth values

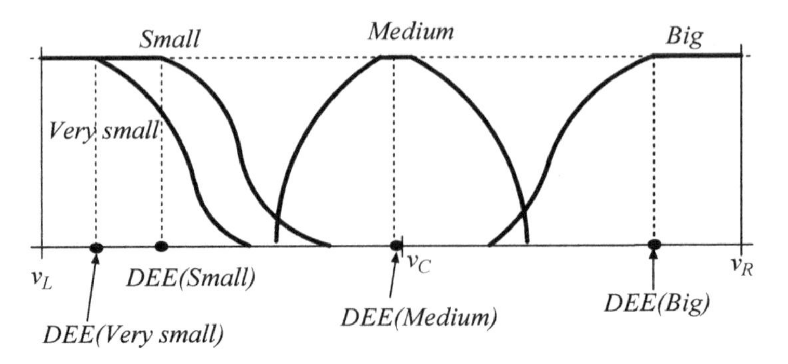

Fig. 8. Scheme of the DEE defuzzification method

$$\mathrm{Int}(\mathcal{R}_1) = \mathrm{Int}(X \text{ is } \mathcal{A}_1) \Rightarrow \mathrm{Int}(Y \text{ is } \mathcal{B}_1),$$

$$\cdots\cdots\cdots\cdots\cdots\cdots\cdots\cdots$$

$$\mathrm{Int}(\mathcal{R}_m) = \mathrm{Int}(X \text{ is } \mathcal{A}_m) \Rightarrow \mathrm{Int}(Y \text{ is } \mathcal{B}_m).$$

Furthermore, we will consider a context $w \in W$ for the variable X and $w' \in W$ for the variable Y. Let an observation $X = x_0 \in w$ in the context w be given. Using (14), we assign x_0 a perception $\mathrm{Int}(X \text{ is } \mathcal{A}_{i_0})$. On the basis of that the following *rule of perception-based deduction* is valid (see [25, 28]):

$$r_{PbLD} : \frac{LPerc^{LD}(x_0, w) = \mathrm{Int}(X \text{ is } \mathcal{A}_i), \quad LD}{Eval(\hat{y}, w', \mathcal{B}_i)} \tag{16}$$

where $\mathrm{Int}(X \text{ is } \mathcal{A}_i) \in Topic^{LD}$, $\mathrm{Int}(Y \text{ is } \mathcal{B}_i) \in Focus^{LD}$ and

$$\hat{y}_i = \mathrm{DEE}\left(\left\{\mathrm{Ext}_w(X \text{ is } \mathcal{A}_i)(x_0) \to \mathrm{Ext}_{w'}(Y \text{ is } \mathcal{B}_i)(y)/y \mid y \in w'\right\}\right). \tag{17}$$

Informal explanation of (16) is the following: from the perception 'X is $\mathcal{A}_i$' of the observation x_0 and the given linguistic description LD we derive a specific element $\hat{y}$ that is surely evaluated by the evaluating expression $\mathcal{B}_i$.

Note that $\mathrm{Ext}_w(X \text{ is } \mathcal{A}_i)(x_0) \to \mathrm{Ext}_{w'}(Y \text{ is } \mathcal{B}_i)(y)$, $y \in w'$ is a truth value of the statement

$$\text{'}X = x_0 \text{ is } \mathcal{A}_i\text{' implies '}Y = y \text{ is } \mathcal{B}_i\text{'} \tag{18}$$

which is a logical characterization of the dependence between X and Y characterized by the rule $\mathcal{R}_i$. The value $\hat{y}_i$ in (17) is the worst possible value of Y implied by x_0, which is evaluated by the expression $\mathcal{B}_i$ in the best possible way.

Implementation

This method has been implemented in the software system LFLC 2000 developed at the University of Ostrava in the Czech Republic (see [8]). Figure 9

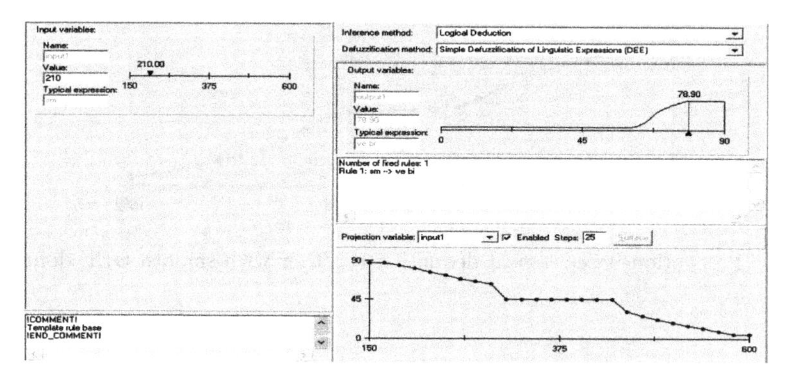

Fig. 9. Demonstration of perception-based logical deduction

shows a screen with the results of perception-based logical deduction run on the data from Example 4. Some of other results are presented in the table below

x_0	y_0
160	87
180	85.2
210	78.9
560	8.7
580	5.1
595	4.5

where input values can be interactively set in the line depicted in *left upper* part of Fig. 9 (the given observation is $x_0 = 210$). The *right upper* part shows the result of

$$\text{DEE}\left(\left\{\text{Ext}_w(X \textsf{ is small})(210) \rightarrow \text{Ext}_{w'}(Y \textsf{ is very big})(y)/y \mid y \in w'\right\}\right).$$

The *right lower* part shows the behavior of the outputs based on 2 rules from Example 4. Note that a decrease of small values of X leads to increase of Y and vice-versa. The reason is that the output of Y must be in correspondence with the truth degree of the statement

$$X = x_0 \textsf{ is very big}$$

by rule $\mathcal{R}_1$ or '$X = x_0$ is small' by rule $\mathcal{R}_2$. The greater this degree, the greater must be the truth of the statement '$Y = y_0$ is $\mathcal{B}_{i_0}$'. Consequently, if $\mathcal{B}_{i_0}$ has the form '⟨linguistic hedge⟩ small' then the value of y_0 decreases (it is still more and more true that y_0 is '⟨linguistic hedge⟩ small'). And vice-versa, if $\mathcal{B}_{i_0}$ has the form '⟨linguistic hedge⟩ big' then the value of y_0' increases. For comparison, we also present the result of PbLD when using the smooth DEE defuzzification in Fig. 10.

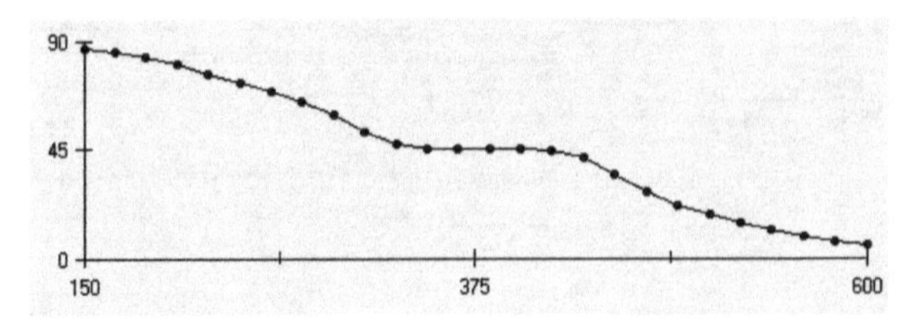

Fig. 10. Perception-based logical deduction together with smooth DEE defuzzification

In the rest of this chapter, we present several kinds of applications in which natural language played a crucial role. We have applied the above described model of the meaning of evaluating linguistic expressions and perception-based logical deduction.

5 Applications of The Model of Evaluating Expressions

5.1 Application to Geology

The application described below is a part of wider application of fuzzy logic in geology which was described in detail in [22]. A specific feature of it is the use of evaluating linguistic expressions for branching inside a specific algorithm.

Geological Problem

The geologist has at disposal a section illustrating changes in stacking patterns of limestone. Without going into the geological details, we present his problem in a simple abstract way as follows.

The given section consists of a set of sequences of 8 types of rocks in Fig. 11. Each of them is a result of one geological cycle during which the rocks are obtained by depositing a certain organic material. In the ideal situation, each sequence begins with type 1 and ends with type 8. The problem is, however, that the behavior of the nature is far from being ideal. Therefore, the data consist of sequences of the rock types that may have the form demonstrated in Table 1. The first, third, fifth and seventh columns of this table contain numbers representing rock types in the order given by the increasing drill depth. The second, fourth, sixth, and eighth columns contain the corresponding rock thickness (in m).

The Algorithm and Results

Our goal is to determine boundaries of the sequences according to expert rules formulated in natural language as follows:

1. Ends of sequences are usually rocks types 6, 7, or 8, if they are followed by rock type 1,2, or 3. If the given rock has lower number followed again by 6, 7, or 8 and it is *too thin* then it is ignored.

Table 1. Typical data with sequences of rock types

Rock type	Thick	Rock type	Thick	Rock type	Thick	Rock type	Thick
4	1.2	5	1.0	4	0.4	3	0.5
1	0.4	4	0.2	1	0.5	4	0.2
3	0.1	5	0.6	2	1.5	5	1.3
4	5.5	2	0.4	1	0.1	3	8.2
3	0.1	4	4.3	3	2.7	5	1.0
4	3.0	1	0.3	4	1.5	3	1.5
2	0.7	2	0.6	3	0.3	5	5.0
4	0.3	3	0.8	5	1.0	6	2.0

2. Check whether the obtained sequences are *sufficiently thick.* If the given sequence is *too thin* then it is joined with the following one, provided that the resulting sequence does not become *too thick.*
3. If the sequence is *too thick* then it is further divided: check all rock types 4 and mark them as ends of a new sequence provided that the new sequence is not *too thin*; mark the new sequence only if it is *sufficiently thick.*

We have tested two sets of the data from real drilling sections. The first data were 75 meters of section illustrating changes in stacking patterns of the Conococheague Limestone. It contained 226 rock units. The geologist determined 44 sequences from them, while our algorithm determined 47 sequences. Of these, 36 coincide with the geologist's solution, i.e. 81% agreement. The second data (Lower Ordovician El Paso Group observed in the Franklin Mountains of west Texas) contained 272 rock units. The geologist determined 99 rock sequences while our algorithm determined 97 sequences. Of these, 87 are coincident with the geologist's solution which is 88% agreement. The geologist confirmed that *our entire solution* was acceptable because the differences in evaluation of sequences were disputable and the geologist finally agreed with the results of our algorithm.

5.2 Application to Fuzzy Control

An important group of applications based especially on the above described perception-based logical deduction are applications in fuzzy control. Let us remind the reader of the original idea of fuzzy control: to utilize expert knowledge of the way of control in situations when the successful control is known but its precise description does not exist, or is too rough, or obtaining it is too expensive.

We have developed such a control and using LFLC 2000 realized several tens of simulations of various kinds of processes including 8 real applications (see [17]). In this subsection, we show simulation of control of inverted pendulum using a PD-fuzzy controller.

The context of variables is the following: *Error* (rad): $\langle 0, 0.52, 1.3\rangle$, *Change of error*: $\langle 0, 3.2, 8\rangle$, *Control action* (impuls): $\langle 0, 36, 90\rangle$.

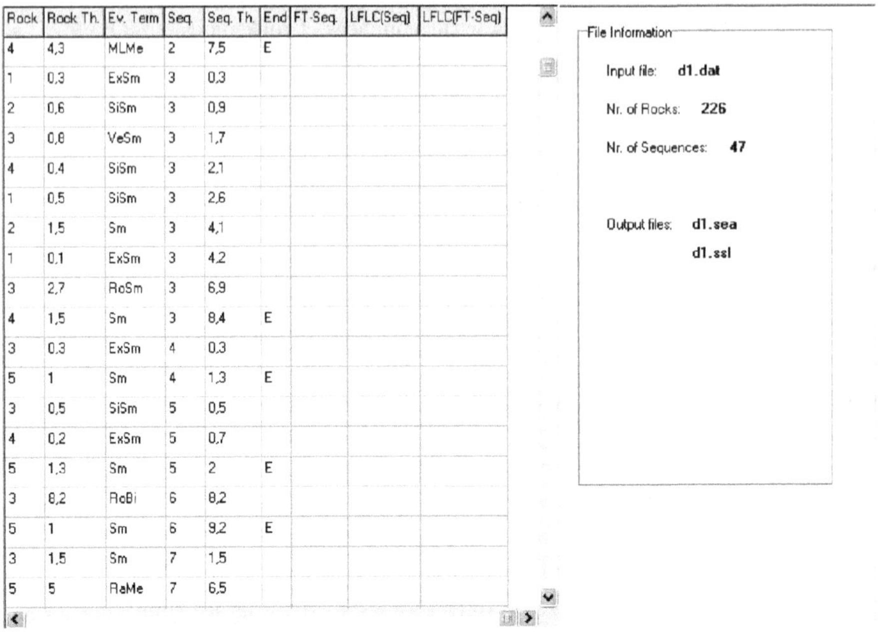

Rock	Rock Th.	Ev. Term	Seq.	Seq. Th.	End	FT-Seq.	LFLC(Seq)	LFLC(FT-Seq)
4	4,3	MLMe	2	7,5	E			
1	0,3	ExSm	3	0,3				
2	0,6	SiSm	3	0,9				
3	0,8	VeSm	3	1,7				
4	0,4	SiSm	3	2,1				
1	0,5	SiSm	3	2,6				
2	1,5	Sm	3	4,1				
1	0,1	ExSm	3	4,2				
3	2,7	RoSm	3	6,9				
4	1,5	Sm	3	8,4	E			
3	0,3	ExSm	4	0,3				
5	1	Sm	4	1,3	E			
3	0,5	SiSm	5	0,5				
4	0,2	ExSm	5	0,7				
5	1,3	Sm	5	2	E			
3	8,2	RoBi	6	8,2				
5	1	Sm	6	9,2	E			
3	1,5	Sm	7	1,5				
5	5	RaMe	7	6,5				

File Information

Input file: **d1.dat**

Nr. of Rocks: **226**

Nr. of Sequences: **47**

Output files: **d1.sea**

d1.ssl

Fig. 11. Results of rock sequences determination based on the theory of evaluating expressions. Ends of sequences are denoted by 'E'

The corresponding linguistic description is presented in Table 2. For example, Rule 12 should be read as

$$\text{IF } E \text{ is } +\textsf{small} \text{ AND } dE \text{ is } \textsf{roughly zero} \text{ THEN } U \text{ is } -\textsf{very small}.$$

One can see that the number of rules is rather high. This disadvantage, however, is balanced by many advantages, namely: the description is immediately understandable to everybody even after years; there is no necessity to know shapes of fuzzy sets*. Furthermore, the same description can be used for control of very different processes without any change except for the context of variables**.

* In fact, this linguistic description had been designed on pure linguistic basis when the expert had the meaning of rules in mind and did not know about fuzzy sets behind.

** We have a lot of concrete experiences with it. We have even developed an almost general linguistic description for PI-fuzzy control which can control several kinds of processes of completely different physical nature, from stable to non-stable ones. The only thing that must be specified in each process is the context of variables.

Table 2. Full linguistic description of a PD-fuzzy controller. The meaning of shorts is *Sm-small, Me-medium, Bi-big, Ze-zero, No-not, Ex-extremely, Si-significantly, Ve-very, ML-more or less, Ro-roughly, QR-quite roughly*

No.	E	dE ⇒	U	No.	E	dE ⇒	U
1.	+bi	+no ze	−bi	21.	ro ze	+ve sm	−si sm
2.	+bi	ro ze	−me	22.	ro ze	−ve sm	+si sm
3.	+bi	−sm	−me	23.	−no ze	−ex bi	+ex bi
4.	+bi	−no sm	ze	24.	−ve sm	+sm	−sm
5.	+me	+no sm	−bi	25.	−ve sm	+ve sm	ze
6.	+me	+sm	−me	26.	−sm	+no sm	−ve sm
7.	+me	ro ze	−me	27.	−sm	+sm	+ex sm
8.	+me	−sm	−ex sm	28.	−sm	+ve sm	+si sm
9.	+me	−no sm	+ve sm	29.	−sm	ro ze	+ve sm
10.	+sm	+no sm	−me	30.	−sm	−sm	+sm
11.	+sm	+sm	−sm	31.	−sm	−no sm	+me
12.	+sm	ro ze	−ve sm	32.	−me	+no sm	−ve sm
13.	+sm	−ve sm	−si sm	33.	−me	+sm	+ex sm
14.	+sm	−sm	−ex sm	34.	−me	ro ze	+me
15.	+sm	−no sm	+ve sm	35.	−me	−sm	+me
16.	+ve sm	+ve sm	−sm	36.	−me	−no sm	+bi
17.	+ve sm	−ve sm	ze	37.	−bi	+no sm	ze
18.	+ve sm	−sm	+sm	38.	−bi	+sm	+me
19.	+no ze	+ex bi	−ex bi	39.	−bi	ro ze	+me
20.	ze	ze	ze	40.	−bi	−no ze	+bi

Our concept of linguistic description has great learning abilities (see [4,20]). First of all, it is possible to automatically learn the context of variables. Furthermore, there exists an algorithm checking how near the controlled condition is to the set point. In case it is sufficiently near, the context is modified to obtain a more precise control. We speak about the *changing context* algorithm. It makes it possible to realize very precise control. Of course, the system also reacts on various kinds of disturbances so that the control remains successful.

A comparison of fuzzy control of an inverted pendulum with and without the context changing algorithm is demonstrated in Fig. 12. Besides the context, it is also possible to learn the linguistic description. For the details see [4].

If we do not insist on such precise control, it is possible to simplify the linguistic description from Table 2 to contain only 7 rules; the result is in Table 3. The corresponding fuzzy control of the inverted pendulum both with and without the context changing algorithm is demonstrated in Fig. 13.

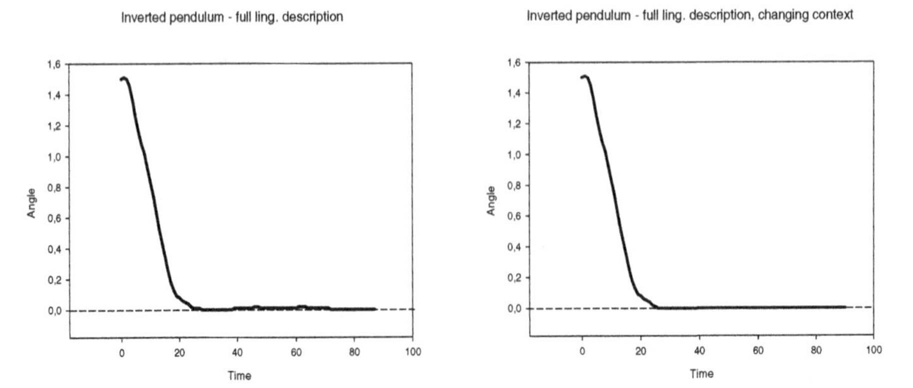

Fig. 12. Fuzzy control of inverted pendulum without and with change of the context of variables

Table 3. A simplified linguistic description of a PD-fuzzy controller

No.	E	dE⇒	U
1.	ze	ze	ze
2.	ro ze	−sm	+ve sm
3.	ro ze	+sm	−ve sm
4.	-no ze	−no ze	+me
5.	+no ze	+no ze	−me
6.	-no sm	+no ze	+sm
7.	+no sm	−no ze	−sm

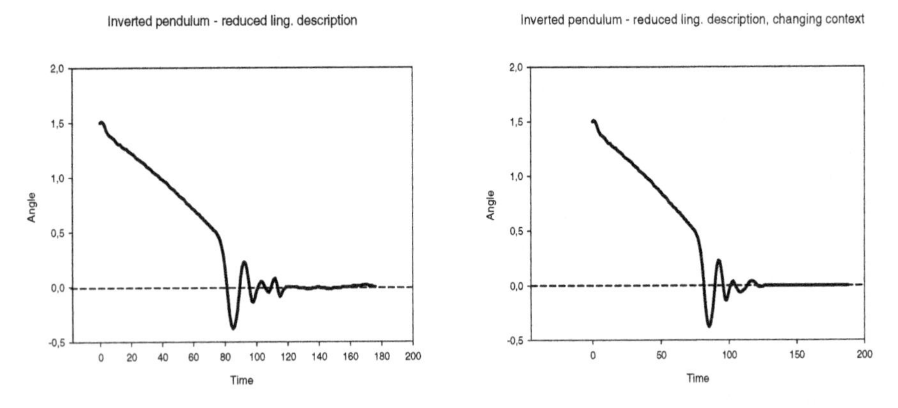

Fig. 13. Fuzzy control of inverted pendulum using a simplified linguistic description without and with change of the context of variables

5.3 Application to Decision-Making

General Approach

Other areas of applications where the linguistic approach has a great potential is decision making. The linguistic description can be used for an apt description of the decision situation when various degrees of fulfilment of the decision criteria, including their importance, can be distinguished sufficiently subtly and possible discrepancies can be overcome. Among advantages of this approach, we also see the fact that it is unnecessary to assign weights to the criteria. This is good because methods for weights assignment, though sophisticated, are rather intricate and still very subjective (though they are seemingly objective). Of course, linguistic formulation of the problem using fuzzy IF-THEN rules is subjective, as well. However, this subjectivity is directed and formulated using natural language which is understandable to everybody. Therefore, people (experts) may more easily find agreement on the optimal formulation.

A problem is that the rules in the linguistic description should not contain more than about 4–5 variables since a human being may hardly manage more of them at the same time. The solution is to form a multicriterial decision-making problem as a *hierarchical system* of linguistic descriptions.

Let $C_1, \ldots, C_n$ be criteria on the basis of which we are to decide about variants $v_1, \ldots, v_m$. Each criterion is associated with a scale that is used to measure the degree of its fulfillment by each variant. This is accomplished using the linguistic description consisting of the system of rules of the form

$$\begin{aligned} \mathcal{R}_1 &:= \text{IF } C_1 \text{ is } \mathcal{A}_{11} \text{ AND } \ldots \text{ AND } C_n \text{ is } \mathcal{A}_{1n} \text{ THEN } H \text{ is } \mathcal{B}_1 \\ &\cdots\cdots\cdots\cdots\cdots\cdots\cdots\cdots\cdots\cdots \\ \mathcal{R}_p &:= \text{IF } C_1 \text{ is } \mathcal{A}_{p1} \text{ AND } \ldots \text{ AND } C_n \text{ is } \mathcal{A}_{pn} \text{ THEN } H \text{ is } \mathcal{B}_p \end{aligned} \tag{19}$$

where $C_1, \ldots, C_n$ are the criteria, H is a global evaluation of variants and $\mathcal{A}_{ji}, \mathcal{B}_{ji}$, $j = 1, \ldots, m$, $i = 1, \ldots, p$ are the above discussed evaluating linguistic expressions. The global evaluation is usually a spaceless characteristics with the context $\langle 0, 0.4, 1 \rangle$. The value $H = 1$ means that the given variant is the best while $H = 0$ means that it is the worst.

Since the number n can be large, such a linguistic description can hardly be formed directly by experts and even if it could be, it would hardly be understandable. Therefore, we will formally divide the criteria $C_1, \ldots, C_n$ into r groups $H_1, \ldots, H_r$. Let us denote

$$\mathcal{H}_k = \{C_{k1}, \ldots, C_{kn(k)}\}, \qquad k = 1, \ldots, r.$$

Then the above introduced linguistic description can be transformed into a hierarchical system of linguistic descriptions

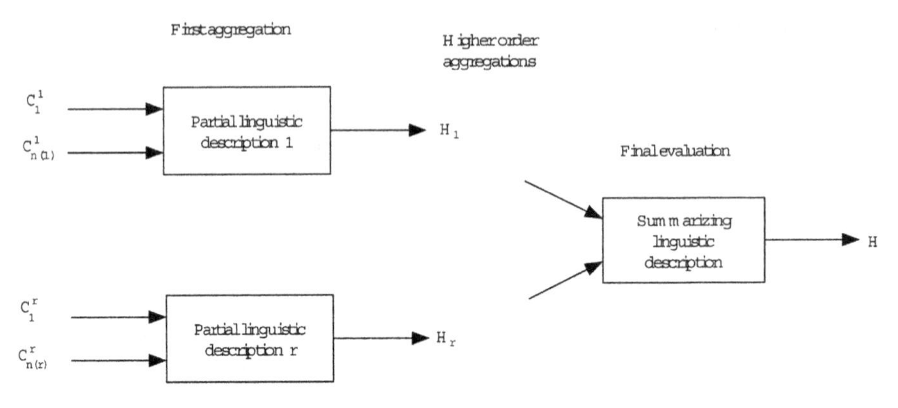

Fig. 14. Scheme of hierarchical fuzzy decision-making system.

$$
\begin{aligned}
\mathcal{R}_{1k} &= \text{IF } C_{k1} \text{ is } \mathcal{A}_{11} \text{ AND } \dots \text{ AND } C_{kn(k)} \text{ is } \mathcal{A}_{1n(k)} \text{ THEN } H_k \text{ is } \mathcal{B}_1 \\
& \quad \dots\dots\dots\dots\dots\dots\dots\dots\dots\dots\dots\dots\dots\dots\dots\dots \qquad (20) \\
\mathcal{R}_{1p(k)} &= \text{IF } C_{k1} \text{ is } \mathcal{A}_{p(k)1} \text{ AND } \dots \text{ AND } C_{kn(k)} \text{ is } \mathcal{A}_{p(k)n(k)} \\
& \qquad\qquad \text{THEN } H_k \text{ is } \mathcal{B}_{p(k)}
\end{aligned}
$$

$k = 1, \dots, r$ and

$$
\begin{aligned}
\mathcal{R}_1 &= \text{IF } H_1 \text{ is } \mathcal{A}_{11} \text{ AND } \dots \text{ AND } H_r \text{ is } \mathcal{A}_{1r} \text{ THEN } H \text{ is } \mathcal{B}_1 \\
& \quad \dots\dots\dots\dots\dots\dots\dots\dots\dots\dots\dots\dots\dots\dots\dots\dots \qquad (21) \\
\mathcal{R}_s &= \text{IF } H_1 \text{ is } \mathcal{A}_{s1} \text{ AND } \dots \text{ AND } H_r \text{ is } \mathcal{A}_{sr} \text{ THEN } H \text{ is } \mathcal{B}_s.
\end{aligned}
$$

The last description provides a final evaluation from which the proper decision is made. The situation is graphically demonstrated in Fig. 14.

The final decision is made using the system of linguistic descriptions (20), (21) (they replace the linguistic description (19)) for each variant v_j, $j = 1, \dots, m$ after inserting the corresponding values of the criteria C_i, $i = 1, \dots, n$ as observations into them. The value of the resulting global evaluation H provides ranking of the variants (the higher the value of H, the better the given variant).

Demonstration

Let us demonstrate this approach on an example, which was simulated using the software system LFLC 2000.

> *A small company has to decide about leasing a new car since the old car costs are still increasing and its reliability is decreasing. The decision is influenced by economical development of the company and several characteristics, which are very difficult to quantify.*

Variants to be decided are a *car type and its leasing offer*, namely Toyota Avensis, Toyota Yaris and Volvo with 25% or 50% advance payment offered by four companies. The decision is influenced by the following characteristics:

1. Economical development of the company over 4 years. For its analysis, we will use the F-transform technique (see I. Perfilieva [33]).
2. Old car costs, representability of a new car (creditworthiness of the company).
3. Offer of a leasing company: This will be evaluated from various points of view. The following information will be of special importance:
   (a) Trustworthiness of a leasing company.
   (b) Additional offers, for example whether the company also offers insurance as part of payments, etc.
   (c) Advance payment, monthly payment and total surplus.

The decision situation is characterized by hierarchical subtasks, as described below.

Evaluation of the Economical Development

Economical development of the company is characterized by a month-time series $(z_t)_{t \in T_M}$ of economical results (e.g. a gain) where each z_t can be split into trend r_t and remainder q_t, i.e.

$$z_t = r_t + q_t, \qquad t \in T_M.$$

The global trend $(r_t)_{t \in T_M}$ is analyzed using the F-transform technique. The economical development of the company will be evaluated using a linguistic description in Table 4 which takes into account interannual trend differences, i.e.

$$\Delta r_t = r_t - r_{t-1}, \qquad t \in T_R$$

where T_R is the year periodicity derived from month one. The economical development is characterized on the basis of the last three inter-annual trend

Table 4. Linguistic description for evaluation of the economical development

	Δr_{t-2}	Δr_{t-1}	Δr_t	$\Rightarrow$ V
1	+Bi	+Bi	+Bi	VeBi
2	+NoSm	+NoSm	+QRBi	Bi
3	+Sm	+Sm	+Sm	Me
4	−NoBi	RoZe	+Sm	RoSm
5	RoZe	RoZe	RoZe	VeSm
6	−NoZe	−NoZe	−NoSm	Ze
7	+Me	+Me	+Me	RoBi
8	−NoZe	−NoZe	+Sm	VeSm
9	+Bi	+Me	+Sm	MLBi

Table 5. Linguistic description for evaluation of the leasing offer

	R	A	C	⇒ N
1	Sm	Sm	+Sm	Bi
2	Sm	Me	+Sm	RaBi
3	NoBi	Sm	+Me	Me
4	Bi	Bi	+Bi	VeSm
5	NoBi	RoMe	+RoSm	RoBi
6	Me	Me	+Me	MLSm
7	VeSm	Me	+Sm	VeBi
8	Sm	Bi	−NoZe	VeBi
9	Sm	RoMe	+RoMe	MLBi
10	Sm	Bi	+Sm	RoSm
11	MLMe	VeBi	+VeSm	VeSm

differences using a spaceless characteristics V (its context is $\langle 0, 0.4, 1 \rangle$). If $V = 1$ then it is extremely good, and if $V = 0$ then it is critical. Note that values of V also appear in other linguistic descriptions.

Evaluation of the Leasing Offer

The leasing offer by the given company is characterized by a spaceless characteristics N. This is obtained using the linguistic description in Table 5 where R is a spaceless characteristic of trustworthiness of a leasing company. The context of both characteristics is $\langle 0, 0.4, 1 \rangle$.

Naturally, $N = 1$ expresses that the company is trustworthy while $N = 0$ means that it is untrustworthy. This information can be based, for example, on the size of the company, its popularity, experiences with it, etc. We can derive it using a special linguistic description, which may include many criteria

Table 6. Linguistic description for the global evaluation

	V	L	Z	N	⇒ H
1	Bi	Sm	Bi	Bi	Bi
2	Sm	Bi	ignor	Sm	Sm
3	VeSm	NoSm	ignor	ignor	ExSm
4	Me	Me	NoSm	NoSm	RoBi
5	NoSm	RoMe	RoBi	Sm	RoSm
6	NoSm	RoMe	RoBi	Bi	Me
7	VeBi	RaSm	VeBi	Bi	VeBi
8	RoMe	Me	Bi	Bi	RaBi
9	RoBi	VeBi	Bi	RoSm	RaSm
10	RoBi	Bi	Bi	RoBi	MLBi
11	RoBi	RoSm	Sm	RoBi	RoSm
12	RoBi	RoSm	Sm	RoSm	VeSm

having an objective as well as a subjective character. Finally, A is the advance payment and C is the total surplus computed from the car price, advance payment, fee, leasing period, month payment and insurance.

Global Evaluation of Variants

On the basis of the results obtained using the above linguistic descriptions, we may now evaluate each variant globally. For this, we will use the linguistic description in Table 6 where V and N are the above described characteristics of the economic development and quality of the leasing offer, respectively. The variable L characterizes the global leasing cost with respect to the economic

Car	Price	Repres.	Leas. comp.	Length	Glob. eval.	Gain devel.	M. cost	Risc	Init. cost	Surplus	Offer
Avensis-25	875000	0,70	RT Torax	36	0,20 (RoSm)	0,72 (MLBi)	32405	0,50	227186	188070	0,12 (Sm)
			Toyota Leasing	42	0,70 (RoBi)		22903	0,10	231342	-33818	0,95 (SiBi)
			RT Torax	42	0,20 (RoSm)		28985	0,50	227186	221628	0,12 (Sm)
			Toyota Leasing	36	0,70 (RoBi)		26407	0,10	231342	-27842	0,95 (SiBi)
			CAC Leasing	48	0,62 (QRBi)		21519	0,20	219000	19920	0,67 (RoBi)
			CAC Leasing	36	0,62 (QRBi)		27537	0,20	219000	12832	0,67 (RoBi)
Avensis-50	875000	0,70	Toyota Leasing	36	0,70 (RoBi)		25773	0,05	450342	-50690	0,98 (ExBi)
			Toyota Leasing	42	0,70 (RoBi)		22262	0,05	450342	-60728	0,98 (ExBi)
			RT Torax	48	0,20 (RoSm)		25047	0,50	446186	189234	0,12 (Sm)
			RT Torax	36	0,28 (QRSm)		31471	0,50	446186	139446	0,38 (RaMe)
			CAC Leasing	36	0,70 (RoBi)		26819	0,10	438000	-13004	0,95 (SiBi)
			CAC Leasing	48	0,70 (RoBi)		20789	0,10	438000	-15144	0,95 (SiBi)
Yaris-25	370000	0,25	Toyota Leasing	36	0,42 (RaMe)		11212	0,10	97641	-24259	0,95 (SiBi)
			Toyota Leasing	48	0,42 (RaMe)		8570	0,10	97641	-35831	0,95 (SiBi)
			CAC Leasing	36	0,42 (RaMe)		11626	0,20	92000	-9368	0,81 (Bi)
			CAC Leasing	48	0,42 (RaMe)		9091	0,20	92000	-10832	0,81 (Bi)
			RT Torax	36	0,25 (RoSm)		12215	0,50	95780	11824	0,64 (RoBi)
Yaris-50	370000	0,25	Toyota Leasing	36	0,42 (RaMe)		10903	0,10	190641	-35407	0,95 (SiBi)
			Toyota Leasing	48	0,26 (QRSm)		8305	0,10	190641	28625	0,33 (VRSm)
			CAC Leasing	48	0,42 (RaMe)		8437	0,20	185000	-42216	0,81 (Bi)
			CAC Leasing	36	0,18 (MLSm)		12591	0,20	185000	25384	0,31 (VRSm)
			RT Torax	36	0,06 (VeSm)		11739	0,50	188780	5300	0,06 (VeSm)
Volvo-25	1045000	0,85	CSOB Leasing	36	0,73 (MLBi)		33038	0,15	272222	13870	0,65 (RoBi)
			CAC Leasing	36	0,75 (RaBi)		32651	0,25	261500	-56	0,67 (RoBi)
			RT Torax	48	0,15 (RaSm)		30887	0,50	271500	259556	0,12 (Sm)
			CSOB Leasing	48	0,75 (RaBi)		25649	0,25	261500	8132	0,67 (RoBi)
Volvo-50	1045000	0,85	CSOB Leasing	36	0,76 (RaBi)		31945	0,15	533472	-25480	0,86 (Bi)
			CAC Leasing	36	0,23 (RoSm)		34695	0,25	522500	73516	0,32 (VRSm)

Fig. 15. Results of decision-making

results of the company and Z is a coefficient of the new car representability which, analogous to the trustworthiness of the leasing company, can be determined as a subjective value from the interval $[0, 1]$. Results of the demonstration example are in Fig. 15. The global evaluation obtained using the last linguistic expression provides us with information about the quality of the given alternative. The highest global evaluation $H = 0.76$ has been assigned to Volvo with 50% of advance payment from the CSOB Leasing Company. The global evaluations are for better understandability assigned perceptions in the context $\langle 0, 0.4, 1\rangle$ using the function (13) (in the case of the chosen variant, it is *RaBi (rather big)*). Note that all the three variants have been evaluated by the same expression. This may be useful in situations where the management prefers words to numbers and wants to have some space for final decisions.

6 Summary

In this chapter, we demonstrated that mathematical fuzzy logic is a strong, well developed formal tool for modeling of the vagueness phenomenon and that it is a good basis for construction of PNL (namely, the theory of evaluating linguistic expressions and the theory of fuzzy IF-THEN rules as special sentences of natural language). We are convinced that this opens way to the development of mathematical model of natural human reasoning and that it has a great power for applications including learning.

Among problems that should be solved in further research is the development of mathematical FLb and deepening the theory of PNL by introducing generalized quantifiers, involving time, developing the theory of the meaning of verbs, extending special reasoning schemes, forming the world knowledge and finding methods for automation of its learning from Internet.

References

1. P. Andrews, An Introduction to Mathematical Logic and Type Theory: To Truth Through Proof, Kluwer, Dordrecht, 2002.
2. D. Boixander, J. Jacas, J. Recasens, Fuzzy equivalence relations: advanced material, in: D. Dubois, H. Prade (Eds.), Fundamentals of Fuzzy Sets, Kluwer, Dordrecht, 2000, pp. 261–290.
3. B. Bouchon-Meunier, Fuzzy logic and knowledge representation using linguistic modifiers, in: L. A. Zadeh, J. Kacprzyk (Eds.), Fuzzy Logic for the Management of Uncertainty, J. Wiley, New York, 1992, pp. 399–414.
4. R. Bělohlávek, V. Novák, Learning rule base of the linguistic expert systems, Soft Computing 7 (2002) 79–88.
5. R. Carnap, Meaning and Necessity: a Study in Semantics and Modal Logic, University of Chicago Press, Chicago, 1947.

6. A. De Soto, R. Recasens, Modeling a linguistic variable as a hierarchical family of partitions induced by an indistinguishability operator, Fuzzy Sets and Systems 121 (2001) 427–437.
7. B. Diaz, T. Takagi, PNL applied to economics, in: Proc. Int. Conf. FUZZ-IEEE'2005, Reno, USA, 2005, pp. 149–154.
8. A. Dvořák, H. Habiballa, V. Novák, V. Pavliska, The software package LFLC 2000 - its specificity, recent and perspective applications, Computers in Industry 51 (2003) 269–280.
9. A. Dvořák, V. Novák, Fuzzy logic model of detective reasoning, in: Proc. 11th Int. Conf. IPMU, Paris, July 2006, Vol. 2, Éditions EDK, Les Cordeliers, Paris, 2006, pp. 1890–1897.
10. A. Dvořák, V. Novák, A model of complex human reasoning in fuzzy logic in broader sense, Fuzzy Sets and Systems.
11. F. Esteva, L. Godo, Monoidal t-norm based logic: towards a logic for left-continuous t-norms, Fuzzy Sets and Systems 124 (2001) 271–288.
12. P. Hájek, Metamathematics of Fuzzy Logic, Kluwer, Dordrecht, 1998.
13. P. Hájek, V. Novák, The sorites paradox and fuzzy logic, International Journal of General Systems 32 (2003) 373–383.
14. U. Höhle, Fuzzy Sets and Sheaves, Bergische Universität, Wuppertal, Germany, 2003.
15. G. Lakoff, Hedges: A study in meaning criteria and logic of fuzzy concepts, Journal of Philosophical Logic 2 (1973) 458–508.
16. P. Materna, Concepts and Objects, Acta Philosophica Fennica 63, Helsinki, 1998.
17. V. Novák, , J. Kovář, Linguistic if-then rules in large scale application of fuzzy control, in: R. Da, E. Kerre (Eds.), Fuzzy If-Then Rules in Computational Intelligence: Theory and Applications, Kluwer Academic Publishers, Boston, 2000, pp. 223–241.
18. V. Novák, Fuzzy Sets and Their Applications, Adam Hilger, BristoYork, 1989.
19. V. Novák, Towards formalized integrated theory of fuzzy logic, in: Z. Bien, K. Min (Eds.), Fuzzy Logic and Its Applications to Engineering, Information Sciences, and Intelligent Systems, Kluwer, Dordrecht, 1995, pp. 353–363.
20. V. Novák, Fuzzy if-then rules in logical deduction and their learning form data, in: Proc. 4th Czech-Japanese Seminar on Data Analysis and Decision Making under Uncertainty, Hyogo, Japan, 2001, pp. 96–107.
21. V. Novák, Descriptions in the full fuzzy type theory, Neural Network World 5 (2003) 559–565.
22. V. Novák, Fuzzy logic deduction with words applied to ancient sea level estimation, in: R. Demicco, G. Klir (Eds.), Fuzzy logic in geology, Academic Press, Amsterdam, 2003, pp. 301–336.
23. V. Novák, Fuzzy type theory as higher order fuzzy logic, in: Proc. 6th Int. Conference on Intelligent Technologies (InTech'05), Dec. 14-16, 2005, Fac. of Science and Technology, Assumption University, Bangkok, Thailand, 2005, pp. 21–26.
24. V. Novák, On fuzzy type theory, Fuzzy Sets and Systems 149 (2005) 235–273.
25. V. Novák, Perception-based logical deduction, in: B. Reusch (Ed.), Computational Intelligence, Theory and Applications, Springer, Berlin, 2005, pp. 237–250.

26. V. Novák, Which logic is the real fuzzy logic?, Fuzzy Sets and Systems 157 (2006) 635–641.
27. V. Novák, A comprehensive theory of evaluating linguistic expressions, in: S. Gottwald, E. P. Klement (Eds.), Proc. of 26th Linz Seminar on Fuzzy Set Theory, Springer, Berlin, (to appear), research Report No. 71, http://irafm.osu.cz/irafm.
28. V. Novák, S. Lehmke, Logical structure of fuzzy IF-THEN rules, Fuzzy Sets and Systems.
29. V. Novák, I. Perfilieva, On the semantics of perception-based fuzzy logic deduction, International Journal of Intelligent Systems 19 (2004) 1007–1031.
30. V. Novák, I. Perfilieva, Dvořák, A. Chen, GQ., Q. Wei, P. Yan, Mining linguistic information from numerical databases, in: Fuzzy Logic, Soft Computing and Computational Intelligence I. Eleventh IFSA World Congress, Tsinghua Univ. Press and Springer, Beijing, China, 2005, pp. 546–551.
31. V. Novák, I. Perfilieva, J. Močkoř, Mathematical Principles of Fuzzy Logic, Kluwer, Boston, 1999.
32. J. Pavelka, On fuzzy logic I, II, III, Zeitschrift für Mathematische Logik und Grundlagen der Mathematik 25 (1979) 45–52, 119–134, 447–464.
33. I. Perfilieva, Fuzzy transforms: theory and applications, Fuzzy Sets and Systems 157 (2006) 993–1023.
34. P. Vopěnka, Mathematics in the Alternative Set Theory, Teubner, Leipzig, 1979.
35. M. Ying, B. Bouchon-Meunier, Quantifiers, modifiers and qualifiers in fuzzy logic, Journal of Applied Non-Classical Logics 7 (1997) 335–342.
36. L. A. Zadeh, The concept of a linguistic variable and its application to approximate reasoning I, II, III, Information Sciences 8-9 (1975) 199–257, 301–357, 43–80.
37. L. A. Zadeh, Precisiated natural language, AI Magazine 25 (2004) 74–91.

Analytical Theory of Fuzzy IF-THEN Rules with Compositional Rule of Inference *

Irina Perfilieva

Abstract. A system of fuzzy IF-THEN rules is considered as a knowledge-base system where inference is made on the basis of three rules of inference, namely *Compositional Rule of Inference*, *Modus Ponens* and *Generalized Modus Ponens*. The problem of characterizing models of such systems is investigated. We propose an analytical theory of fuzzy IF-THEN rules where all the above mentioned rules of inference work properly. *Modus Ponens* and *Generalized Modus Ponens* formally express interpolation between fuzzy nodes and continuity at fuzzy nodes, respectively. We show that the validity of *Modus Ponens* and thus, interpolation between fuzzy nodes lead to the problem of solvability of fuzzy relation equations. We prove that if the computation is based on sup $-*$ composition then the validity of *Modus Ponens* implies the validity of *Generalized Modus Ponens*.

1 Introduction

We consider knowledge-base systems where the knowledge is given by a set of fuzzy IF-THEN rules

$$\begin{array}{c}\mathsf{IF}\ x\ \mathsf{is}\ A_1\ \mathsf{THEN}\ y\ \mathsf{is}\ B_1\\ \dots\dots\dots\dots\dots\dots\dots\dots\\ \mathsf{IF}\ x\ \mathsf{is}\ A_n\ \mathsf{THEN}\ y\ \mathsf{is}\ B_n\end{array} \tag{1}$$

and reasoning is based on the following rules of inference: *Modus Ponens* (**MP**), *Generalized Modus Ponens* (**GMP**), *Compositional Rule of Inference* (**CRI**). A knowledge-base system (rule-base system) is always used when a user aims at arriving to a conclusion on the basis of preliminary discovered rules and new facts or observations. The following are advantages and disadvantages of a general rule-base system.

* This paper has been partially supported by the grant 201/04/1033 of GA ČR and partially by the research project MSM 6198898701 of MŠMT ČR

Advantages:

(i) simple creation and interpretation of rules,
(ii) simple extension and modification,
(iii) simple deduction technique.

Disadvantages:

(a) relationship among different rules is unclear,
(b) low flexibility in deduction (*Modus Ponens* requires coincidence between antecedent of some of the rules and a new incoming fact),
(c) low effectiveness of processing (search for the above mentioned coincidence is time consuming).

We claim that the disadvantages listed above are not relevant if a properly constructed model of fuzzy rule-based system is chosen. By the "properly constructed" model we mean a *fuzzy function* F_{Rules} which is defined at given nodes (fuzzy sets) $A_1, \ldots, A_n$ by the respective values (fuzzy sets) $B_1, \ldots, B_n$, i.e. $F_{Rules}(A_i) = B_i$, $i = 1, \ldots, n$, and moreover fulfils the following conditions:

M1. F_{Rules} is represented by a fuzzy relation R such that its value at arbitrary node (fuzzy set) A (which may be equal to any of A_i, $i = 1, \ldots, n$) can be computed on the basis of the *Compositional Rule of Inference*

$$F_{Rules}(A) = A \circ R.$$

Recall that **CRI** has been introduced by L. A. Zadeh in [14] in the following form:

$$\begin{array}{l} (x, y) \text{ are } R \\ x \text{ is } A \\ \hline y \text{ is } A \circ R. \end{array} \tag{2}$$

M2. F_{Rules} is continuous at nodes $A_1, \ldots, A_n$. This means that (informally speaking) if a fuzzy set A is close to A_i then $F_{Rules}(A)$ is close to B_i, $i = 1, \ldots, n$.

Let us show how the disadvantages listed above can be overcome.

Ad (a): Relationship among rules

We argue that only rules which characterize local extremes of the corresponding function F_{Rules} are necessary. The extremal values should be characterized by fuzzy sets $B_1, \ldots, B_n$. All intermediate values can be computed with the help of **CRI**.

Ad (c): Effectiveness of processing

Fuzzy IF-THEN rules processing (deduction) is based on **CRI** and therefore, it requires a routine computation only. We conclude that it is *effective.*

Ad (b): Flexibility in deduction

The deduction is *flexible* because **M2.** is fulfilled.

In this paper we propose an *analytical theory of fuzzy rule-base systems* in which the requirements **M1.**–**M2.** are fulfilled. At the beginning, we will summarize the up-to-date state of knowledge in this field and then indicate where is the novelty of our investigation.

Systems of fuzzy IF-THEN rules were extensively studied theoretically attempting at creation of a methodology for analysis of knowledge-base systems. It is worth of reminding some papers which noticeably contributed to the progress of this topic [2–4,6,8,11]. The leading idea was to elaborate a special theory of fuzzy IF-THEN rules in a first or higher order fuzzy logic. This was explicitly expressed by L. A. Zadeh when he characterized the agenda of fuzzy logic. However, a lot of attempts to build a rigorous formal logical theory of fuzzy IF-THEN rules failed. Some exceptions are worth to be remarked — the V. Novák's formal logical theory of fuzzy IF-THEN rules based on the fuzzy logic with evaluated syntax [12], and S. Gottwald's theory of fuzzy relation equations [4]. In the author's opinion, the main source of difficulties stems from a copying purely logical approach to the analysis of fuzzy IF-THEN rules. To overcome this restrictive processing we propose to base the analysis of fuzzy IF-THEN rules on their functional origin.

In parallel with the logical approach, systems of fuzzy IF-THEN rules were also extensively studied in connection with their numerous applications in fuzzy control, identification of dynamic systems, prediction of dynamic systems, decision-making, etc. (see e.g. [1, 4]). The success in this area can be explained by the proposed calculus [14, 18] for making "deductions" with the help of a number of semantically adjusted rules of inference. Two of them have been used most frequently: the *Compositional Rule of Inference* and the *Generalized Modus Ponens.* Each rule of inference being applied to a system of fuzzy IF-THEN rules induced investigations regarding its utilization: **CRI** generated the problem of representation of a system of fuzzy IF-THEN rules by a single fuzzy relation, and **GMP** generated the problem of establishing a relationship between A and A_i. Furthermore, **CRI** generated the problem of solvability of a system of fuzzy relation equations, while **GMP** focused on the problem of characterizing a closeness between two fuzzy sets. However, only few attempts (see e.g. [3]) have been focused on the analysis of *compatibility of both rules.* It turned out that if deductions are made on the basis of **CRI** then **GMP** and even **MP** may not be valid (see Sect. 7 with Illustrations). The problem of *characterization of models of fuzzy IF-THEN rules where all three inference rules listed above work properly* arises. In [3] this problem has been considered with the restriction to one fuzzy IF-THEN rule only. The proposed solution put limiting conditions on operations which are used in the interpretation of logical connectives.

In this paper we solve the above given problem in general, i.e. for a fixed algebra of operations we found conditions on fuzzy parameters A_i, B_i, $i = 1, \dots, n$, which guarantee the validity of **CRI**, **MP** and **GMP**. Our principal position on fuzzy rule-based systems is that their analysis can be done without logical framework. We propose to consider a system of fuzzy IF-THEN rules as a partial description of a *fuzzy function* whose domain and range are universes of fuzzy sets. In this treatment, **CRI** plus **MP** imply that such function must be represented by a fuzzy relation and this relation must solve the respective system of fuzzy relation equations (see Sect. 4 for details).

Moreover, we proved (Sect. 6) that the following relationship is valid:

If **CRI** *is accepted then* **GMP** *is equivalent to* **MP**.

Therefore, **GMP** automatically holds true if fuzzy IF-THEN rules processing is based on **CRI** and **MP** rules of inference.

The analysis of fuzzy rule-base systems is made on the basis of a single algebra that forms a complete *residuated lattice* [9]. This makes the proposed theory well-balanced and mathematically elegant.

2 Functional Versus Logical Approach

In this section we emphasize the functional origin of a model of fuzzy IF-THEN rules, contrary to the logical one which is otherwise automatically assumed. In both cases a model of fuzzy IF-THEN rules is, or is based on a fuzzy relation.

The logical approach emphasizes how a fuzzy relation (as a model of IF-THEN rules) can be represented by a formula. The functional approach puts the accent on the way how a fuzzy relation can be used in computation of dependent values of a corresponding fuzzy function. Let us discuss this principal difference in general. A new definition of a model of fuzzy IF-THEN rules will be given below, in Sect. 4.

Aiming at creation of a model of the given system of IF-THEN rules, we will distinguish two different approaches:

(i) The first one called a *logical approach* forms a model in a way similar to that producing interpretation of some formula (see, e.g. [8, 11]). This approach (in general) agrees on interpretation of atomic expressions and connectives and recursively constructs an interpretation of any properly written formula. In the case of a system of fuzzy IF-THEN rules (2), the corresponding formula is

$$((A_1(x) \rightarrow B_1(y)) \wedge \cdots \wedge (A_1(x) \rightarrow B_1(y))) \tag{3}$$

provided that the connective "IF-THEN" is expressed by implication $\rightarrow$. In this formula, the fuzzy constraints A_i, B_i are atomic expressions. If we agree to model them by fuzzy sets A_i, B_i (we will denote them by the same symbols as the fuzzy constraints) then formula (3) (and the corresponding system of fuzzy IF-THEN rules) is modeled in a class of fuzzy relations. Therefore, the logical approach assigns a unique fuzzy relation (in each structure) to a system of fuzzy IF-THEN rules on the basis of precisely formulated procedure.

(ii) We suggest another approach and call it *functional* having in mind the way of making computations using a given system of fuzzy IF-THEN rules. In this approach, fuzzy constraints A_i, B_i are again modeled by fuzzy sets A_i, B_i, but the respective system of IF-THEN rules is modeled by a fuzzy function F_{Rules} which is defined at nodes $A_1, \dots, A_n$ so that $F_{Rules}(A_i) = B_i$, $i = 1, \dots, n$. In order to extend this function to other fuzzy nodes different from $A_1, \dots, A_n$, we assume that F_{Rules} has an *adjoint* fuzzy relation, say R, so that $F_{Rules}(A) = A \circ R$ for arbitrary fuzzy node A (see Sect. 3 for the details). If we take $A = A_i$, $i = 1, \dots, n$, then we obtain restrictions on R in a form of a system of fuzzy relation equations

$$\begin{cases} A_1 \circ R = B_1, \\ \dots\dots\dots\dots \\ A_n \circ R = B_n \end{cases} \tag{4}$$

We agree that any solution to this system can be taken as an underlying fuzzy relation which determines the corresponding fuzzy function F_{Rules}.

As can be seen, the functional approach also uses a fuzzy relation for specifying a model of fuzzy IF-THEN rules, but this relation need not be unique and moreover, need not be represented by any logical formula.

In the forgoing text we will denote function F_{Rules} by F_R in order to emphasize that it has an adjoint fuzzy relation R. Let us remark that the proposed functional approach models also rules of inference **CRI** and **MP** which are based on the given IF-THEN rules (see the details below).

3 Fuzzy Sets, Fuzzy Relations and Fuzzy Functions

In this section, we will step aside and give precise definitions and explanations to all the notions mentioned above. For this purpose, we need to choose an appropriate algebra of operations over fuzzy sets. A complete residuated lattice $\mathcal{L}$ on $[0,1]$ will be sufficient to meet all our requirements where

$$\mathcal{L} = \langle [0,1], \vee, \wedge, *, \rightarrow, \mathbf{0}, \mathbf{1} \rangle \tag{5}$$

is a structure with four binary operations and two constants (see [9] for details). We extend $\mathcal{L}$ by the derived binary equivalence operation

$$x \leftrightarrow y = (x \rightarrow y) \wedge (y \rightarrow x).$$

Recall that $\leftrightarrow$ is reflexive, symmetric and transitive with respect to the multiplication $*$. The latter means that

$$\forall x, y, z \in [0,1] \quad (x \leftrightarrow y) * (y \leftrightarrow z) \leq (x \leftrightarrow z).$$

Let us remark that if a residuated lattice has the support $[0,1]$ then the multiplication operation $*$ is a *left continuous t-norm.*

The universe of fuzzy sets on $\mathbf{X}$ (respectively, on $\mathbf{Y}$) is introduced as a universe of functions (membership functions) as follows:

$$\mathcal{F}(\mathbf{X}) = [0,1]^{\mathbf{X}}$$

(as well as $\mathcal{F}(\mathbf{Y}) = [0,1]^{\mathbf{Y}}$). No special requirements on $\mathbf{X}$, $\mathbf{Y}$ and membership functions are imposed. For two fuzzy sets A and B on $\mathbf{X}$ we write $A = B$ or $A \leq B$ if $A(x) = B(x)$ or $A(x) \leq B(x)$ holds for all $x \in \mathbf{X}$, respectively.

The algebra of operations over fuzzy sets of $\mathbf{X}$ is introduced as an induced residuated lattice on $\mathcal{F}(\mathbf{X})$. This means that each operation from $\mathcal{L}$ induces the corresponding operation on $\mathcal{F}(\mathbf{X})$ taken pointwise. Obviously, operations over fuzzy sets fulfil the same properties as operations in the respective residuated lattice.

A (binary) *fuzzy relation* on $\mathbf{X} \times \mathbf{Y}$ is a fuzzy set of the Cartesian product. $\mathcal{F}(\mathbf{X} \times \mathbf{Y})$ denotes the set of all binary fuzzy relations on $\mathbf{X} \times \mathbf{Y}$. Analogously, an n-ary fuzzy relation can be introduced.

Let $R \in \mathcal{F}(\mathbf{X} \times \mathbf{Y})$ and $S \in \mathcal{F}(\mathbf{Y} \times \mathbf{Z})$. Then the fuzzy relation T on $\mathbf{X} \times \mathbf{Z}$

$$T(x,z) = \bigvee_{y \in \mathbf{Y}} (R(x,y) * S(y,z))$$

is a result of the *composition* (or sup $-*$ composition) of R and S and denoted by

$$T = R \circ S.$$

In particular, if A is a unary fuzzy relation on $\mathbf{X}$ or simply a fuzzy set of $\mathbf{X}$ then the result of the sup $-*$ composition between A and $R \in \mathcal{F}(\mathbf{X} \times \mathbf{Y})$ is the fuzzy set of $\mathbf{Y}$ defined by

$$(A \circ R)(y) = \bigvee_{x \in \mathbf{X}} (A(x) * R(x,y)).$$

Let a fuzzy relation $R \in \mathcal{F}(\mathbf{X} \times \mathbf{Y})$ be given. Then it can be used for a representation of the *adjoint fuzzy function* $F_R : \mathcal{F}(\mathbf{X}) \longrightarrow \mathcal{F}(\mathbf{Y})$:

$$F_R(A) = \bigvee_{x \in \mathbf{X}} (A(x) * R(x,y)) = A \circ R \tag{6}$$

where A is an arbitrary fuzzy set from $\mathcal{F}(\mathbf{X})$ (see more in [15]). The computed value $F_R(A)$ is a fuzzy set from $\mathcal{F}(\mathbf{Y})$ and it is the value of F_R given argument A. In general, if the connection between a fuzzy function $F_R : \mathcal{F}(\mathbf{X}) \longrightarrow \mathcal{F}(\mathbf{Y})$ and a fuzzy relation R is given by (6) then we say that they are *adjoint*. We will use this term both for a fuzzy function as well as a fuzzy relation.

We say that the adjoint fuzzy function F_R *interpolates* among pairs of fuzzy sets (A_i, B_i), $i = 1, \ldots, n$, where $A_i \in \mathcal{F}(\mathbf{X}), B_i \in \mathcal{F}(\mathbf{Y})$ if for each i

$$B_i = F_R(A_i) = A_i \circ R.$$

4 Model of Fuzzy IF-THEN Rules: Guaranty of MP

In this section we give a definition of a model of fuzzy IF-THEN rules which has been informally discussed above. We assume that if fuzzy IF-THEN rules are given then the fuzzy sets A_i, B_i, $i = 1, \dots, n$, are given by their membership functions $A_i \in \mathcal{F}(\mathbf{X}), B_i \in \mathcal{F}(\mathbf{Y})$ in accordance with our specification in Sect. 3. Therefore, a model of fuzzy IF-THEN rules specifies the way how are these parameters connected. This will be realized using operations from a complete residuated lattice $\mathcal{L}$ on $[0,1]$ by specifying a way of computation (6). Consequently, the discussed parameters form a *structure*

$$\mathcal{S} = \langle \mathbf{X}, \mathbf{Y}, \{A_i, B_i\}_{i=1,\dots,n}, \mathcal{L}, \circ \rangle. \tag{7}$$

Definition 1. *We say that a fuzzy function $F_R : \mathcal{F}(\mathbf{X}) \longrightarrow \mathcal{F}(\mathbf{Y})$ is a model of fuzzy IF-THEN rules (2) in a structure $\mathcal{S}$ given by (7) if for all $i = 1, \dots, n$*

$$F_R(A_i) = A_i \circ R = B_i \tag{8}$$

where $R \in \mathcal{F}(\mathbf{X} \times \mathbf{Y})$ is an adjoint fuzzy relation.

Note that Definition 1 takes into account requirement **M1.** on a model of fuzzy IF-THEN rules, which has been discussed in the Introduction. Therefore, a model of fuzzy IF-THEN rules realizes the rule of inference (**MP**) which is taken with respect to these rules.

If we rewrite (8) as a system of fuzzy relation equations (cf. 4)

$$\begin{cases} A_1 \circ R = B_1, \\ \dots\dots\dots\dots \\ A_n \circ R = B_n \end{cases}$$

where R is an unknown fuzzy relation then from Definition 1, R is a solution to (4) if and only if it is the underlying fuzzy relation of a model F_R of fuzzy IF-THEN rules (2) in the structure $\mathcal{S}$.

We may refer to different criteria of solvability (see [4,13,13]) which tell us under which conditions we may obtain a solution to a system of fuzzy relation equations. As a side result of our main theorem (Sect. 6), we will present a new criterion of solvability (Theorem 3).

Remark 1. It follows from Definition 1 that a choice of a structure determines whether a set of fuzzy IF-THEN rules has a model or not. For example, it may happen that (4) is not solvable in some structure $\mathcal{S}$ and therefore, fuzzy IF-THEN rules do not have a model in $\mathcal{S}$. In this case, some parameters in $\mathcal{S}$, e.g. fuzzy sets A_i and B_i should be changed in order to assure the solvability. The change of A_i and B_i leads to another structure where fuzzy IF-THEN rules can be successfully modeled (see Sect. 7 with illustrations). In [13,14] different necessary and sufficient conditions of solvability of (4), putting restrictions on fuzzy sets A_i and B_i, have been proposed.

5 Model of Fuzzy IF-THEN Rules: Guaranty of GMP

The property (8) takes into account the requirement **M1.** on a model of fuzzy IF-THEN rules and realizes the classical inference rule of **MP**:

$$\begin{array}{l} \mathsf{IF}\ x\ \mathsf{is}\ A_i\ \mathsf{THEN}\ y\ \mathsf{is}\ B_i, \quad i = 1, \ldots, n \\ x\ \mathsf{is}\ A_i \\ \hline y\ \mathsf{is}\ B_i. \end{array} \tag{9}$$

If we relax the assumption 'x is A_i' and replace it by 'x is close to A_i' then by a similar relaxation of the conclusion, we will obtain the inference rule of **GMP**:

$$\begin{array}{l} \mathsf{IF}\ x\ \mathsf{is}\ A_i\ \mathsf{THEN}\ y\ \mathsf{is}\ B_i, \quad i = 1, \ldots, n \\ x\ \mathsf{is\ close\ to}\ A_i \\ \hline y\ \mathsf{is\ close\ to}\ B_i. \end{array} \tag{10}$$

GMP is often assumed (see e.g. [3]) when dealing with a set of fuzzy IF-THEN rules. Informally, we have formulated this rule in our requirement **M2.** Below we give the precise formalization of **GMP**.

Definition 2. *We say that a fuzzy function $F_R : \mathcal{F}(\mathbf{X}) \longrightarrow \mathcal{F}(\mathbf{Y})$ realizes* **GMP** *in a structure $\mathcal{S}$ with respect to fuzzy IF-THEN rules (2) if for each i and for each fuzzy set $A \in \mathcal{F}(\mathbf{X})$ the following inequality holds true:*

$$\bigwedge_{y \in \mathbf{Y}} (B_i(y) \leftrightarrow F_R(A)(y)) \geq \bigwedge_{x \in \mathbf{X}} (A_i(x) \leftrightarrow A(x)). \tag{11}$$

Inequality (11) can be rewritten in terms of an adjoint fuzzy relation R:

$$\bigwedge_{y \in \mathbf{Y}} (B_i(y) \leftrightarrow (A \circ R)(y)) \geq \bigwedge_{x \in \mathbf{X}} (A_i(x) \leftrightarrow A(x)). \tag{12}$$

If (12) is fulfilled then we will also say that the fuzzy relation R realizes **GMP** in a structure $\mathcal{S}$ with respect to fuzzy IF-THEN rules (2).

In this section, we will explain why (11) is used for the specification of **GMP**. The closeness between fuzzy sets in (11) is measured using the operation $\leftrightarrow$. Let us characterize the connection between $\leftrightarrow$ and the pseudo-metric on $\mathcal{F}(\mathbf{X})$ induced by the former.

Let us assume that the multiplication $*$ is a continuous Archimedean t-norm. Recall that in this case it has an additive generator $g : [0,1] \longrightarrow [0, +\infty]$ which is a strictly decreasing function, continuous on $[0,1]$ and fulfilling two conditions (see [10]):

1. $g(1) = 0$,
2. $x * y = g^{-1}(\min(g(0), g(x) + g(y)))$.

Let us remark that the range of g is the extended non-negative part of the real line and therefore, the equality $g(0) = +\infty$ is possible. We will use the following representation of the operation $\leftrightarrow$ ([10]):

$$x \leftrightarrow y = g^{-1}(|g(x) - g(y)|) \tag{13}$$

where $g^{-1} : [0, +\infty] \longrightarrow [0, 1]$ is the inverse function, and where in the case $g(0) = +\infty$ we additionally define $g(0) - g(0) = 0$.

A continuous Archimedean t-norm $*$ determines the metric space $(\mathcal{F}(\mathbf{X}), D_g)$ where

$$D_g(A, B) = \bigvee_{x \in \mathbf{X}} |g(A(x)) - g(B(x))|.$$

Analogously, $*$ determines the metric space $(\mathcal{F}(\mathbf{Y}), D_g)$.

Theorem 1. *Let $\mathcal{L}$ be a residuated lattice on $[0, 1]$ and a structure $\mathcal{S}$ be given by $\langle \mathbf{X}, \mathbf{Y}, \{A_i, B_i\}_{i=1,\dots,n}, \mathcal{L}, \circ \rangle$. Let, moreover, the t-norm $*$ be a continuous Archimedean t-norm with a continuous additive generator g. A fuzzy function $F_R : \mathcal{F}(\mathbf{X}) \longrightarrow \mathcal{F}(\mathbf{Y})$ realizes* **GMP** *in the structure $\mathcal{S}$ with respect to fuzzy IF-THEN rules if and only if*

$$D_g(B_i, F_R(A)) \le D_g(A_i, A)$$

for each i and each fuzzy set $A \in \mathcal{F}(\mathbf{X})$.

Proof. Assume that a fuzzy function $F_R : \mathcal{F}(\mathbf{X}) \longrightarrow \mathcal{F}(\mathbf{Y})$ realizes **GMP** with respect to fuzzy IF-THEN rules in $\mathcal{S}$ and rewrite (11) having in mind (13):

$$\bigwedge_{y \in \mathbf{Y}} (B_i(y) \leftrightarrow F_R(A)(y)) \ge \bigwedge_{x \in \mathbf{X}} (A_i(x) \leftrightarrow A(x)) \Leftrightarrow$$

$$\bigwedge_{y \in \mathbf{Y}} g^{-1}(|g(B_i(y)) - g(F_R(A)(y))|) \ge \bigwedge_{x \in \mathbf{X}} g^{-1}(|g(A_i(x)) - g(A(x))|).$$

Based on the fact that g^{-1} is a continuous and strictly decreasing function on $[0, +\infty]$, we come to the conclusion of the lemma. Indeed,

$$\bigwedge_{y \in \mathbf{Y}} g^{-1}(|g(B_i(y)) - g(F_R(A)(y))|) \ge \bigwedge_{x \in \mathbf{X}} g^{-1}(|g(A_i(x)) - g(A(x))|) \Leftrightarrow$$

$$\bigvee_{y \in \mathbf{Y}} (|g(B_i(y)) - g(F_R(A)(y))|) \le \bigvee_{x \in \mathbf{X}} (|g(A_i(x)) - g(A(x))|) \Leftrightarrow$$

$$D_g(B_i, F_R(A)) \le D_g(A_i, A).$$

It is worth noticing that validity of **GMP** with respect to fuzzy IF-THEN rules in its functional representation (11) means that a model of those rules (fuzzy function F_R) is *continuous* at the fuzzy nodes $A_1, \dots, A_n$.

6 MP Is Equivalent to GMP

Let us consider two rules of inference simultaneously: **MP** (9) and **GMP** (10), both with respect to the same set of fuzzy IF-THEN rules.

We will show that **GMP** in its functional representation (11) holds true if and only if F_R is a model of the respective fuzzy IF-THEN rules (and equivalently, F_R realizes the rule of inference (**MP**)). Therefore, if we accept the requirement **M1.** on a model of fuzzy IF-THEN rules then the requirement **M2.** is fulfilled automatically. This will be proved in the main theorem of this section which relies on the following lemma.

Lemma 1. *Let a structure* $\mathcal{S} = \langle \mathbf{X}, \mathbf{Y}, \{A_i, B_i\}_{i=1,\dots,n}, \mathcal{L}, \circ\rangle$ *be given. Let* $R \in \mathcal{F}(\mathbf{X}\times\mathbf{Y})$ *be a fuzzy relation and* $F_R : \mathcal{F}(\mathbf{X}) \longrightarrow \mathcal{F}(\mathbf{Y})$ *be the adjoint fuzzy function. Then for any* $A \in \mathcal{F}(\mathbf{X})$ *and all* $i = 1,\dots,n$ *and* $y \in \mathbf{Y}$ *it is true that*

$$B_i(y) \leftrightarrow F_R(A)(y) \geq \delta_{R,i}(y) * \bigwedge_{x\in\mathbf{X}} (A_i(x) \leftrightarrow A(x)) \tag{14}$$

where

$$\delta_{R,i}(y) = B_i(y) \leftrightarrow F_R(A_i)(y).$$

Proof. Let us denote

$$B = F_R(A) \tag{15}$$

and observe, that $B \in \mathcal{F}(\mathbf{Y})$. By the property of transitivity of the operation $\leftrightarrow$ with respect to $*$, we obtain

$$B \leftrightarrow B_i \geq (B \leftrightarrow F_R(A)) * (F_R(A) \leftrightarrow F_R(A_i)) * (F_R(A_i) \leftrightarrow B_i) \tag{16}$$

where $i \in \{1,\dots,n\}$.

Let us estimate each of three multiplicands in the right-hand side of (16). By (15), the first one is equal to 1:

$$B \leftrightarrow F_R(A) \equiv 1.$$

The second one may be estimated from below for arbitrary $y \in \mathbf{Y}$ as follows:

$$F_R(A)(y) \leftrightarrow F_R(A_i)(y) = (\bigvee_{x\in\mathbf{X}} (A(x) * R(x,y))) \leftrightarrow \bigvee_{x\in\mathbf{X}} (A_i(x) * R(x,y)) \geq$$
$$\bigwedge_{x\in\mathbf{X}} (A(x) * R(x,y) \leftrightarrow A_i(x) * R(x,y)) \geq \bigwedge_{x\in\mathbf{X}} (A(x) \leftrightarrow A_i(x)).$$

Here, we used the facts

$$\bigwedge_{i\in I}(a_i \leftrightarrow b_i) \leq (\bigvee_{i\in I} a_i \leftrightarrow \bigvee_{i\in I} b_i)$$

and

$$(a \leftrightarrow b) * (c \leftrightarrow d) \leq (a * c \leftrightarrow b * d)$$

that are valid in any complete residuated lattice.

The third multiplicand

$$F_R(A_i)(y) \leftrightarrow B_i(y)$$

is equal to $\delta_{R,i}(y)$. Summarizing all three estimations, we obtain the conclusion of the lemma.

Lemma 1 estimates a deviation (via equivalence) of a value $F_R(A)$ from the respective fuzzy set B_i, presented on the right-hand side of the rules. The estimation uses the deviation between fuzzy sets A and A_i (again expressed via equivalence) and the deviation expressed by $\delta_{R,i}(y)$. The latter estimates whether R is a solution to the i-th equation of the system (4). If this holds then for all $y \in \mathbf{Y}$, $\delta_{R,i}(y) = 1$ and by (11), the fuzzy function F_R realizes **GMP** with respect to the i-th fuzzy relation equation.

Theorem 2 (Main theorem). *Let $\mathcal{S} = \langle \mathbf{X}, \mathbf{Y}, \{A_i, B_i\}_{i=1,\dots,n}, \mathcal{L}, \circ \rangle$ be a structure. A fuzzy function $F_R : \mathcal{F}(\mathbf{X}) \longrightarrow \mathcal{F}(\mathbf{Y})$ is a model of fuzzy IF-THEN rules (2) in the structure $\mathcal{S}$ if and only if it realizes* **GMP** *in $\mathcal{S}$ with respect to all fuzzy IF-THEN rules.*

Proof. $\Rightarrow$ Suppose that function $F_R : \mathcal{F}(\mathbf{X}) \longrightarrow \mathcal{F}(\mathbf{Y})$ is a model of fuzzy IF-THEN rules in the structure $\mathcal{S}$. Then the adjoint fuzzy relation R solves the system (8) and therefore, $\delta_{R,i}(y) = 1$ for all $i = 1, \dots, n$ and for all $y \in \mathbf{Y}$. By (14),

$$B_i(y) \leftrightarrow (A \circ R)(y) \geq \bigwedge_{x \in \mathbf{X}} (A_i(x) \leftrightarrow A(x))$$

and therefore,

$$B_i(y) \leftrightarrow F_R(A)(y) \geq \bigwedge_{x \in \mathbf{X}} (A_i(x) \leftrightarrow A(x)).$$

This means that F_R realizes **GMP** in $\mathcal{S}$ (cf. (11)) with respect to fuzzy IF-THEN rules.

$\Leftarrow$ Suppose that a fuzzy function $F_R : \mathcal{F}(\mathbf{X}) \longrightarrow \mathcal{F}(\mathbf{Y})$ realizes **GMP** in $\mathcal{S}$ with respect to fuzzy IF-THEN rules. Then (11) holds for each $i = 1, \dots, n$ and for each $A \in \mathcal{F}(\mathbf{X})$. Let us take an arbitrary $i \in \{1, \dots, n\}$ and apply (11) for $A = A_i$. We obtain

$$\bigwedge_{y \in \mathbf{Y}} (B_i(y) \leftrightarrow F_R(A_i)(y)) \geq 1$$

which implies that for each $y \in \mathbf{Y}$, $B_i(y) \leftrightarrow F_R(A_i)(y) = 1$ or that

$$F_R(A_i) = B_i.$$

Therefore, F_R is a model of fuzzy IF-THEN rules in $\mathcal{S}$ by Definition 1.

It is worth noticing that besides the equivalence between the property of being a model and the property of realizing **GMP** with respect to fuzzy IF-THEN rules, Theorem 2 establishes a new solvability criterion of a system of fuzzy relation equations.

Theorem 3 (Criterion of Solvability of a System of Fuzzy Relation Equations). *A fuzzy relation R gives a solution to the system of fuzzy relation equations (4) if and only if for any $i = 1, \ldots, n$ and any $A \in \mathcal{F}(\mathbf{X})$ it satisfies (12), i.e.*

$$\bigwedge_{y \in \mathbf{Y}} (B_i(y) \leftrightarrow (A \circ R)(y)) \geq \bigwedge_{x \in \mathbf{X}} (A_i(x) \leftrightarrow A(x)).$$

Proof. Let R gives a solution to the system of fuzzy relation equations (4) where fuzzy sets A_i, B_i are given by some structure $\mathcal{S}$. Then by Definition 1, it is true if and only if R is the underlying fuzzy relation of a model F_R of fuzzy IF-THEN rules (2) in $\mathcal{S}$. By theorem 2, F_R is a model of fuzzy IF-THEN rules in $\mathcal{S}$ if and only if F_R realizes **GMP** in $\mathcal{S}$. By Definition 2, the last is equivalent to (11) which is an equivalent reformulation of (12).

Summarizing all what has been proved in this Section with the previously established results we can claim that the following statements are equivalent:

(a) a system of fuzzy IF-THEN rules (2) has a model in a structure $\mathcal{S}$,
1. there is a fuzzy function F_R which interpolates among pairs of fuzzy sets (A_i, B_i), $i = 1, \ldots, n$, which are given by $\mathcal{S}$,

(b) the adjoint fuzzy relation R gives a solution to the system of fuzzy relation equations (4) where fuzzy sets A_i, B_i are given by $\mathcal{S}$,
(c) the fuzzy function F_R realizes **MP** in $\mathcal{S}$ with respect to fuzzy IF-THEN rules,
(d) the fuzzy function F_R realizes **GMP** in $\mathcal{S}$ with respect to fuzzy IF-THEN rules.

7 Illustrations

We will illustrate our theory on two sets of pictures. Both demonstrate results of computation on the basis of the Compositional Rule of Inference in two different structures $\mathcal{S}_1$ and $\mathcal{S}_2$. Both structures use standard Łukasiewicz algebra Ł on $[0, 1]$ as a complete residuated lattice and sets with two fuzzy IF-THEN rules. The pictures differ in the fuzzy parameters A_1, A_2, B_1, B_2 as well as in the fuzzy relations $R(x, y)$ computed by the formula

$$R(x, y) = \bigvee_{i=1}^{2} A_i(x) * B_i(y). \tag{17}$$

On Fig. 1, the fuzzy sets A_1, A_2, B_1, B_2 from the structure $\mathcal{S}_1$ and the fuzzy relation R given by (17) are shown. On Fig. 2–5, the adjoint function F_R realizes **GMP** in $\mathcal{S}_1$. This is illustrated by choosing different input fuzzy sets (denoted by A) and shifting them towards the fuzzy set A_1. We observe that the computed outputs (fuzzy sets denoted by B) are getting closer and closer to the fuzzy set B_1. This means that the considered set of fuzzy IF-THEN rules has a model in the structure $\mathcal{S}_1$, and that this model can be given by the fuzzy function F_R. Moreover, the fuzzy relation R (see Fig. 1) solves the system of fuzzy relation equations where the fuzzy sets A_i, B_i, $i = 1, 2$, are given by $\mathcal{S}_1$.

On Fig. 6, the fuzzy sets A_1, A_2, B_1, B_2 from the structure $\mathcal{S}_2$ and the fuzzy relation R given by (17) are shown.

On the second set of Fig. 7–10, the adjoint function F_R does not realize **GMP** in $\mathcal{S}_2$. This is illustrated by choosing different input fuzzy sets (denoted by A) and shifting them towards the fuzzy set A_1. We observe that the computed outputs (fuzzy sets denoted by B) do not come closer to the fuzzy set B_1. Moreover, in the case $A = A_1$, the computed result $A \circ R$ does not coincide with B_1. This means that the considered set of fuzzy IF-THEN rules does not have a model in the structure $\mathcal{S}_2$. Moreover, the fuzzy relation R in Fig. 6 does not solve the system of fuzzy relation equations where fuzzy sets A_i, B_i, $i = 1, 2$, are given by $\mathcal{S}_2$.

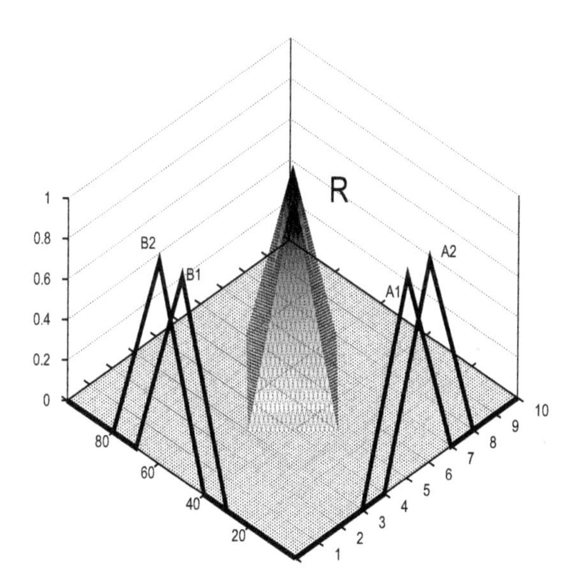

Fig. 1. The fuzzy sets A_1, A_2, B_1, B_2 and the fuzzy relation R from the structure $\mathcal{S}_1$

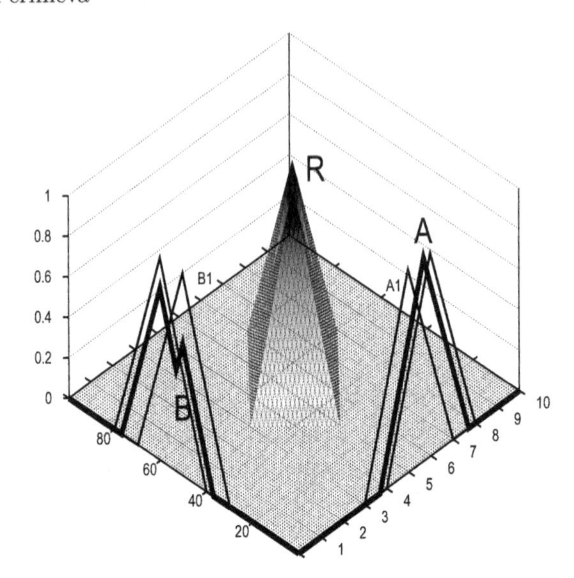

Fig. 2. *Structure* $\mathcal{S}_1$. The fuzzy input A is "far" from A_1 and the same holds for the fuzzy outputs B and B_1

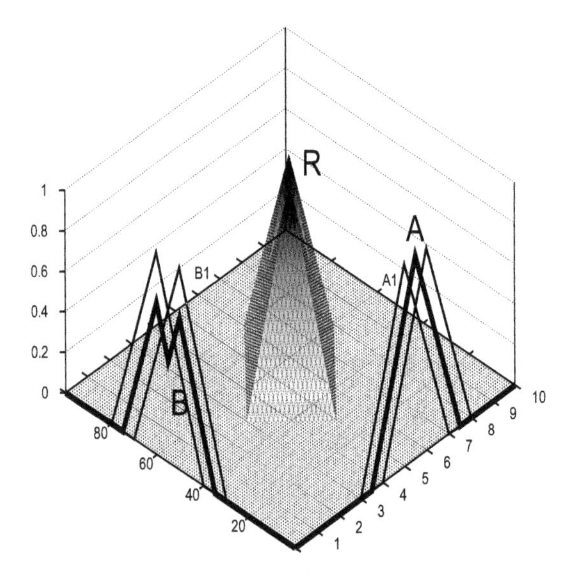

Fig. 3. *Structure* $\mathcal{S}_1$. The fuzzy input A is closer to A_1 and the same holds for the fuzzy output B with respect to B_1

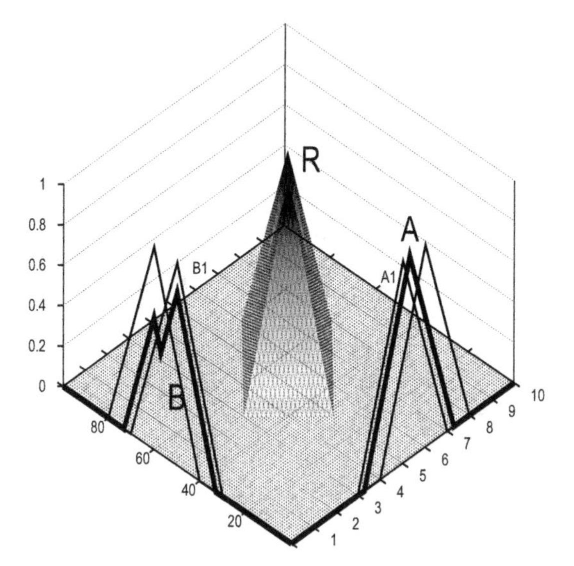

Fig. 4. *Structure* $\mathcal{S}_1$. The fuzzy input A is still closer to A_1 than above and the same holds for the fuzzy output B with respect to B_1

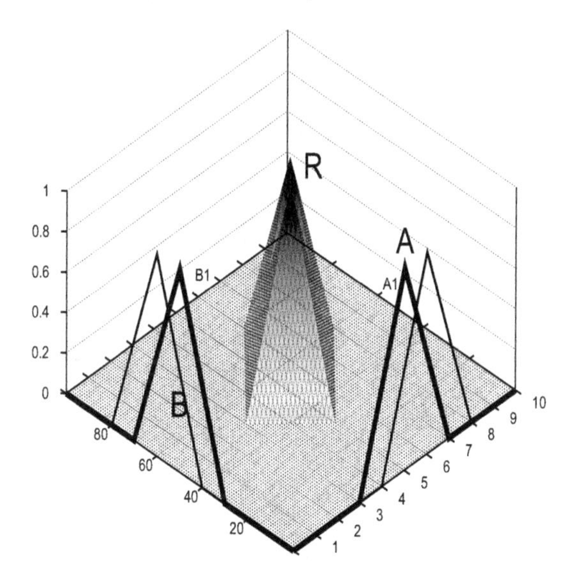

Fig. 5. *Structure* $\mathcal{S}_1$. The fuzzy input A is equal to A_1 and the same holds for the fuzzy outputs B and B_1

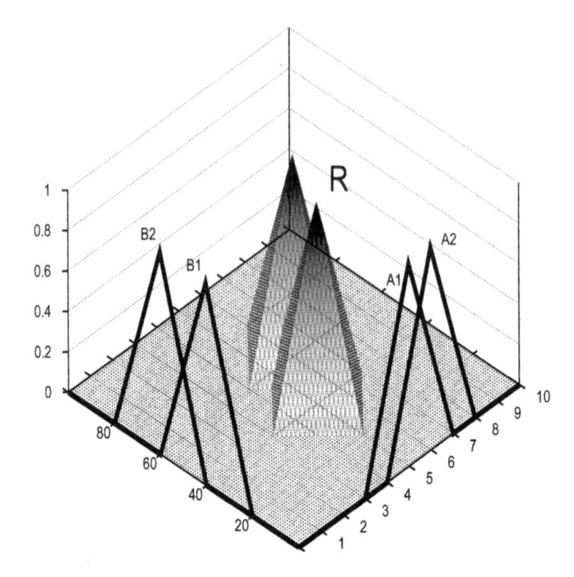

Fig. 6. The fuzzy sets A_1, A_2, B_1, B_2 and the fuzzy relation R from the structure $\mathcal{S}_2$

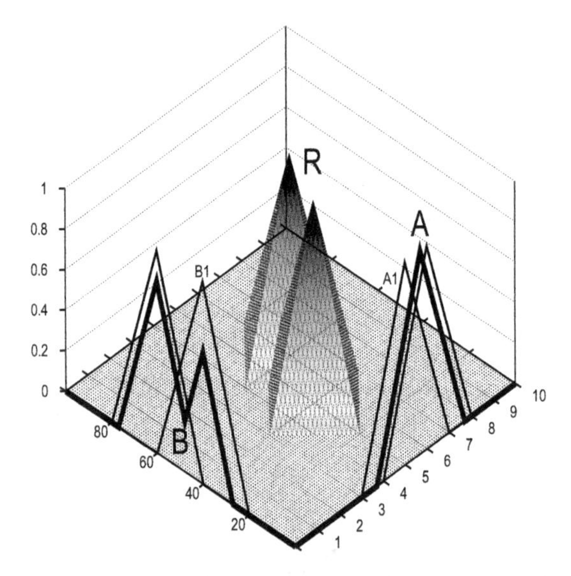

Fig. 7. *Structure* $\mathcal{S}_2$. The fuzzy input A is "far" from A_1 and the same holds for the fuzzy outputs B and B_1

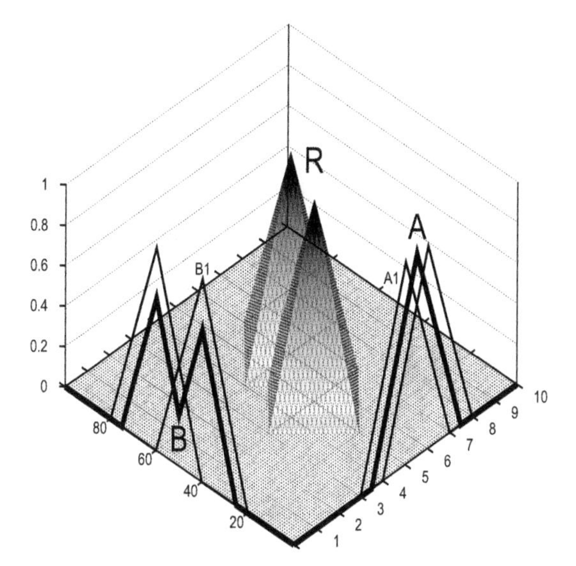

Fig. 8. *Structure* $\mathcal{S}_2$. The fuzzy input A is closer to A_1, but this does not hold for the fuzzy output B with respect to B_1

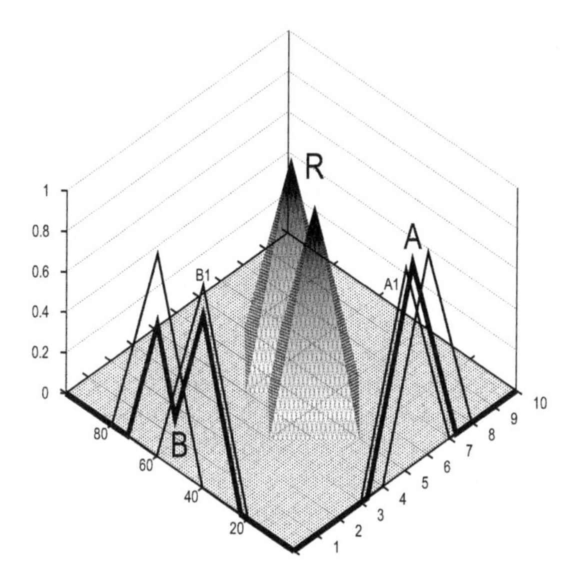

Fig. 9. *Structure* $\mathcal{S}_2$. The fuzzy input A is still closer to A_1 than above, but this does not hold for the fuzzy output B with respect to B_1

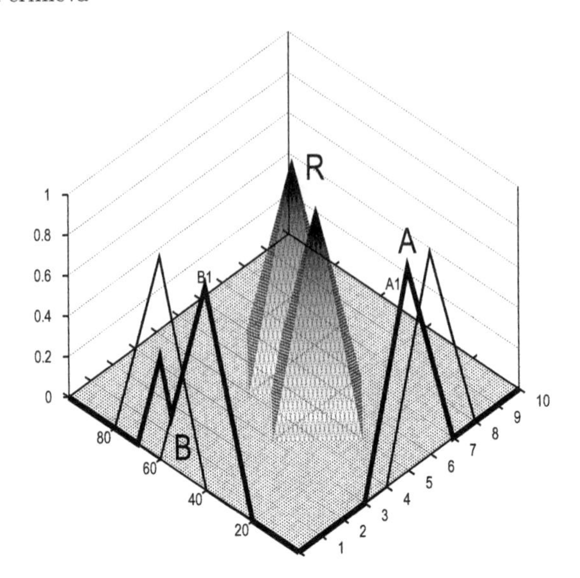

Fig. 10. *Structure $\mathcal{S}_2$.* The fuzzy input A is equal to A_1, but the fuzzy outputs B and B_1 are different

8 Conclusions

We proposed the analytical theory of fuzzy rule-base systems which is based on fuzzy IF-THEN rules considered together with the inference rules *Compositional Rule of Inference*, *Modus Ponens* and *Generalized Modus Ponens*. In this theory a system of fuzzy IF-THEN rules is considered as a partial description of a *fuzzy function* F_R whose domain and range are universes of fuzzy sets. *Compositional Rule of Inference* together with *Modus Ponens* imply that such function must be represented by a fuzzy relation and this relation must solve the respective system of fuzzy relation equations. We proved that *Generalized Modus Ponens* in its functional representation holds true if and only if F_R is a model of the respective fuzzy IF-THEN rules (and equivalently, F_R realizes the rule of inference (**MP**). Summarizing all these facts, our theory is free of disadvantages generally known for rule-based systems. As a by product, we have also established a new criterion of solvability of a system of fuzzy relation equations.

References

1. Baldwin J F and Guild N C F (1980) Modelling controllers using fuzzy relations, Kybernetes, 9: 223–229
2. Bouchon B. (1987) Fuzzy inferences and conditional possibility distributions, Fuzzy Sets and Systems, 23: 33–41

3. Bouchon-Meunier B, Mesiar R, Marsala C, Rifqi M (2003) Compositional rule of inference as as analogical scheme, Fuzzy Sets and Systems, 138: 53–65
4. Di Nola A, Sessa S, Pedrycz W, Sanchez E (1989) Fuzzy Relation Equations and Their Applications to Knowledge Engineering. Kluwer, Dordrecht
5. Dubois D, Prade H (1996) What are fuzzy rules and how to use them, Fuzzy Sets and Systems, 84: 169–185
6. Godo L, Jacas J, Valverde L (1991) Fuzzy values in fuzzy logic, Int. J. Intelligent Systems, 6: 191–212
7. Gottwald S (1993) Fuzzy Sets and Fuzzy Logic. The Foundations of Application – from a Mathematical Point of View. Vieweg, Braunschweig
8. Hájek P (1998) Metamathematics of fuzzy logic. Kluwer, Dordrecht
9. Höhle U (1995) Commutative residuated l-monoids. In: Höhle U, Klement E P (eds) Non-Classical Logics and Their Applications to Fuzzy Subsets. A Handbook of the Mathematical Foundations of Fuzzy Set Theory. Kluwer, Dordrecht, 53–106
10. Klement P, Mesiar R, Pap E (2001) Triangular Norms. Kluwer, Dordrecht
11. Novák V, Lehmke S (2006) Logical structure of fuzzy IF-THEN rules. Fuzzy Sets and Systems, to appear
12. Novák V, Perfilieva I, Močkoř J (1999) Mathematical Principles of Fuzzy Logic. Kluwer, Boston/Dordrecht
13. Perfilieva I, Tonis A (2000) Compatibility of systems of fuzzy relation equations. Internat. J. General Systems 29:511–528
14. Perfilieva I (2003) Solvability of a system of fuzzy relation equations: easy to check conditions. Neural Network World 13:571–580
15. Perfilieva I.(2004): Fuzzy function as an approximate solution to a system of fuzzy relation equations, Fuzzy Sets and Systems, 147:363–383
16. Sanchez E (1976) Resolution of composite fuzzy relation equations. Information and Control 30:38–48
17. Zadeh L A (1973) Outline of a new approach to the analysis of complex systems and decision processes. IEEE Trans. Systems, Man and Cybernet. SMC-3: 28–44
18. Zadeh L A (1975) The concept of a linguistic variable and its applications to approximate reasoning. Information Sciences, 9, 43–80

Fuzzy Logic and Ontology-based Information Retrieval

Mustapha Baziz, Mohand Boughanem, Yannick Loiseau, and Henri Prade

Abstract. Most of information retrieval (IR) approaches relies on the hypothesis that keywords extracted from a document are sufficient to evaluate the relevance of that document with respect to the query. Such an approach may insufficiently lay bare the semantic contents of the documents. In addition to keywords, automatic indexing methods need external knowledge such as thesauri and ontologies for improving the representation of documents or for expanding queries to related keywords. Moreover, ontologies may be combined with a view of for estimating the relevance of documents, the "proximity" between words, or for expressing flexible queries. In this chapter, we survey several recent approaches. Then, two types of methods are discussed in detail. The first one uses a symbolic pattern matching approach, which is based on possibilistic ontologies (where qualitative necessity and possibility degrees estimate to what extent two terms refer to the same thing). The second type of approaches projects fuzzy set representations of queries and documents on a classical ontology, and compare these projections for rank ordering the documents according to a retrieval status value.

1 Introduction

Information Retrieval (IR) is concerned with defining retrieval models that accurately discriminate between relevant and non-relevant documents. The retrieval model specifies how documents and queries are represented, and how these representations are compared in order to produce relevance estimates [1].

Many information retrieval systems represent documents and queries as sets of "bags of words". These approaches assume that the contents of documents themselves are enough for identifying the representative keywords. This simplistic view inevitably leads to many problems caused by the ambiguity of natural language due to polysemy and synonymy. Indeed, the vocabulary extracted in this way from documents is far to be complete, which may introduce an heterogeneity problem with respect to other documents or with respect to queries (e.g. when documents come from different sources). For instance, documents written by different authors, even if they deal with the same field, may use slightly different vocabularies. Moreover, queries specifying user needs may be stated in yet another vocabulary, since the user does not know a priori which terms are used in the documents collection. This issue is still more obvious when using documents written in different languages, as it is the case in multilingual information retrieval [2]. This raises the problem of putting into correspondence the terms used in the different vocabularies. Indeed, natural language words may have several meanings (polysemy), and different words may have the same or very similar meanings (synonymy). Word sense disambiguation attempts to solve this ambiguity by pinpointing which meaning (or concept) is represented by a word or phrase in a given context.

To cope with these issues, various approaches have been developed, in particular *concept-based information retrieval*, which has received much attention in recent years. Concept-based IR approaches (as defined, e.g. in [3–5]) promise to increase the quality of responses since they aim at capturing some part of the semantics of documents. The main idea is that the meaning of a text depends on conceptual relationships to objects in the world rather than to linguistic relations found in texts or dictionaries [6]. The key issue is to identify appropriate *concepts* that describe and characterize the document contents. Thus, the challenge is to make sure that irrelevant concepts will not be kept, and that relevant concepts will not be discarded. Concepts are mental structures, while words and phrases are the linguistic representatives of concepts. Since concepts are abstract entities, representing them is another problem.

One way to facilitate the identification of related concepts and their linguistic representatives given a key concept is to use *ontologies*. They help to build document representations at a higher level of granularity, trying to describe the topical content of documents. These semantic ontologies define relations between terms or concepts, such as *synonymy* or *hypernymy*. However, these relations are generally strict ones, and can therefore be too limited for describing the real world, with all its ambiguity and vagueness, as previously mentioned. Moreover, they have to interact with flexible queries, where the user express preferences (e.g. [7–9]). Therefore, *fuzzy ontologies* in which relations between terms are weighted or encoded by different kinds of measures, may be used. Many other works use fuzzy logic in IR to solve the ambiguity and vagueness issues, by defining flexible queries or fuzzy indexes (e.g. [7,10–12]). However, we will focus here on the use of fuzzy logic in ontologies to represent fuzzy relational knowledge about words or phrases.

The rest of the chapter is organized as follows. First of all, recent approaches in concept-based information retrieval and fuzzy ontologies are surveyed (Sect. 2). Then two particular approaches are overviewed and discussed.

The first approach, presented in Sect. 3, defines a fuzzy ontology assessing the strengths of specialization and approximate synonymy relations in terms of possibility and necessity measures. This ontology is then used in a symbolic pattern matching framework to estimate the relevance of documents with respect to a flexible query. The documents and the query are indexed using the ontology, allowing its use to propagate the relations between terms in documents and query terms.

In the second approach presented in Sect. 4, both documents and queries are represented by means of weighted trees. The evaluation of a conjunctive query is then interpreted as computing a degree of inclusion between sub-trees (the sub-tree of the query and the ones related to each document of the collection). The ontology-based description of the contents of the documents takes into account the semantic equivalences between expressions, as well as the basic principle stating that if a document heavily includes some terms, it also concerns to some extent more general concepts. This latter point is handled at the technical level by a completion procedure that assesses positive weights also to terms that do not appear directly in the documents.

2 Concept-based IR and Fuzzy Logic

In order to improve the relevance of retrieved documents, IR systems often make use of linguistic resources, to expand queries, represent queries and/or documents. These resources can be weighted to introduce fuzzy relations in the matching process, or this matching process can itself be fuzzy in nature. This section presents recent approaches using such resources.

2.1 Linguistic Resources

Several kinds of knowledge or linguistic resources, such as controlled vocabularies, taxonomies, thesauri or ontologies, are commonly used in information retrieval in order to improve performances with respect to simple statistical text analysis. These resources can be classified according to their level of abstraction or the knowledge they contain.

A *controlled vocabulary* is a list of unambiguous valid terms (i.e. words or expressions). Terms that may have several meanings are qualified in order to resolve the ambiguity.

A *taxonomy* is a hierarchical controlled vocabulary. Several hierarchical relations can be found, and depend on the context of use of the resource (type–instance, whole–part,. . .).

A *thesaurus* is a controlled vocabulary where terms are associated. In the context of information retrieval, this association can be automatically computed by analyzing cooccurrences of terms in a learning corpus, or by defining a similarity measure between terms.

An *ontology* introduces the notion of concept, corresponding to ideas or meanings that can be represented by several related terms. Many different kinds of semantic or linguistic relations can be defined between terms or concepts, such as synonymy, hypernymy (is-a), meronymy (part-of) relations. These relations and their representations are more formal than in a taxonomy, since ontologies are generally used to model complex knowledge about the real world, and to infer additional knowledge as it is the case in the semantic web. These kinds of resources are commonly used in information retrieval for expanding queries using related terms (e.g. [13]) or for indexing both documents and queries in a common controlled space.

An example of widely used ontology is WordNet [14]. This english ontology contains general terms organized in *synsets* (sets of synonymous terms) related using semantic relations. A short description is given to help identifying the concept represented by the synset.

Since a word can have several meanings, it can pertains to different concepts. This introduce the main problem of automatic use of linguistic resources in information retrieval: the ambiguity of natural language. Many approaches try to disambiguate words in documents and queries by using linguistic resources, as in query expansion, or to use them as in concept-based indexing.

2.2 Classical Ontologies and IR

Concept-based IR approaches try to increase the quality of responses of information retrieval systems by extracting semantics from documents, often using ontologies. Two key contributions can then be distinguished: query re-engineering and document representations. In both tasks, the expected advantage is to get a richer and more precise meaning representation in order to obtain a more powerful identification of relevant documents meeting the query. In concept-based IR, sets of words, names, noun phrases are mapped into the concepts they encode [15]. By these models, a document is represented as a set of concepts: to this aim a crucial component is a conceptual structure for mapping concepts to document representations. These conceptual structures can be general or domain specific. They can be either manually or automatically generated or they may pre-exist [16]. A conceptual structure can be represented using distinct data structures: trees (ex. [3] or [17] in the medical domain), semantic networks [18, 19] or conceptual graphs [6, 20–22].

In [3], Woods proposed a conceptual indexing method by mapping words and phrases onto conceptual taxonomies. As proposed in the project by Sun Microsystems, this approach brings 60% accuracy when applied to index short collections (about 10MB of UNIX manual pages). However, as shown in [23], this is not a limit; conceptual indexing can also be applied successfully to larger collections, and possibly used for the exploration of the Web. Nevertheless, one can notice that in practice, works reporting on experiments dealing with large collections are rare. Only some ones using Latent Semantic Indexing (LSI) [24, 25] are carried out, even though this technique (LSI) is not essentially a conceptual approach since it is based on cooccurrence to bring closer the words appearing in the same context. In [6] and [21], the use of conceptual graphs for representing documents and queries is discussed. The authors proposed a method for measuring the similarity of phrases represented as conceptual graphs. Some investigation are also done in the understanding of conceptual graph expression power for IR and the integration of the facet notion into the theoretical model (see e.g. [22]).

Gonzalo et al. [15] proposed an indexing method based on WordNet synsets. Each document is indexed by the identifiers of the synsets to which its keywords belong. These synsets are determined by manually disambiguating the documents keywords. The authors reported a gain of 29% in retrieval accuracy compared to a classical keyword-based approach. However, their experiments showed that this improvement is quite sensitive to disambiguation errors.

Other authors used methods for measuring the similarity between concepts on the basis of the semantic network of an ontology, e.g. a domain ontology for sport [5], or WordNet synsets [26] when dealing respectively with audio sport transcriptions and XML data. These similarities are mainly used for managing the expansion of queries. Navigli et al. [27] presented in their system (OntoLearn) a method called *structural semantic interconnection* in order to disambiguate words in queries using WordNet glosses (definitions with possibly example(s) from real world). Moreover, ontology concepts that are semantically related to the query can be added in order to expand the query and to select more documents, namely those using a different vocabulary but dealing with the same topic [5, 16, 28, 29]. For instance, the *Ontoseek* information retrieval system [16] uses WordNet to expand queries.

In most of the previous approaches, a strict comparison is used to match document concepts with concepts belonging to a crisp ontology. However, this may be too rough in practice. To allow for a more flexible use of ontologies, fuzzy logic may be introduced in the processing of the knowledge contained in these resources. The fuzzification can be introduced in the ontology itself by weighting the relations between concepts, or by representing concepts themselves as fuzzy sets of terms. Weights may also be used in the matching process between document concepts and query concepts in IR.

Ontologies are also used for improving the clustering of users' profiles [30]. These ontologies represent knowledge on the domains of interest of the users. While relations between concepts in the ontologies are crisp, the users' profiles are then linked through the ontology, which is used to compute a similarity measure between them in order to define a weighted graph that allows for a more accurate clustering system.

2.3 Fuzzy Ontologies

Some authors considered relevance rather than similarity between terms [28], where degrees representing terms specialization and generalization are introduced. These degrees are asymmetric, generalization being less favored (from a relevance point of view) than specialization. As an example, *poodle* specializes *dog* at 0.9 whereas *dog* generalizes *poodle* at 0.4. In [31], a fuzzy ontology is used to summarize news. This ontology is built by fuzzifying an existing ontology using a fuzzy inference mechanism, based on several similarity measures between terms. These measures are computed by textual analysis of corpora. The fuzzy inference inputs are a *part-of-speech distance*, a *term word similarity* that counts the number of common Chinese ideograms in expressions or phrases, and a *semantic distance similarity* based on the distances in the crisp ontology.

Statistical analysis of texts can be used to estimate similarities between terms in order to define thesauri that may be interpreted in a fuzzy way, as presented in [32], where fuzzy sets of terms similar to a given term are viewed as representing concepts. These thesauri are used to reformulate queries in order to retrieve more relevant documents. Another approach [33] uses a weighted multilingual ontology to exploit multilingual documents in a translation and search process. In this system, the multilingual ontology is used to translate and expand the query. The query is stated as a subgraph of the ontology, by selecting concepts judged to be relevant. The relevance of each concept can be weighted by the user. The matching is done by computing the inclusion of the query representation in the document representation. A similar approach is presented in [34], where authors used an automatically built fuzzy ontology to expand and refine the user query. The ontology is built using WordNet to extract keywords from a collection of documents, the fuzzy relations being computed as in [32], a statistical analysis of cooccurrences of terms gives a "'fuzzy"' value of there hierarchical relations. The ontology is then pruned to eliminate redundant and meaningless relations.

Query expansion using ontologies is also presented in [35], where authors used a context sensitive semantic query expansion based on a fuzzy concept hierarchy. This hierarchy is based on a fuzzy inclusion relation, computed as the transitive closure of both *specialization* (is-a) and *composition* (part-of) fuzzy relations. Each word in a query is therefore expanded by all the concepts that are included in it according to the fuzzy hierarchy.

To support semantic search in documents collection, [36] have proposed an ontology-based retrieval model exploiting a hand-made domain ontology and a knowledge base. They adapted the classical vector-space model in order to use a representation of documents and queries based on the ontology. Documents are therefore semantically indexed by *annotating* them using instances from the knowledge base that are classified in the ontology. This semi-automatic process is done by searching representative labels of the concept in the document. Annotations are weighted by an adapted version of the classical $tf * idf$ measure*, substituting the ontology concepts to the classical keywords index. The query evaluation is therefore done by selecting instances of the knowledge

*The $tf * idf$ is a weight often used in information retrieval. The term frequency (tf) in the given document gives a measure of the importance of the term t_i within the particular document. The inverse document frequency (idf) is a measure of the general importance of the term (it is the logarithm of the number of all documents divided by the number of documents containing the term) base that satisfy the query, and retrieving documents annotated by these instances by computing the similarity between the two vectors of concepts representing the query and the document. Annotations weights are then used to rank results. This work is extended in [37] to deal with vague queries based on user's preferences represented by weighted ontology concepts.

These approaches use ontologies in a possibly fuzzy way, either in the ontologies representation or in its use. The following sections present other approaches using ontologies for matching queries and documents, using either a fuzzy ontology as an inference framework, or a crisp one to estimate a fuzzy logic degree of matching.

3 Symbolic Pattern Matching on Possibilistic Ontologies

The fuzzy pattern matching framework [38] provides a tool for evaluating flexible queries in face of possibly imprecise data, where each linguistic label, both in the data or in the query, is represented by a fuzzy set. It is used to formulate flexible queries by fuzzy sets, evaluated on imprecise or fuzzy data also represented by fuzzy sets. This technique estimates to what extent it is possible and to what extent it is certain that data represented by imprecise attributes fulfill a flexible query representing the user needs and preferences. Fuzzy sets can interface numerical values with linguistic terms, using membership functions, and the comparison of terms can be evaluated by fuzzy pattern matching. However, the terms used for describing information need to be represented by a fuzzy set on a clearly identified domain, which is a severe limitation for information retrieval purposes.

In order to deal with more general linguistic terms, the fuzzy pattern matching idea has been adapted to symbolic labels [39, 40] recently. The matching between query and document no longer requires the identity of terms, but rather a *semantic similarity* is computed between these terms, in a qualitative matching process.

Similarity measures between words have been extensively studied in information retrieval literature, using for instance distance between nodes in a taxonomy, or based on a common information probability (e.g. [41]). Besides, a usual method when a query fails is to replace it by similar queries generated using ontologies or thesauri (e.g. [42]). Qualitative pattern matching remedies limitations of fuzzy pattern matching by enlarging the meaning of a term through the fuzzy set of its similar terms.

In the following, the relations between terms are provided by means of an ontology estimating approximate synonymy and hypernymy relations between terms by means of possibility and necessity degrees.

3.1 Possibilistic Ontology

In these ontologies, the vocabulary of a domain i is defined by a set of terms $\mathcal{T}_i = \{t_i^j, j = 1, \ldots, n(i)\}$, where t_i^j is a label (e.g. *hotel*), that can be used in order to describe a piece of information (here, an accommodation place). Their meanings are related through the corresponding ontology O_i where relations are modeled by possibility and necessity degrees. For two labels t_i^j and t_i^k:

- $\Pi(t_i^j, t_i^k) = \Pi(t_i^k, t_i^j)$ represents to what extent t_i^j and t_i^k can describe the same thing. A zero possibility means that the two labels never represent the same thing. A positive possibility lesser than 1 expresses that the two terms may mean the same thing, but it is not always the case. If the two meanings overlap, but are not true synonyms nor specializations: $\Pi(t_i^j, t_i^k) = 1$ and $N(t_i^j, t_i^k) = N(t_i^k, t_i^j) = 0$;
- $N(t_i^j, t_i^k)$ estimates to what extent it is certain that t_i^k is a specialization of t_i^j. Moreover, $N(t_i^j, t_i^k) = N(t_i^k, t_i^j) = 1$ represents genuine synonymy. If t_i^k is a perfect specialization of t_i^j, then $N(t_i^j, t_i^k) = 1$. However, a zero value only means a total lack of certainty for the specialization relation between these terms. This relation is asymmetric.

If the meanings of the two terms are clearly distinct, we have $\Pi(t_i^j, t_i^k) = N(t_i^j, t_i^k) = N(t_i^k, t_i^j) = 0$.

These measures must satisfy the following properties:

- Reflexivity: $\Pi(t_i^j, t_i^j) = 1$.
- Symmetry: $\Pi(t_i^j, t_i^k) = \Pi(t_i^k, t_i^j)$.
- $\Pi(t_i^j, t_i^k) \geq N(t_i^j, t_i^k)$, since specialization entails that the meanings overlap.
- $N(t_i^j, t_i^k) > 0 \Rightarrow \Pi(t_i^j, t_i^k) = 1$. If it is somewhat certain that t_i^k specialize t_i^j, then it must be fully possible that they are used for referring to the same thing.

The degrees specified in the ontology are actually only defined on a subset of the Cartesian product of the vocabulary $\mathcal{T}_i \times \mathcal{T}_i$. They can be completed using previous properties and the two following forms of transitivity:

$$N(t_i^j, t_i^h) \geq \min\left(N(t_i^j, t_i^k), N(t_i^k, t_i^h)\right) , \quad (1)$$

$$\Pi(t_i^j, t_i^h) \geq N(t_i^j, t_i^k) * \Pi(t_i^k, t_i^h) . \quad (2)$$

where $*$ is defined as:

$$a * b = \begin{cases} b \text{ if } b > 1 - a , \\ 0 \text{ otherwise.} \end{cases}$$

Equation (1) represents the specialization transitivity [43]. The "hybrid transitivity" (2) states that if t_i^k specializes t_i^j and if t_i^k and t_i^h may refer to the same thing, then the meanings of t_i^j and t_i^h should overlap as well; see [44] for a proof of formulas (1) and (2).

Therefore, values that are not specified can be deduced from existing ones using the previous properties and relations. Values that cannot be inferred are supposed to be zero. The fact that default possibility is zero corresponds to a closed world hypothesis, since it is supposed that two terms cannot overlap if it is not specified. From a practical point of view and to simplify the use of the degrees, evaluations will be estimated "at worst," and the $\geq$ will be generally taken as an equality. It is therefore possible to estimate the relevance degrees between the data and the query even if the searched terms are not directly present in the information representation without using any explicit query expansion stage, as usually proposed in IR.

As a matter of illustration, Fig. 1 presents a fragment of a simple ontology for accommodation places. This graph is a simplified representation of how an agent may perceive similarity relations between terms. Since it is possible to deduce implicit values for relations between terms, using properties and constraints of the used degrees, only direct links need to be given and are represented here.

In Fig. 1, note that words like *lodge* and *inn* are only considered as *possible* synonyms, or as entity that can provide the same services. Nothing can be inferred for the necessity from the possibility degree, and it can exist *lodges* that are not *inns*. On the other hand, both necessity degrees between *motel* and *motor inn* being 1, these terms are considered as genuine synonyms.

The values of degrees present in such ontologies are qualitative in nature, and estimates semantic relations between terms. As an example (in Fig. 1), $N(hotel, motel) = 0.6$ means that it may exist *motels* that are not considered as *hotel*, but that *generally*, *motels* are a kind of *hotels*. Thus, despite the use of numerical values, only the relative ordering between these values is significant, as the purpose is to rank-order query results and not to assess an absolute similarity degree. Practically, only few levels should be used, e.g. $\{0, 0.4, 0.6, 1\}$. These values may be associated for convenience with linguistic labels, such as "very similar," "rather similar," etc. to specify the strength of the relations.

As in the approach presented in [28] (see Sect. 2), two kinds of degrees are used, but with a different meaning. The possibility degree is symmetrical, and a positive necessity for $N(t_i^j, t_i^k)$ implies nothing for $N(t_i^k, t_i^j)$, contrary

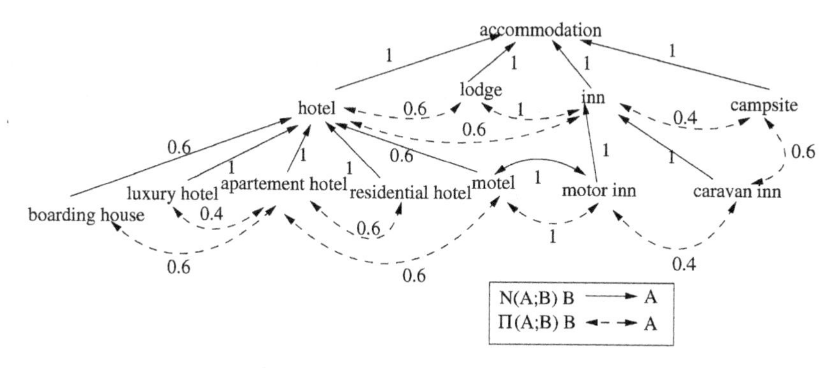

Fig. 1. Accommodation ontology

to specialization and generalization degrees of two reversed pairs which are simultaneously strictly positive as in the above example. Moreover, the product used as a transitivity operator in [28] leads to a weakening of association weights between terms with the distance in the ontology. Here, the *min* operator implies that the matching is independent from the ontology granularity (inserting a new term between two terms cannot change their similarity).

Building such ontologies is a complex task, specially if done manually even on a task domain. Ontologies such as WordNet [14] can be used as a starting point. For instance, typed relations used in these ontologies, such as hypernymy can be matched with necessity degrees. Other relations such as "being a part of," e.g. a *room* and a *hotel*, can be interpreted in terms of possibility degrees. Besides, statistical ontologies can also be built from corpora analysis, extracting relations from terms co-occurrences (e.g. [45]). This may provide a basis for assessing values in possibilistic ontologies using both crisp semantic ontologies and possibilistic rescaling of probabilities. Subparts of general ontologies can be identified in order to obtain domain specific ontologies.

3.2 Possibilistic Indexing

Documents are viewed as a set of keywords, obtained by lemmatizing their significant terms and forgetting the stop-words. Keywords correspond to a unique multiple-valued attribute, in which terms pertain to the same global domain $\mathcal{T}$. To be homogeneous with the ontology model, the association between documents and the ontology nodes must be stated using the same possibility and necessity degrees, taking into account the statistical weights of the terms in the documents. Classically, the significance weight ρ_i^j for a given term t_i with respect to a document D_j is computed by combining its frequency tf_{ij} in D_j and its inverse frequency $idf_i = \log(d/df_i)$, where df_i is the number of documents containing t_i and d is the number of documents in the collection. The weights ρ_i^j are assumed to be rescaled between 0 and 1. The document D_j is therefore represented by the fuzzy set of its significant terms [9, 46]:

$$D_j = \{(\rho_i^j, t_i), i = 1, \ldots, n\}$$

where n is the number of terms in the ontology.

Assuming that the ρ_i^j is an intermediary degree between the possibility and the necessity that the term describes the document, a simple and parametrized way to assess the Π and N degrees from the ρ_i^j is to use the following piecewise linear transformation, based on [47]:

$$\Pi(t_i, D_j) = \begin{cases} 0 & \text{if } \rho_i^j = 0 \\ 1 & \text{if } \rho_i^j \geq \alpha \\ \frac{\rho_i^j}{\alpha} & \text{otherwise} \end{cases} \tag{3}$$

$$N(t_i, D_j) = \begin{cases} 1 & \text{if } \rho_i^j = 1 \\ \frac{\rho_i^j - \alpha}{1-\alpha} & \text{if } \alpha < 1 \text{ and } \rho_i^j \geq \alpha \\ 0 & \text{otherwise} \end{cases} \tag{4}$$

The intuition underlying (3–4) is that a sufficiently frequent term in the document is necessarily somewhat relevant, while a less frequent term is only possibly relevant. It has been shown in [48] that the best value for α depends on the terms' weight repartition in the collection.

3.3 Symbolic Pattern Matching

Queries are conjunctions of disjunctions of possibly weighted terms. It is assumed that all the terms used in the query are in the ontology. This can be practically achieved by enforcing the user to choose query terms in the ontology. Thus a query R may be viewed as a conjunction of fuzzy sets R_i representing a disjunction of flexible user needs. Moreover, weights ω are introduced at the conjunctions level in order to express the relative *importance* of the elementary requirements of the query. Namely,

$$R = \bigwedge_k \left(\omega_k, \bigvee_j (\lambda_k^j, t_k^j) \right) , \text{ with } t_k^j \in \mathcal{T} .$$

The weights $\lambda_i^j \in [0, 1]$ reflects how satisfactory this term is for the user (i.e. how well it corresponds to his/her request). It is assumed that $\max_j \lambda_i^j = 1$, i.e. at least one query term reflects the exact user requirement. The importance weighting ω_k obey the same constraint as the weights λ_k^j. In practice, disjunctions are between terms which are more or less interchangeable for the user (the weight λ_k^j expressing his/her preference between them). The weight ω_k expresses how compulsory is each elementary requirement in the conjunction. Note that weighted disjunctions allow us to define new concepts. As an example, a user can specify its own definition of a *cosy lodging* as:

$$(0.5, \mathit{lodge}) \vee (0.7, \mathit{motel}) \vee (0.8, \mathit{apartment\ hotel}) \vee (1, \mathit{luxury\ hotel}) .$$

Given a collection of documents indexed using an ontology as described in Sect. 3.2, the possibilistic query evaluation consists in retrieving all documents D such that the possibility of relevance $\Pi(R, D)$ or the necessity of relevance $N(R, D)$ are non-zero. Taking into account the possibility and the certainty of significance as given by (3–4), leads for a query R and a document D, to the following relevance status value (rsv):

$$rsv(R, D) = (\Pi(R, D), N(R, D)) ,$$

where relevance degrees are given by:

$$\Pi(R, D) = \min_k \max \left(1 - \omega_k, \max_{i,j} \min(\lambda_k^j, \Pi(t_k^j, t_i), \Pi(t_i, D)) \right) , \quad (5)$$

$$N(R, D) = \min_k \max \left(1 - \omega_k, \max_{i,j} \min(\lambda_k^j, N(t_k^j, t_i), N(t_i, D)) \right) . \quad (6)$$

The $\max_{ij}$ parts are weighted disjunctions corresponding to those in the query (where a fuzzy set of more or less satisfactory labels expresses a disjunctive requirement inside the same domain). In the same way, as the query is a conjunction of elementary requirements, and since the ω_k's are importance weights, the weighted min operator is used in the final aggregation. Therefore, having an importance weight ω_k less than 1 leads to retrieve results violating the corresponding elementary requirement with a degree at most equal to $1 - \omega_k$. Note that if R contains a disjunction of redundant terms, that is, $R = t \vee t'$ and $N(t, t') = 1$ in the ontology, it can be checked that evaluating t and $t \vee t'$ leads to the same result.

$\Pi(R, D)$ and $N(R, D)$ values estimate to what extent the document D corresponds possibly and certainly to the query R. Results are sorted first using decreasing values of $N(R, D)$, then decreasing values of $\Pi(R, D)$ for documents having the same necessity value.

This matching process can be applied to databases, classical or fuzzy ones, or adapted to information retrieval for collections of sentences or keywords, as presented and evaluated in [49]. Besides, in the above formulas, the aggregation of the evaluations associated with each elementary requirement is performed by means of the conjunction *min*. It is well known in information retrieval that the minimum operation is often too restrictive in practice, and is usually outperformed by other operations such as the sum. However, it has been shown in a recent work [48,50], that it is possible to refine the minimum operation (using a leximin ordering on ordered sets of values to be compared), and to obtain results as good or even better than with the sum. Such a refinement can be applied also in the above approach.

3.4 Discussion

The approach described in this section is an adaptation of fuzzy pattern matching to purely linguistic terms. The main idea is to retrieve information containing terms that may not match exactly those of the query. To cope with this point, a *possibilistic ontology* is used, where the relations between therms are stated by the possibility and the certainty that their meanings refer to the same thing. This allows to specify semantic relations, such as synonymy or specialization and generalization of meanings. Thanks to the transitivity properties of possibilistic ontologies, relations that are not explicitly stated can be deduced. A property of this model is the independence of the similarity with respect to the hierarchical distance of terms in the ontology, and therefore to the granularity of the vocabulary. Since the matching is qualitative in nature, results can be rank-ordered even though no document matches exactly the query without reformulation. However, due to the implicit query expansion, the ontology is an important aspect of the system efficiency, and therefore this approach depends on the quality of the ontology used. By introducing weights in the query, the user can represent his/her preferences and priorities, as well as take into account some knowledge about the collection.

4 IR Model Based on Projecting Documents on Ontologies

This section presents a recently proposed approach to information retrieval based on the use of a fuzzy conceptual structure (ontology) that is used both for indexing documents and expressing user queries [51–53]. The conceptual structure is hierarchical and encodes the knowledge of the topical domain of the considered documents. This structure is formally represented as a weighted tree. In this approach, the evaluation of conjunctive queries is based on the comparison of minimal sub-trees containing the two sets of nodes corresponding to the concepts expressed in the document and the query respectively. The comparison is based on the computation of a multiple-valued degree of inclusion. Some candidate implications are discussed on the basis of their respective semantics. The proposed approach generalizes standard fuzzy information retrieval.

This part is organized into six main subsections. The first one (Sect. 4.1) gives a synthetic overview of the proposed approach and the remaining sections develop the approach as follows. In Sect. 4.2, the proposed concept-based representations of documents and queries are introduced. In Sect. 4.3, the process of the conceptual query evaluation is described. Sections 4.4 and 4.5 give some details concerning the use of the approach in practice. They describe respectively the concept detection stage and the proposed empirical method for pruning abstract nodes from the sub-trees. Finally, some concluding remarks are given in Sect. 4.6.

4.1 Summary of the Approach

The principle of the proposed approach aims at representing both documents and queries by means of sub-trees. A document/query is then represented by a set of concepts corresponding to nodes in the hierarchical structure of the ontology. Each node in the resulted sub-trees corresponds to a disambiguated term from a document/query that matches one concept of the ontology. Sub-trees are obtained by considering only the *subsumption* relation represented in our case by a classical $is-a$ relation (hypernymy). The idea behind this representation is to complete the document/query description by possibly adding intermediate nodes in order to complete these representations by concepts that do not appear explicitly in a document and/or a query but that deal somewhat with the same topic.

Roughly, as it can be seen in Fig. 2, in order to estimate the relevance of a document from the archive compared to a given query, we proceed as follows. Initially both the document and the query are projected onto ontology and a corresponding subtree is built for each of them. At this stage, intermediate nodes are possibly added to the two representations in order to get subtree structures. In a second step, a minimal subtree containing the two initial ones is built, then both the subtree representing the document and the query

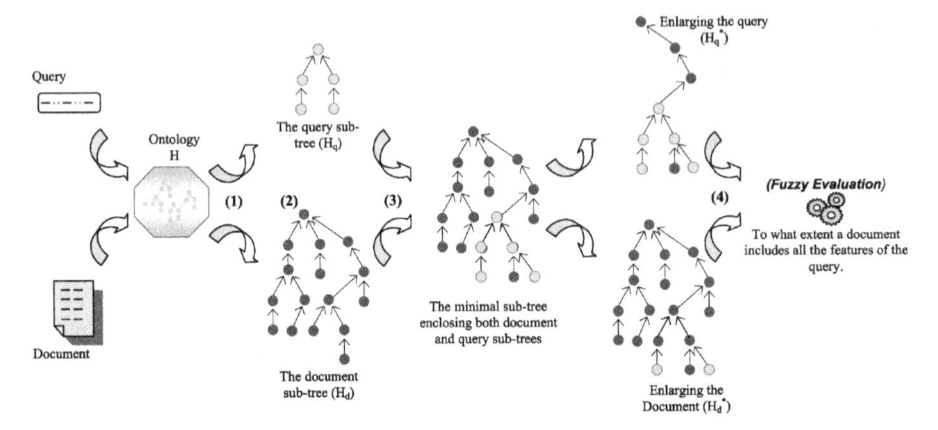

Fig. 2. Synoptic scheme of the approach

are enlarged by adding new intermediate nodes from the minimal subtree. Finally, the two enlarged subtree representations are compared using fuzzy connective implications to determine to which extend the document includes all the features of the query. More details of this approach are given in the next sections.

4.2 Concept-based Representation of Documents and Queries

In the proposed approach, the query evaluation of textual documents is supposed to be mediated by an ontology made by a unique tree-like hierarchy H of concepts, which are supposed to be sufficient for describing the contents of the considered documents with an appropriate level of accuracy. Leaves in H can be thought as simple keywords, i.e. keywords expressing specialized concepts, while other nodes refer to keywords which are labels of more general concepts. Edges in this hierarchy represent the classical *is-a* link. Both documents and queries are supposed to be interpreted or expressed in terms of labels of nodes of H, possibly in association with weights.

Let d be a document. Each document d is identified by means of a set of pairs $R_d = (w_i, \alpha_i), i = 1, \ldots, k(d)$ where w_i is a key word or phrase taken from d that corresponds in a univocal way to a concept c_i from H and α_i is its importance weight (index term weight) computed by an occurrence-based indexing function described in [54, 55], $k(d)$ is the number of terms in document d. We then compute the projection H_d of d on H, in the following way:

1. a weighted subset $N(H, d)$ of nodes of H, namely $N(H, d) = (n_j, \gamma_j), j = 1, \ldots, m(d)$ where for any node n_j, there exists $w_i \in d = \{(w_i, \alpha_i), i = 1, \ldots, k(d)\}$ such that $n_j = w_i$, and n_j is known as the most appropriate concept (or node) [56] that represent better w_i

in the conceptual structure H, and then we take $\gamma_j = \alpha_i$. $m(d)$ is the number of nodes in H that are equivalent to some terms in R_d. When several equivalent expressions w_i in H exist, the longest term is retained as described in Sect. 4.4.
2. H_d is the minimal sub-tree of H which contains $N(H,d)$, where the weights associated with the nodes are those obtained at step 1 if the nodes belong to $N(H,d)$ and are 0 otherwise.

Remark 1. If necessary, one may restrict $N(H,d)$ to those expressions for which γ_j is sufficiently high, thus specifying an acceptance threshold.

Let q be a query obtained by selecting a collection of labels (concepts) in H, with possibly an importance weighting, namely as a set $(l_k, \delta_k), k = 1, \ldots, r(q)$. We assume that the query is viewed as a (weighted) conjunction of the concept labels. A query q is also modeled by a sub-tree H_q of H. Namely H_q is the minimal weighted sub-tree of H containing $(l_k, \delta_k), k = 1, \ldots, r(q)$, keeping the weights k, and putting 0 on the other nodes of H_q.

4.3 Query Evaluation

Query evaluation is based on the comparison of two weighted subsets of nodes of H, one corresponding to the query and the other to the current document. First, a minimal sub-tree of H containing both subsets is determined, then an inclusion degree is computed for evaluating to what extent a document includes all the features of the query. The use of aggregation functions and various implication connectives in the definition of the inclusion degree is discussed. Lastly, some possible procedures for completing the description of the document or extending the query, by propagating weights in H, are presented.

Comparison Based on the Minimal Common Sub-tree

Let H_E be the minimal non-weighted sub-tree which contains both H_d and H_q. Let H_d^* and H_q^* be the extensions of H_d and H_q on H_E putting zero weights on the nodes of $H_E - H_d$ and $H_E - H_q$ respectively. The evaluation of a query q with respect to a document d, is performed in terms of a degree of relevance $rel_c(d;q)$ of d with respect to q computed as degree of inclusion of H_q into H_d, namely

$$rel_c(d;q) = \min_{n \in H_E} \mu_{H_q^*}(n) \rightarrow H_d^*(n) \tag{7}$$

where $\mu_{H_d^*}(n)$ (resp. $\mu_{H_q^*}(n)$) is the weight associated with node n in H_d^* (resp. H_q^*), and $\rightarrow$ is a multiple-valued implication connective expressing that all the concepts of the query should appear in the description of the document. One may think of introducing equivalence connectives in place of implications in

(7) for requiring that the topic of the document corresponds exactly to the topic of the query.

However, note that looking for exact matching may be dangerous: suppose we are looking for documents dealing with a topic $A(q = A)$ but there does not exist any document dealing with A without $B(d = A, B)$; in such a case the exact matching strategy will give nothing. However, a strict equivalence could be relaxed into an approximate similarity by weakening the equivalence connective by means of a similarity relation. A strict conjunctive evaluation as in (7) may be too requiring, and in information retrieval, "best matching" is usually preferred to exact matching. Therefore, a simple function that is known to allow best matching is the *sum* computed as follows:

$$Sum : rel_d(d; q) = \sum_{n \in H_E} \mu_{H_q^*}(n) \rightarrow H_d^*(n) \tag{8}$$

Other refinements of the *sum* could also be used in order to perform the ranking function. This means that not only the whole *sum* computed in (8) is considered for rank-ordering documents, but also partial summations of terms appearing in (7). The idea is the following.

Definition 1 (Lexisum strict, Lexisum1, Lexisum2). *Let us denote* $rel_d(d; q) = \sum_n t_n$ *(where* $t_n = \mu_{H_q^*}(n) \rightarrow H_d^*(n)$*). First, rank-order the* t_i*'s increasingly: assume* $t_1 \leq t_2 \leq \ldots \leq t_k$*. Then build the vector:* $(t_k + \ldots + t_1, t_k + \ldots + t_2, \ldots, t_k)$*. Lastly, the documents are rank-ordered on the basis of the lexicographic ordering of the vector, i.e.,* $(a_1, \ldots, a_k) >_{lex} (b_1, \ldots, b_k)$*. If* $\exists i < k \; \forall j \leq i \; a_j = b_j$ *and* $a_i + 1 > b_i + 1$*.*

When ranking, we may use a strict equality $a_j = b_j$ *or an approximate one* $(\frac{|a_i + b_j|}{a_i} \leq 5\%)$*. The second refinement follows the same procedure, except that the vector is now* $(t_k + \ldots + t_1, t_{k-1} + \ldots + t_1, \ldots, t_1)$*. These ranking procedures are denoting* Lexisum1, Lexisum2 *respectively if approximate equality is used. The word "strict" is added if the strict equality is used. The above idea of* Lexisum *is related to the notion of Lorenz dominance (see e.g. [57]).*

A comparative study concerning the impact of using Sum function, Lexisum strict, Lexisum1, Lexisum2 and other basic functions such as Conjunctive (8) in IR is discussed in [58]. The experiments were done by adapting an existing IR system [59] to support these new relevance functions. The used test collection is issued from the MuchMore project ** [60]. It contains about eight thousand documents (medical papers abstracts) obtained from the Springer Link web site, 25 topics from which the queries are extracted and a relevance judgments file which determines for each topic its set of relevant documents. These assessments were established by domain experts from Carnegie Mellon University, LT Institute.

Two main results come out from this report. The first one shows that Sum and its refinements allow to rank-order more documents for relevance

** http://muchmore.dfki.de/

than the strict Conjunctive function. In fact, Sum and its refinements which performs a best matching are more suitable to information retrieval, whilst Conjunctive function which performs an exact matching is mostly adapted to Data Retrieval (a document is retrieved if and only if it contains all query terms) [61].

The second result concerns the comparison between Sum and its refinements. It shows that the refinements of the sum outperform the simple Sum namely when an approximate equality (Lexisum1 and Lexisum2)is used to compare the partial sums (the used equality threshold=5%).

Choice of an Implication Connective

Several choices can be considered for the implication $\rightarrow$ used in (7), depending on the intended semantics of the weights in the query. Among the possible choices we just consider the Lukasiewicz implication connective. More details concerning the use of more implications connective that have clear semantics in a retrieval context are discussed in [52] and [62] (another comparison is also described in [63] in a database context):
Lukasiewicz implication $a \rightarrow b = \min(1, 1 - a + b)$, namely $a \rightarrow b = 1$ if $a \leq b$ and $a \rightarrow b = 1 - a$ if $b > 0$.

As explained in [64], by delocalizing the weights and using the implication, we can easily build an ordered weighted minimum (Owmin) aggregation, and model a query asking for the satisfaction of most of the terms of the query (rather than all).

Remark 2. Note that a prototypical document can be directly used in this approach as a query. If the document is itself present in the collection, it will be retrieved with the maximal estimated degree of relevance 1. Thus, the approach is appropriate for handling a case-based querying process (for example, to seek documents closer to the one given by the user).

Remark 3. The evaluation of a *disjunctive* query q by a degree $rel_d(d;q)$ of a non-empty intersection between H_d and H_q can be computed as follows:

$$rel_d(d;q) = \max_{n \in H_E} \min(\mu_{H_d^*}(n), \mu_{H_q^*}(n))$$

This expresses that at least one of the important concepts of the query is somewhat relevant to the document.

Completing the Description of a Document/Enlarging the Query

In the above procedure, the weights both in H_d^* and H_q^* have been used as they are. However, it may be advisable to modify some of the zero weights both in the description of the document, and in the query, due to different reasons. Indeed, regarding document d, if a node has a non-zero weight in

H_d^*, we may think that a node which is an ancestor of the node in H is also somewhat relevant for the description of the document (even if its own weight in H_d^*, is zero or small). Then, we may think of "completing" H_d^* by computing updated weights in the following way. Let α_i^s and α_i^{s+1} denote weights at level s and $s+1$ in the hierarchy (the root is at level 0). The idea is to recursively update the weights of the nodes starting from the leaves by having the revised weights computed as:

$$\alpha_{i,rev}^s = \max(\alpha_i^s, (\max_i \alpha_{i,rev}^{s+1}) * disc(s))$$

where $disc(s)$ is a discounting factor possibly depending on level s. Indeed, if a document includes many instances of the word *cat*, it clearly deals with *pets* (the "father" of *cat*), but to a smaller extent if the word *pets* (or its synonyms) do not appear as much in the document. In order to control the number of nodes to be added to document/query descriptions, only the common ancestors of couples or triples of co-occurring words in a same document might be considered in the completion procedure. Regarding the queries, the completion procedure of the weights may be motivated by a potential enlargement of the query to less specific terms. Here, the use of a discounting factor will reflect the fact that documents dealing directly with the terms initially chosen should be preferred to more general documents. A similar idea has been used in [65] when dealing with fuzzy conceptual graphs for handling possibilistic information and fuzzy queries (however with a different interpretation for the weights in the fuzzy conceptual graphs leading to a different evaluation procedure).

Remark 4. One might also think of enlarging the query by introducing children of nodes present in q. In fact, this makes the query more demanding (at least if we keep unchanged the levels of importance for the labels present in the original conjunctive query).

4.4 Detecting Concepts from Text

The first stage of the proposed approach aims at identifying terms from a document to connect to concepts (nodes) of ontology. This stage includes tokenizing a text into sentences; parsing each sentence; extracting from the parsing results all terms (singles and compounds) that belong to at least one entry of ontology and then selecting for each term, the appropriate entry (concept). Terms could be noun phrases like *academy_of_motion_picture_arts_and_sciences* or proper nouns like *henry_kenneth_alfred_russell*. It may arise that a given term possibly corresponds to several entries (concepts) in ontology (polysemy problem). In this case, the appropriate concept is selected according to a contextual disambiguation algorithm described in [56] by carrying out similarity measures between all candidate concepts.

Roughly when detecting concepts from documents, two alternative ways can be distinguished. The first one consists in *projecting the ontology on the document* by extracting all multi-word concepts (compound terms) from the ontology and then identifying those occurring in the document. This method has the advantage to be rapid and makes it possible to have a reusable resource even though the corpus changes. Its drawback is the possibility to omit some concepts, which appear in the source text and in the ontology with different forms. For example if the ontology contains a compound concept *solar battery*, a simple comparison does not recognize in the text the same concept appearing in its plural form *solar batteries*.

The second way, which we adopt in this chapter, follows the reverse path, projecting the document onto ontology: for each candidate concept formed by combining adjacent words in text phrases, we first question the ontology using these words just as they are, and then we use their base forms, if necessary to resolve the problem of word forms.

Concerning word combination, we select the longest term for which a concept is detected. For instance, in the case of WordNet and if we consider the example shown on Fig. 3, the sentence contains five (7) different concepts (named synsets in WordNet terminology): *four-stroke internal-combustion engine*, *internal-combustion engine*, *four-stroke engine*, in addition to the four single terms. The first compound concept *four-stroke internal-combustion engine* and the third one *four-stroke engine* are synonyms. So, they belong to the same WordNet synset designed also by the term *node*, their definition is:

***four-stroke internal-combustion engine**, four-stroke engine – (an internal-combustion engine in which an explosive mixture is drawn into the cylinder on the first stroke and is compressed and ignited on the second stroke; work is done on the third stroke and the products of combustion are exhausted on the fourth stroke).*

The second recognized concept is defined as follow:

***internal-combustion engine**, ICE – (a heat engine in which combustion occurs inside the engine rather than in a separate furnace; heat expands a gas that either moves a piston or turns a gas turbine).*

The selected concept is associated with the longest multi-word *basal body temperature method of family planning*, which corresponds to the correct sense in the sentence. Remind that in word combination, the order must be respected (left to right) otherwise we could be confronted to the syntactic variation problem (*science library* is different from *library science*).

The extracted concepts are then weighted as following. The weight of a concept (node) n_i in a document d_j is:

$$Weight(n_i, d_j) = tf/max\_tf\_c,$$

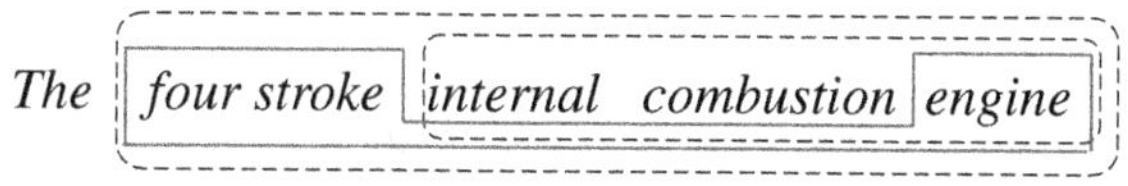

The *four stroke* *internal combustion* *engine*

Fig. 3. Example of text with different concepts

where tf is the frequency of a concept n_i in a document d_j and max_tf_c is the maximal frequency of all the nodes overall the collection.

This weighting scheme can be seen as a very simplistic way to weight concept. Other methods for concept weighting are proposed in the literature, they use in general statistical and/or syntactical analysis [1, 55, 56]. Roughly, they add single words frequencies, multiply them or multiply the number of concept occurrences by the number of single words belonging to the concept.

4.5 Pruning Abstract Nodes

In practice, when completing a document or a query, not all intermediate nodes are added to the document (query) representation. Indeed, nodes located in the high level of the hierarchy are removed as they represent abstract concepts. This additional stage, which consists in a pruning method, uses two pieces of information in order to decide whether a node could be added to a document (query) representation. The first one is the position in the hierarchy (*depth*) of the head node of the sub-tree containing the original nodes and the second one is the length (number of nodes) of the current sub-tree branch containing the candidate intermediate nodes to be added. So, for a branch B_i of a given sub-tree, the number Nb of extra nodes to be added is given by the following formula:

$$Nb(B_i) = \min\left((length(B_i) - 1 + depth)/2, length(B_i) - 1\right)$$

A high value of *depth* means that a node is located near leaves and allows adding specific intermediate nodes, while a low value of *depth* permits to prune abstract nodes, as they are located in the immediate vicinity of the root.

Example

In order to show how abstract intermediate nodes are pruned from the sub-tree, let us consider two concepts (nodes) extracted from a document (named Arthroskopie.00130047 in the used collection): the sense 1 of the name alternative (*alternative#n#1*) and the sense 1 of the name amount (*amount#n#1*). The sub-tree the two nodes belong to is as represented in Fig. 4. Depth of the head node of the branches equals 1 as the head node in this case corresponds to the root of the hierarchy. Let B_1 and B_2 the two branches in Fig. 4:

$B_1 = alternative\#n\#1 \rightarrow decision\_making\#n\#1 \rightarrow higher\_cognitive\_process$ $\#n\#1 \rightarrow process\#n\#3 \rightarrow cognition\#n\#1 \rightarrow psychological\_feature\#n\#1$
$B_2 = amount\#n\#1 \rightarrow magnitude\#n\#1 \rightarrow property\#n\#3 \rightarrow attribute\#n\#2$ $\rightarrow abstraction\#n\#6$

Thus, $length(B_1) = 6$ and $length(B_2) = 5$. So the number of nodes to be added (starting from the leaf) for the first branch B_1 is:

$$Nb(B_1) = min\left((length(B_1) - 1 + depth)/2, length(B_1) - 1\right) = 3 \ nodes$$

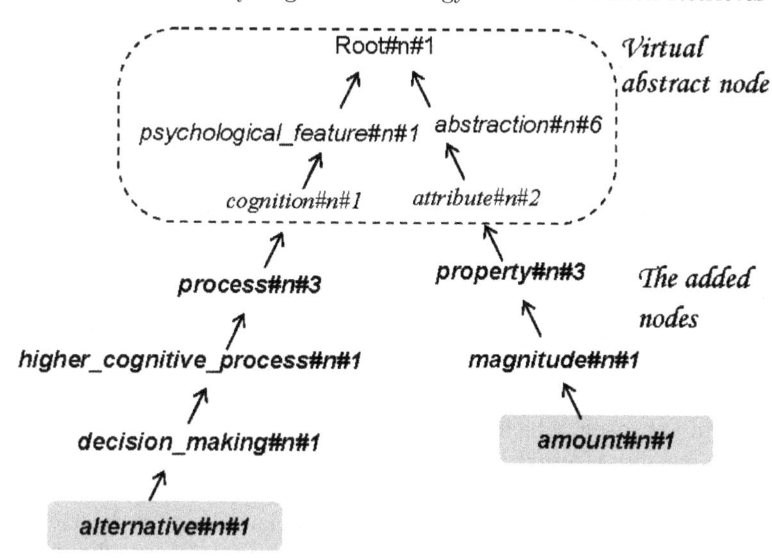

Fig. 4. A sub-tree containing the nodes labeled *alternative#n#1* and *amount#n#1*

which are : *decision_making#n#1*, *higher_cognitive_process#n#1* and *process#n#3*. And from the second branch B_2:

$$Nb(B_2) = min\left((length(B_2) - 1 + depth)/2, length(B_2) - 1\right) = 2\ nodes$$

which are: *magnitude#n#1* and *property#n#3*. The remaining nodes are effectively more abstract and could be pruned from the document representation. They form with the Root node what we call a *virtual abstract node.*

4.6 Discussion

Modern information retrieval systems need the capability to represent and to reason with the knowledge conveyed by text bases. This knowledge can be represented through the use of ontologies. The approach presented in this section lies within the scope of the use of ontologies to concept-based indexing in information retrieval. The introduced approach models both documents and queries as tree-like ontologies where nodes are weighted. The query evaluation process uses fuzzy connectives.

Another direct application of this work may concern the use of the approach at inter-document level. Indeed, the sub-trees resulting from the projection of documents onto ontology could be compared for document clustering. Thus, one can assume that documents with closest sub-trees could be regarded as covering the same subject. It would require the definition of an intersection function between two sub-trees.

5 Concluding Remarks

This chapter discussed different uses of ontologies in fuzzy set-based information retrieval. After a brief survey of recent related works, two approaches have been more particularly presented. They respectively rely on the use of weighted (possibilistic) ontologies, and of a standard ontology (WordNet) processed in a fuzzy logic way.

The first approach based on possibilitic ontologies is motivated by the idea of introducing uncertainty levels and modalities in the assessment of (partial) synonymy and hypernymy relations between terms. In practice, the number of levels used should remain small (even if they are numerically encoded), since their assessments should be qualitative.

The second approach, which is based on the evaluation of degrees of inclusion between weighted subtrees, is oriented towards the need of expanding requests as well as the semantic descriptions of documents contents, using a standard ontology. Fuzzy logic is then used for acknowledging the fact that the terms added in the request, or in the document description, should have weights smaller than those present in the original request or in the document.

The preliminary experiments carried out on IR collections indicate that the two proposed approaches are viable in IR (see [49] for the evaluation of the first approach on document titles and [52] for the evaluation of the second approach when applied to a benchmark collection). Concerning the second approach for instance, the results showed that the concept-based approach outperforms the classical one based on a vector space model, namely when the completion procedure (Sect. 4.3) is used to extend both documents and query sub-trees. Future works should focus on the evaluation of the two approaches on larger collections such as the TREC collection [66]. Another perspective is the use of alternate techniques for rank-ordering query results by different aggregation strategies, as suggested in Sect. 3.

References

1. Croft, W.B., Turtle, H.R., Lewis, D.D.: The use of phrases and structured queries in information retrieval. In Bookstein, A., Chiaramella, Y., Salton, G., Raghavan, V.V., eds.: Proc. of the 4th Annual Intern. Conf. on Research and Development in IR, ACM-SIGIR, Chicago, Illinois (1991) 32–45
2. Grefenstette, G., ed.: Cross-Language Information Retrieval. Kluwer Academic, Boston (1998)
3. Woods, W.A.: Conceptual indexing: A better way to organize knowledge. Technical Report TR-97-61, Sun Microsystems Laboratories (1997)
4. Woods, W.A.: Conceptual indexing: Practical large-scale ai for efficient information access. In: Proc. of the 17th National Conf. on Artificial Intelligence and 12th Conf. on Innovative Applications of Artificial Intelligence, AAAI Press / The MIT Press (2000) 1180–1185

5. Khan, L., McLeod, D., Hovy, E.H.: Retrieval effectiveness of an ontology-based model for information selection. Int. Journal on Very Large Data Bases **13** (2004) 71–85
6. Sprck Jones, K.: Further reflections on TREC. Information Processing and Management **36** (2000) 37–85
7. Bordogna, G., Pasi, G.: A fuzzy linguistic approach generalizing boolean information retrieval: a model and its evaluation. Journal of the American Society for Information Science **44** (1993) 70–82
8. Andreasen, T., Christiansen, H., Larsen, H., eds.: Flexible Query Answering Systems. Kluwer (1997)
9. Kraft, D.H., Bordogna, G., Pasi, G.: Fuzzy set techniques in information retrieval. In Bezdek, J.C., Dubois, D., Prade, H., eds.: Fuzzy Sets in Approximate Reasoning and Information Systems. Kluwer Academic Publishers (1999) 469–510
10. Bordogna, G., Pasi, G.: Controling information retieval through a user adaptive representation of documents. International Journal of Approximate Reasoning **12** (1995) 317–339
11. Yager, R.R.: A note on weighted queries in information retrieval systems. Journal of the American Society for Information Science **38** (1987) 23–24
12. Buell, D.: A problem in information retrieval with fuzzy sets. Journal of the American Society for Information Science **36** (1985) 398–401
13. Voorhees, E.M.: Query expansion using lexical-semantic relations. In: Proc. of the 17th Annual Intern. Conf. on Research and Development in Information Retrieval, ACM-SIGIR, Dublin, Ireland, Springer (1994) 61–69
14. Miller, G., Beckwith, R., C.Fellbaum, Gross, D., Miller, K.: Introduction to wordnet: An on-line lexical database. Journal of Lexicography **3** (1990) 235–244
15. Gonzalo, J., Verdejo, F., Chugur, I., Cigarrn, J.: Indexing with wordnet synsets can improve text retrieval. In: Proc. of the COLING/ACL '98 Workshop on Usage of WordNet for Natural Language Processing. (1998)
16. Guarino, N., Masolo, C., Vetere, G.: Ontoseek: Content-based access to the web. IEEE Intelligent Systems **14** (1999) 70–80
17. Berrut, C.: Indexing medical reports: The rime approach. Information Processing and Management **26** (1990) 93–109
18. Crestani, F.: Application of spreading activation techniques in information retrieval. Artif. Intell. Rev. **11** (1997) 453–482
19. Lee, J.H., Kim, M.H., Lee, Y.J.: Information retrieval based on a conceptual distance in IS-A heirarchy. Journal of Documentation **49** (1993) 188–207
20. Sowa, J.F.: Conceptual Structures: Information Processing in Mind and Machine. Addison-Wesley (1984)
21. Montes-y Gmez, M., Lpez-Lpez, A., Gelbukh, A.: Information retrieval with conceptual graph matching. In: Proc. of the 11th Int. Conf. on Database and Expert Systems Applications, DEXA-2000, Greenwich, England, Springer-Verlag (2000) 312–321
22. Ounis, I., Chevallet, J.P.: Using conceptual graphs in a multifaceted logical model for information retrieval. In Wagner, R., Thoma, H., eds.: Proc. of the 7th Int. Conf. on Database and Expert Systems Applications. Volume 1134 of LNCS., Springer (1996) 812–823
23. Ambroziak, J.R.: Conceptually assisted web browsing. In: 6th International World Wide Web conference. (1997)

24. Dumais, S.T.: Latent semantic indexing (LSI): TREC-3 report. In: Proc. of the 3rd Text Retrieval Conf., NIST special publiscation (1994) 319–330
25. Berry, M.W., Browne, M.: Understanding Search Engines: Mathematical Modeling and Text Retrieval. 2 edn. Society for Industrial and Applied Mathematics (2005)
26. Cinque, L., Malizia, A., Navigli, R.: A semantic-based system for querying personal digital libraries. In Marinai, S., Dengel, A., eds.: Proc. of the 6th International Workshop on Document Analysis Systems. Volume 3163 of LNCS., Springer (2004) 39–46
27. Navigli, R., Velardi, P., Cucchiarelli, A., Neri, F.: Extending and enriching wordnet with ontolearn. In Sojka, P., Pala, K., Smrz, P., Fellbaum, C., Vossen, P., eds.: Proc. of The 2nd Global Wordnet Conference, GWC 2004, Brno, Czech Republic, Masaryk University (2004) 279–284
28. Bulskov, H., Knappe, R., Andreasen, T.: On measuring similarity for conceptual querying. In: Flexible Query Answering Systems, LNAI 2522. Springer (2002) 100–111
29. Baziz, M., Aussenac-Gilles, N., Boughanem, M.: Dsambigusation et expansion de requtes dans un sri : Etude de l'apport des liens smantiques. Revue des Sciences et Technologies de l'Information **8** (2003) 113–136
30. Smirnov, A., Pashkin, M., Chilov, N., Levashova, T., Krizhanovsky, A., Kashevnik, A.: Ontology-based user and requests clustering in customer service management system. In Gorodetsky, V., Liu, J., Skormin, V., eds.: Autonomous Intelligent Systems: Agent and Data Mining, Int. Workshop , AIS-ADM 2005, Springer-Verlag (2005) 231–246
31. Lee, C.S., Jian, Z.W., Huang, L.K.: A fuzzy ontology and its application to news summarization. IEEE Trans. on Systems, Man and Cybernetics **35** (2005) 859–880
32. Miyamoto, S.: Fuzzy sets in Information Retrieval and Cluster Analysis. Kluwer Academic Publisher (1990)
33. Cross, V., Voss, C.: Fuzzy ontologies for multilingual document exploitation. In: Proc. of the 18th Conference of NAFIPS, New York City, IEEE Computer Society Press (1999) 392–397
34. Widyantoro, D.H., Yen, J.: A fuzzy ontology-based abstract search engine and its user studies. In: Proc. of the 10th Intern. Conf. on Fuzzy Systems. Volume 2., IEEE, Melbourne, Australia (2001) 1291–1294
35. Akrivas, G., Wallace, M., Andreou, G., Stamou, G., Kollias, S.: Context-sensitive semantic query expansion. In: Proc. of the Intern. Conf. on Artificial Intelligence Systems, ICAIS, IEEE, Divnomorskoe, Russia (2002)
36. Vallet, D., Fernández, M., Castells, P.: An ontology-based information retrieval model. In Gómez-Pérez, A., Euzenat, J., eds.: Proc. 2nd European Semantic Web Conference. Volume 3532 of LNCS., ESWC 2005, Heraklion, Crete, Greece, Springer (2005) 455–470
37. Castells, P., Miriam Fernndez, M., Vallet, D., Mylonas, P., Avrithis, Y.: Self-tuning personalized information retrieval in an ontology-based framework. In: OTM Confederated Intern. Conf. Volume 3761 of LNCS., ODBASE 2005, Agia Napa, Cyprus, Springer (2005) 977–986
38. Dubois, D., Prade, H., Testemale, C.: Weighted fuzzy pattern matching. Fuzzy Sets and Systems **28** (1988) 313–331

39. Boughanem, M., Loiseau, Y., Prade, H.: Graded pattern matching in a multilingual context. In: Proc. 7th Meeting Euro Working Group on Fuzzy Sets, Eurofuse, Varena (2002) 121–126
40. Loiseau, Y., Prade, H., Boughanem, M.: Qualitative pattern matching with linguistic terms. Ai Communications, The European Journal on Artificial Intelligence **17** (2004) 25–34
41. Resnik, P.: Semantic similarity in a taxonomy: an information-based measure and its application to problem of ambiguity in natural language. J. Artif. Intellig. Res. **11** (1999) 95–130
42. Bidault, A., Froidevaux, C., Safar, B.: Similarity between queries in a mediator. In: Proc. 15th European Conference on Artificial Intelligence, ECAI'02, Lyon (2002) 235–239
43. Rossazza, J., Dubois, D., Prade, H.: A hierarchical model of fuzzy classes. In Caluwe, R.D., ed.: Fuzzy and Uncertain Object-Oriented Databases. World Pub. Co. (1997) 21–62
44. Dubois, D., Prade, H.: Resolution principles in possibilistic logic. Int. Jour. of Approximate Reasoning **4** (1990) 1–21
45. Crouch, C.: An approach to the automatic construction of global thesauri. Information Processing and Management **26** (1990) 629–640
46. Buell, D.: An analysis of some fuzzy subset applications to information retrieval systems. Fuzzy Sets and Systems **7** (1982) 35–42
47. Prade, H., Testemale, C.: Application of possibility and necessity measures to documentary information retrieval. LNCS **286** (1987) 265–275
48. Boughanem, M., Loiseau, Y., Prade, H.: Rank-ordering documents according to their relevance in information retrieval using refinements of ordered-weighted aggregations. In Detyniecki, M., Jose, J.M., Nrnberger, A., van Rijsbergen, C., eds.: 3rd Int. Workshop on Adaptive Multimedia Retrieval. Volume 3877 of LNCS., AMR'05, Glasgow (UK), Springer-Verlag (2005) 44–54
49. Loiseau, Y., Boughanem, M., Prade, H.: Evaluation of term-based queries using possibilistic ontologies. In Herrera-Viedma, E., Pasi, G., Crestani, F., eds.: Soft Computing in Web Information Retrieval: Models and Applications. Volume 197 of Studies in Fuzziness and Soft Computing. Springer (2006) 135–160
50. Boughanem, M., Loiseau, Y., Prade, H.: Refining aggregation functions for improving document ranking in information retrieval. Int. J. Appl. Math. Comput. Sci. - Soft Computing for Information Management on the Web (2006, submitted)
51. Boughanem, M., Pasi, G., Prade, H.: A Fuzzy set approach to concept-based Information Retrieval . In: 10th International Conference IPMU'04 , Perugia (Italy), 04/07/04-09/07/04, IPMU (2004) 1775–1782
52. Baziz, M., Boughanem, M., Pasi, G., Prade, H.: A fuzzy set approach to concept-based information retrieval. In: 4th Conf. of the Euro. Soc. for Fuzzy Logic and Tech. and 11me Rencontres Francophones sur la Logique Floue et ses Applications, EUSFLAT-LFA 2005, Barcelona, Spain (2005) 1287–1292
53. Baziz, M., Boughanem, M., Pasi, G., Prade, H.: A fuzzy logic approach to information retrieval using an ontology-based representation of documents. In Sanchez, E., ed.: Fuzzy Logic and the Semantic Web. Elsevier (2006) 363–377
54. Salton, G., McGill, M.: Introduction to modern information retrieval. McGraw-Hill, New York (1983)

55. Huang, X., Robertson, S.E.: Comparisons of probabilistic compound unit weighting methods. In: Proc. of the ICDM'01 Workshop on Text Mining, San Jose, USA (2001) 1–15
56. Baziz, M., Boughanem, M., Aussenac-Gilles, N., Chrisment, C.: Semantic cores for representing documents in ir. In: SAC'2005 - 20th ACM Symp. on Applied Computing. Volume 2., Santa Fe, USA. (2005) 1011–1017
57. Dubois, D., Prade, H.: On different ways of ordering conjoint evaluations. In: Proc. of the 25th Linz seminar on Fuzzy Set Theory, Linz, Austria (2004) 42–46
58. (Technical report)
59. Boughanem, M., Dkaki, T., Mothe, J., Soule-Dupuy, C.: Mercure at TREC-7. In: Proc. of the 7th Text Retrieval Conf., NIST special publiscation (1997) 135–141
60. Buitelaar, P., Steffen, D., Volk, M., Widdows, D., Sacaleanu, B., Vintar, S., Peters, S., Uszkoreit, H.: Evaluation resources for concept-based cross-lingual information retrieval in the medical domain. In: Proc. of the 4th Int. Conf. on Language Resources and Evaluation, LREC2004, Lissabon, Portugal (2004)
61. van Rijsbergen, C.: Information Retrieval. Butterworths & Co., Ltd, London (1979)
62. Pasi, G.: A logical formulation of the boolean model and of weighted boolean models. In: Workshop on Logical and Uncertainty Models for Information Systems, LUMIS 99), University College London, UK (1999)
63. Dubois, D., Nakata, M., Prade, H.: Extended divisions for flexible queries in relational databases. In Pons, O., Vila, M.A., Kacprzyk, J., eds.: Knowledge Management in Fuzzy Databases. Physica-Verlag (1999) 105–121
64. Dubois, D., Prade, H.: Semantic of quotient operators in fuzzy relational databases. Fuzzy Sets and Systems **78** (1996) 89–93
65. Thomopoulos, R., Buche, P., Haemmerl, O.: Representation of weakly structured imprecise data for fuzzy querying. Fuzzy Sets and Systems **140** (2003) 111–128
66. Vorhees, E.M., Harman, D.: Overview of the sixth text retrieval conference (TREC-6). In Vorhees, E.M., Karman, D.K., eds.: Proc. of the Sixth Text Retrieval Conference (TREC-6). (1998)

Real-World Fuzzy Logic Applications in Data Mining and Information Retrieval

Bernadette Bouchon-Meunier, Marcin Detyniecki, Marie-Jeanne Lesot, Christophe Marsala, and Maria Rifqi

Abstract. This chapter focuses on real-world applications of fuzzy techniques for information retrieval and data mining. It gives a presentation of the theoretical background common to all applications, lying on two main elements: the concept of similarity and the fuzzy machine learning framework. It then describes a panel of real-world applications covering several domains namely medical, educational, chemical and multimedia.

1 Introduction

Information retrieval and data mining are two components of a same problem, the search of information and knowledge extraction from large amounts of data, very large databases or data warehouses.

In information retrieval, the user knows approximately what he looks for, for instance an answer to a question, or documents corresponding to a given requirement in a database. The search is performed in text, multimedia documents (images, videos, sound) or in web pages. Transmedia information retrieval takes advantage of the existence of several media to focus on a more specific piece of information, for instance using sound and speech to help retrieving sequences in a video. The main difficulty lies in the identification of relevant information, i.e. the closest or the most similar to the user's need or expectation. The concept of relevance is very difficult to deal with, mainly because it is strongly dependent on the context of the search and the purpose of the action launched on the basis of such expected relevant information. Asking the user to elicit what he looks for is not an easy task and, the more flexible the query-answer process, the more efficient the retrieval. This is a first reason to use fuzzy sets in knowledge representation to enable the user to express his expectations in a language not far from natural. The second reason lies in the approximate matching between the user's query and existing elements in the database, on the basis of similarities and degrees of satisfiability.

In data mining, the user looks for new knowledge, such as relations between variables or general rules for instance. The search is performed in databases or data warehouses. The purpose is to find homogeneous categories, prototypical behaviors, general associations, important features for the recognition of a class of data. In this case again, using fuzzy sets brings flexibility in knowledge representation, interpretability in the obtained results, in rules or in characterizations of prototypes. Looking for too strict a relation between variables may be impossible because of the variability of descriptions in the database, while looking for an imprecise relation between variables or to a crisp relation between approximate values of variables may lead to a solution. The expressiveness of fuzzy rules or fuzzy values of attributes in a simplified natural language is a major quality for the interaction with the final user.

The main problems in information retrieval and data mining lie in the large scale of databases, especially when dealing with video or web resources, in the heterogeneous data of various types, numerical or symbolic, precise or imprecise, ambiguous, approximate, with incomplete files, uncertain because of the poor reliability of sources or the difficulties of measurement of observation. Another source of problems is the complexity of the user's requests, expressed in natural language or involving various criteria for instance. The necessity to create cooperative systems, friendly and user-oriented, adapted to the user's needs or capabilities, providing a personalized information, leads to soft approaches to man-machine interaction and to on-line or off-line learning of the best way to satisfy the demand. Fuzzy logic is very useful in this matter because of its capability to represent miscellaneous data in a synthetic way, its robustness with regard to changes of the parameters of the user's environment, and obviously its unique expressiveness.

This chapter focuses on real-world applications of fuzzy techniques for information retrieval and data mining. It first gives a brief presentation of the theoretical background common to all applications (Sect. 2), decomposed into two main elements: the notion of similarity and the fuzzy machine learning techniques that are applied in the described applications (Sect. 3). Indeed, similarity, or more generally comparison measures are used at all levels of the data mining and information retrieval tasks: at the lowest level, they are used for the matching between a query to a database and the elements it contains, for the extraction of relevant data. Then similarity and dissimilarity measures can be used in the process of cleaning and management of missing data to create a training set. In the various techniques to generalize particular information contained in this training set, dissimilarity measures are used in the case of inductive learning, similarity measures for case-based reasoning or clustering tasks. Eventually, similarities are used to interpret results of the learning process into an expressible form of knowledge, for instance through the definition of prototypes. Section 2.1 presents the similarity notion more formally.

Section 2.2 considers a complementary component of similarity, the fuzzy learning techniques in which they can be used. It describes methods used in the applications presented in Sect. 3, namely fuzzy decision trees, that perform fuzzy inductive learning, fuzzy prototype extraction, that provides flexible characterization of data sets, and fuzzy clustering, that identifies relevant subgroups in data sets.

Finally we describe in Sect. 3 real-world applications exploiting these methods and belonging both to the data mining and information retrieval fields. They cover several domains, such as medical (Sect. 3.1), educational (Sect. 3.2), chemical (Sect. 3.3), and multimedia (Sect. 3.4).

2 Theoretical Background

In this section, we recall the theoretical background common to the applications presented in Sect. 3, considering successively the notion of similarity (Sect. 2.1) and fuzzy machine learning techniques (Sect. 2.2).

2.1 Similarity

The notion of similarity, or more generally of comparison measures, is central for all real-world applications: it aims at quantifying the extent to which two objects are similar, or dissimilar, one to another, providing a numerical value for this comparison.

Similarities and dissimilarities between objects are generally evaluated from values of their attributes or variables characterizing these objects. It is the case in various domains, such as statistics and data analysis, psychology and pattern recognition for instance. Dissimilarites are classically defined from distances. Similarities and dissimilarities are often expressed from each other: the more similar two objects are, the less dissimilar they are, the smaller their distance. Weights can be associated with variables, according to the semantics of the application or the importance of the variables. It appears that some quantities are used in various environments, with different forms, based on the same principles.

For instance, the most classic dissimilarity measures between two objects with continuous numerical attributes are the Euclidian distance, the Manhattan distance, and more generally Minkowski distances. In the case of binary attributes, coefficients introduced by Russel-Rao, Jaccard, Dice or Ochiai are very popular. For more details, see [6, 47, 52].

Tversky's Model

A more theoretical form of similarity measure has been introduced by Tversky in a psychometrical framework within the so-called Tversky's contrast model [52]: given two objects described by the sets of their characteristics, respectively denoted A and B, this model defines a similarity measure $s(A, B)$ as a function of the common features of the two objects ($A \cap B$) and their respective distinct features ($A - B$ and $B - A$) that verifies the properties of monotonicity, independence, solvability and invariance [52].

Using this axiomatic, Tversky shows that the similarity takes the form

$$s(A,B) = \theta f(A \cap B) - \alpha f(A - B) - \beta f(B - A)$$

where f is a non-negative scale and α, β and θ are non-negative parameters.

Tversky also proposed non-linear similarity measures, in the so-called ratio model that takes the form

$$s(A,B) = \frac{f(A \cap B)}{f(A \cap B) + \alpha f(A - B) + \beta f(B - A)}$$

Fuzzy Comparison Measures

Considering Tversky's approach and the large variety of measures existing in the literature, we proposed a unified vision of measures of comparison [6], considering both measures of similarity and dissimilarity, in the case of fuzzy set-valued variables (crisp values can be handled as well with such measures).

Given a universe X, let $F(X)$ denote its fuzzy power set. Let M be a fuzzy set measure $M : F(X) \to \mathbb{R}^+$ and a difference $\Theta : F(X) \times F(X) \to F(X)$. For instance, denoting f_A the membership function of a fuzzy set A, we can choose $M(A) = \int_X f_A(x)dx$ and $f_{A\Theta B}(x) = \max(0, f_A(x) - f_B(x))$ for all $x \in X$.

A *measure of comparison* is then defined [6] on X as a function S verifying

$$\begin{aligned} S:\ & F(X) \times F(X) \to [0,1] \\ & \text{such that } S(A,B) = F_s(M(A \cap B), M(B\Theta A), M(A\Theta B)) \end{aligned}$$

where F_s is a function $F_s : \mathbb{R}^3 \to [0,1]$. Depending on the properties required from the F_s, different particular cases of comparison measures can be distinguished:

- A *measure of similitude* is such that $F_s(u,v,w)$ is non-decreasing in u and non-increasing in v and w.
- A *measure of resemblance* is a particular measure of similitude, reflexive (i.e. $S(A,A) = 1$ for any A) and symmetrical (i.e. $S(A,B) = S(B,A)$). For instance, one can consider

$$S(A,B) = \frac{\displaystyle\int_X f_{A\cap B}(u)du}{\displaystyle\int_X f_{A\cup B}(u)du}$$

- A *measure of dissimilarity* is such that $F_S(u,v,w)$ is non-increasing in u, non-decreasing in v and w.

2.2 Fuzzy Machine Learning

The second part of the theoretical background common to all applications described in Sect. 3 concerns the fuzzy machine learning techniques, that use the previous similarity measures. Machine learning is an important way to extract knowledge from sets of cases, specially in large scale databases. In this section, we consider only the fuzzy machine learning methods that are used in the applications described in Sect. 3, leaving aside other techniques as for instance fuzzy case based reasoning or fuzzy association rules (for a complete review on fuzzy learning methods, the interested reader is referred to [20]).

Three methods are successively considered: fuzzy decision trees, fuzzy prototypes and fuzzy clustering. The first two belong to the supervised learning framework, i.e. they consider that each data point is associated with a category. Fuzzy clustering belongs to the unsupervised learning framework, i.e. no a priori decomposition of the data set into categories is available.

Fuzzy Decision Trees

Fuzzy decision trees (FDT) are particularly interesting for data mining and information retrieval because they enable the user to take into account imprecise descriptions of the cases, or heterogeneous values (symbolic, numerical, or fuzzy) [22,44,54,56]. Moreover, they are appreciated for their interpretability, because they provide a linguistic description of the relations between descriptions of the cases and decision to make or class to assign [20]. The rules obtained through FDT make it easier for the user to interact with the system or the expert to understand, confirm or amend his own knowledge. Another quality of FDT is their robustness, since a small variation of descriptions does not drastically change the decision or the class associated with a case, which guarantees a resistance to measurement errors and avoids sharp differences for close values of the descriptions.

For these reasons, FDT have been extensively used in the past years. What they mainly provide for knowledge extraction is, first a ranking of attributes bringing information about the importance of various criteria in the assignment of decision or class, secondly rules establishing a link between descriptions and decision.

Most algorithms to construct decision trees proceed in the same way, the so-called *Top Down Induction of Decision Tree* (TDIDT) method. They build a tree from the root to the leaves, by successive partitioning of the training set into subsets. Each partition is done by means of a test on an attribute and leads to the definition of a node of the tree. An attribute is selected thanks to a *measure of discrimination* H (in classic decision tree, the Shannon entropy is generally used [8,43], it comes from Information Theory [2]). Such a measure makes it possible to order the attributes according to an increasing accuracy when splitting the training set. The discriminating power of each attribute is valued with regard to the classes. The attribute with the highest discriminating power is selected to construct a node in the decision tree.

Methods to construct decision trees, whether crisp or fuzzy, differ mainly in their choice of H [34]; in the fuzzy case, two main families can be distinguished. The first one deals with methods based on a generalized Shannon entropy: the *entropy of fuzzy events* as a measure of discrimination [22, 44, 54, 56]. It corresponds to the Shannon entropy extended to fuzzy events by substituting probabilities of fuzzy events to classic probabilities. The second one deals with methods based on another family of fuzzy measures [7,11,38,53], namely a *measure of classification ambiguity*, defined from both a measure of fuzzy subsethood and a measure of non-specificity.

For the practical construction of fuzzy decision trees for the applications in Sect. 3, we used our software *Salammbô* [33]: this system implements the previous tree learning method in a flexible framework, allowing the user to choose the measure of discrimination among the previous possibilities, as well as the splitting strategy and the stopping criterion. Furthermore, it offers an internal method for the construction of fuzzy values, deduced from the universe of values of continuous attributes related to the distribution of the classes [32].

Fuzzy Prototype Construction

Fuzzy prototypes [30, 46, 57] constitute another approach to the characterization of data categories: they provide descriptions or interpretable summarizations of data sets, so as to help a user to better apprehend their contents: a prototype is an element chosen to represent a group of data, to summarize it and underline its most characteristic features. It can be defined from a statistic point of view, for instance as the data mean or the median; more complex representatives can also be used, as the Most Typical Value [17] for instance.

The prototype notion was also studied from a cognitive science point of view, and specific properties were pointed out [48, 49]: it was shown that a prototype underlines the common features of the category members, but also their distinctive features as opposed to other categories, underlining the specificity of the group. Furthermore, prototypes were related to the typicality notion, i.e. the fact that all data do not have the same status as regards the group: some members of the group are better examples, more representative or more characteristic than others. It was also shown [48, 49] that the typicality of a point depends both on its resemblance to other members of the group (internal resemblance), and on its dissimilarity to members of other groups (external dissimilarity).

These definitions were exploited by Rifqi [46] who proposed a construction method implementing these principles and exploiting the similarity measure framework presented in Sect. 2.1. More precisely, the method consists in first computing internal resemblance and external dissimilarity for each data point: they are respectively defined as the aggregation (mean or median e.g.) of the ressemblance to the other members of the group, and as the aggregation of the dissimilarity to members of other groups, for a given choice of the resemblance and dissimilarity measures (see Sect 2.1). In a following step, a typicality degree is computed for each data point as the aggregation of its internal resemblance and external dissimilarity. In a last step, the prototype itself is defined, as the aggregation of the most typical category members.

Fuzzy prototypes are defined as the application of this method to fuzzy data [46], or to crisp data that are aggregated into fuzzy sets [30]. Former case, illustrated in Sect. 3.1, corresponds to data whose attribute values are not numerical values but fuzzy subsets. Latter case, illustrated in Sect. 3.2, makes it possible to model the intrinsic imprecise nature of the prototype: it is more natural to say "the typical French person measures around 1.70m", rather than "the typical French person measures 1.6985" (fictitious values). This implies the prototype is best described by an imprecise linguistic expression "around 1.70m" than a crisp numerical value. Now the aggregation step that builds prototypes from the most typical data can build a fuzzy set, derived from the typicality degree distribution [30]. Such fuzzy prototypes then characterize data sets, underlining both the common features of the group members and their distinctive features, and modeling their unsharp boundaries.

Fuzzy Clustering

Contrary to fuzzy decision trees and fuzzy prototype construction methods, clustering algorithms belong to the unsupervised learning framework, i.e. they do not consider that a decomposition of the data set into categories is available. They perform data mining as the identification of relevant subgroups of the data, determining subsets of similar data and thus highlighting the underlying structure of the data set. More precisely relevant subgroups are such that points within a group are more similar one to another than to points assigned to a different subgroup. Thus, as the previous learning methods, they rely on comparison measures as presented in Sect. 2.1.

The clustering aim can also be expressed as the decomposition of the data set into subgroups that are both homogeneous and distinct: the fact that clusters are homogeneous implies that points in the same subgroup indeed resemble one another, which justifies their grouping. The fact that they are distinct justifies the individual existence of each cluster that captures different characteristics of the data. Through this decomposition, clustering leads to a simplified representation of the data set that can be summarized by a reduced number of clusters instead of considering each individual data point.

The fuzzy set theory proves its advantage in this framework through the notion of membership degrees: in crisp clustering algorithms, such as the k-means or hierarchical methods (see e.g. [21]), a point is assigned to a single cluster. Now this is not adapted to the frequent case where clusters overlap and points have partial memberships to several subgroups. Ruspini [50] first proposed to exploit fuzzy set theory to represent clusters, so as to model unclear assignments and clusters with unsharp boundaries. Dunn [16] proposed the first fuzzy clustering algorithm, called fuzzy c-means (FCM), that was generalized by Bezdek [3, 4]. Since then, many variants have been proposed to address specific aims (e.g. noise handling [12, 26], adapting to other data or cluster types [18, 24, 55], or considering fuzzy clustering at a more formal level [23]), leading to the vast fuzzy clustering domain (see e.g. [19]).

3 Real-World Applications

In this section, a panel of real-world fuzzy logic applications is presented, based on the similarity framework and the fuzzy learning methods described in the previous section. They belong both to the data mining and information retrieval fields, and cover several domains, namely medical (Sect. 3.1), educational (Sect. 3.2), chemical (Sect. 3.3) and multimedia (Sect. 3.4). For each application, the objective of the task, the considered data, the applied method and the obtained results are successively described.

3.1 Medical Applications

Medical applications are good cases where Fuzzy Set Theory can bring out enhancement as compared to classic algorithms because most of the attributes used here to characterize cases are associated with imprecise values. In this section, we present three applications of data mining, respectively to prevent cardio-vascular diseases, to measure asthma severity and to detect malign microcalcifications in mammographies.

Data Mining to Prevent Cardio-Vascular Diseases

This project was done thanks to financial supports by INSERM and was led by M.-C. Jaulent (INSERM ERM 0202). Researchers from several French universities collaborated with a medical scientist on a well-known database to prevent cardio-vascular diseases.

Objective

The main objective here was to find discriminating features in order to prevent cardio-vascular diseases. Predictions should help medical scientists to detect and prevent cardio-vascular diseases for hypertensive patients.

Data

The used data were the INDANA (INdividual Data ANalysis of Antihypertensive intervention) database. This database is composed of ten therapeutic samples based on cardio-vascular risks for patients. For each patient, a set of classic features is combined with a set of medical measurements on several years and leads to the conclusion for this patient (death or not).

The particular database used for our experiments was composed by the features for patients. In this set, 107 patients died of a cardio-vascular disease, and 2132 were alive at the end of the experiment.

Method

One of the main problems to solve here comes from the fact that the proportion of classes is heavily unbalanced, which makes it impossible to use classic decision tree algorithms.

We focused on the use of fuzzy decision trees in order to obtain an interpretable set of fuzzy rules. The working plan was the following.

First of all, one hundred pairs of sets (dual training set and test set) was generated randomly from the initial data set. In a second step, each training set was used to construct a fuzzy decision tree, as detailed in the sequel. Afterwards, each fuzzy decision tree was used to classify the corresponding test set. In the last step an aggregation was performed: for each case of the initial set, a set of classification degrees was obtained (from the 100 fuzzy decision trees). The final class associated with the case was the one that obtained the highest aggregation degree.

The construction of the fuzzy decision trees was done as presented in Sect. 2.2: we used the *Salammbô* software [33] (see Sect. 2.2), choosing star entropy as discrimination measure, $\alpha-$cut strategy as splitting strategy and a criterion based on the Shannon information measure and the number of examples in the local training set [33] as stopping criterion.

Results

The obtained global good classification rate for the cardio-vascular rate was around 70%. More interesting was the analysis of the details of the classification: it showed the existence of high disparities among the patients. Indeed, it highlighted the existence of patient subsets that were very hard to classify and others that obtained 100% good classification.

Data Mining for Assessing Asthma Severity

Objective

The main objective in this study [31], conducted in collaboration with Dr Alain Lurie[1], was to identify variables and decision pathways patients use

[1](1): Service de Pneumologie, Hôpital Cochin, Assistance Publique-Hôpitaux de Paris, 27 rue du Faubourg Saint-Jacques, F-75679 Paris Cedex 14,
(2): Laboratoire d'Explorations Fonctionnelles Respiratoires et du Sommeil, Hertford British Hospital, 3 rue Barbès, F-92300 Levallois-Perret

to determine the severity of their asthma (perceived severity of the asthma). In a second step, the identified variables were compared to those involved in the assessment of asthma severity according to the National Asthma Education and Prevention Program (NAEPP) Guidelines (objective severity of the asthma).

Data

The database was composed by a set of 113 outpatients (51 men, 62 women), with (% patients) mild intermittent (6.2), mild persistent (15.9), moderate (65.5) and severe (12.4) asthma. A questionnaire was filled for each patient with several features, among which the patient's sociodemographic characteristics and his asthma characteristics. For the latter, two parts are to be distinguished, respectively assessed by the patient himself and by the doctors: the patient had to assess his perceived asthma severity (rated as mild intermittent, mild persistent, moderate or severe), the response to treatment (perceived treatment efficiency), the quality of his life, and the rating of its medical adherence by the patient. The doctor part concerned the objective asthma severity (also rated as mild intermittent, mild persistent, moderate or severe) derived from medical criteria and the valuation of the respiratory functions.

All these variables were pooled, and considered as potential variables patients might use to determine the perceived severity of their asthma.

Method

A fuzzy decision tree was constructed thanks to the *Salammbô* software (see Sect. 2.2) in order to obtain a fuzzy rule set to identify a set of variables and decision pathways patients use to determine the severity of their asthma.

Results

A 4-fold cross validation was undertaken to validate the obtained model. This cross-validation evaluated the accuracy and the robustness of the model, and highlighted the usefulness of the decision tree for patients outside the training set. A fuzzy decision tree predicted the decision class with a mean rate of 73%.

Learning to Detect Malign Microcalcifications in Mammographies

One woman in 8 in the United States and one woman in 10 in Europe has a breast cancer during her life. Nowadays, mammography is the primary diagnostic procedure for the early detection of breast cancer.

Until recently, all information in mammography was acquired, fused and evaluated by the doctor. Today however, with the increasing performance of image processing techniques, we are at a point where doctors using a computer aided detection (CAD) system perform better in some aspects than the doctors by themselves. The strong points of CAD systems are their precision and repeatability. Logically, efforts are being made to combine the expert knowledge of the doctor with the precision and repeatability of the CAD system.

Microcalcification[2] clusters are an important element in the detection of breast cancer. This kind of finding is the direct expression of pathologies which may be benign or malignant.

Objective

The objective of the project, conducted in collaboration with General Electric Medical Systems, was to provide an automatic classification of microcalcifications. Such a tool provides to the radiologist:

- an objective description,
- a systematic classification of every microcalcification in the image.

These two characteristics are the foundations for a good diagnosis. Furthermore, providing a good description of the relevant classes to the radiologists enables them to improve their performances.

Fuzzy logic is a powerful tool for the formulation of expert knowledge and the combination of imprecise information from different sources. To achieve meaningful results the imprecision in all information used to come to a conclusion should be taken into account.

Data

The description of microcalcifications is not an easy task, even for an expert. If some of them are easy to detect and to identify, some others are more ambiguous. The texture of the image, the small size of objects to be detected (less than one millimeter), the various aspects they have, the radiological noise, are parameters which impact the detection and the characterization tasks.

More generally, mammographic images present two kinds of ambiguity: *imprecision* and *uncertainty*. The *imprecision* on the contour of an object comes from the fuzzy aspect of the borders: the expert can approximately define the contour but certainly not with a high spatial precision. The *uncertainty* comes from the microcalcification superimpositions: because objects are built from the superimpositions of several 3D structures on a single image, we may have a doubt about the contour position.

The first step consists in finding automatically the contours of microcalcifications. This segmentation is also realized thanks to a fuzzy representation of imprecision and uncertainty (more details in [45]). Each microcalcification

[2] The microcalcifications are small depositions of radiologically very opaque materials that can be seen on mammography exams as small bright spots

is then described by means of 5 fuzzy attributes computed from its fuzzy contour. These attributes enable us to describe more precisely:

- the shape (3 attributes): elongation (minimal diameter/maximal diameter), compactness1, compactness2.
- the dimension (2 attributes): surface, perimeter.

Figure 1 shows an example of the membership functions of the values taken by a detected microcalcification. One can notice that the membership functions are not "standard" in the sense that they are not triangular or trapezoidal (as it is often the case in the literature) and this is because of the automatic generation of fuzzy values (we will not go into details here, interested readers may refer to [5]).

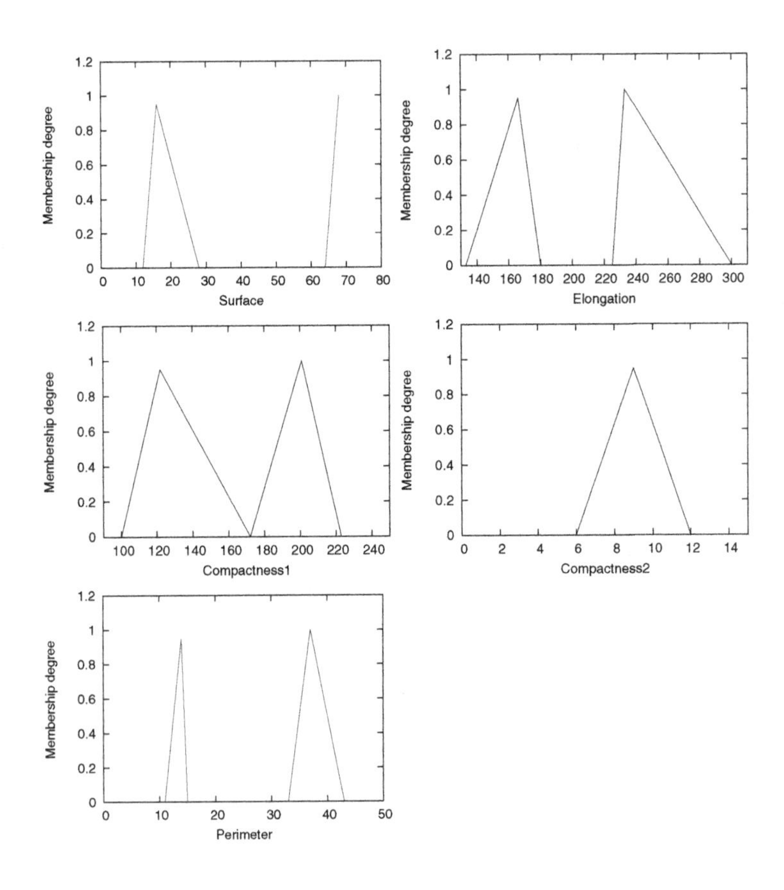

Fig. 1. Description of a microcalcification by means of fuzzy values

Method

Experts categorised microcalcifications into 2 classes: *round* microcalcifications and *not round*, because this property is important to qualify the malignancy of the microcalcification. Each class was described by means of fuzzy prototypes, as described in Sect. 2.2, with the median as the aggregation operator for both internal similarity and external dissimilarity. Details about the specific similarity and dissimilarity are given in [45].

Results

Figure 2 gives the obtained fuzzy prototypes describing the two classes. It can be seen that on the attributes elongation, compactness1 or compactness2, the typical values of the two classes round and not round, are quite different: the intersection between them is low. This can be interpreted in the following way: *a round microcalcification has typically an elongation approximatly between 100 and 150 whereas a not round microcalcification has typically an elongation approximatly between 150 and 200*, etc. For the attributes *surface* and *perimeter*, at the opposite of the previously attributes, the typical values of the two classes are superimposed, it means that these attributes are not typical.

3.2 Educational Applications

Providing Interpretable Characterizations of Students

In this section, we consider another domain application for fuzzy machine learning methods, namely the educational domain. The presented application was performed in the framework of a project with the schoolbook publisher Bordas-Nathan.

Objective

The considered task consists in characterizing students, through the identification of relevant groups of students having the same characteristics, and the comparison of several student classes, to determine whether the classes present the same characteristics or not. Of special importance is the interpretability of the results, to enable a teacher to exploit the information and the structure identified in the student data.

Data

The considered data are descriptions of students, each student being represented as the vector of its results to several exams. No category information is provided, i.e. no knowledge about the decomposition of the data into categories is available.

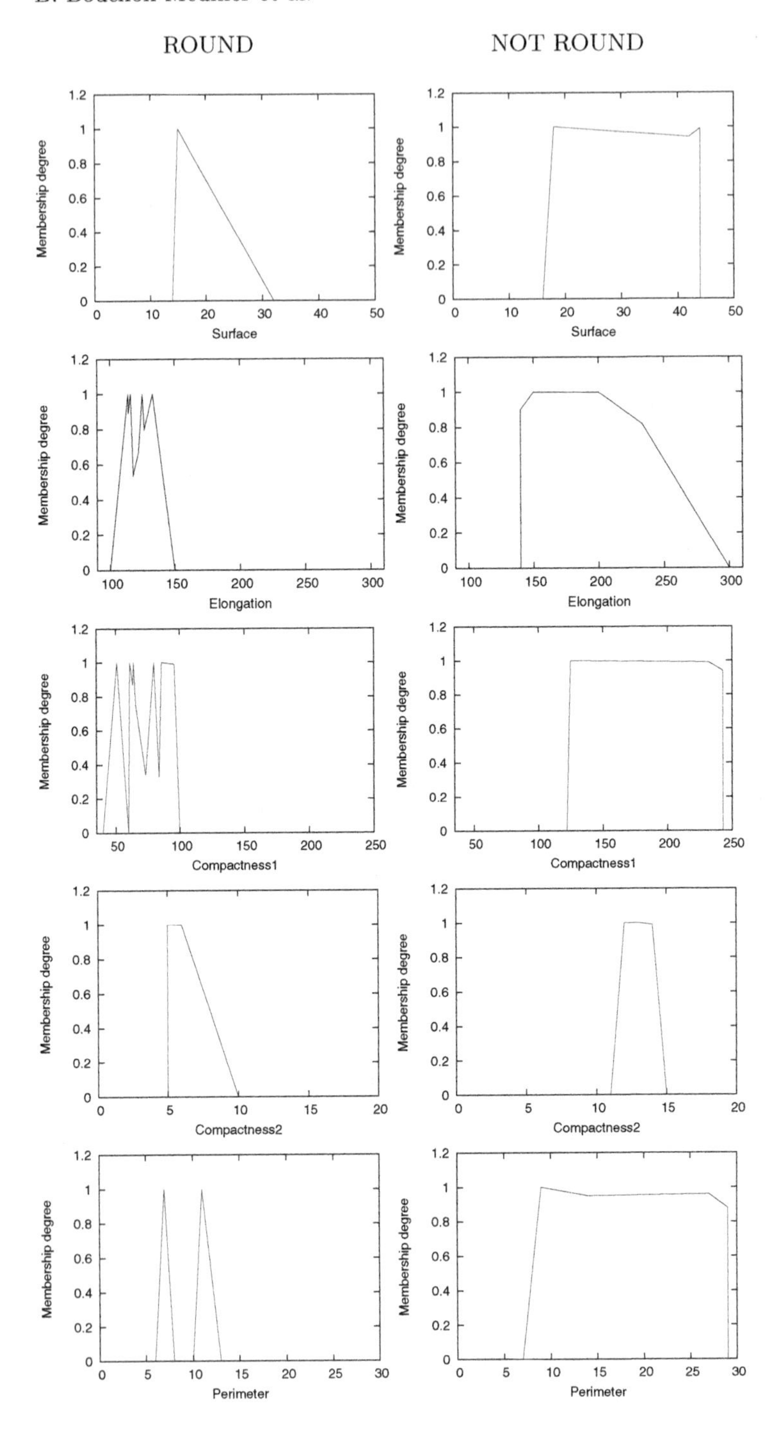

Fig. 2. Fuzzy prototypes of the classes "round" and "not round"

For the dissimilarity measure, we simply use the Euclidean distance between the vectors representing the students. In the case where more information is available (for instance information about the relationships between the exams), more sophisticated similarity measures can be used [27].

Method

To perform data mining with these data, several methods were combined, namely fuzzy prototypes, fuzzy clustering and exceptionality coefficients. Fuzzy prototypes, as detailed in Sect. 2.2, correspond to flexible data characterizations, that underline the common features of the group members, as well as their distinctive features as compared to other groups.

Fuzzy clustering (see Sect. 2.2) is applied to identify relevant subgroups in the student set; the chosen algorithm is the Outlier Preserving Clustering Algorithm, OPCA [29]. This combination of FCM and hierarchical clustering with single linkage offers a particular outlier handling: contrary to the general case, it does not consider outliers as noisy points or aberrations that must be identified and excluded from the data, but interprets them as specific cases, that are as significant as classic clusters to summarize and describe the data. Thus it handles outliers as clusters reduced to a single point, and preserves them in the final data decomposition.

Finally, exceptionality coefficients [28] are tools to further characterize clusters, indicating the extent to which a cluster is exceptional or representative of the data: it makes it possible to distinguish major trends and atypical behaviors and models intuitive descriptions: consider for instance a device having three modes, described as "high", "low" and "abnormally low". The exceptional case "abnormally low" is part of the system description, which is indeed necessary, but the adverb "abnormally" underlines its specificity. Exceptionality coefficients quantifies this notion, allowing to identify the most representative clusters for the whole data set.

Results

First, we consider the task of characterizing a set of students, at a semantic and interpretable level, using fuzzy prototypes: Figure 3 shows the level lines of the fuzzy prototypes describing the data set containing the results obtained by 150 students at 2 exams. It was decomposed into 5 categories by the FCM algorithm: the central cluster corresponds to students having average results for both exams, the 4 peripheral clusters correspond to the 4 combinations success/failure for the 2 exams. The obtained fuzzy prototypes capture the semantics of the subgroups: they are approximately centered around the group means, but take into account the discriminative features of the clusters and underline their specificity. Indeed, they influence each other and are further apart than the averages would be. In the case of the lower left cluster for instance, the student having obtained twice the mark 0 totally belongs to the prototypes, which corresponds to the group interpretation as students having

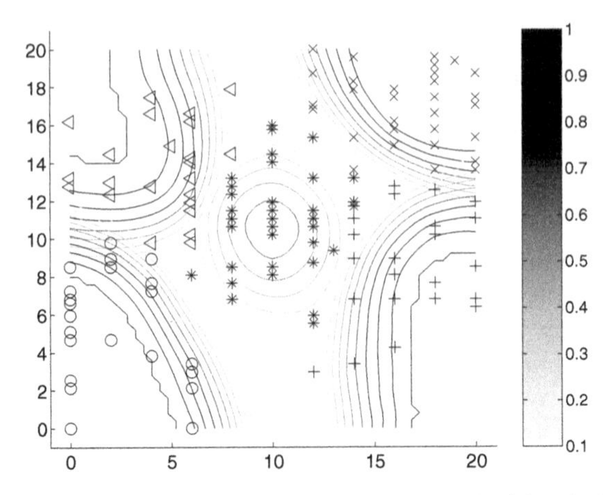

Fig. 3. Fuzzy prototypes characterizing students described by their results on two exams [28]

failed at both exams and underlines its specificity. Furthermore, the fuzzy properties of these prototypes indicate the transitions between the groups and model their unsharp boundaries, providing richer information than a single numerical value.

Figure 4 illustrates the task of class comparison: two classes were evaluated for the same exam and are to be compared. To that aim, the two classes are first decomposed using the OPCA algorithm, clusters are then represented through their fuzzy profiles [29], as shown on the first row of Fig. 4, and

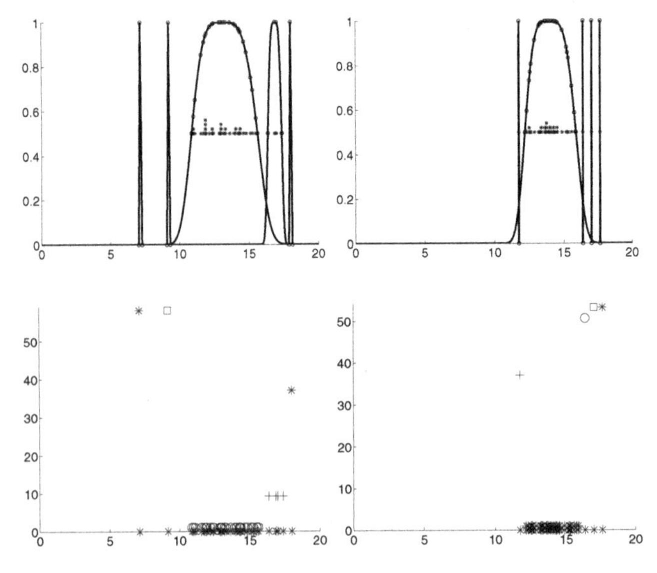

Fig. 4. Fuzzy profiles and exceptionality coefficients for two student classes [28]

finally enriched by exceptionality coefficients, as shown on the second row. As expected, OPCA identifies both classic clusters and one-point cluster corresponding to specific cases: for instance in the left case, two students having especially big difficulties are identified; in the right case, three levels of especially brilliant students are detected. The fuzzy profiles show that the two classes have the same global structure, made of a large middle group, with some smaller groups. The class represented on the left appears less homogeneous, as it contains two students having difficulty but it has a more important best student group. Exceptionality coefficients enrich this description and modify the impression, showing that the two classes are actually different: the left class can be globally characterized as having better results. In both cases, the group with the lower exceptionality coefficient is the middle group, which indeed is the most representative cluster for the whole class. Yet differences are to be observed for the other groups: in the right-hand case, the best students have high exceptionality coefficients, i.e. they appear as outliers. In the left-hand case, the exceptions correspond to the lower results, and the high ones have lower exceptionality coefficients, indicating they are more characteristic of the whole class.

3.3 Chemical Applications

Management of Sensorial Information

This project was conducted in collaboration with the chemistry and computer science departments of the Science and Technique faculty of the Mohammedia university (Morocco).

Objective

In order to be discerned, volatile compounds, pure chemicals or miscellanies, are elated by the air to the nose where they are warmed and humidified. The birth of olfactory message results from the adsorption of the arriving smelling molecules to the nasal mucus. Several theories of olfaction were suggested but, until now, the mechanism of perception stays unrecognized although progress have been made in different disciplines related to the chimioreception. Chemists attempt to contribute to the understanding of the mechanisms implied in the olfaction by the investigation of the links that bind the chemical structure of the smelling to the quality of their odors. To this academic goal, a convenient and commercial interest can be added as the rational design of new molecules which could be used in the industry of fragrance, aromas and cosmetic.

Thus the aim of the study [37] is to extract relationships between the molecule properties and its odor.

The first difficulty of this task deals with the translation of the structure of chemicals in a set of parameters containing the usable information to solve the problem. Among other methods, the auto correlation [58] and the fragmentation method [9,25] have been applied and obtained a satisfactory degree of success. More recently, artificial neural networks have been applied [10]. Yet they allow neither to establish mathematical equations between the studied odor and the used descriptors, nor to give off any rules allowing to know the role and the influence of every used descriptor. While descriptors generally used in this domain are of precise nature -continuous or discrete- the made decisions are shaded. Experts always classify the quality of an odor by some symbolic descriptors on some numeric scales such as "strong or weak odor of musk", or "absence or presence of studied odor". Classic systems do not allow the treatment of this double description of the odor.

Data

The set of studied compounds is constituted by 99 aliphatic alcohols. The odor of these alcohols was described in details by Schnabel et al. [51] using 16 symbolic descriptors on an active scale valued from 1 to 5. We focus here on the odor of camphor, considering as camphoraceous the 58 compounds that have a score between 1 and 5 on the Schnabel's scale, and as non-camphoraceous the 41 compounds with score 0.

The molecules are represented using a descriptive model obtained by the GESDEM method (Generation and selection of descriptor and pattern elaboration) [59], combined to geometric and physico-chemical characteristics. The GESDEM methodology uses as descriptors the groups of atoms held together by bonds and constituting specific subgraphs of the considered molecule (see [37] for more details). The geometric and physico-chemical information is made of 10 attributes such as the molecule length, its width, its depth, its molecular refraction, its density, its aqueous solubility, an indication of its relative lipophilicity, and some ratios between these quantities.

Method

The method used here was the construction and the use of fuzzy decision trees thanks to the *Salammbô* software (see Sect. 2.2).

Results

First a 4-fold cross-validation test using all attributes was conducted, leading to 88.0% good classification rate, and showing an improvement when compared to crisp decision trees. Figure 5 shows the obtained tree: within each node of the tree, the number of molecules pertaining to this node is given. On each vertex going out of a node, a test on the value of an attribute is given. This test can be a crisp one (e.g. ≤ 1) or a fuzzy one (e.g. *small*), in this case, the membership partition constructed by the *Salammbô* software is shown. The root of the tree tests the 4D4 attribute, which is related to bonds of the type C-C(C)-C: it indicates the number of atoms of tertiary carbons and quaternary carbons and gives an idea on the global shape of the molecule.

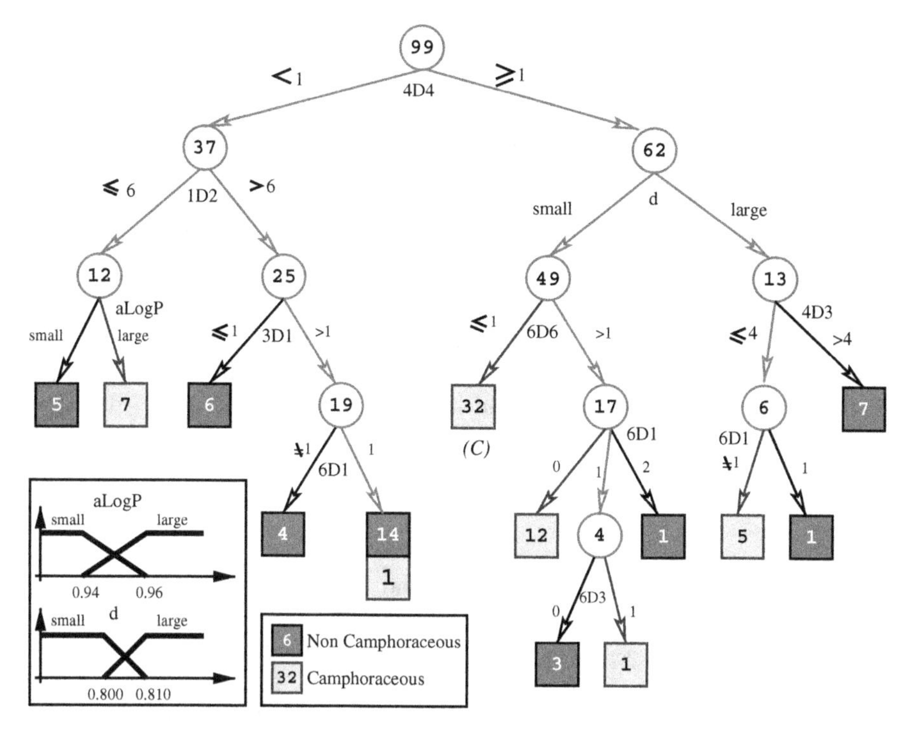

Fig. 5. Fuzzy decision tree to classify camphoraceous data

Tests were also performed using only the attributes of the descriptive model of molecules or only the geometric and physico-chemical attributes. They showed that the descriptive model is efficient to determine attributes to predict the camphoraceous odor of a molecule. Moreover, these attributes are more easily evaluated than geometric attributes which are more complex to measure and that do not improve greatly the prediction of the odor.

3.4 Multimedia Applications

In this section, we consider applications in data mining and information retrieval in the multimedia field. We describe first an image retrieval application based on a visual similarity navigation paradigm and second a learning approach for a semantic annotation of a video signal, based on some examples.

Searching in a Clothes Catalogue by Visual Similarity

This work is part of the results of the ITEA European project: KLIMT - KnowLedge InterMedation Technology. Although in this project several laboratories and industry partners were involved to accomplish what follows, the company Sinequa played a significant role.

Objective

The main objective of this work was to enhance a classic text search engine of an on-line clothes catalogue, with an image search tool. A prototype was developed that illustrated the complementarities between the two navigation schemas: text queries and visual-similarity browsing.

Data

The data was provided by an on-line store client of Sinequa: La Redoute. For the development of the prototype we disposed of a database of 5000 products. For each of them, we had one or two images and the associated text description. The text was semi-structured in an XML format (title, price, description, etc) and the images were generally of low quality and of mid-size (around 300 x 300).

Method

We were responsible for the image search and the fusion of the information. The text search was based on Sinequas engine: Intuition. It works very similarly any other text based engine, with a strong focus on linguistics and semantics.

When searching based on visual features, it is extremely difficult to specify the query. In fact, we do not have any keywords that could be found, in a way or another, inside a targeted text. One idea is to use an example as starting point and then look for images similar to this initial one. In order to achieve this, we need two keystones: on the one hand, we need to define a numerical description of the image and a similarity that will translate the fact that two images are visually similar.

Traditionally histograms of colors, textures and forms are used to describe the image; then a classic distance (usually Euclidean) is used to compare them. The main disadvantage of using distances is that they are very sensitive to the number of dimensions used to describe the histogram. Therefore we used Tversky's ratio model (see Sect. 2.1), which is not only stable in this respect, but is also based on a psychological framework.

Searching visually-similar clothes based on an example is interesting as long as we can isolate the clothes from the rest of the image. To achieve this, we developed a segmentation algorithm, which is fast and robust with respect to textures (prints on the clothes, etc). During the indexation phase (off-line), we applied the segmentation algorithm to all images and we described all segments based on their colors, forms and position. In a second phase (on-line), we compute on the fly the similarities of the example segment with all other segments. This computation for the size of the database is instantaneous.

Results

As a result we developed an image and text search engine, with a look and feel of the La Redoute on-line store. On Fig. 6 we can see part of the interface, where the user chooses the region of the image of his interest.

Although no usability tests were driven, we observed that users, when looking for something specific, prefer to use text queries. But when lounging around they do like the possibility to find visually similar clothes. This is particularly true for descriptions that are difficult to translate into text, as for instance specific color patterns or prints.

More details about this specific work and further results on text and image fusion can be found in [39]. The reader interested on further results on image search based on segments can read [40] and if interested on the theoretical consequences of using Tversky's similarities in the general case of information retrieval refer to [41].

Learning to Label a Video Signal with High Level Features

This application was conducted for the TREC Video Retrieval 2005 challenge organized yearly by the NIST institute [1].

Objective

In order to retrieve specific information in a video, we need to index it. The index consists of a set of labels and attached to each label a set of time locator. These labels may be of different type and we focus here on high level features, as for instance: the presence of a map in the segment, the presence of

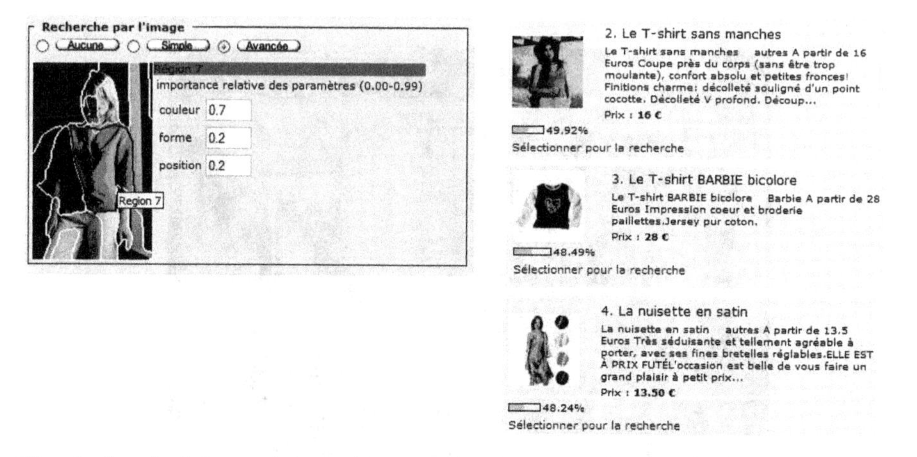

Fig. 6. On the left we observe the part of the interface allowing the user to choose the region he would like to focus on and the weights he would give to the color, form and position aspects. On the right we see the result

a particular person (politician, sportsman,. . .), the occurrence of an explosion, the fact that the segment has been recorded outside or inside a building, etc. Our approach consists in learning, from a set of examples, how to label a video signal, with high level features, in particular here the presence of a map on the screen.

The TRECVID challenge is decomposed into 2 steps: a model has to be built according to a training set of video, and this model should enable the ranking of a set of test videos. The objective is to propose a ranking of all the segments of the test videos according to the presence of a given high level feature.

Data

The TRECVID challenge offers a video database in order to train and test each method. The data are composed of more than 200 video news (each of around 30 minutes length) that represent more than 100 hours of video data. Moreover, some additional data is proposed: a set of XML files that describes the cutting into shot of each video, the set of all of the image files representing keyframes, and the set of annotations files for development keyframes [1,42].

Each keyframe is represented by two sets of numerical values, the *Visual Information Descriptors* and the *Video Information Descriptors*. The *Visual Information Descriptors* are obtained directly from the keyframes. To obtain visual spatial-related information from the keyframe, the image is cut into 5 pieces (see Fig. 7). Each piece corresponds to a spatial part of the keyframe: top, bottom, left, right, and middle. The five regions do not have the same size in order to reflect the importance of the contained information based on its position. Afterwards, for each region, the associated histogram in the HSV

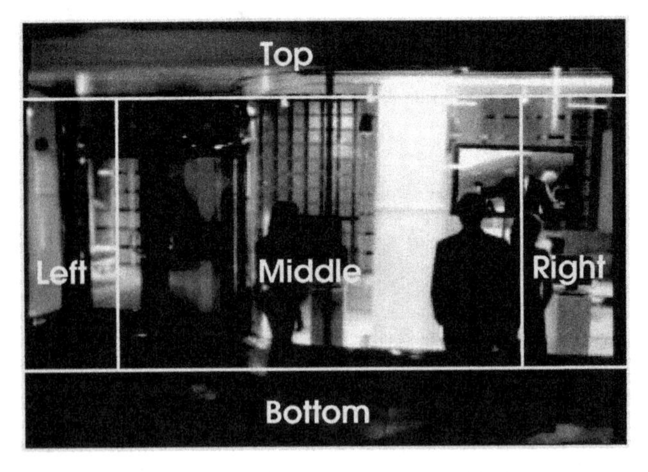

Fig. 7. Spatial decomposition of a keyframe

space was computed. Depending on the area of the region, the histogram is defined more or less precise (based on the number of bins): 6x3x3 for Middle, Top, and Bottom, 4x2x2 for Left and Right. At the end, a first set of numerical values (each one ranging from 0 to 1) is obtained that characterizes every keyframe: the Visual Information Descriptors.

The *Video Information Descriptors* are obtained from the information associated with the video and given by means of the shot detection process. They correspond to the temporal information associated with the shot from which the keyframe was extracted. For a given keyframe, these descriptors are extracted from the XML file associated with a video and obtained from shot detection process. The XML tags associated with each shot enables us to obtain the following information for every keyframe:

- the name of the keyframe and its kind
- the timecode of the keyframe in the video
- the timecode of the beginning of the shot containing the keyframe
- the duration of the shot containing the keyframe

At the end, a second set of numerical values is obtained that characterizes the keyframe and the shot to which it belongs: the Video Information Descriptors.

Finally, each keyframe is associated to a *Class Descriptor*, obtained from the indexation of the video. It corresponds to the feature(s) that should be associated with a shot. It is extracted from the file obtained from the (human) indexation process of the training videos. A keyframe can be associated with more than one feature depending on the result of the indexation process. The set of experiments that has been conducted this year focuses on the detection of the presence of a map in a segment.

Method

The method used in the NIST TRECVID'2005 evaluation task is based on the use of Fuzzy Decision Trees (see Sect. 2.2 and [13–15, 35]). For more details on the approach, please refer to [36].

First of all, a training set was constructed by means of a set of descriptors that can be extracted from the video data. It enabled us to obtain a classifier used afterwards to classify and rank the test keyframes. In order to use the FDT learning method, which is a supervised learning method, the training set must contain cases with the feature to be recognized and cases that do not possess that feature. Moreover, decision tree construction methods are based on the hypothesis that the value for the class is equally distributed. This hypothesis is not valid when considering the TRECVID'05 data set: for instance, for the Map feature, in the whole development set of indexed keyframes, there are 940 keyframes with the Map feature and 61273 keyframes without the Map feature. Thus, to have a valid training set for the construction of a fuzzy decision tree, the number of keyframes of each class have to be balanced. In our experiment, 900 keyframes with each class (with the Map feature, or without the Map feature) have been selected in order to build a training set with 1800 keyframes.

The *Salammbô* software (see Sect. 2.2) was used to train and test the data. A cross validation was conducted, as follows:

Step 1 The training set is composed of 900 keyframes with the Map feature and 900 keyframes without the Map feature. Each of these keyframes is randomly selected in the corresponding set of keyframes,

Step 2 An evaluation set is composed using the rest of the keyframes (40 with the Map feature, and 60373 without the Map feature).

Step 3 A FDT is constructed by means of the training set, and is used to classify the evaluation set (the presence of the feature for a keyframe is predicted by means of the FDT, and the result is compared with the indexation of this keyframe).

These 3 steps are renewed 3 times in order to obtain several results, which are then averaged.

The final FDT were obtained by using the whole set of training keyframes with the Map feature:

Step 1 The training set is composed of 940 keyframes with the Map feature and 940 keyframes without the Map feature. The keyframes without the Map feature are randomly selected in the whole set of keyframes without the Map feature,

Step 2 A FDT is constructed by means of the training set, and is used to classify the whole test set of keyframes from video 1 to video 140.

As the random selection of a subset of 940 keyframes without the Map feature from the set of 61273 ones enables us to obtain several subsets, these 2 steps were renewed 5 times in order to obtain several runs.

After the construction of the FDT as explained in Sect. 2.2, each FDT is used to classify the whole test set of keyframes.

First of all, Visual Information Descriptors and Video Information Descriptors are extracted for all the keyframes from the test set. This enables us to obtain vectors of numerical data that can be classified with the FDT.

By means of the classification, each keyframe e from the test set is associated with a membership degree $\texttt{FDeg}(c_e)$ to the Map feature. At the end, each shot from the test video set is associated with a membership degree $\texttt{FDeg}(c_e)$ of its keyframe e. All test shots can thus be ranked by means of these membership degrees. We assumed that the higher the membership degree, the more confident the FDT is of the presence of the feature in the shot. This ranking method is the one used for all runs submitted to TRECVID 2005.

Results

The global results obtained using extremely simple visual description and out-of-the-box fuzzy decision tree software were encouraging. The use of this type of algorithm is a novelty on this kind of application. This approach provides classification rules which are human understandable, thus allowing further developments. The presented runs are an underestimation of what could be easily obtained.

In fact, the FDT optimizes the classification of all the examples and not the ranking of the results. It appeared in the results below: 21 teams participated in the NIST TRECVID'2005 Challenge [1]. For the feature "Maps", the results obtained by our approach were:

- Average precision: *0.163* (range of the whole results: [0.001, 0.526], mean: 0.24).
- Number of hits within the 100 first shots: *69 hits* (range of the whole results: [0, 100], mean: 81).
- Number of hits within the 1000 first shots: *411 hits* (range of the whole results: [32, 897], mean: 491).
- Number of hits within the 2000 first shots: *683 hits* (range of the whole results: [49, 1095], mean: 651).

Further developments on the adaptation of fuzzy decision trees to ranking problems (instead of just classification) are under study.

4 Conclusion

Real world applications address a double challenge. On the one hand they are responses to specific problems with their specific constraints and on the other hand they have to build on solid and general theoretical foundations. In this paper we briefly present first two of these essential pillars: fuzzy comparison measures and fuzzy machine learning. Then, based on these two strongly interrelated bases, a set of applications in domains ranging from medical to educational but also chemical and multimedia domains illustrate specific solutions.

All these applications focus on information retrieval or data mining, which are in fact, as we saw in this chapter, two components of a same challenge: the search or the extraction of information and knowledge from large amounts of data. For each of the presented solutions, we focused only on how the theory supported the application, ignoring a large set of other difficulties, appearing when dealing with real world challenges: as for instance technical issues, solutions for fast execution (essential in the case of large data sets), management of large data bases, corrupted data, etc. For more details on each of theses applications please refer to the corresponding publications.

Finally, what makes all these applications unique is the use of fuzzy logic. In fact, their success is due, to some extent, to its capability to represent diverse types of data in a synthetic way, its robustness with regard to changes or noise, and obviously its unique expressiveness, crucial for the understanding of the results.

References

1. Guidelines for the TRECVID 2005 evaluation - National Institute of Standards and Technology, 2005. http://www-nlpir.nist.gov/projects/tv2005/tv2005.html.
2. J. Aczel. Entropies, characterizations, applications and some history. In *Modern Information Processing, from Theory to applications*. Elsevier, 2006.
3. J. Bezdek. *Fuzzy mathematics in pattern classification*. PhD thesis, Applied Mathematical Center, Cornell University, 1973.
4. J. Bezdek. *Pattern Recognition with Fuzzy Objective Function Algorithm*. Plenum, New York, 1981.
5. S. Bothorel, B. Bouchon-Meunier, and S. Muller. Fuzzy logic-based approach for semiological analysis of microcalcification in mammographic images. *International Journal of Intelligent Systems*, 12:814–843, 1997.
6. B. Bouchon-Meunier, M. Rifqi, and S. Bothorel. Towards general measures of comparison of objects. *Fuzzy sets and systems*, 84(2):143–153, 1996.
7. X. Boyen and L. Wehenkel. Automatic induction of fuzzy decision tree and its application to power system security assessment. *Fuzzy Sets and Systems*, 102(1):3–19, 1999.
8. L. Breiman, J.H. Friedman, R.A. Olshen, and C.J. Stone. *Classification And Regression Trees*. Chapman and Hall, New York, 1984.
9. W. E. Brugger and P. C. Jurs. Extraction of important molecular features of musk compounds using pattern recognition techniques. *J. Agric. Food Chem.*, 25(5):1158–1164, 1977.
10. M. Chastrette, D. Zakarya, and J. P. Peyraud. Structure-musk odor relationships for tetralins and indans using neural networks. *Eur. J. Med. Chem.*, 29:343–348, 1994.
11. K.J. Cios and L.M. Sztandera. Continuous ID3 algorithm with fuzzy entropy measures. In *Proceedings of the first International IEEE Conference on Fuzzy Systems*, San Diego, 1992.
12. R. Davé. Characterization and detection of noise in clustering. *Pattern Recognition Letters*, 12:657–664, 1991.
13. M. Detyniecki and C. Marsala. Fuzzy inductive learning for multimedia mining. In *Proc. of the EUSFLAT'01 conference*, pages 390–393, Leicester (UK), September 2001.
14. M. Detyniecki and C. Marsala. Fuzzy multimedia mining applied to video news. In *Proc. of the 9th IPMU'00 Conf.*, pages 1001–1008, Annecy, France, July 2002.
15. M. Detyniecki and C. Marsala. Discovering knowledge for better video indexing based on colors. In *Proc. of the Fuzz-IEEE'03 conference*, pages 1177–1181, St Louis (USA), May 2003.
16. J.C. Dunn. A fuzzy relative of the isodata process and its use in detecting compact well-separated clusters. *Journal of Cybernetics*, 3:32–57, 1973.
17. M. Friedman, M. Ming, and A. Kandel. On the theory of typicality. *International Journal of Uncertainty, Fuzzyness and Knowledge-Based Systems*, 3(2):127–142, 1995.
18. E. Gustafson and W. Kessel. Fuzzy clustering with a fuzzy covariance matrix. In *Proc. of IEEE CDC*, pages 761–766, 1979.
19. F. Höppner, F. Klawonn, R. Kruse, and T. Runkler. *Fuzzy Cluster Analysis, Methods for classification, data analysis and image recognition*. Wiley, 2000.

20. E. Hüllermeier. Fuzzy methods in machine learning and data mining: Status and prospects. *Fuzzy Sets and Systems*, 156(3):387–406, 2005.
21. A. Jain, M. Murty, and P. Flynn. Data clustering: a review. *ACM Computing survey*, 31(3):264–323, 1999.
22. C.Z. Janikow. Fuzzy decision trees: Issues and methods. *IEEE Transactions on Systems, Man and Cybernetics*, 28(1):1–14, 1998.
23. F. Klawonn. Understanding the membership degrees in fuzzy clustering. In *Proc. of the 29th Annual Conference of the German Classification Society, GfKl 2005*, pages 446–454. Springer, 2006.
24. F. Klawonn, R. Kruse, and H. Timm. Fuzzy shell cluster analysis. In *Learning, networks and statistics*, pages 105–120. Springer, 1997.
25. G. Klopman and D. Ptchelintsev. Application of the computer automated structure evaluation methodology to a QSAR study of chemoreception- aromatic musky odorants. *J. Agric. Food Chem.*, 40:2244–2251, 1992.
26. R. Krishnapuram and J. Keller. A possibilistic approach to clustering. *IEEE Transactions on fuzzy systems*, 1:98–110, 1993.
27. M.-J. Lesot. Kernel-based outlier preserving clustering with representativity coefficients. In B. Bouchon-Meunier, G. Coletti, and R. Yager, editors, *Modern Information Processing: From Theory to Applications*, pages 183–194. Elsevier, 2005.
28. M.-J. Lesot and B. Bouchon-Meunier. Cluster characterization through a representativity measure. In *Proc. of Flexible Query Answering Systems, FQAS'04*, pages 446–458. Springer, 2004.
29. M.-J. Lesot and B. Bouchon-Meunier. Descriptive concept extraction with exceptions by hybrid clustering. In *Proc. of the IEEE Int. Conf. on Fuzzy Systems, Fuzz-IEEE'04*, pages 389–394. IEEE Press, 2004.
30. M.-J. Lesot, L. Mouillet, and B. Bouchon-Meunier. Fuzzy prototypes based on typicality degrees. In *Proc. of the 8th Fuzzy Days 2004*, pages 125–138. Springer, 2005.
31. A. Lurie, C. Marsala, S. Hartley, B. Bouchon-Meunier, F. Guillemin, and D. Dusser. Patients' perception of asthma severity. *(to be published)*.
32. C. Marsala. Fuzzy partitioning methods. In W. Pedrycz, editor, *Granular Computing: an Emerging Paradigm*, Studies in Fuzziness and Soft Computing, pages 163–186. Springer-Verlag, 2001.
33. C. Marsala and B. Bouchon-Meunier. An adaptable system to construct fuzzy decision trees. In *Proc. of the NAFIPS'99 (North American Fuzzy Information Processing Society)*, pages 223–227, New York, USA, June 1999.
34. C. Marsala, B. Bouchon-Meunier, and A. Ramer. Hierarchical model for discrimination measures. In *Proc. of the IFSA'99 World Congress*, pages 339–343, Taiwan, 1999.
35. C. Marsala and M. Detyniecki. Fuzzy data mining for video. In *Proc. of the EUSFLAT'03 conference*, pages 73–78, Zittau, (Germany), September 2003.
36. C. Marsala and M. Detyniecki. University of Paris 6 at TRECVID 2005: High-level feature extraction. In *TREC Video Retrieval Evaluation Online Proceedings*, 2005. http://www-nlpir.nist.gov/projects/tvpubs/tv.pubs.org.html.
37. C. Marsala, M. Ramdani, D. Zakaria, and M. Toullabi. Fuzzy decision trees to extract features of odorous molecules. In B. Bouchon-Meunier, R.R. Yager, and L.A. Zadeh, editors, *Uncertainty in Intelligent and Information Systems*, volume 20 of *Advances in Fuzzy Systems - Applications and Theory*, pages 235–249. World Scientific, 2000.

38. C. Olaru and L. Wehenkel. A complete fuzzy decision tree technique. *Fuzzy Sets and Systems*, 138(2):221–254, 2003.
39. J.-F. Omhover and M. Detyniecki. Combining text and image retrieval. In *Proceedings of the EUROFUSE Workshop on Data and Knowledge Engineering*, pages 388–398, 2004.
40. J.-F. Omhover, M. Detyniecki, and B. Bouchon-Meunier. A region-similarity-based image retrieval system,. In B. Bouchon-Meunier, G. Coletti, and R. Yager, editors, *Modern Information Processing: From Theory to Applications.* Elsevier, 2005.
41. J.-F. Omhover, M. Detyniecki, M. Rifqi, and B. Bouchon-Meunier. Image retrieval using fuzzy similarity: measure equivalence based on invariance in ranking. In *Proc. of the IEEE Int. Conf. on Fuzzy Systems - Fuzz-IEEE'04*, pages 1367–1372, 2004.
42. C. Petersohn. Fraunhofer HHI at TRECVID 2004: Shot boundary detection system. Technical report, TREC Video Retrieval Evaluation Online Proceedings, TRECVID, 2004. URL: www-nlpir.nist.gov/projects/tvpubs/tvpapers04/fraunhofer.pdf.
43. J. Ross Quinlan. Induction of decision trees. *Machine Learning*, 1(1):86–106, 1986.
44. M. Ramdani. Une approche floue pour traiter les valeurs numériques en apprentissage. In *Journées Francophones d'apprentissage et d'explication des connaissances*, 1992.
45. A. Rick, S. Bothorel, B. Bouchon-Meunier, S. Muller, and M. Rifqi. Fuzzy techniques in mammographic image processing. In Etienne Kerre and Mike Nachtegael, editors, *Fuzzy Techniques in Image Processing*, Studies in Fuzziness and Soft Computing, pages 308–336. Springer Verlag, 2000.
46. M. Rifqi. Constructing prototypes from large databases. In *Proc. of IPMU'96*, 1996.
47. M. Rifqi. *Mesure de comparaison, typicalité et classification d'objets flous : théorie et pratique.* PhD thesis, Université de Paris VI, 1996.
48. E. Rosch. Principles of categorization. In E. Rosch and B. Lloyd, editors, *Cognition and categorization*, pages 27–48. Lawrence Erlbaum associates, 1978.
49. E. Rosch and C. Mervis. Family resemblance: studies of the internal structure of categories. *Cognitive psychology*, 7:573–605, 1975.
50. E. Ruspini. A new approach to clustering. *Information control*, 1(15):22–32, 1969.
51. K.O. Schnabel, H.D. Belitz, and C. Ranson. Untersuchungen zur Struktur-Activität-Beziehung bei Geruchsstoffen. *Z. Lebensm Unters Forsch*, 187: 215–233, 1988.
52. A. Tversky. Features of similarity. *Psychological Review*, 84:327–352, 1977.
53. X. Wang, B. Chen, G. Qian, and F. Ye. On the optimization of fuzzy decision trees. *Fuzzy Sets and Systems*, 112(1):117–125, May 2000.
54. R. Weber. Fuzzy-ID3: A class of methods for automatic knowledge acquisition. In *IIZUKA'92 Proceedings of the 2nd International Conference on Fuzzy Logic*, pages 265–268, 1992.
55. Z. Wu, W. Xie, and J. Yu. Fuzzy c-means clustering algorithm based on kernel method. In *Proc. of ICCIMA'03*, pages 1–6, 2003.
56. Y. Yuan and M.J. Shaw. Induction of fuzzy decision trees. *Fuzzy Sets and systems*, 69:125–139, 1995.

57. L.A. Zadeh. A note on prototype theory and fuzzy sets. *Cognition*, 12:291–297, 1982.
58. D. Zakarya. Use of autocorrelation components and Wiener index in the evaluation of the odor threshold of aliphatic alcohols. *New J. Chem.*, 16:1039–1042, 1992.
59. D. Zakarya, M. Chastrette, M. Tollabi, and S. Fkih-Tetouani. Structure-camphor odor relationships using the generation and selection of pertinent descriptors approach. *Chemometrics and Intelligent Laboratory Systems*, 48: 35–46, 1999.

Gene Regulatory Network Modeling: A Data Driven Approach

Yingjun Cao, Paul P. Wang, and Alade Tokuta

Abstract. In this chapter, a novel gene regulatory network inference algorithm based on the fuzzy logic network is proposed and tested. The algorithm is intuitive and robust. The key motivation for this algorithm is that genes with regulatory relationships can be modeled via fuzzy logic, and the degrees of regulations can be represented as the accumulated distance during a period of time intervals. One unique feature of this algorithm is that it makes very limited prior assumptions concerning the modeling; hence the algorithm is categorized as a data-driven algorithm. As a non-parametric model, the algorithm is very useful when only limited a priori knowledge on the target microarray exists. Another characteristic of this algorithm is that the time-series microarray data have been treated as a dynamic gene regulatory network, and the fuzzification of gene expression values makes the algorithm more agreeable to reality. We have deduced the dynamic properties of the FLN using the anneal approximation, and dynamic equations of the FLN have been analyzed. Based upon previous investigation results that in yeast protein-protein networks, as well as in the Internet and social networks, the distribution of connectivity follows the Zipf's law, the criteria of parameter quantifications for the algorithm have been achieved. The algorithm was applied on the yeast cell-cycle dataset from Stanford *Saccharomyces cerevisiae* database which produced pleasing results. The computation also showed that a triplet search is suitable and efficient for the inference of gene regulatory networks.

The chapter is organized as follows: first, the broader context of gene regulatory networks research is addressed. Then the definition of the FLN is given, and the dynamic properties of the FLN are deduced using the annealed approximation in Sect. 2. Combined with the Zipf's law, the parameter quantification guidelines are achieved at the end of this section. In Sect. 3, the algorithm is illustrated in detail with its data-driven rationale. Then, the algorithm's inference results on *Saccharomyces cerevisiae* dataset are presented and analyzed in Sect. 4. Future research work with the algorithm is discussed at the end of this chapter.

1 Introduction

With the development of molecular biology and genetic technologies, huge amounts of data will be measured for a big genome with relatively high accuracy. This, in return, arouses the concurrent development of computational bioinformatics tools to analyze the data from genetic laboratories. One of the most important problems in bioinformatics is to discover how genes interregulate in a systematic manner which results in different translated protein products and phenotypes. To study the causal pathways that control complex biological functions, some studies concerning gene regulations have focused on the group properties of clustered genes via proximity index [1–3]. Despite some of the important discoveries on genetic features inside a class of genes, this methodology, indeed, makes the use of static methods for this dynamic process. In the mean time, more researchers have begun to realize that gene regulatory mechanisms should be modeled as a network topologically in order to gain more detailed insight [4]. The reason for the networking idea is that normal regulation pathways are composed of regulations resulting from many genes, ribonucleic acids (RNAs) and transcription factors (TFs), and these controlling chemical complexes serve as the driving force in maintaining normal organism functions. In addition, the network model enables us to process spatial and temporal gene interactions simultaneously. Furthermore, the large amount of data should not pose many adverse effects on the stability, the behavior, and the processing time.

Based on the network representation of gene regulations, a number of network inference models have been proposed. Boolean network was the first proposed model with extensive studies and application results [5–8]. Due to the binary limitation inherent in Boolean values, the exact properties of gene regulations cannot be expressed in relative detail by this model. Other mathematical models and approaches have also been adopted to model the gene regulatory mechanism, such as differential equations [9–11], Bayesian networks [12–16], hybrid Petri net [17], growth network [18,19], circuit construction via biochemical perturbations [20], weighted matrix [21], neural networks [22,23], and hybrid networks [24,25]. These models, however, have stressed different aspects of the behavior and each model has contributed good inference results in certain aspects. The ongoing research on those models is focusing on the challenges of data integration or information fusion. The reason is, unlike the lower level analysis (sequence analysis), gene expression data, thus far, have not been able to provide the information accurate enough to construct the exact regulatory network. Thus other types of information, like DNA sequence data, should be integrated into the inference model; hence causes the fusion of different types of data. Other aspects of regulatory networks research include non-linear data processing, noise tolerance, synchronization and model over fitting [26].

Research on genome wide gene regulations has used dynamic microarray data, which offers a throughput measurement of genomic expression levels at each sample time. Given a series of microarray data, researchers have been trying to locate the spatial and temporal modes of regulations regarding different conditions or different stages of cell-cycle on different species [26–28]. But because of the hybridization process and the synchronization issues of time-series microarray, the data, very often, contain missing, noisy or unsynchronized subsets. Other techniques like serial analysis of gene expression (SAGE) [29,30], TF mappings [31,32], and antibody arrays [33] have appeared in the literature to discover the regulatory mechanism.

As the simplest yet the mostly used model for gene regulatory networks, the NK Boolean network was first proposed by Kauffman [5, 7, 34, 35]. The fundamental theory of the NK Boolean network can be simply illustrated as the following.

Gene A and gene B interact during some time intervals, and their interactions will determine or regulate the status of another gene C. In the NK Boolean network, their regulating mechanism can be modeled by a logical function

$$C(t+1) = Logic(A(t), B(t))$$

It means the current state of A and B will regulate the next state of gene C through certain intrinsic logical function. If numerous genetic regulations occur simultaneously, the participating genes with their unique logical functions constitute a gene regulatory network. The network will be self-evolving, and eventually reach certain final states. In this model, N is the total number of genes in the network, and K is the average number of regulating genes. According to the theoretical studies and simulations based on this model, the NK Boolean network seems to fit the biological systems best when K equals to 2. This constraint on the connectivity of the network enables the network to mimic biological mutations and evolution phenomena by the so-called chaotic dynamics of the system [36–40]. With the properties of stability, redundancy, and periodicity, further research on Boolean network model involves with the mutual information inference and probabilistic Boolean models as extensions [8]. These hybrid models tend to make a Boolean network more realistic and pragmatic; hence the network will yield more power.

In this research, a novel network model, fuzzy logic network (FLN), is proposed and thoroughly examined. The possibility of fitting gene regulation into this model will also be addressed. In a sense, FLN is the generalization of Boolean network, but it is capable of overcoming the unrealistic constraint of Boolean value (ON/OFF symbolically). Fuzzy logic has evolved as a powerful tool over more than 40 years, and its applications are widely available in scientific research and engineering literature. This research aims to build up the basic model of FLN, and study the characteristics of this network in detail. In this study, the gene regulation mechanism has been proved to be suitably modeled as a FLN. The inference algorithm for gene regulatory networks is devised to locate the optimal network structure purely based on data. Further, the FLN is able to inherit all the good properties of Boolean network especially the causal property, but FLN is expected to be a more effective model with the nuance of membership function adjustment, flexible inference rules, and the ability to model highly non-linear relationships and periodicity.

With the distinctive properties in processing real life incomplete data and uncertainties, the gene regulation analysis based on fuzzy logic theory has appeared in open literature [41–45] with good inference results. The proposed FLN has inferred excellent regulatory networks for yeast cell-cycle [46–48]. With the power of modeling complex network, the FLN developed in this work is expected to evolve into one of the important tools for gene regulatory networks inference.

2 Fuzzy Logic Network Theory

The proposed fuzzy logic network theory is novel, and is presented in the following manner. In the first part, the basic definitions of FLN and their appropriate meanings are given. Secondly, the dynamic process of the network is deduced using the approach of annealed approximation. Combined with the Zipf's law, the critical conditions that a network must satisfy in order to model regulatory mechanisms are presented in the end.

2.1 Definitions

1). Fuzzy logic Network

Given a set of N variables (genes),
$\mathbf{F(t)} = (F_1(t), F_2(t), \ldots, F_N(t)), F_i(t) \in [0,1]$ $(i = 1, 2, \ldots, N)$, t represents time; the variables are to be updated by the dynamic equations:

$$F_i(t+1) = \Lambda_i(F_{i_1}(t), F_{i_2}(t), \ldots, F_{i_K}(t)) \quad (1)$$

where Λ_i is a randomly chosen fuzzy logical function.

For an FLN, the logical functions can be constructed using the combination of AND ($\wedge$), OR ($\vee$) and NOT (-). The total number of choices for fuzzy logical functions is decided only by the number of inputs. If a node has K $(1 \leq K \leq N)$ inputs, then there are 2^K different logical functions. In the definition of FLN, each node, $F_i(t)$, has K inputs. But this fixed connectivity will be relaxed later.

2). Membership functions

The membership function is defined as a function $\Lambda_u : U \to [0,1]$ where Λ_u is the degree of the membership. Usually, the membership function has to satisfy the so called t-norm/t-co-norm, which is a binary operation that satisfies the identity, commutative, associative, and increasing properties. Table 1 is a list of commonly used fuzzy logical functions [49].

3). Quenched update

If all fuzzy logical functions, Λ_i $(i = 1, 2, \ldots, N)$, and their related variable sets, $\{F_{i_1}(t),\ F_{i_2}(t),\ \ldots,\ F_{i_K}(t)\}$, chosen at the first step of the system remain the same throughout the whole dynamic process, then the system is defined as a quenched updated network.

Table 1. Four commonly used fuzzy logical functions including their AND ($\wedge$), OR ($\vee$) and NOT (-)

logical functions	$a \wedge b$	$a \vee b$	$\overline{a}$
Max-Min	$\min(a, b)$	$\max(a, b)$	1-a
GC	$a \times b$	$\min(1,\ a + b)$	1-a
MV	$\max(0,\ a + b$ -1)	$\min(1,\ a + b)$	1-a
Probabilistic	$a \times b$	$a + b - a \times b$	1-a

4). Synchronous update
If all the variables, $F_i(t)$ $(i = 1, 2,\ \ldots,\ N)$, are updated at the same time, then the system is called synchronously updated; otherwise, it is asynchronously updated. In this chapter, the FLN is assumed to be synchronously updated.

5). Annealed approximation
If fuzzy logical functions are randomly chosen in each update, then the network is called an annealed network. The annealed network can be thought as the lower-bound approximation of the quenched network. Although it neglects the quenched property, one can expect that the properties deduced from annealed approximations will have good agreement with numerical simulations of quenched system, quantitatively and qualitatively alike. The reason is that the most important difference between quenched and annealed networks is annealed approximation cannot produce cycles. Since not all FLNs have cycles as will be shown later, we believe that annealed approximation is capable of giving very good analytical results on the dynamics of the FLN [50, 51].

6). Basin of attraction
It is the set of points in the system state space, such that initial conditions chosen in this set dynamically evolve toward a particular steady state.

7). Attractor
It is a set of states invariant under the dynamic progress, toward which the neighboring states in a given basin of attraction asymptotically approach in the course of dynamic evolution. It can also be defined as the smallest unit which cannot be decomposed into two or more attractors with distinct basins of attraction.

8). Limit cycle
It is an attracting set of state vectors to which orbits or trajectories converge, and upon which their trajectories are periodic.

9). Length of a limit cycle
It is the number of states contained within the cycle, and it also represents the fundamental period of that limit cycle.

10). Basin number
It is the number of reachable states by a limit cycle or attractor.

11). Stable variables
It is the set of state variables that evolve to a constant state eventually, independent of the initial configurations. The stable variables in Boolean networks

are usually those variables that are always give 1 or 0. However, in the FLN, the steady variables should be rare. On the other hand, variables that are not stable are called unstable variables.

12). Relevant variables
It is a set of unstable variables that control some other unstable variables. The FLN should have much more relevant variables than Boolean network.

2.2 Theorems

The majority of this work has focused on the dynamical aspects of the FLN. The reason is, on one hand, not all FLNs have limit cycles or attractors as strictly as in the Boolean network. On the other hand, the properties of cycles are not well reflected in the real network. A cycle forms when the system revisits some previously covered configurations. But the demand of entering a cycle is not a single constraint, but a constraint with N dimensions [52]. Excellent work has been done on the characteristics of cycles in the Boolean network [53–55], but it has been shown that power law appears when the system has exponentially short cycles. The length of cycles and the number of cycles are heavily affected by the chaotic property and for biological systems, the most important and desirable property is robustness. The meaning of robustness is that small variations of system parameters should not dramatically change the system behaviors. Hence, we believe that the dynamic properties of the FLN are more important.

Theorem 1 *A quenched FLN using the Max-Min fuzzy logical function must reach limit cycles or attractors.*

Proof. Given initial conditions of the network,

$$\mathbf{F(0)} = (F_1(0), F_2(0), \ldots, F_N(0)) \tag{2}$$

and if the Max-Min logical function is used, it is obvious that the possible value of any variable at any time t, $F_i(t)$, can be only selected from the following set.

$$\{F_1(0), 1 - F_1(0), F_2(0), 1 - F_2(0) \ldots, F_N(0), 1 - F_N(0)\}$$

So the state space, in the beginning, includes maximally $2N$ possible values (some values out of $2N$ may be the same, so $2N$ is the upper limit). Since the FLN is quenched, the initial configurations will remain the same throughout the whole dynamic process. So the state space remains the same, which is all the possible iterations of $2N$ values on an $N \times 1$ vector space. Therefore, the state space maximally includes $(2N)^N = 2^N \cdot N^N$ different vectors.

After $2^N \cdot N^N$ updates at most, the network must have reached a state where it has already been visited. Hence the network must have limit cycles or attractors. □

This property is only valid for the quenched network using the Max-Min logical function. If other types of logical functions (GC, MV or Probabilistic shown in Table 1) are used, then the network cannot be guaranteed to reach limit cycles or attractors.

Take the GC fuzzy logical function as an example. Given a simple two variable network, $\{F_1(t), F_2(t)\}$, the update rules are

$$\begin{aligned} F_1(t+1) &= F_1(t) \wedge F_2(t) \\ F_2(t+1) &= F_2(t) \end{aligned} \quad (3)$$

Suppose the initial values are $F_1(0) = 0.2,\ F_2(0) = 0.5$, then the network will evolve through the following states:
$(0.2, 0.5) \rightarrow (0.2 \times 0.5, 0.5) \rightarrow (0.2 \times 0.5^2, 0.5) \rightarrow \ldots (0.2 \times 0.5^i, 0.5) \ldots$

As can be seen, it will never reach a previously visited state because the value of the first variable at the current time is always different from any of its ancestors. Although some FLNs will not reach the exact steady state, the networks, in fact, can be thought as reaching a pseudo-steady state asymptotically. For the previous example, the pseudo steady state can be thought as (0, 0.5).

Theorem 2 *For a quenched FLN using the Max-Min logical function, values of all variables at the end of the process have a lower bound of* $min\{F_1(0), 1 - F_1(0), F_2(0), 1 - F_2(0) \ldots, F_N(0), 1 - F_N(0)\}$*, and an upper bound of* $max\{F_1(0), 1 - F_1(0), F_2(0), 1 - F_2(0) \ldots, F_N(0), 1 - F_N(0)\}$.

Proof. Suppose at time t the system reaches its steady state, then $\forall F_i(t)$, we can trace it back to the initial configurations due to the quenched property.

$$\begin{aligned} F_i(t) &= \Lambda_i(F_{i_1}(t-1), F_{i_2}(t-1), \ldots, F_{i_K}(t-1)) \\ F_{i_j}(t-1) &= \Lambda_{i_j}(F_{i_{j_1}}(t-2), F_{i_{j_2}}(t-2), \ \ldots, \ F_{i_{j_K}}(t-2)), \ (1 \leq j \leq K) \\ &\vdots \end{aligned} \quad (4)$$

After t steps of tracing back shown in (4), we can trace the value of $F_i(t)$ as the composite of K^t membership functions applied on the initial conditions. For any membership function based on the Max-Min fuzzy logical function, it can be decomposed as the conjunction of disjunctions (same as minterm presentations in Boolean logic). Since the Max-Min logical function preserves its initial values, so each disjunction will preserve its input values. From the definition of composite functions, the composite of disjunctions will also preserve the input values.

Thus, we have proved that the initial values will be channeled to the steady state. The value of any variable in the steady state has to be chosen from the set

$$\{F_1(0), 1 - F_1(0), F_2(0), 1 - F_2(0) \ldots, F_N(0), 1 - F_N(0)\}$$

So the lower bound of the variable values in the steady state is $min\{F_1(0), 1 - F_1(0), F_2(0), 1 - F_2(0) \ldots, F_N(0), 1 - F_N(0)\}$ and the upper bound is $max\{F_1(0), 1 - F_1(0), F_2(0), 1 - F_2(0) \ldots, F_N(0), 1 - F_N(0)\}$. □

2.3 Critical Connectivity and Dynamics of the FLN

In this section, the annealed approximation is used to deduce the important equations concerning the connectivity and the dynamics of the FLN. First consider two configurations of a quenched FLN at time t: $\Sigma(t), \widetilde{\Sigma}(t)$, where

$$\begin{aligned} \Sigma &= \{\sigma_1(t), \sigma_2(t), \ldots, \sigma_N(t)\} \\ \widetilde{\Sigma} &= \{\widetilde{\sigma_1}(t), \widetilde{\sigma_2}(t), \ldots, \widetilde{\sigma_N}(t)\} \end{aligned} \tag{5}$$

The distance between the two configurations at time t, $D(t)$, can be computed as the accumulated Hamming distance (AHD),

$$D(t) = \sum_{i=1}^{N} Hamming(\sigma_i(t), \widetilde{\sigma_i}(t)) \tag{6}$$

and

$$Hamming(\sigma_i(t), \widetilde{\sigma_i}(t)) = \begin{cases} 1 & \text{if } |\sigma_i(t) - \widetilde{\sigma_i}(t)| \geq \delta \\ 0 & \text{if } |\sigma_i(t) - \widetilde{\sigma_i}(t)| < \delta \end{cases} \tag{7}$$

The Hamming distance uses $\delta \in [0,1]$ as a parameter to differentiate the closeness of two fuzzy values. As a matter of fact, the distance between two Boolean values can also be computed using (7) with $\delta \equiv 0$. So the AHD of the FLN is the extension of Boolean distance. One can easily see that the maximum distance between $\Sigma(t)$ and $\widetilde{\Sigma}(t)$ is N, and the minimum distance is 0.

Similar to the Boolean network, we may designate another quantity

$$a_t = 1 - \frac{D(t)}{N} \tag{8}$$

as the normalized coverage of the two configurations. As can be seen, $a_t \in [0,1]$, and the higher a_t is, the higher degree of similarity exists between the two configurations.

Suppose at time t, $\Sigma(t)$ and $\widetilde{\Sigma}(t)$ are at a distance l_t, then we can compute the probability that the two configurations have a distance l_{t+1} at time $t+1$. Let's denote this probability as $P(l_{t+1}, l_t)$. Suppose $\Sigma(t)$ and $\widetilde{\Sigma}(t)$ have the same logical function selections but different initial values for each variable. We assume that the variables in the system can select one out of S values. The probability of selecting each one of the S values is the same, i.e. $\frac{1}{S}$. The proposed FLN uses fuzzy membership functions, but they can still be decomposed into the minterm form. So the number of membership functions in the FLN is the same as that in the NK Boolean network except the fuzzy variable can take any value in the closed interval $[0,1]$. In other words, the number of values that a variable can take in the FLN is actually infinite. But for the Max-Min logical function, we have shown that a variable can take $2N$ values. It is finite although N can in the order of thousands. In this situation, we can analyze the system as the following [53, 56–58].

Suppose A is the set of variables which are identical in $\Sigma(t)$ and $\widetilde{\Sigma}(t)$ at time t. Obviously set A has $N - l_t$ variables. Define $Q(N_0)$ as the probability that N_0 variables have all their K parents from the set A. Then $Q(N_0)$ is a discrete random variable following binomial distribution with a parameter $\left(\frac{N-l_t}{N}\right)^K$. By definition,

$$\frac{N - l_t}{N} = 1 - \frac{l_t}{N} = a_t \tag{9}$$

Then

$$\begin{aligned} Q(N_0) &= \binom{N}{N_0} \left[\left(\frac{N - l_t}{N}\right)^K\right]^{N_0} \left[1 - \left(\frac{N - l_t}{N}\right)^K\right]^{N-N_0} \\ &= \binom{N}{N_0} \left[a_t{}^K\right]^{N_0} \left[1 - a_t{}^K\right]^{N-N_0} \end{aligned} \tag{10}$$

It is obvious that these N_0 variables will remain the same at step $t+1$ in both $\Sigma(t+1)$ and $\widetilde{\Sigma}(t+1)$. For the remaining $N - N_0$ variables, since at least one of their parents will be different, there is a probability of $p = \frac{S(S-1)}{S^2}$ that a variable will be different in the next step while $1 - p$ is the probability that it will be the same in two networks. Using conditional probability and binomial theorem,

$$\begin{aligned} P(l_{t+1}, l_t) &= \sum_{N_0=0}^{N-l_{t+1}} Q(N_0) \binom{N - N_0}{l_{t+1}} \left(\frac{S(S-1)}{S^2}\right)^{l_{t+1}} \\ &\quad \left(1 - \frac{S(S-1)}{S^2}\right)^{N-N_0-l_{t+1}} \\ &= \frac{N!}{l_{t+1}!(N - l_{t+1})!}(1 - a_t^K)^{l_{t+1}}(\frac{S-1}{S})^{l_{t+1}}(\frac{1}{S})^{N-l_{t+1}} \\ &\quad \sum_{N_0=0}^{N-l_{t+1}} \frac{(N - l_{t+1})!}{N_0!(N - l_{t+1} - N_0)!}(Sa_t^K)^{N_0}(1 - a_t^K)^{N-l_{t+1}-N_0} \\ &= \frac{N!}{l_{t+1}!(N - l_{t+1})!}\left[\frac{S-1}{S}(1 - a_t^K)\right]^{l_{t+1}}(\frac{1}{S})^{N-l_{t+1}} \\ &\quad (Sa_t^K + 1 - a_t^K)^{N-l_{t+1}} \\ &= \frac{N!}{l_{t+1}!(N - l_{t+1})!}P^{l_{t+1}}(1 - P)^{N-l_{t+1}} \end{aligned} \tag{11}$$

where $P = 1 - a_t^K - \frac{1}{S}(1 - a_t^K)$.

Equation (11) obviously follows the binomial distribution, and the mean of this distribution is

$$N\left(1 - a_t^K - \frac{1}{S}(1 - a_t^K)\right) \tag{12}$$

Since the distribution reaches it peak at the mean, we can obtain that

$$l_{t+1} = N\left(1 - a_t^K - \frac{1}{S}(1 - a_t^K)\right) \tag{13}$$

Because $\frac{l_{t+1}}{N} = 1 - a_{t+1}$, we have a very important dynamic equation for the two systems,

$$a_{t+1} = a_t^K + \frac{1}{S}(1 - a_t^K) \tag{14}$$

The dynamical behaviors of $\Sigma(t)$ and $\widetilde{\Sigma}(t)$ are governed by (14). If the networks are stable, the coverage, a_t, should eventually approach 1 because it is assumed that the two networks have the same configurations of logical functions except different initial values. The stability issue also requires that the coverage does not decrease in the steady state. So the following criteria should be imposed:

$$\frac{\partial a_{t+1}}{\partial a_t} < 1 \tag{15}$$

$$\lim_{t\to\infty} a_t = 1 \tag{16}$$

Applying (15) and (16) to (14), we have the critical requirement for the connectivity K:

$$K < \frac{S}{S-1} \tag{17}$$

Luque et al. [56] have discussed the stability equation for a system whose variable can have one out of S values. They assumed that the probability of a function producing 0 is p, and all other states have the same probability. Through some probabilistic deduction, they located the critical condition for connectivity K. In this paper, we proposed another method to yield some interesting result.

The probability that the variable takes 0 is p, and all the other values have the same probability of being taken by the variable, i.e. $\frac{1-p}{S-1}$. Then the dynamic equation about the updated coverage can be written as:

$$a_{t+1} = p[1 - (1-p)(1 - {a_t}^K)] + \sum_{i=1}^{S-1} \frac{1-p}{S-1}[1 - (1 - \frac{1-p}{S-1})(1 - {a_t}^K)] \tag{18}$$

The first part on the right side of (18) describes the probability that the current coverage is based on the smallest value taken by the variables; the second part describes the probability that the current coverage is based on any other values taken by the variables.

Re-arranging (18), we get the same result with Luque et al. If annealed approximation is used, then it is obtained that

$$a_{t+1} = {a_t}^K + [p^2 + \frac{(1-p)^2}{S-1}](1 - {a_t}^K) \tag{19}$$

So the critical connectivity turns out to be:

$$K = \frac{1}{1 - p^2 - \frac{(1-p)^2}{S-1}} \tag{20}$$

The dynamic equation resulted from the anneal approximation can be expressed a more standard way. Let $\overline{P}$ be the probability that a function produces different values given different inputs. Then the annealed approximation (as shown in (11)) can applied as:

$$\begin{aligned}
P(l_{t+1}, l_t) &= \sum_{N_0=0}^{N-l_{t+1}} Q(N_0) \binom{N - N_0}{l_{t+1}} \overline{P}^{l_{t+1}} (1 - \overline{P})^{N-N_0-l_{t+1}} \\
&= \frac{N!}{l_{t+1}!(N - l_{t+1})!} (1 - a_t^K)^{l_{t+1}} (\overline{P})^{l_{t+1}} (1 - \overline{P})^{N-l_{t+1}} \\
&\quad \sum_{N_0=0}^{N-l_{t+1}} \frac{(N - l_{t+1})!}{N_0!(N - l_{t+1} - N_0)!} (\frac{a_t^K}{1 - \overline{P}})^{N_0} (1 - a_t^K)^{N-l_{t+1}-N_0} \\
&= \frac{N!}{l_{t+1}!(N - l_{t+1})!} \left[\overline{P}(1 - a_t^K)\right]^{l_{t+1}} (\overline{P})^{N-l_{t+1}} \\
&\quad (\frac{a_t^K}{1 - \overline{P}} + 1 - a_t^K)^{N-l_{t+1}} \\
&= \frac{N!}{l_{t+1}!(N - l_{t+1})!} \left(\overline{P}(1 - a_t^K)\right)^{l_{t+1}} \left[1 - \overline{P}(1 - a_t^K)\right]^{N-l_{t+1}}
\end{aligned} \tag{21}$$

So the dynamic recursive equation can then be found as:

$$a_{t+1} = 1 - \overline{P}(1 - a_t{}^K) \tag{22}$$

It is obvious that for the situation described in [56],

$$\overline{P} = p(S - 1)\frac{1 - p}{S - 1} + (S - 1)\frac{1 - p}{S - 1}(1 - \frac{1 - p}{S - 1}) \tag{23}$$

It can also be easily shown that this generalization is also valid for Boolean networks ($S \equiv 2$).

Now we consider the critical connectivity of a general FLN. Given the *pdf* that a fuzzy logical function produces a fuzzy value in the two networks, $f_i(x)$, we can compute the probability that a function produces different values from different inputs. Since any node in $\Sigma(t)$, and $\widetilde{\Sigma}(t)$ uses the same logical function and the difference of two networks is AHD, the probability $\overline{P}$ can be expressed as:

$$\begin{aligned}
\overline{P} &= P(|F_i(t) - F_j(t)| \geq \delta) \\
&= \int\int_A [f_i(\sigma_i(t)) \cdot f_i(\widetilde{\sigma}_i(t))] d(\sigma_i(t)) d(\widetilde{\sigma}_i(t))
\end{aligned} \tag{24}$$

where A is the slant integral area in the unit box with δ as a parameter.

A simple case is considered where the distribution is uniform for all $x \in [0,1]$, the difference probability can be computed via (24), which yields:

$$\overline{P} = (1-\delta)^2 \tag{25}$$

So the connectivity is:

$$K = \frac{1}{(1-\delta)^2} \tag{26}$$

Another generalization is to relax the requirement of uniform connectivity throughout the whole network. We can consider a network where the nodes can have different number of connections with a discrete distribution ρ_k, where

$$\rho_k = Prob(\text{a node has a connectivity of k}) \text{ and } \sum_{k=1}^{N} \rho_k = 1 \tag{27}$$

The general dynamics, (22), will change to

$$a_{t+1} = 1 - \sum_{k=1}^{N} \rho_k \overline{P}(1 - {a_t}^k) \tag{28}$$

Then the mean connectivity can be found using the marginal stability:

$$< K >= \frac{1}{\overline{P}} \tag{29}$$

One important question we ask is how the critical connectivity affects the FLN's application to model gene regulations. It has been found that in yeast protein-protein networks, Internet and social networks, the distribution of connectivity follows Zipf's law [59, 60], i.e.

$$P((\text{number of inputs}) = K) \propto \frac{1}{K^\gamma}, 1 \le K \le N \tag{30}$$

where γ is an integer usually between 2 and 3.

Then the mean connectivity is:

$$< K >= \epsilon \sum_{K=1}^{N} K \frac{1}{K^\gamma} = \epsilon \sum_{K=1}^{N} \frac{1}{K^{\gamma-1}} \tag{31}$$

where ϵ is a constant to guarantee the sum equals 1. It is obvious that

$$\epsilon = \frac{1}{\sum_{K=1}^{N} \frac{1}{K^\gamma}} \tag{32}$$

Let

$$H_N^{(\gamma)} = \sum_{i=1}^{N} \frac{1}{K^\gamma} \tag{33}$$

as the partial sum of the generalized harmonic series. Then we can obtain that

$$\epsilon = \frac{1}{H_N^{(\gamma)}} \text{ and } < K >= \frac{H_N^{(\gamma-1)}}{H_N^{(\gamma)}} \tag{34}$$

Since there is no general formula for the sum of this series, we can approximate the series sum if N is large enough, which is definitely true for the gene regulatory network. The approximation of $H_N^{(\gamma)}$ is

$$H_N^{(\gamma)} \approx \begin{cases} \infty & \text{if } \gamma = 1 \\ \frac{\pi^2}{6} & \text{if } \gamma = 2 \\ 1.202 & \text{if } \gamma = 3 \\ \frac{\pi^4}{90} & \text{if } \gamma = 4 \\ 1.036 & \text{if } \gamma = 5 \\ \frac{\pi^6}{945} & \text{if } \gamma = 6 \end{cases} \tag{35}$$

Combined with (34), the mean connectivity is computed as:

$$< K > \approx \begin{cases} \infty & \text{if } \gamma = 2 \\ 1.3685 & \text{if } \gamma = 3 \\ 1.1106 & \text{if } \gamma = 4 \\ 1.0447 & \text{if } \gamma = 5 \\ 1.0183 & \text{if } \gamma = 6 \end{cases} \tag{36}$$

Then we can find the critical conditions of S for the FLN. From the expression of critical probability, the relationship between S and $< K >$ is

$$S \geq \frac{< K >}{< K > -1}$$

So from (36), the number of values, S, has a lower bound as shown in the following:

$$S \geq \begin{cases} 1 & \text{if } \gamma = 2 \\ 3.7137 & \text{if } \gamma = 3 \\ 10.0416 & \text{if } \gamma = 4 \\ 23.3714 & \text{if } \gamma = 5 \\ 55.6448 & \text{if } \gamma = 6 \end{cases} \tag{37}$$

Considering that a network should usually be on the chaotic side of critical connectivity, we observe that a small S might not be a good choice for the network. This, in return, triggers the question on whether a crisp Boolean network is powerful enough to infer the genetic network structure on the edge of chaos.

If we compare K in (36) with (26), we can compute the necessary δ to satisfy the requirement,

$$\delta > \begin{cases} 0.1452 & \text{if } \gamma = 3 \\ 0.0511 & \text{if } \gamma = 4 \\ 0.0216 & \text{if } \gamma = 5 \\ 0.0090 & \text{if } \gamma = 6 \end{cases} \tag{38}$$

As can be seen, when γ increases, the network has to adjust itself by adopting stricter criteria. The result also agrees with the relationship between S and γ.

3 Inference Algorithm Design

The algorithm searches all possible regulations among all genes by measuring the AHD throughout the microarray time span. The key idea of the algorithm is that the AHD among genes according to all possible logic combinations should be able to represent gene regulatory relationships. If some genes have a regulatory relationship, their AHD should be smaller with respect to that of the others. Besides, the algorithm is based on fuzzy set modeling which allows it to capture the non-linear relationship in gene regulatory processes, and the algorithm under investigation is indeed very robust to capture genetic activities in different situations.

Let $\boldsymbol{G} \in R^{n \times m}$ be the time-series microarray data where n is the number of genes in the data and m is the number of time slots in the microarray set. The algorithm first randomly selects $G_r = (G_{r_1}, G_{r_2}, ..., G_{r_i}, ...G_{r_K})$, a group of regulators that regulates another gene G_t $(G_{r_i}, G_t \in R^{1 \times m}, t \neq r_1, r_2, ...r_K)$. Then the algorithm filters the regulators through a fuzzy logic mask, $FLogic$[3], to generate a pseudo-gene-time-series, $G_s \in R^{1 \times m}$ where

$$G_s^j = FLogic(G_{r_1}^j, G_{r_2}^j, ..., G_{r_i}^j, ...G_{r_K}^j), 1 \leq j \leq m \tag{39}$$

The AHD between G_t and G_s is, then, computed using the Hamming distance metric:

$$AHD(G_t, G_s) = \sum_{j=1}^{m-1} Hamming(G_t^{j+1}, G_s^j) \tag{40}$$

where

[3] $FLogic$ is one of the possible fuzzy logical functions that is applied on K variables

$$Hamming(G_t^j, G_s^{j+1}) = \begin{cases} 1, & if|G_t^{j+1} - G_s^j| > T_h \\ 0, & if|G_t^{j+1} - G_s^j| \leq T_h \end{cases} \tag{41}$$

T_h in (41) is the Hamming threshold (same as δ in (7)) to discern the difference of two close numbers. The AHD between G_t and G_s is computed with a time shift assuming that regulations happen with one time delay.

For each group of possible regulators and regulated gene, the algorithm can determine its AHD, and record it. In the end, the algorithm will infer the appropriate groups with AHD less than the inference threshold, T_i, $(0 < m \times T_i < 1)$. The flow chart of the algorithm is depicted in Fig.1

Generally speaking, the complexity of the algorithm is $o(n^K)$. As shown in (36), the mean connectivity should be a value between 1.1106 and 1.1385. Thus the complexity of the algorithm is almost linear with the number of the genes. In this paper, the connectivity K is first assumed to be 2. A greater K $(K = 3)$ other than a triplet search has also been investigated on the dataset.

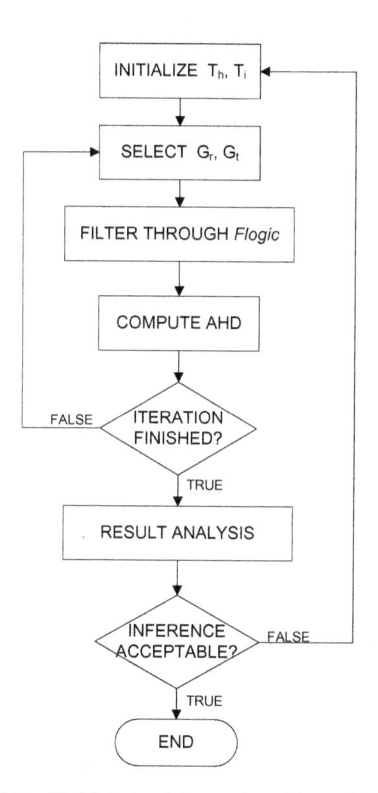

Fig. 1. The structure of the FLN data-driven algorithm. T_h is the Hamming threshold and T_i is the inference threshold. After the selection of a group of regulators G_r and the regulated gene G_t, the AHD is computed. After all iterations are finished, the inference results are analyzed

4 Computation Results

4.1 Data Pre-processing

The chosen dataset includes 6178 commonly accepted opening reading frames (ORFs) in *Saccharomyces cerevisiae* genome [61]. A subset of the time-series microarray, α-arrest set, was chosen to test the algorithm. α-factor is a polypeptide mating pheromone secreted into the medium by mating-type protein MATα. It causes MATα cells to arrest cell division at the initial states, and also causes morphological changes. α factor also has other physiological functions such as the induction of agglutinin on the surface, and biomedical changes in the cell wall. In this subset, there are 18 time slots. To reduce errors introduced by the noise in the microarray, the dataset was pre-processed with three criteria. Any gene failing to satisfy any one of the three criteria was deleted. The criteria are stated as follows:

- Only genes with more than $\frac{2}{3}$ valid time slots with respect to the total time slots are considered above the noise level. The reason for this requirement is that some genes do not stand out from the background noise, and accordingly, the dataset has blanks at certain time slots. If more than $\frac{1}{3}$ of the time slots of a gene are blanks, that gene's expressions are interpreted as noisy, and hence, removed.
- The maximum value of each gene's expression levels must be at least 3 times bigger than the minimum in the gene's time series. If not, the gene is deleted from the dataset. This requirement guarantees that genes processed by the algorithm have enough dynamic changes in their expressions, and the computational time is also reduced by limiting the search space.
- Genes with spikes in the time series are not included. The signal-noise ratio of the spike was selected as 5.

After the pre-processing and filtering, 680 genes survived the cut in α-arrest set. Following gene deletions, the values of genes were normalized into [0, 1] interval throughout the time series. Firstly, the values in the dataset were changed from log-odds into real values. Then for each gene, the maximum throughout the series was found, and it was used to divide each expression value of that gene in the time series. After these steps, the expression levels of each gene in the series have been normalized into [0, 1] interval. The algorithm was programmed in C language and run at Pittsburgh Supercomputer Center using parallel computing.

4.2 Triplet Search ($K = 2$)

All four different logical functions in Table 1 were tested with same parameter settings ($T_h = 0.01, T_i = 21\%$), and they have inferred four different regulatory networks. The inferred networks are included in Fig. 3, Appendix A, and B. Despite the difference among the networks, they share some common features.

- All four networks contain two common center genes, FIG1 and PRM1. Both genes have been confirmed to affect the cell-cell mating mechanism of yeast.
- All four networks have certain dominating regulatory logical functions. Fig. 2 shows the distribution of inferred regulatory logic. As can be seen, AND logic and promoter/represser cases dominate the regulatory logic found by Max-Min, GC and Probabilistic logical functions while the MV logical function finds that the combinatorial of AND logic or promoter/represser situations are more prevalent. In fact, GC and Probabilistic logical functions yield the same network because all their discoveries are AND logic, and they share the same rules for AND. Comparing all three inferred networks, the MV logical function shows the robustness to small changes of T_h. So the final inference network, Fig. 3, is based on this logical function.

The algorithm used the MV logical function to locate 54 groups of regulations (out of 4.3×10^9 possible groups) involving 26 genes. As can be seen from Table 3, 14 out of the 26 genes in the network have been proved to be important

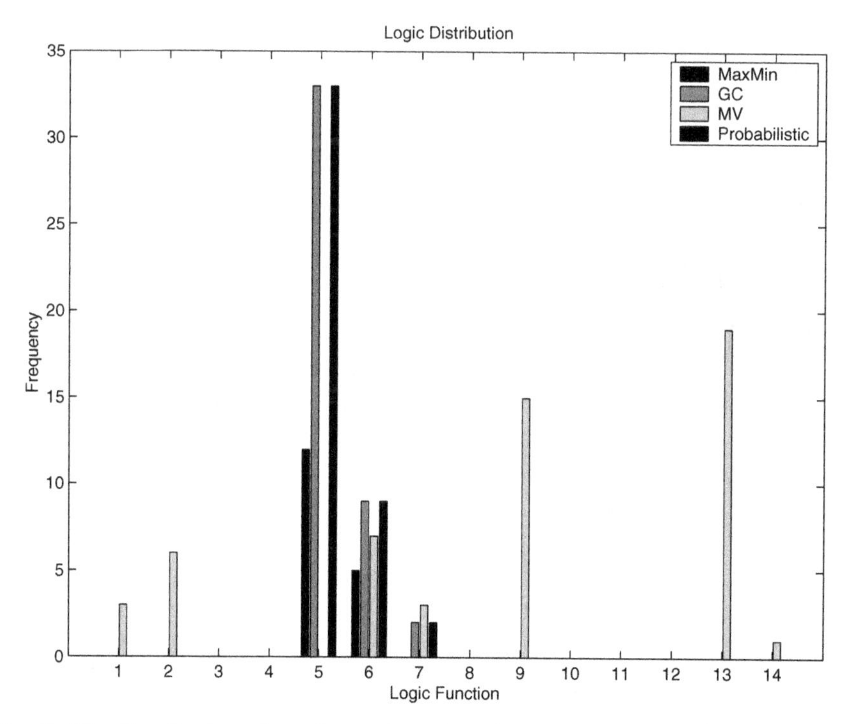

Fig. 2. The distribution of regulatory logic based on the four fuzzy logical functions in the inferred gene regulatory networks with $K = 2$. X-axis represents the logic functions coded from 1 to 14 according to Table 2, and y-axis is the frequency. Each color bar represents the frequency of a specific fuzzy logical function

for the mating or cell cycle in yeast. In addition, the backbone of the network (nodes with high connectivity) is made up of 9 out of these 14 genes. From the functions of the genes involved in the network, we can see that the algorithm indeed located the correct genes that inter-regulate. The inferred network also shows the important network motif that have been previously studied [62, 63]. The network includes 7 feed-forward loops, 3 single-input modules as well as the dense overlapping modules. Those modules share different evolutionary properties and advantages. For example, the feed forward loop is believed to play a functional role in information processing. This motif can serve as a circuit-like function to activate output only if the input signal is persistent, and also allows a rapid deactivation if the input is off. Furthermore, the inferred network includes two internal cycles (FIG1 $\longleftrightarrow$ PRM1, FIG1 $\longleftrightarrow$ FIG2) and one feedback loop among FIG1, FIG2, and ASG7. All the genes in the cycle or feedback loop are involved in the signaling for yeast mating and the close regulations among them are integral to yeast mating. Although network motif studies of *E. Coli* did not find cyclic structures [63], the feedback loop is believed to be the most important regulatory mechanism for cells to adapt into new conditions. The inferred network shows a striking similarity of regulation mechanisms across different species while preserving specific regulatory mechanisms.

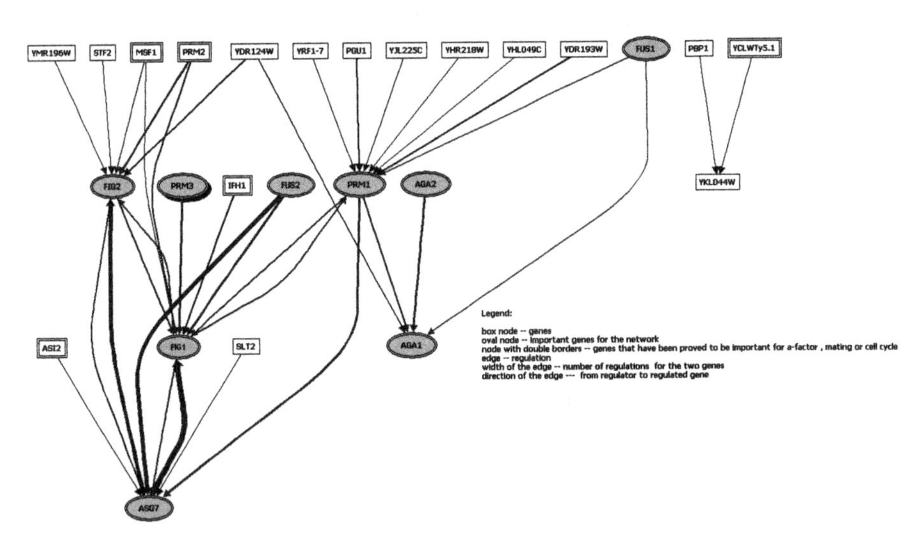

Fig. 3. The α-factor gene regulatory network using the MV fuzzy logical function for $K = 2$. There are 26 genes in the network with 14 biologically proved genes, and 9 genes are the backbone of the network. Each node represents a gene, and double-border nodes represent genes with biological functions for the mating or cell-cycle in yeast. Each edge represents a regulation, and the width of the edge is the regulatory strength of the regulation. The direction of the edge is from regulator to regulated genes. The backbone of the network is represented as: oval nodes with green color and red edges. The figure legends for appendix A, B, and C are the same

Table 2. All possible regulatory logic with $K = 2$

1. $G_1 \to G_t$	2. $G_2 \to G_t$	3. $\overline{G_1} \to G_t$
4. $\overline{G_2} \to G_t$	5. $G_1 \wedge G_2 \to G_t$	6. $\overline{G_1} \wedge G_2 \to G_t$
7. $G_1 \wedge \overline{G_2} \to G_t$	8. $\overline{G_1} \wedge \overline{G_2} \to G_t$	9. $G_1 \vee G_2 \to G_t$
10. $\overline{G_1} \vee G_2 \to G_t$	11. $G_1 \vee \overline{G_2} \to G_t$	12. $\overline{G_1} \vee \overline{G_2} \to G_t$
13. $G_1 \oplus G_2 \to G_t$	14. $\overline{G_1} \oplus G_2 \to G_t$	

4.3 More Complex Quadruplet Search ($K = 3$)

If there are 3 regulators for a target gene, then the total possible number of regulatory logical functions is $2^8 - 2 = 254$ (exclude all-zero and all-one logical functions). The computation complexity is also affected by the total number of genes in the microarray. Therefore, the computation load of $K = 3$ is about 12,000 times larger than the triplet search. The MV logical function was used, and logical functions were implemented by considering all combinations of AND logic through Karnaugh maps. According to the FLN theory, $K = 3$ will result in a much more complex system and it is confirmed by our computations. The total computation time is about 600 hours on a 12-cpu cluster in a parallel manner (each computation node has a CPU of 1GHz and 1G memory). Using the same parameters as for $K = 2$, the algorithm located 556 groups of regulations (from 5.4×10^{13} possible groups) involving 272 genes. The network is very complex as it should because 3-regulator scenario is more sensitive to the parameters. To get a better understanding of $K = 3$, T_h is reduced to 0.08 and this time, the network located 105 regulation groups involving 98 genes to form a sparse network [46]. Appendix C shows the inferred network.

The backbone of the network is similar to that of Fig. 3 but gives more insightful details. 44 out 98 discovered genes have been proved by previous experiments to be involved in the mating of yeast. The network is quite complicated, and provides many different sub-modules existing only within the network backbones. The relationships among AGA1, ASG7, FIG1, FIG2, FUS2, AGA1, AGA2, PRM1, and PRM2 are very complex. These 7 genes are almost fully connected, which yields 10 feed forward loops and 7 cycles. This high connectivity implies that those 7 genes are highly dependent upon each other.

Another important discovery of this computation is the regulatory logical functions. Table 4 shows the distribution of regulatory logical functions. As can be seen, most of the discovered logical functions involve one or two genes in the form of possible alternative paths. This finding confirms the result for $K = 2$. In conclusion, the quadruplet-search gives a more detailed structure of genetic inter-regulations, while the logical functions are still within the grasp of $K = 2$. Considering the huge computational increase from $K = 2$

Table 3. Genetic function of genes in Fig. 3

Name	Function
FIG1	Integral membrane protein for efficient mating; regulate intracellular signaling & fusion
FIG2	Cell wall adhesin specifically for mating to maintain cell wall integrity during mating
FUS1	Membrane protein required for cell fusion; regulated by mating pheromone; Cdc28p substrate
FUS2	Cytoplasmic protein for the alignment of parental nuclei before nuclear fusion during mating
AGA1	Anchorage subunit of a-agglutinin of a-cells; C-terminal signal for addition of GPI anchor to cell wall
AGA2	Adhesion subunit of a-agglutinin of a-cells, C-terminal sequence as Sag1p during agglutination
ASG7	Protein that regulates signaling from Ste4p; specific to a-cells and induced by alpha-factor
PRM1	Pheromone-regulated protein for membrane fusion during mating; regulated by Ste12p
PRM2	Pheromone-regulated protein regulated by Ste12p
PRM3	Pheromone-regulated protein required for karyogamy
STF2	ATPase stabilizing factor
PBP1	Poly(A)-binding protein binding protein
SLT2	Suppressor of lyt2
PGU1	Endo-polygalacturonase
MSF1	Mitochondrial phenylalanyl-tRNA synthetase alpha subunit
IFH1	Essential protein with a highly acidic N-terminal domain; Cdc28p substrate
ASI2	Predicted membrane protein negatively regulating amino acid uptake
YRF1-7	YRF1 Helicase encoded by the Y' element of subtelomeric regions; phosphorylated by Cdc28p
YCLWTy5.1	Encodes fragment of Ty Pol protein
YDR124W	Uncharacterized ORF
YHL049C	Uncharacterized ORF
YMR196W	Uncharacterized ORF
YJL225C	Uncharacterized ORF
YHR218W	Uncharacterized ORF
YDR193W	Dubious ORF
YKL044W	Dubious ORF

Table 4. Logic distribution of α-factor network using the MV logical function with $K = 3$

Regulatory logic	Frequency
$G_1 \wedge G_3 \to G_t$	3
$(G_1 \wedge G_2) \vee (G_1 \wedge G_3) \to G_t$	21
$(G_1 \wedge \overline{G_3}) \vee (G_1 \wedge \overline{G_2}) \to G_t$	1
$G_1 \to G_t$	1
$(G_1 \wedge \overline{G_2} \wedge G_3) \vee (\overline{G_1} \wedge G_2 \wedge G_3) \to G_t$	13
$(G_1 \wedge G_2 \wedge \overline{G_3}) \vee (\overline{G_1} \wedge G_2 \wedge G_3) \to G_t$	5
$G_2 \wedge G_3 \to G_t$	2
$(G_1 \wedge G_2) \vee (G_2 \wedge G_3) \to G_t$	17
$(G_1 \wedge G_3) \vee (G_2 \wedge G_3) \to G_t$	20
$G_1 \wedge \overline{G_2} \wedge G_3 \to G_t$	1
$G_1 \wedge G_2 \wedge \overline{G_3} \to G_t$	1
$(\overline{G_1} \wedge \overline{G_2} \wedge G_3) \vee (G_1 \wedge G_2 \wedge G_3) \to G_t$	1
$G_1 \wedge G_2 \to G_t$	2
$(G_1 \wedge G_2 \wedge \overline{G_3}) \vee (G_1 \wedge \overline{G_2} \wedge G_3) \to G_t$	17

to $K = 3$, it might be wise to first conduct $K = 2$ to discover some major inter-regulations, and screen out unrelated genes. Then a more general search can be performed as desired.

5 Conclusions

This work describes a completely novel data-driven algorithm to infer gene regulatory networks from time-series microarray data. It is based on the FLN, and introduces the accumulated Hamming distance as its inference criteria. The availability of a super computing facility in this case is noteworthy, and may certainly hold the key to the success of this research project. Validated by the computations on the yeast dataset, the FLN has been proved to be capable of modeling the regulatory network with pleasing accuracy.

In spite of the findings, there is still room for further fine tuning and improvement. The future effort should focus on the effect of other distance metrics as means of comparing the performance of the modeling. Certainly, the metric used in this paper is one of the best possible choices. We believe other choices must be explored to improve the behavior and results. Further, it is our belief that there is a need to improve computational complexity by adopting heuristic searches, as well as a deeper understanding concerning the chaos generating mechanism. The reason is that biological phenomena of brain information processing, and gene regulations in the cell are two important inquiries in which the chaos phenomena are prominently observed, and we must deal with the chaos accordingly. Finally, this paper is the starting point of a new and powerful mechanic which is capable of handling uncertainty on one hand, and modeling the phenomena of complexity on the other hand.

A Max-Min Inference Network ($K = 2$)

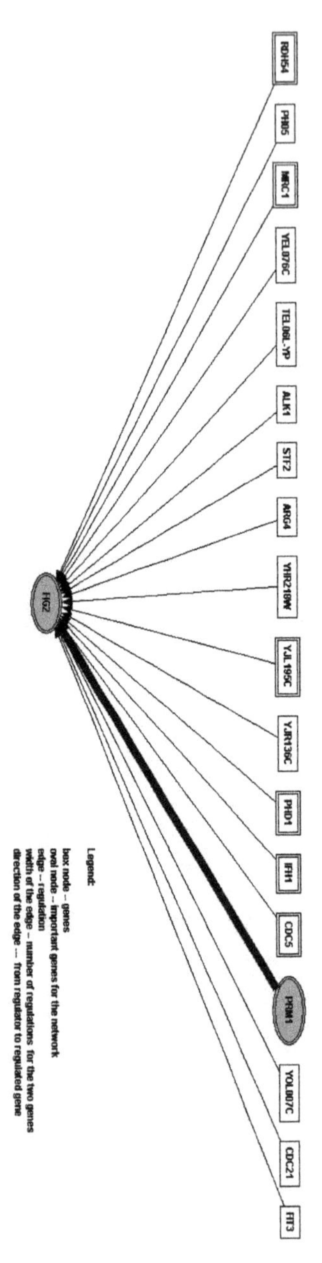

Fig. 4. α-factor inference network using the Max-Min fuzzy logical function with $K = 2$. There are 19 genes in the network with 8 biologically proved genes, and 2 genes are the backbone of the network

B Probabilistic/GC Inference Network ($K = 2$)

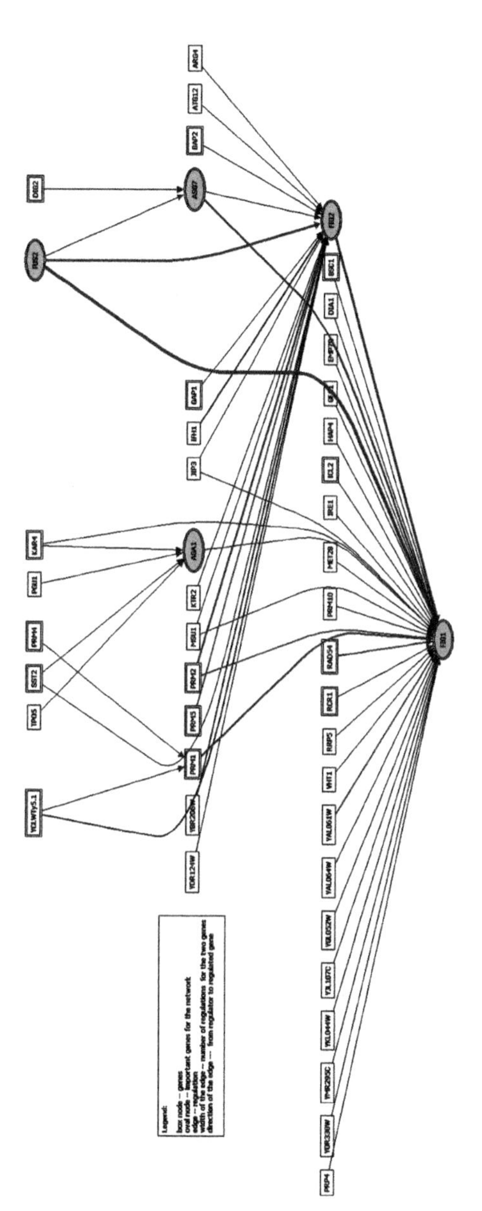

Fig. 5. α-factor inference network using the Probabilistic/GC fuzzy logical functions for $K = 2$. There are 46 genes in the network with 18 biologically proved genes, and 5 genes are the backbone of the network

C Quadruplet Inference Network ($K = 3$)

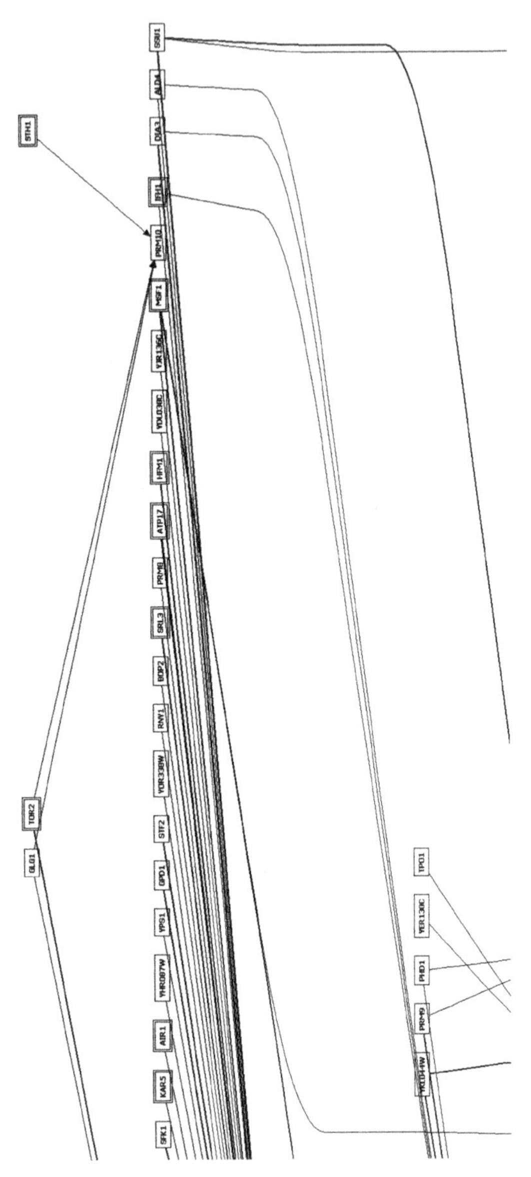

Fig. 6. The detailed structure for the α-factor network using the MV decomposition ($K = 3$). The four sub-figures are listed from top to bottom, and from left to right. This is the upper-left section

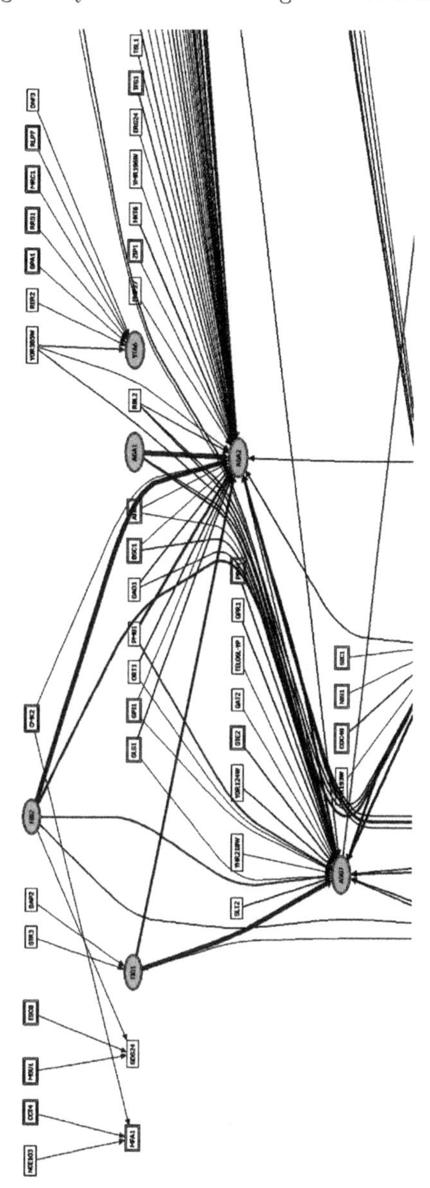

Fig. 6. The detailed structure for the α-factor network using the MV decomposition ($K = 3$). The four sub-figures are listed from top to bottom, and from left to right. This is the lower-left section

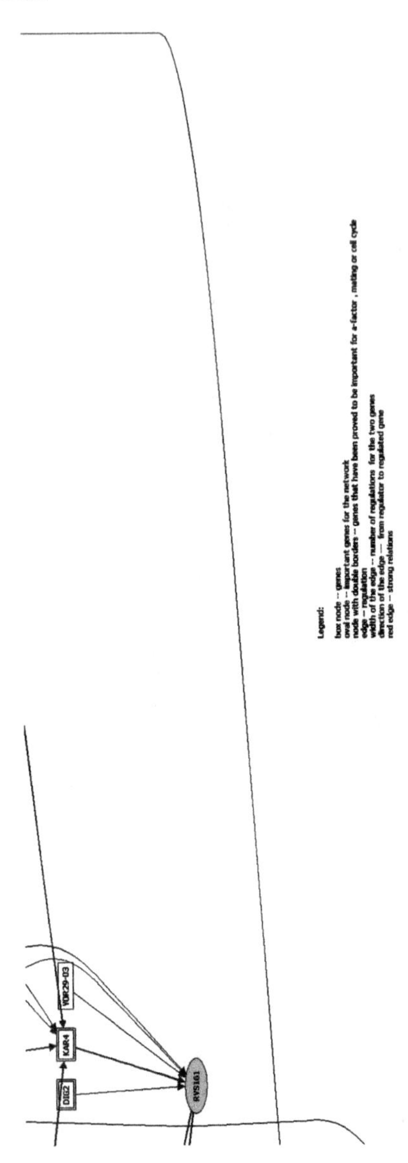

Fig. 6. The detailed structure for the α-factor network using the MV decomposition ($K = 3$). The four sub-figures are listed from top to bottom, and from left to right. This is the upper-right section

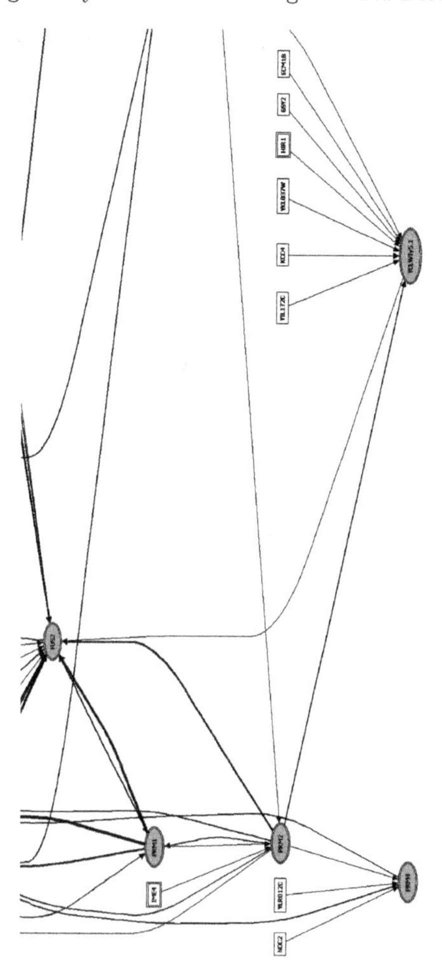

Fig. 6. The detailed structure for the α-factor network using the MV decomposition ($K = 3$). The four sub-figures are listed from top to bottom, and from left to right. This is the lower-right section

References

1. Costa, I., F. Carvalho, and M. Souto (2003). Comparative study on proximity indices for cluster analysis of gene expression time series. Journal of Intelligent and Fuzzy Systems 13(2-4), 133–142.
2. Gulob, T., D. Slonim, F. Huard, M. Gaasenbeek, J. Mesirov, H. Coller, M. Loh, J. Downing, M. Caligiuri, C. Bloomfield, and E. Lander (1999). Molecular classification of cancer: class discovery and class prediction by gene expression monitoring. Science 286, 531–537.
3. Krishnapuram, B., L. Carin, and A. Hartemink (2002). Applying logistic regression and RVM to achieve accurate probabilistic cancer diagnosis from gene expression profiles. IEEE Workshop on Genomic Signal Processing and Statistics.
4. Strogatz, S. (2001). Exploring complex networks. Nature 410, 268–276.
5. Kauffman, S. (1993). Origins of order: self-organization and selection in evolution. New York, NY: Oxford University Press.
6. Yuh, C., H. Bolouri, and E. Davision (1998). Genomic cis-regulatory logic: experimental and computational analysis of a sea urchin gene. Science 279(5358), 1896–1902.
7. Akutsu, T., S. Miyano, and S. Kuhara (2000). Inferring qualitative relations in genetic networks and metabolic pathways. Bioinformatics 16(8), 727–734.
8. Shmulevich, I., E. Fougherty, S. Kim, and W. Zhang (2002). Probabilistic Boolean networks: a rule-based uncertainty model for gene regulatory networks. Bioinformatics 18(2), 261–274.
9. Plahte, E., T. Mestl, and S. Omholt (1994). Global analysis of steady points for systems of differential equations with sigmoid interactions. Dynamics and Stability of Systems 9(4), 275–291.
10. Chen, T., H. He, and G. Church (1999). Modeling gene expression with differential equations. Pacific Symposium on Biocomputing, 29–40.
11. Gibson, M. and E. Mjolsness (2001). Modeling the activity of single genes, Chapter 1, pp. 1–48. Computational Methods for Molecular and Cellular Biology. MIT Press.
12. Murphy, K. and S. Mian (1999). Modeling gene expression data using dynamic Bayesian networks. Technical report, University of California at Berkeley, Berkeley, CA.
13. Friedman, N., M. Linial, I. Nachman, and D. Peer (2000). Learning Bayesian network structure of dynamic probabilistic network. Journal of Computational Biology 7(3-4), 601–620.
14. Moler, E., D. C. Radisky, and I. Mian (2000). Integrating naive Bayes models and external knowledge to examine copper and iron homestasis in s. cerevisiae. Physiological Genomics 4, 127–135.
15. Nachman, I., A. Regev, and N. Friedman (2004). Inferring quantitative models of regulatory networks from expression data. Bioinformatics 20(Suppl. 1), i248–i256.
16. Beal, M., F. Falciani, Z. Ghahramani, C. Rangel, and D. Wild (2005). A Bayesian approach to reconstructing genetic regulatory networks with hidden factors. Bioinformatics 21(5543), 349–356.
17. Matsuno, H., A. Doi, M. Nagasaki, and S. Miyano (2000). Hybrid Petri net representation of gene regulatory network. Pacific Symposium on Biocomputing 5, 341–352.

18. Hashimoto, R., S. Kim, I.Shmulevich, W. Zhang, M. L. Bittner, and E. Dougherty (2004). Growing genetic regulatory networks from seed genes. Bioinformatics 20(8), 1241–1247.
19. TeichMann, S. and M. Babu (2004). Gene regulatory network growth by duplication. Nature Genetics 36(5), 492–496.
20. Sprinzak, D. and M. Elowitz (2005). Reconstruction of genetic circuits. Nature 438(7067), 443–448.
21. Weaver, D., C. Workman, and G. D. Stormo (1999). Modeling regulatory networks with weight matrices. Pacific Symposium on Biocomputing, 112–123.
22. Vohradsky, J. (2001). Neural network model of gene expression. The FASEB Journal 15, 846–854.
23. Khan, J., J. Wei, M. Ringner, L. Saal, M. Ladanyi, F. Westermann, F. Berthold, M. Schwab, C. Antonescu, C. Peterson, and P. Meltzer (2001). Classification and diagnostic prediction of cancers using gene expression profiling and artificial neural networks. Nature Medicine 7(6), 658–659.
24. Mestl, T., E. Plahte, and S. Omholt (1995). A mathematical framework for describing and analyzing gene regulatory networks. Journal of Theoretical Biology 176(2), 291–300.
25. Kwon, A., H. Hoos, and R. Ng (2003). Inference of transcriptional regulation relationships from gene expression data. Bioinformatics 19(8), 905–912.
26. Joseph, Z. (2004). Analyzing time series gene expression data. Bioinformatics 20(16), 2493–2503.
27. Jong, H. (2002). Modeling and simulation of genetic regulatory systems: A literature review. Journal of Computational Biology 9(1), 67–103.
28. John, W. (2002). Deciphering gene expression regulatory networks. Current Opinion in Genetics and Development 12, 130–136.
29. Velculescu, V., L. Zhang, B. Vogelstein, and K. Kinzler (1995). Serial analysis of gene expression. Science 270(5235), 484–487.
30. Georgantas, R., V. Tanadve, M. Malehorn, S. Heimfeld, C. Chen, L. Carr, F. Martinez-Murillo, G. Riggins, J. Kowalski, and C. Civin (2004). Microarray and serial analysis of gene expression analysis identify known and novel transcripts overexpressed in hematopoietic stem cells. Cancer Research 64, 4434–4441.
31. Lee, T., N. Rinaldi, F. Robert, D. Odom, Z. Joseph, G. Gerber, N. Hannettt, C. T. Harbinson, C. M. Thompson, I. Simon, J. Zeitlinger, E. Jennings, H. Murray, D. B. Gordon, B. Ren, J. Wyrick, J. Tagne, T. Volkert, E. Fraenkel, D. Gifford, and R. Young (2002). Transcriptional regulatory networks in Saccharomyces cerevisiae. Science 298, 799–804.
32. J.Qian, J. Lin, N. Luscombe, H. Yu, and M. Gerstein (2003). Prediction of regulatory networks: genome-wide identification of transcription factor targets from gene expression data. Bioinformatics 19(15), 1917–1926.
33. Fung, E., V. Thulasiraman, S. R. Weinberger, and E. Dalmasso (2001). Protein biochips for differential profiling. Current Opinion in Biotechnology 12, 65–69.
34. Huang, S. and D. Ingber (2000). Shape-dependent control of cell growth, differentiation and apoptosis: switching between attractors in cell regulatory networks. Experimental Cell Research 261, 91–103.
35. Akutsu, T. and S. Miyano (2000). Algorithm for inferring qualitative models of biological networks. Pacific Symposium on BioComputing, 293–304.

36. Waelbroeck, H. and F. Zertuche (1999). Discrete chaos. Journal of Physics A: Mathematical and General 32, 175–189.
37. Wang, P., Y. Cao, J. Robinson, and A. Tokuta (2003). A study of the two gene network - the simplest special case of SORE (self organizable regulating engine). Proceedings of 7th Joint Conference on Information Science, 1716–1720.
38. Cao, Y., P. Wang, and A. Tokuta (2004). SORE (self organizable regulating engine) - a possible building block for a "biogizing" control system. Proceedings of 4th International Symposium on Intelligent Manufacturing, 42–48.
39. Cao, Y., P. Wang, and A. Tokuta (2003). A study of two gene network. BISC FLINT-CIBI 7th Joint Workshop on Soft Computing for Internet and Bioinformatics, UC Berkeley.
40. Zawidzki, T. (1998). Competing models of stability in complex, evolving systems: Kauffman vs. Simon. Biology and Philosophy 13(4), 541–554.
41. Woolf, P. and Y. Wang (2000). A fuzzy logic approach to analyzing gene expression data. Physiological Genomics 3, 9–15.
42. Chen, H. C., H. Lee, T. Lin, W. Li, and B. Chen (2004). Quantitative characterization of the transcriptional regulatory network in the yeast cell cycle. Bioinformatics 20(12), 1914–1927.
43. Sokhansanj, B., J. Fitch, J. Quong, and A. Quong (2004). Linear fuzzy gene network models obtained from microarray data by exhaustive search. BMC Bioinformatics 5(108).
44. Sokhansanj, B. and J. Fitch (2001). URC fuzzy modeling and simulation of gene regulation. In 23rd Annual International Conference of the IEEE Engineering in Medicine and Biology Society, Instanbul, Turkey.
45. Cao, Y., P. Wang, and A. Tokuta (2005). Gene Regulating Network Discovery, Volume 5 of Studies in Computational Intelligence, Chapter 3, pp. 49–78. Springer-Verlag GmbH.
46. Cao, Y., P. Wang, and A. Tokuta (2006). S. pombe regulatory network construction using the fuzzy logic network. Poster, LSS Computational Systems Bioinformatics Conference.
47. Resconi, G., Y. Cao, and P. Wang (2006). Fuzzy Biology. Proceedings of 5th International Symposium on Intelligent Manufacturing Systems. 29–31.
48. Cao, Y., H. Clark, and P. Wang (2006). Fuzzy logic network (FLN) on gene regulatory network modeling. Proceedings of International Conference on Intelligent systems and Knowledge Engineering.
49. Reiter, C. (2002). Fuzzy automata and life. Complexity 3(7), 19–29.
50. Derrida, B. and Y. Pomeau (1986). Random networks of automata: a simple annealed approximation. Europhysics Letters 1, 45–49.
51. Derrida, B. and D. Stauffer (1986). Phase transitions in two-dimensional Kauffman cellular automata. Europhysics Letters 2, 739–745.
52. Qu, X., M. Aldana, and L. Kadanoff (2002). Numerical and theoretical studies of noise effects in the Kauffman model. Journal of Statistical Physics 109(5-6), 967–986.
53. Luque, B. and R. Sole (1997a). Controlling chaos in random Boolean networks. Europhysics Letter 37(9), 597–602.
54. Somogyvari, Z. and S. Payrits (2000). Length of state cycles of random Boolean networks: an analytic study. Journal of Physics A: Mathematical and General 33, 6699–6706.
55. Thieffry, D. and D. Romero (1999). The modularity of biological regulatory networks. Biosystems 50(11), 49–59.

56. Luque, B. and R. Sole (1997). Phase transitions in random networks: simple analytic determination of critical points. Physical Review E 55(1), 257–260.
57. Sole, R. and B. Luque (1994). Phase transitions and antichaos in generalized Kauffman networks. Physical Letters A 196(1-2), 331–334.
58. Sole, R., B. Luque, and S. Kauffman (2000). Phase transition in random network with multiple states. Technical report, Santa Fe Institute.
59. Kauffman, S., C. Peterson, B. Samuelsson, and C. Troein (2004, December). Genetic networks with canalyzing Boolean rules are always stable. PNAS 101(49), 17102–17107.
60. Uetz, P., L. Giot, G. Cagney, T. Mansfield, R. Judson, J. R. Knight, D. Lockshon, V. Narayan, M. Srinivasan, P. Pochart, A. Qureshi-Emili, Y. Li, B. Godwin, D. Conover, T. Kalbfleisch, G. Vijayadamodar, M. Yang, M. Johnston, S. Fields, and J. Rothberg (2000). A comprehensive analysis of protein-protein interactions in saccharomyces cerevisiae. Nature 403, 623–627.
61. Spellman, P., G. Sherlock, M. Zhang, V. Iyer, K. Anders, M. Eisen, P. Brown, D. Botstein, and B. Futcher (1998). Comprehensive identification of cell cycle-regulated genes of the yeast saccharomyces cerevisiae by microarray hybridization, molecular biology of the cell. Molecular Biology of the Cell 9, 3273–3297.
62. Milo, R., S. Itzkovitz, N. Kashtan, R. Levitt, S. Shen-Orr, I. Ayzenshtat, M.Sheffer, and U. Alon (2004). Superfamilies of evolved and designed networks. Science 303(5663), 1538–1542.
63. Shen-Orr, S., R. Milo, S. Mangan, and U. Alon (2002). Network motifs in the transcriptional regulation network of Escherichia coli. Nature Genetics 31, 64–68.

An Abstract Approach Toward the Evaluation of Fuzzy Rule Systems

Siegfried Gottwald

Abstract. The general mathematical problem of fuzzy control is an interpolation problem: a list of fuzzy input-output data, usually provided by a list of linguistic control rules, should be realized as argument-value pairs for a suitably chosen fuzzy function. However, contrary to the usual understanding of interpolation, in the actual approaches this interpolation problem is considered as a global one: one uniformly and globally defined function should realize all the fuzzy input-output data.

In standard classes of functions thus this interpolation problem often becomes unsolvable. Hence it becomes intertwined with an approximation problem which allows that the given fuzzy input-output data are realized only approximately by argument-value pairs.

In this context the paper discusses some quite general sufficient conditions for the solvability of the interpolation problem, as well as similar conditions for suitably modified data, i.e. for a quite controlled approximation to the original argument-value pairs.

1 Introduction

The standard paradigm of rule based fuzzy control is that one supposes to have given, as an incomplete and fuzzy description of a control function Φ from an input space $\mathbf{X}$ to an output space $\mathbf{Y}$, a finite family

$$\mathcal{D} = (\langle A_i, B_i \rangle)_{1 \le i \le n} \tag{1}$$

of (fuzzy) input-output data pairs to characterize this function Φ.

Let us assume for simplicity that all the input data A_i are *normal*, i.e. that there is a point x_0 in the universe of discourse with $A_i(x_0) = 1$. Sometimes even weak normality would suffice, i.e. that the supremum over all the membership degrees of the A_i equals one; but we do not indent to discuss this in detail.

In the usual approaches such a family of input-output data pairs is provided by a finite list

$$\text{IF } x \text{ is } A_i \quad \text{THEN } y \text{ is } B_i, \qquad i = 1, \ldots, n \tag{2}$$

of linguistic control rules, also called fuzzy IF-THEN rules.

The main mathematical problem of fuzzy control, besides the engineering problem to get a suitable list of linguistic control rules for the actual control problem, is therefore the interpolation problem to find a function $\Phi^* : \mathcal{F}(\mathbf{X}) \longrightarrow \mathcal{F}(\mathbf{Y})$ which interpolates these data, i.e. which satisfies

$$\Phi^*(A_i) = B_i \quad \text{for each } i = 1, \ldots, n, \tag{3}$$

and which, in this way, gives a fuzzy representation for the control function Φ.

Actually the standard approach is to look for *one* single function, more precisely: for some uniformly defined function, which should interpolate all these data, and which should be globally defined over (X), or at least over a suitably chosen sufficiently large subclass of (X).

The additional approximation idea of the compositional rule of inference (CRI for short), as proposed by Zadeh [14], is to approximate Φ^* by a fuzzy function $\Psi^* : \mathcal{F}(\mathbf{X}) \longrightarrow \mathcal{F}(\mathbf{Y})$ determined for all $A \in \mathcal{F}(\mathbf{X})$, or for all A of that subclass, by the relationship

$$\Psi^*(A) = A \circ R \tag{4}$$

which refers to some suitable fuzzy relation $R \in \mathcal{F}(\mathbf{X} \times \mathbf{Y})$, and understands the operation $\circ$ as "sup-T-composition", with T a triangular norm.

In general we shall call functions which can, according to (4), be represented by a fuzzy relation R simply *CRI-representable.*

This "global" interpolation problem, presented by such a finite family (1) of input-output data only, in general has different solutions. However, the main approach toward this global interpolation problem is to search for a solution in a restricted class $\mathcal{IF}$ of "interpolating" functions, e.g. inside the class of CRI-representable functions. And such a restriction of the class of interpolating functions offers also the possibility that within such a class $\mathcal{IF}$ of interpolating functions the interpolation problem becomes unsolvable.

Interpolation in numerical mathematics for functions f over the reals, i.e. over a linearly ordered domain, or also for k-ary functions over the $\mathbb{R}^k$, i.e. over a partially ordered domain, usually is understood as a *local* representation of the given data via functions of some predetermined type, e.g. by polynomials, which are supposed to realize some "few" neighboring argument-value pairs for f. And this means that some such "few" neighboring argument-value pairs determine the domains of these local representations of the interpolating function. Contrary to this standard usage, in fuzzy control up to now one has not discussed any version of a "localized" interpolation approach. One of the reasons for this situation may be that in the context of fuzzy sets a suitable notion of "localization" is missing, which for real functions is given by the standard metric.

Instead, in the fuzzy context the global interpolation problem becomes in a natural way intertwined with an *approximation problem*: one may be interested to look for a function $\Psi^* \in \mathcal{IF}$ which does not really interpolate, but which "realizes" the given fuzzy input-output data "suitably well".

Such an approximative approach is completely reasonable if one has in mind that even a true solution Φ^* of the interpolation problem (3) only gives a fuzzy representation for the crisp control function Φ.

2 Two Standard Interpolation Strategies

More or less the standard theoretical understanding for the design of a fuzzy controller is the reference to the *compositional rule of inference* (CRI) first discussed by Zadeh [14].

A suitable general context for the structure of the corresponding membership degrees, which at the same time are truth degrees of a corresponding many-valued logic, is a lattice ordered abelian monoid enriched with a further operation $\rightarrowtail$, which is connected with the semigroup operation $*$ by the adjointness condition

$$x * z \leq y \quad \text{iff} \quad z \leq (x \rightarrowtail y)\,.$$

The resulting structure often is called a *residuated lattice*. Its corresponding formalized language has besides the (idempotent) conjunction $\wedge$ which is provided by the lattice meet a further (in general not idempotent) "strong" conjunction &, which has the semigroup operation $*$ as its truth degree function.

For a full formalization one therefore would embed these considerations into the context of the basic fuzzy logic BL or the monoidal t-norm logic MTL, both explained e.g. in [7].

The previously mentioned formalized language may be further enlarged by a suitable class term notation for fuzzy sets by writing $\{x \,\|\, H(x)\}$ to denote that one fuzzy set A which has as its membership degree $A(a)$ in the point a of the universe of discourse just the truth degree $[\![H(a)]\!]$ of the formula $H(a)$; or written more straightforward: $\{x \,\|\, H(x)\}(a) = [\![H(a)]\!]$.

This context yields for the CRI-based strategy, which was first applied to a control problem by Mamdani/Assilian [10], the following formulation:

From the data (A_i, B_i) one determines a fuzzy relation R in such a way that the approximating function Ψ_R^* for Φ^* becomes "describable" as

$$\Psi_R^*(A) = A \circ R = \{y \,\|\, \exists x (A(x) \mathbin{\&} R(x,y))\}\,, \tag{5}$$

which means, for the membership degrees, the well known definition

$$\Psi_R^*(A)(y) = \sup_{x \in \mathbf{X}} \big(A(x) * R(x,y)\big)\,.$$

Of course, the most preferable situation would be that the function Ψ_R^* really interpolates the given input-output-data.

A closer look at fuzzy control applications shows that one has, besides this approach via CRI-representable functions and a final application of the CRI to fuzzy input data, also a competing approach: the *method of activation degrees* which first was used by Holmblad/Ostergaard [9] in their fuzzy control algorithm for a cement kiln. In a slightly modified form this approach has more recently also been discussed in [12].

This method of activation degrees changes the previous CRI-based approach in the following way:

For each actual input fuzzy set A and each single input-output data pair (A_k, B_k) one determines a modification B_k^* of the corresponding "local" output B_k, characterized only by the local data (A_k, B_k) and the actual input A, and finally aggregates all these modified "local" outputs into one global output:

$$\Xi^*(A) = \bigcup_{i=1}^{n} B_i^* . \tag{6}$$

The particular choice of Holmblad/Ostergaard for B_k^* has been

$$B_k^*(y) = \mathsf{hgt}\,(A \cap A_k) \cdot B_k(y) , \tag{7}$$

with the rule output modifier $\mathsf{hgt}\,(A \cap A_k)$ *understood as the degree of activation of the k-th rule by the input A.*

In general terms, this modification of the first mentioned approach does not only offer one particular diverging approach toward the general interpolation problem, it also indicates that besides those both CRI-related approaches other ones with different inference and perhaps also with different aggregation operations could be of interest – as long as they are determined by finite lists of input-output data (A_i, B_i) and realize mappings from $\mathcal{F}(\mathbf{X})$ to $\mathcal{F}(\mathbf{Y})$.

This has not been done up to now in sufficient generality. Further on in this paper we shall present some considerations which point in this direction.

3 Modifying the Initial Data

The standard strategy to "solve" such a system of relation equations is, following [10], to refer to its Mamdani-Assilian relation R_{MA}, determined by the membership degrees

$$R_{\mathrm{MA}}(x, y) = \bigvee_{i=1}^{n} A_i(x) * B_i(y) , \tag{8}$$

and to apply, for a given fuzzy input A, the *compositional rule of inference*, i.e., to treat the fuzzy set $A \circ R_{\mathrm{MA}}$ as the corresponding, "right" output.

Similarly one can "solve" the system of relation equations, following [13], with reference to its Sanchez-relation $\widehat{R}$, which is the largest solution in the case of solvability and determined by the membership degrees

$$\widehat{R}(x,y) = \bigwedge_{i=1}^{n} (A_i(x) \rightarrowtail B_i(y)), \tag{9}$$

and again to the CRI, which means to treat for any fuzzy input A the fuzzy set $A \circ \widehat{R}$ as its "right" output.

But both these "solution" strategies have the (at least theoretical) disadvantage that they may give insufficient results, at least for the predetermined input sets.

Thus R_{MA} and $\widehat{R}$ may be considered as *pseudo-solutions*. Call $\widehat{R}$ the *maximal* or *S-pseudo-solution*, and R_{MA} the *MA-pseudo-solution*.

As is well known and explained e.g. in [4,6,8], these pseudo-solutions R_{MA} and $\widehat{R}$ are upper and lower approximations for the realizations of the linguistic control rules:

$$A_i \circ \widehat{R} \subseteq B_i \subseteq A_i \circ R_{\mathrm{MA}} . \tag{10}$$

Now one may equally well look for new pseudo-solutions, e.g. by some *iteration* of these pseudo-solutions in the way, that for the *next iteration step* in such an iteration process the system of relation equations is changed such that its (new) output sets become the *real output* of the *former* iteration step. This has been done in [8].

To formulate the dependence of the pseudo-solutions R_{MA} and $\widehat{R}$ from the input and output data, we denote the "original" pseudo-solutions with the input-output data (A_i, B_i) in another way and write

$$R_{\mathrm{MA}}[B_k] \text{ for } R_{\mathrm{MA}}, \qquad \widehat{R}[B_k] \text{ for } \widehat{R}.$$

Using the fact that for a given solvable system of relation equations its maximal pseudo-solution $\widehat{R}$ is really a solution one immediately gets

Proposition 1. *For any fuzzy relation S one has for the modified system with. input-output data $(A_k, A_k \circ S)_{1 \le k \le n}$ for all i:*

$$A_i \circ \widehat{R}[A_k \circ S] = A_i \circ S .$$

The reason is simply that the modified system is solvable according to the chosen modification.

Hence it does not give a new pseudo-solution if one iterates the solution strategy of the maximal, i.e. S-pseudo-solution after some (other) pseudo-solution.

The situation changes if one uses the MA-solution strategy after an other pseudo-solution strategy. Because R_{MA} has the superset property, one should use it for an iteration step which follows a pseudo-solution step w.r.t. a fuzzy relation which has the subset property, e.g. after the S-strategy using $\widehat{R}$. This gives, cf. again [8]:

Proposition 2. *If one first applies the S-pseudo-solution strategy, and then the RA-pseudo-solution strategy to the modified system, then one has always, similar to* (10), *the inclusions*

$$A_i \circ \widehat{R}[B_k] \subseteq A_i \circ R_{\mathrm{MA}}[A_k \circ \widehat{R}[B_k]] \subseteq A_i \circ R_{\mathrm{MA}}[B_k]\,.$$

Thus the iterated relation $R_{\mathrm{MA}}[A_k \circ \widehat{R}]$ is a *better* pseudo-solution as each one of R_{MA} and $\widehat{R}$.

We shall reconsider this approach to modify the initial data in some suitable form later on in Sect. 7.

4 Interpolation Strategies and Aggregation Operators

There is the well known distinction between FATI and FITA strategies to evaluate systems of linguistic control rules w.r.t. arbitrary fuzzy inputs from $\mathcal{F}(\mathbf{X})$.

The core idea of a FITA strategy is that it is a strategy which **F**irst **I**nfers (by reference to the single rules) and **T**hen **A**ggregates starting from the actual input information A. Contrary to that, a FATI strategy is a strategy which **F**irst **A**ggregates (the information in all the rules into one fuzzy relation) and **T**hen **I**nfers starting from the actual input information A.

From the two standard interpolation strategies of the last section, obviously (5) offers a FATI strategy, and (6) provides a FITA strategy.

Both these strategies use the set theoretic union as their aggregation operator. Furthermore, both of them refer to the compositional rule of inference as their core tool of inference.

In this particular case it is under mild restrictions even possible to transform an approach which follows one of these strategies into an equivalent one which uses the other strategy, cf. e.g. [5], but this–of course–does not hold in general.

In general, however, the interpolation operators we intend to consider depend more generally upon some inference operator(s) as well as upon some aggregation operator.

Definition 1. *By an* inference operator *we mean here simply a mapping from the fuzzy subsets of the input space to the fuzzy subsets of the output space.*[1]

[1] This terminology has its historical roots in the fuzzy control community. There is no relationship at all with the logical notion of inference intended and supposed here; but–of course–also not ruled out

Definition 2. *An* aggregation operator $\mathbf{A}$, *as explained e.g. in* [1, 2], *is a family* $(f^n)_{n \in \mathbb{N}}$ *of ("aggregation") operations, each* f^n *an* n*-ary one, over some partially ordered set* $\mathbf{M}$, *with ordering* $\leqslant$, *with a bottom element* $\mathbf{0}$ *and a top element* $\mathbf{1}$, *such that each operation* f^n *is non-decreasing, maps the bottom to the bottom:* $f^n(\mathbf{0}, \ldots, \mathbf{0}) = \mathbf{0}$, *and the top to the top:* $f^n(\mathbf{1}, \ldots, \mathbf{1}) = \mathbf{1}$.

Such an aggregation operator $\mathbf{A} = (f^n)_{n \in \mathbb{N}}$ is a *commutative* one iff each operation f^n is commutative. And $\mathbf{A}$ is an *associative* aggregation operator iff e.g. for $n = k + l$ one always has

$$f^n(a_1, \ldots, a_n) = f^2(f^k(a_1, \ldots, a_k), f^l(a_{k+1}, \ldots, a_n))$$

and in general

$$f^n(a_1, \ldots, a_n) = f^r(f^{k_1}(a_1, \ldots, a_{k_1}), \ldots, f^{k_r}(a_{m+1}, \ldots, a_n))$$

for $n = \sum_{i=1}^{r} k_i$ and $m = \sum_{i=1}^{r-1} k_i$.

Our aggregation operators further on are supposed to be commutative as well as associative ones.[2]

Because the general associativity condition reduces for the case $k = l = 1$ to the condition

$$f^2(x, y) = f^2(f^1(x), f^1(y))$$

it is reasonable to assume that one additionally has $f^1 = \mathsf{id}$, i.e. always $f^1(a) = a$.

Observe that an associative aggregation operator $\mathbf{A} = (f^n)_{n \in \mathbb{N}}$ is essentially determined by its binary aggregation function f^2; more precisely: by its subfamily $(f^n)_{n \leq 2}$.

For these aggregation operators $\mathbf{A}$ we need some more properties. First we consider such ones which characterize the behavior of $\mathbf{A}$ in the cases that additional information has to be aggregated. And second we look at the behavior of $\mathbf{A}$ if only one piece of information has to be aggregated.

Definition 3. *We call an aggregation operator* $\mathbf{A} = (f^n)_{n \in \mathbb{N}}$ additive *iff always* $f^1(b) \leqslant f^2(b, c)$*; we call it* multiplicative *iff always* $f^2(b, c) \leqslant f^1(b)$*; we call it* idempotent *iff always* $f^1(b) = f^2(b, b)$*; and we call it* sup-dominated *iff always* $f^2(a, b) \leqslant \sup\{a, b\}$.

Definition 4. *We call an aggregation operator* $\mathbf{A} = (f^n)_{n \in \mathbb{N}}$ optimistic *iff always* $b \leqslant f^1(b)$*; we call it* pessimistic *iff always* $f^1(b) \leqslant b$*; and we call it* realistic *iff always* $f^1(b) = b$.

Corollary 3. *Let* $\mathbf{A} = (f^n)_{n \in \mathbb{N}}$ *be an aggregation operator.*

(*i*) *If* $\mathbf{A}$ *is idempotent, then one has always* $f^2(\mathbf{0}, b) \leqslant f^2(b, b) = f^1(b)$*; and one has also always* $f^n(b, \ldots, b) = f^1(b)$.

[2] It seems that this is a rather restrictive choice from a theoretical point of view. However, in all the usual cases these restrictions are satisfied

(ii) *$\mathbf{A}$ is additive iff one has always $f^1(b) \leqslant f^2(\mathbf{0}, b)$.*
(iii) *If $\mathbf{A}$ is multiplicative, then one has always $f^2(\mathbf{0}, b) = \mathbf{0}$.*
(iv) *$\mathbf{A}$ is multiplicative iff one has always $f^2(\mathbf{1}, b) \leqslant f^1(b)$.*

If we now consider interpolation operators Ψ of FITA-type and interpolation operators Ξ of FATI-type then they have the abstract forms

$$\Psi_{\mathcal{D}}(A) = \mathbf{A}(\theta_1(A), \dots, \theta_n(A)), \tag{11}$$

$$\Xi_{\mathcal{D}}(A) = \widehat{\mathbf{A}}(\theta_1, \dots, \theta_n)(A). \tag{12}$$

Here we assume that each one of the "local" inference operators θ_i is determined by the single input-output pair $\langle A_i, B_i \rangle$. Therefore we occasionally shall write $\theta_{\langle A_i, B_i \rangle}$ instead of θ_i only. And we have to assume that the aggregation operator $\mathbf{A}$ operates on fuzzy sets, and that the aggregation operator $\widehat{\mathbf{A}}$ operates on inference operators.

With this extended notation the formulas (11), (12) become

$$\Psi_{\mathcal{D}}(A) = \mathbf{A}(\theta_{\langle A_1, B_1 \rangle}(A), \dots, \theta_{\langle A_n, B_n \rangle}(A)), \tag{13}$$

$$\Xi_{\mathcal{D}}(A) = \widehat{\mathbf{A}}(\theta_{\langle A_1, B_1 \rangle}, \dots, \theta_{\langle A_n, B_n \rangle})(A). \tag{14}$$

And it can be observed that a crucial role is played by applications of (aggregations of) inference operators to fuzzy sets. Often these inference operators are determined by fuzzy relations, and the application is given by the CRI methodology. But this is not at all necessary.

5 Some Particular Examples

The first, and most well known example of such an interpolation procedure is the Mamdani/Assilian FATI strategy which uses as local inference operators the fuzzy cartesian products $\theta_i^1 = A_i \times B_i$, their set theoretic union as aggregation operator, and the CRI strategy for application:

$$\Xi_{\mathcal{D}}^1(A) = A \circ \bigcup_{i=1}^{n} (A_i \times B_i).$$

Some other particular cases of these interpolation procedures have been discussed in [11]. These authors consider four different cases which shall now be considered in some detail.

First they look at the FITA-type interpolation

$$\Psi_{\mathcal{D}}^2(A) = \bigcap_{i=1}^{n} \left(A \circ (A_i \rhd B_i)\right), \tag{15}$$

here using as in [4] the notation $A_i \rhd B_i$ to denote the fuzzy relation with membership function

$$(A_i \rhd B_i)(x,y) = A_i(x) \rightarrowtail B_i(y)\,.$$

In this case we obviously have set theoretic intersection, i.e. $\mathbf{A}^2 = \bigcap$, as the aggregation operator (for the local inference results). And the local inference operator for the i-th linguistic control rule is just the CRI w.r.t. the fuzzy relation $A_i \rhd B_i$:

$$\theta^1_{\langle A_i, B_i\rangle}(A) = A \circ (A_i \rhd B_i)\,.$$

Their second example discusses the, in a reasonable sense dual, FATI-type approach given by

$$\Xi^3_{\mathcal{D}}(A) = A \circ \bigcap_{i=1}^{n} ((A_i \rhd B_i))\,. \tag{16}$$

Again here set theoretic intersection, i.e. according to our general notation $\widehat{\mathbf{A}}^3 = \bigcap$, is the aggregation operator for the local inference operators, and these are as in the previous case given by the fuzzy relations $A_i \rhd B_i$. The inference operation is again the CRI application.

Of course, this second approach is thus just the common CRI-based strategy which leads to the S-pseudo-solution, used in this general form already in [3], cf. also [4].

The third example in [11] is again of the FITA-type and determined by

$$\Psi^4_{\mathcal{D}}(A) = \bigcap_{i=1}^{n} \{y \,\|\, \delta(A, A_i) \longrightarrow B_i(y)\}\,. \tag{17}$$

using besides the previously mentioned class term notation for fuzzy sets the activation degree

$$\delta(A, A_i) = \bigwedge_{x \in \mathbf{X}} (A(x) \longrightarrow A_i(x)) \tag{18}$$

which is a degree of subsethood of the actual input fuzzy set A w.r.t. the i-th rule input A_i.

Of course, here the aggregation operator (for the local inference results) is again the set theoretic intersection. But the local inference results

$$\theta^4_{\langle A_i, B_i\rangle}(A) = \{y \,\|\, \delta(A, A_i) \longrightarrow B_i(y)\}$$

are not determined by application of the CRI strategy to the input fuzzy sets and to suitably chosen fuzzy relations. So we have here one of the rare examples for an inference strategy different from the CRI methodology. Another such example has, by the way, been offered in [12].

And the fourth example in [11] is a certain modification and complication of the third one, determined by

$$\Psi^5_{\mathcal{D}}(A) = \bigcap_{\emptyset \neq J \subseteq N} \{y \,\|\, \delta(A, \bigcup_{j \in J} A_j) \longrightarrow \bigcup_{j \in J} B_i(y)\}\,, \tag{19}$$

using $N = \{1, 2, \dots, n\}$.

Here the final aggregation operator for inference results is again the set theoretic intersection. However, what have been aggregated are not inference results which come from the consideration of single rules but of whole rule groups. This is coded in the inference results, because these single inference results

$$\theta^5_{\langle A_i,B_i\rangle}(A) = \{y \,\|\, \delta(A, \bigcup_{j\in J} A_j) \longrightarrow \bigcup_{j\in J} B_i(y)\} \tag{20}$$

come from an aggregation over rule groups—and all possible rule groups are taken into consideration here.

These inference results (20) themselves look like the results of a FATI strategy, with the set union as the aggregation operator. In a certain sense, therefore, we have with this fourth example a kind of a "FATITA" strategy with its "first aggregation" a somehow partial one, then an inference, and then again an aggregation.

But the simplest way to understand this last example seems to be that one (i) modifies the given data set $(A_i, B_i)_{i\in N}$ into the more complicated one

$$\Big(\bigcup_{\emptyset\neq J\subseteq N} A_j, \bigcup_{\emptyset\neq J\subseteq N} B_j \Big),$$

and then (ii) applies the FITA strategy of the previous example to this modified data set.

So only the first four examples remain. And in these examples the main aggregation operators are the set theoretic union and the set theoretic intersection. Both are obviously associative, commutative, and idempotent. Additionally the union is an additive, and the intersection a multiplicative aggregation operator.

6 Stability Conditions for the Given Data

If $\Theta_{\mathcal{D}}$ is a fuzzy inference operator of one of the types (13), (14), then the interpolation property one likes to have realized says that it is satisfied

$$\Theta_{\mathcal{D}}(A_i) = B_i \tag{21}$$

for all the data pairs $\langle A_i, B_i\rangle$. In the particular case that the operator $\Theta_{\mathcal{D}}$ is given by (5), this is just the problem to solve the system (21) of fuzzy relation equations.

Definition 5. *In the present generalized context let us call the property (21) the* $\mathcal{D}$-stability *of the fuzzy inference operator* $\Theta_{\mathcal{D}}$.

To find $\mathcal{D}$-stability conditions on this abstract level seems to be rather difficult in general. However, the restriction to fuzzy inference operators of FITA-type makes things easier.

It is necessary to have a closer look at the aggregation operator $\mathbf{A} = (f^n)_{n\in\mathbb{N}}$ involved in (11) which operates on $\mathcal{F}(\mathbf{Y})$, of course with the set theoretic inclusion as partial ordering.

Definition 6. *Having $B, C \in \mathcal{F}(\mathbf{Y})$ we say that C is* **A**-negligible w.r.t. B *iff $f^2(B,C) = f^1(B)$ holds true.*

The core idea here is that in any aggregation by **A** the presence of the fuzzy set B among the aggregated fuzzy sets makes any presence of C superfluous.

Corollary 4. *(i) If* **A** *is an idempotent and additive aggregation operator then each C with $C \leqslant B$ is* **A***-negligible w.r.t. B.*

(ii) If **A** *is an idempotent and multiplicative aggregation operator then each C with $B \leqslant C$ is* **A***-negligible w.r.t. B.*

(iii) The bottom element $C = \mathbf{0}$ in the domain of an additive and idempotent aggregation operator **A** *is* **A***-negligible w.r.t. any other element of that domain.*

Examples:

1. If **A** is the supremum w.r.t. the partial ordering $\leqslant$ than C is **A**-negligible w.r.t. B iff $C \leqslant B$;
2. particularly C is $\bigcup$-negligible w.r.t. B iff $C \subseteq B$.
3. If **A** is the infimum w.r.t. the partial ordering $\leqslant$ then C is **A**-negligible w.r.t. B iff $B \leqslant C$;
4. particularly C is $\bigcap$-negligible w.r.t. B iff $C \supseteq B$.

Of course, similar results hold true for all idempotent aggregation operators, which are additive or multiplicative, respectively.

Proposition 5. *Consider a fuzzy inference operator of FITA-type*

$$\Psi_{\mathcal{D}} = \mathbf{A}(\theta_{\langle A_1,B_1\rangle}, \ldots, \theta_{\langle A_n,B_n\rangle}) .$$

It is sufficient for the $\mathcal{D}$-stability of $\Psi_{\mathcal{D}}$, i.e. to have

$$\Psi_{\mathcal{D}}(A_k) = B_k \qquad \text{for all } k = 1, \ldots, n \tag{22}$$

that one always has

$$\theta_{\langle A_k,B_k\rangle}(A_k) = B_k \tag{23}$$

and additionally that for each $i \neq k$ the fuzzy set

$$\theta_{\langle A_k,B_k\rangle}(A_i) \quad \text{is } \mathbf{A}\text{-negligible w.r.t.} \quad \theta_{\langle A_k,B_k\rangle}(A_k) . \tag{24}$$

Proof. The result follows immediately from the corresponding definitions.

This result has two quite interesting specializations which themselves generalize well known results about fuzzy relation equations.

Corollary 6. *It is sufficient for the $\mathcal{D}$-stability of a fuzzy inference operator $\Psi_{\mathcal{D}}$ of FITA-type that one has*

$$\Psi_{\mathcal{D}}(A_i) = B_i \qquad \text{for all } 1 \leq i \leq n$$

and that always $\theta_{\langle A_i, B_i \rangle}(A_j)$ is $\mathbf{A}$-negligible w.r.t. $\theta_{\langle A_i, B_i \rangle}(A_i)$.

Corollary 7. *It is sufficient for the $\mathcal{D}$-stability of a fuzzy inference operator $\Psi_{\mathcal{D}}$ of FITA-type, which is based upon an additive, realistic, and idempotent aggregation operator, that one has*

$$\Psi_{\mathcal{D}}(A_i) = B_i \qquad \text{for all } 1 \leq i \leq n$$

and that always $\theta_{\langle A_i, B_i \rangle}(A_j)$ is the bottom element in the domain of the aggregation operator $\mathbf{A}$.

Obviously, this is a direct generalization of the fact that systems of fuzzy relation equations are solvable if their input data form a pairwise disjoint family (w.r.t. the corresponding t-norm based intersection)because in this case one has usually:

$$\begin{aligned} \theta_{\langle A_i, B_i \rangle}(A_j) &= A_j \circ (A_i \times B_i) = \{y \| \exists x (x \varepsilon A_j \mathbin{\&} (x, y) \varepsilon A_i \times B_i)\} \\ &= \{y \| \exists x (x \varepsilon A_j \cap_T A_i \mathbin{\&} y \varepsilon B_i)\}, \end{aligned}$$

with $\cap_T$ the t-norm based intersection operation which refers to the same t-norm T as the conjunction operation &.

To extend these considerations from inference operators (11) of the FITA type to those of the FATI type (12) let us consider the following additional notion.

Definition 7. *Suppose that $\widehat{\mathbf{A}}$ is an aggregation operator for inference operators, and that $\mathbf{A}$ is an aggregation operator for fuzzy sets. Then $(\widehat{\mathbf{A}}, \mathbf{A})$ is an* application distributive *pair of aggregation operators iff*

$$\widehat{\mathbf{A}}(\theta_1, \ldots, \theta_n)(X) = \mathbf{A}(\theta_1(X), \ldots, \theta_n(X)) \tag{25}$$

holds true for arbitrary inference operators $\theta_1, \ldots, \theta_n$ and fuzzy sets X.

Using this notion it is easy to see that one has on the left hand side of (25) a FATI type inference operator, and on the right hand side an associated FITA type inference operator. So one is able to give a reduction of the FATI case to the FITA case.

Proposition 8. *Suppose that $(\widehat{\mathbf{A}}, \mathbf{A})$ is an application distributive pair of aggregation operators. Then a fuzzy inference operator $\Xi_{\mathcal{D}}$ of FATI-type is $\mathcal{D}$-stable iff its associated fuzzy inference operator $\Psi_{\mathcal{D}}$ of FITA-type is $\mathcal{D}$-stable.*

Proof. This follows immediately from the corresponding definitions: equation (25) pulls the $\mathcal{D}$-stability of the FATI type back to the $\mathcal{D}$-stability of the FITA type, and vice versa.

7 Stability Conditions for Modified Data

The combined approximation and interpolation problem, as previously explained, sheds new light on the standard approaches toward fuzzy control via CRI-representable functions originating from the works of Mamdani/Assilian [10] and Sanchez [13] particularly for the case that neither the Mamdani/Assilian relation R_{MA} of (8), nor the Sanchez relation $\widehat{R}$ of (9), offer a solution for the system of fuzzy relation equations. In any case both these fuzzy relations determine CRI-representable fuzzy functions which provide approximate solutions for the interpolation problem.

In other words, the consideration of CRI-representable functions determined by (8) as well as by (9) provides two methods for an approximate solution of the main interpolation problem. As is well known and explained e.g. in [4], the approximating interpolation function CRI-represented by $\widehat{R}$ always gives a lower approximation, and that one CRI-represented by R_{MA} gives an upper approximation for normal input data.

Extending these results, in [8] the iterative combination of these methods has been discussed to get better approximation results. For the iterations there, always the next iteration step consisted in an application of a predetermined one of the two approximation methods to the data family with the original input data and the real, approximating output data which resulted from the application of the former approximation method.

A similar iteration idea was also discussed in [11], and already explained in Sect. 3, however restricted always to the iteration of only one of the approximation methods explained in (15), (16), (17), and (19).

Therefore let us now, in the general context of this paper, discuss the problem of $\mathcal{D}$-stability for a modified operator $\Theta_{\mathcal{D}}^*$ which is determined by the kind of iteration of $\Theta_{\mathcal{D}}$ just explained.

Let us consider the $\Theta_{\mathcal{D}}$*-modified* data set $\mathcal{D}^*$ given as

$$\mathcal{D}^* = (\langle A_i, \Theta_{\mathcal{D}}(A_i)\rangle)_{1\leq i\leq n}\,, \tag{26}$$

and define from it the modified fuzzy inference operator $\Theta_{\mathcal{D}}^*$ as

$$\Theta_{\mathcal{D}}^* = \Theta_{\mathcal{D}^*}\,. \tag{27}$$

For these modifications, the problem of stability reappears. Of course, the new situation here is only a particular case of the former. And it becomes a simpler one in the sense that the stability criteria now refer only to the input data A_i of the data set $\mathcal{D} = (\langle A_i, B_i\rangle)_{1\leq i\leq n}$.

Proposition 9. *It is sufficient for the $\mathcal{D}^*$-stability of a fuzzy inference operator $\Psi_{\mathcal{D}}^*$ of FITA-type that one has*

$$\Psi_{\mathcal{D}}^*(A_i) = \Psi_{\mathcal{D}^*}(A_i) = \Psi_{\mathcal{D}}(A_i) \qquad \text{for all } 1\leq i\leq n \tag{28}$$

and that always $\theta_{\langle A_i, \Psi_{\mathcal{D}}(A_i)\rangle}(A_j)$ is **A***-negligible w.r.t. $\theta_{\langle A_i, \Psi_{\mathcal{D}}(A_i)\rangle}(A_i)$.*

Proof. The modification of the output data obviously satisfies condition (23) of Proposition 5; hence only condition (24) remains.

Let us look separately at the conditions (28) and at the negligibility conditions.

Corollary 10. *The conditions* (28) *are always satisfied if the inference operator* $\Psi^*_{\mathcal{D}}$ *is determined by the standard output-modified system of relation equations* $A_i \circ R[A_k \circ R] = B_i$ *in the notation of* [8].

Corollary 11. *In the case that the aggregation operator* $\mathbf{A}$ *in* (11) *is the set theoretic union, the conditions* (28) *together with the inclusion relationships*

$$\theta_{\langle A_i, \Psi_{\mathcal{D}}(A_i)\rangle}(A_j) \subseteq \theta_{\langle A_i, \Psi_{\mathcal{D}}(A_i)\rangle}(A_i)$$

are sufficient for the $\mathcal{D}^*$*-stability of a fuzzy inference operator* $\Psi^*_{\mathcal{D}}$.

As in Sect. 6 one is able to transfer this result to FATI-type fuzzy inference operators in a straightforward way, simply referring to the application distributivity.

Corollary 12. *Suppose that* $(\widehat{\mathbf{A}}, \mathbf{A})$ *is an application distributive pair of aggregation operators. Then a fuzzy inference operator* $\Phi^*_{\mathcal{D}}$ *of FATI-type is* $\mathcal{D}^*$*-stable iff its associated fuzzy inference operator* $\Psi^*_{\mathcal{D}}$ *of FITA-type is* $\mathcal{D}^*$*-stable.*

8 The Subset Property

Besides the stability problem, which corresponds just to the solvability problem for fuzzy relation equations, in that context it is also of interest—if one does not have solvability— to get information on some approximation behavior. One of the conditions of interest for pseudo-solution of (systems of) fuzzy relation equations is the subset property as discussed e.g. in [6, 8].

Definition 8. *An interpolation operator* $\Phi_{\mathcal{D}}$ *satisfies the* subset property *for a family* (1) *of input-output data iff one has for all* $i = 1, \ldots, n$:

$$\Phi_{\mathcal{D}}(A_i) \subseteq B_i \,.$$

Proposition 13. *The following conditions are sufficient that an interpolation operator* $\Psi_{\mathcal{D}}(A) = \mathbf{A}(\theta_{\langle A_1, B_1\rangle}(A), \ldots, \theta_{\langle A_n, B_n\rangle}(A))$ *of FITA type has the subset property:*

1. *The inference result* $\theta_{\langle A_i, B_i\rangle}(A_j)$ *is always* $\mathbf{A}$*-negligible w.r.t.* $\theta_{\langle A_i, B_i\rangle}(A_i)$*,* $\mathbf{A}$ *is pessimistic, and one has the subset property that* $\theta_{\langle A_i, B_i\rangle}(A_i) \leqslant B_i$ *for each* $i = 1, \ldots, n$.

2. *The aggregation operator* **A** *is idempotent and pessimistic, and one has satisfied for all* $1 \leq i, j \leq n$ *the condition*

$$\theta_{\langle A_j, B_j \rangle}(A_i) \leqslant \theta_{\langle A_i, B_i \rangle}(A_i) \leqslant B_i \, .$$

3. *The aggregation operator* **A** *is sup-dominated and one has always satisfied* $\theta_{\langle A_j, B_j \rangle}(A_i) \leqslant B_i$.

Proof. The results follow by routine calculations in a straightforward way from the corresponding definitions.

However, only the first one of these sufficient conditions seems to be reasonable; the other two are rather strong ones and seem to be quite unrealistic to assume for particular applications.

For the FATI approach suitable results are actually lacking.

9 The Superset Property

Dual to the subset property there is the following superset property.

Definition 9. *An interpolation operator* $\Phi_{\mathcal{D}}$ *satisfies the* superset property *for a family* (1) *of input-output data iff one has for all* $i = 1, \ldots, n$*:*

$$\Phi_{\mathcal{D}}(A_i) \supseteq B_i \, .$$

Proposition 14. *The following conditions are sufficient that an interpolation operator* $\Psi_{\mathcal{D}}(A) = \mathbf{A}(\theta_{\langle A_1, B_1 \rangle}(A), \ldots, \theta_{\langle A_n, B_n \rangle}(A))$ *of FITA type has the superset property:*

1. *if* **A** *is additive and optimistic, and if one always has* $B_i \leqslant \theta_{\langle A_i, B_i \rangle}(A_i)$.
2. *if* $\theta_{\langle A_k, B_k \rangle}(A_i)$ *is always* **A***-negligible w.r.t.* $\theta_{\langle A_i, B_i \rangle}(A_i)$ *and if on always has* $B_i \leqslant \theta_{\langle A_i, B_i \rangle}(A_i)$.
3. *if* **A** *is sup-dominated, if one has the superset property* $B_i \leqslant \theta_{\langle A_i, B_i \rangle}(A_i)$, *and if one always also has*

$$\theta_{\langle A_k, B_k \rangle}(A_i) \leqslant \theta_{\langle A_i, B_i \rangle}(A_i) \, .$$

Proof. The results follow by routine calculations in a straightforward way from the corresponding definitions.

If one considers an aggregation operator **A** which is multiplicative and pessimistic, then one gets as a necessary condition for the superset property of **A** that one always has $B_i \leqslant \theta_{\langle A_k, B_k \rangle}(A_i)$.

But these necessary conditions are again very strong ones, as is the last mentioned sufficient condition of the last proposition.

Again for the FATI type of approach there are less results. In the case that the aggregation operator $\widehat{\mathbf{A}}$ is additive and optimistic, the "local superset" properties

$$B_i \leqslant \theta_{\langle A_i, B_i \rangle}(A_i)$$

are sufficient for the "global" one.

But if one assumes that $\widehat{\mathbf{A}}$ is multiplicative then, even under the additional assumption that $\widehat{\mathbf{A}}$ is optimistic, one needs all the strange conditions

$$B_i \leqslant \theta_{\langle A_k, B_k \rangle}(A_i)$$

as necessary ones.

10 Concluding remarks

The paper intends to offer a (first step toward a) rather general point of view regarding common approaches toward evaluations of fuzzy rule systems. The idea is to look at those evaluations as some combinations of aggregation parts, and of "inference" parts.

One of the interesting observations, actually, is that here only particular properties of the aggregation operators had to be considered, not of the inference operators. Whether this means that the aggregation parts are the more important ones, or whether this only indicates that a deeper understanding of the rôle of the inference operators is lacking, this remains open.

Moreover a lot of details remain open which concern a possible fine tuning of the principal results the paper is aimed to offer.

What also remains outside our present considerations is the problem to discuss other, different interpolation methods for finite (or perhaps even infinite) lists of argument-value pairs which may be of interest to determine functions which connect fuzzy set arguments with fuzzy set values.

And finally let it be mentioned that essentially only the lattice properties of the structure of the membership degrees has been of importance in these considerations. So these considerations may suitably be generalized.

References

1. T. Calvo, G. Mayor, R. Mesiar (eds.): *Aggregation Operators: New Trends and Applications*, Physica-Verlag: Heidelberg, 2002.
2. D. Dubois, H. Prade: On the use of aggregation operations in information fusion processes, *Fuzzy Sets Systems* **142** (2004), 143–161.
3. S. Gottwald: Characterizations of the solvability of fuzzy equations. *Elektron. Informationsverarb. Kybernet.* **22** (1986), 67–91.
4. S. Gottwald: *Fuzzy Sets and Fuzzy Logic.* The Foundations of Application – From a Mathematical Point of View, Vieweg: Braunschweig/Wiesbaden and Teknea: Toulouse, 1993.
5. S. Gottwald: On differences and conformities of different fuzzy control strategies. In: *Proc. 6. Internat. Fuzzy Systems Assoc. World Congress* IFSA '95, *Sao Paulo, July 21 - 28, 1995*, Sao Paulo 1995, vol. 2, 113–116.

6. S. Gottwald: Mathematical Fuzzy Control. A Survey of Some Recent Results. *Logic Journal of IGPL* **13**, no.5 (2005), 525–541.
7. S. Gottwald, P. Hájek: T-norm based mathematical fuzzy logics. In: *Logical, Algebraic, Analytic, and Probabilistic Aspects of Triangular Norms* (E.P. Klement and R. Mesiar, eds.), Elsevier: Dordrecht, 2005, 275–299.
8. S. Gottwald, V. Novák, I. Perfilieva: Fuzzy control and t-norm-based fuzzy logic. Some recent results, in: *Proc. 9th Internat. Conf.* IPMU'2002, ESIA – Université de Savoie, Annecy, 2002, 1087–1094.
9. L.P. Holmblad, J.J. Ostergaard: Control of a cement kiln by fuzzy logic, in: M.M. Gupta/E. Sanchez (eds.), *Fuzzy Information and Decision Processes.* North-Holland Publ. Comp.: Amsterdam, 1982, 389–399.
10. A. Mamdani, S. Assilian, An experiment in linguistic synthesis with a fuzzy logic controller, *Internat. J. Man-Machine Studies* **7** (1975) 1–13.
11. N.N. Morsi, A.A. Fahmy, On generalized modus ponens with multiple rules and a residuated implication, *Fuzzy Sets Systems,* **129** (2002), 267–274.
12. B. Moser, M. Navara, Fuzzy controllers with conditionally firing rules, *IEEE Trans. Fuzzy Systems* **10** (2002), 340–348.
13. E. Sanchez, Resolution of composite fuzzy relation equations, *Information and Control,* **30** (1976) 38–48.
14. L.A. Zadeh, Outline of a new approach to the analysis of complex systems and decision processes. *IEEE Trans. Systems, Man and Cybernet.* **SMC-3** (1973) 28–44.

Nuclear Reactor Power Control Using State Feedback with Fuzzy Logic

Jorge S. Benítez-Read, J. Humberto Pérez-Cruz and Da Ruan

Abstract. A novel control scheme for power regulation in a research nuclear reactor has been developed. The scheme combines different techniques such as state variable feedback, first order numerical integration for state gain estimation, and fuzzy logic to attain a stable power regulation in the reactor. The proposed control strategy attains, in short time and without overshoot, different values of steady-state power departing from a wide range of different values of initial power. The algorithm was subjected to exhaustive simulation tests and the experimental results obtained are highly satisfactory.

1 Introduction

Safety aspects in the control of nuclear power have been always present in every control algorithm design, either explicitly or implicitly. In this sense, some reactivity constraints may be imposed on the control system [3] or, a predefined limiting value of the reactor period can be obtained by applying velocity control to the control rod system [2]. Likewise, controller designs that consider disturbances or uncertainties in the system models tend to reduce the number of unnecessary power excursions and plant scrams [21]. The theory of fuzzy sets and fuzzy logic has been used as the basis of new approaches for the design of nuclear reactor control systems. Thus for instance, human operator linguistic rules have been incorporated in a high temperature nuclear reactor controller [4], and the use of fuzzy logic and neural networks has been a field of study at Penn State University [9]. Computer simulations reported by Moon [13] show a similar or better performance of fuzzy logic control algorithms compared to conventional P-I controllers. Also, a fuzzy logic based controller has been successfully applied to control the power at the Belgian nuclear reactor BR1 [16], and an adaptive fuzzy logic controller has been implemented in real time on a demo model [15].

At the National Nuclear Research Institute of Mexico (ININ), the development of control algorithms for its TRIGA Mark III research nuclear reactor has been focused in obtaining a controlled ascent of power and its regulation for long periods of time, maintaining the reactor period within the safety limits. Thus, for instance, the acquisition of knowledge of the reactor's dynamic behaviour [18] has been used to design some modern control schemes [1], where the simulations show good dynamic characteristics.

A current project at ININ, in collaboration with the Belgian Nuclear Research Centre (SCK•CEN), has the objective of designing an expert neutron power controller, as well as its integration in the reactor control console for real time validation purposes. To this end, a novel control procedure is presented here. The main goal is to investigate the performance of a control scheme that combines state feedback, a first order predictive stage, and a fuzzy logic block to control the ascent of power in the reactor, maintaining at all times the instantaneous period of the reactor within safety limits. The proposed control strategy permits different values of steady-state power be attained from different values of initial power in short time and without overshoot. The algorithm has been subjected to exhaustive simulation tests and the experimental results obtained have been highly satisfactory.

2 Description of the TRIGA Mark III Reactor

This section briefly describes the principle of operation of a nuclear reactor and presents the main features of the ININ's TRIGA Mark III research nuclear reactor, its typical applications and the reasons to implement new control algorithms.

2.1 Nuclear Fission and Control Principle of the Fission Chain Reaction in a Reactor

A nuclear fission may occur when a neutron is absorbed by the atomic nucleus of a heavy element, for instance the Uranium isotope U-235 [6]. As a result, the nucleus is excited and its structure suffers an alteration. In some cases, this structure modification reaches a point in which the nucleus is partitioned, giving as byproducts two or more smaller nuclei (fission fragments), an average between two and three neutrons, and electromagnetic radiation, among others. These fission neutrons can in turn be absorbed by other U-235 nuclei and produce new fissions. If this process is repeated cyclically, it is called a chain reaction (avalanche effect). Each cycle of the chain reaction is called a generation. The principle of nuclear energy is based on the efficient and controlled use of the energy being liberated during the fission process. Precisely, a nuclear reactor is a system in which the fission chain reaction is adequately controlled to transform the atomic energy into a useful form. In each fission, most of the kinetic energy of the fission fragments is transformed into a small amount of calorific energy. In a chain reaction within a reactor, these small amounts of energy are accumulated, and a considerable energy in form of heat is available for its use. The control of the chain reaction is based on introducing a neutron absorber element, such as Boron or Cadmium, inside the volume where the fissions are being generated. When these elements capture neutrons, the fission reactions can be reduced and, with enough absorber material, the nuclear fission can be totally stopped, process known as reactor shut down.

2.2 The TRIGA Mark III Reactor System: Characteristics, Applications, and Safety Considerations

The most known fission nuclear reactors are those aimed to generate electric power; they are called nuclear power plants or NPP. Not less important, however, are the research nuclear reactors whose main purpose is the generation of neutrons used in fundamental experiments in such fields as nuclear physics, nuclear medicine, material sciences, and nuclear activation analysis (NAA), among others. Since 1968 the National Institute for Nuclear Research (ININ) of Mexico owns and operates a TRIGA Mark III research nuclear reactor [7]. The reactor's platform is shown in Fig. 1. This reactor has been mainly used to determine substance composition by different NAA techniques, to study the effect (accelerated aging) of neutron flux on different materials present on NPP, to train personnel on reactor operation and instrumentation, and to produce radioisotopes for industrial, agricultural and, particularly important,

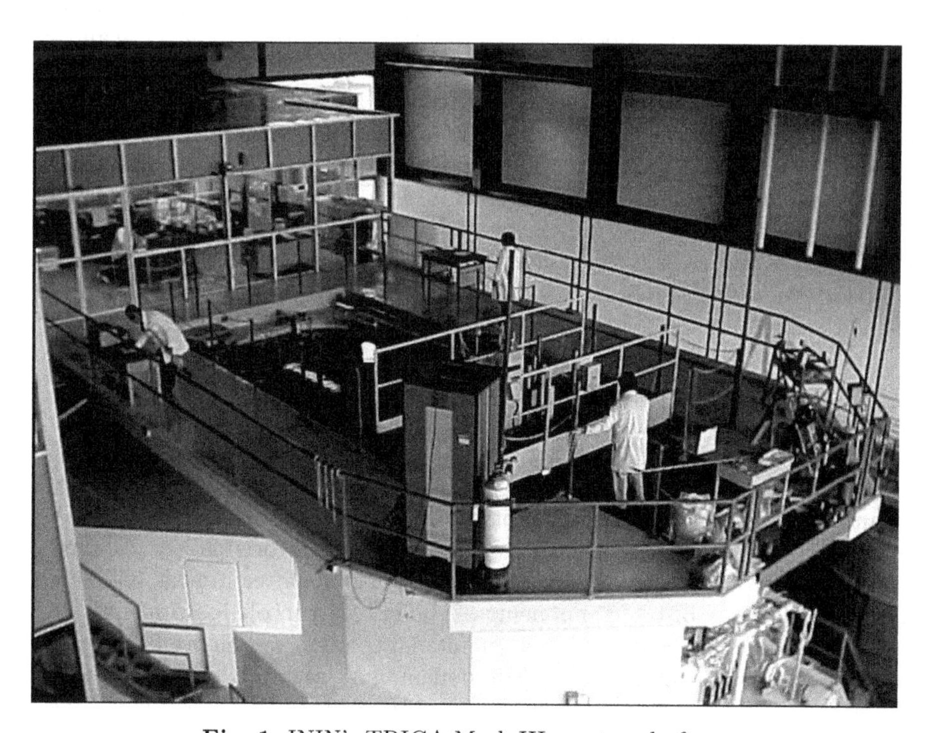

Fig. 1. ININ's TRIGA Mark III reactor platform

medical applications. In this latter area two Samarium-based radio pharmaceutical drugs have been recently produced, ^{153}Sm-ETMP[1] and ^{153}Sm-MH that are used in the treatment of bone cancer (as analgesic) and rheumatoid arthritis, respectively [17].

This reactor can be operated in three alternative modes [7]: 1) Steady state at any power level between 0 and 1 MW with manual or automatic control that use servomechanisms, 2) Rectangular pulse at a power level from 300 to 1000 kW, and 3) Pulse of 1500 MW with a duration of 10 ms approximately. The pulse can be repeated every 5 minutes. The reactor is composed of a core (containing the fuel and the neutron absorber) immersed in a "swimming" pool where the water functions as core coolant and neutron moderator (material used to reduce the speed or energy of the neutrons to a level where the probability of fission is higher). The reactor fuel is a mixture of U-235 with Zirconium hydride (UZrH). The hydrogen present in the nuclear fuel dramatically increases the safety of the reactor to any excursion of reactivity (a measure of neutron population of the current generation with respect to the previous one). Also inside the core, there exist four rods, called control rods, which contain Boron (neutron absorber). Depending on the degree of insertion or withdrawal inside the core of these control rods, it is possible to bring the reactor power to a certain pre-specified level and maintain this power constant for long periods of time, from several hours to several days. In practice, only one of these rods is necessary for power regulation. One of the safety measures, called period scram, during the reactor operation is the automatic shutdown of the reactor if the current reactor period decreases below its safety limit. The reactor period is defined as the time that the reactor would take at any instant of time to increase its current power value by a factor e. In practice, for instance, a period scram occurs if the ratio of power values taken every 10 ms is greater than 1.00333889. Another automatic shutdown occurs when the power overpasses by 10% the demanded final power level. It is clear that to reduce the power overshoot and minimize the occurrence of period scrams, the motion of the control rod has to be governed by an adequate control system.

2.3 Reactor Control: New Console and Convenience of Implementing More Flexible Control Algorithms

As mentioned before, the control of the reactor can be carried out manually or automatically. The automatic control is based on a proportional-integral-derivative (PID) algorithm [8] that was integrated to the original reactor analog console since 1968 by the reactor manufacturer. After almost 35 years of operation, many of the components of the analog console instrumentation were obsolete and discontinued, and maintenance tasks were more frequently required due to the aging of the components. As a result, a new console, a digital one, was designed, built, and installed by ININ's personnel. The original control algorithm was implemented on the new console and works satisfactorily. However, given the great flexibility provided by the new console to implement different controllers, the strategy of ININ, in collaboration with personnel from the Belgian Nuclear Research Centre (SCK-CEN), is to design and test computational intelligence based control algorithms to increase the efficiency and to provide new capabilities for the reactor operation. For instance, some of the algorithms could incorporate operators' experience or could adjust their parameters according to different operating conditions. In fact, different pure fuzzy [1] and input-to-output linearizing control [12] algorithms have been designed and simulated. Some of the limitations of these controllers are mentioned in Sect. 3 and, to overcome them, a scheme is presented here that combines different techniques, each solving a specific problem. First, the availability of the dynamic model of the plant allows the proposition of a control law based on the theory of state feedback [8]. The drawback concerning the slow response of the closed loop system is coped by means of a first order estimator derived from the theory of numerical methods for solving ordinary differential equations [10]. Then, fuzzy logic principles are applied to solve the perturbations of power close to the demanded level due to the nonlinearities of the system.

3 Statement of the Problem

Considering the reactor as a black box, the external reactivity $\rho_{ext}\ (t)$ (with adimensional values between 0 and 3.9) is identified as the system's excitation signal. This control input is related to the control bar displacement. Three states are identified: The neutron power $n\ (t)$, from 1W to 2.988 MW; the delayed neutron precursor concentration $C\ (t)$, from 420 W to 1.257 GW; and the internal reactivity $\rho_{int}\ (t)$, dimensional, from 0 to 3.9 units.

Starting the operation of the reactor at a power level between 1 and 50 W (power provided by an external neutron source), a typical final level is 1 MW for many of the reactor's applications such as radioisotope production, neutron activation analysis, and material testing. In theory, this steady-state power can be easily obtained by applying 1.305 units of external reactivity (stepwise). However, under this hypothetical situation, the simulation of the reactor's point kinetics equations predicts an overshoot of $8.5{*}10^6\%$ (see Fig. 2).

On the other hand, the reactor is subjected to period scram, which consists in an automatic shutdown of the plant every time the power does not satisfy the following inequality:

$$n(t_i + h) < n(t_i)e^{\frac{h}{T}} \tag{1}$$

where $h = 10\,\text{ms}$ and $T = 3\,\text{s}$. Evidently, the adequate extraction of the reactor control rod to get the final power (1 MW) in a short time but without overshoot and without period scram is very important. Since h represents a time increment, normally of 10 ms, and T the reactor period lower limit, usually of 3 seconds, the period scram can also be considered as the automatic

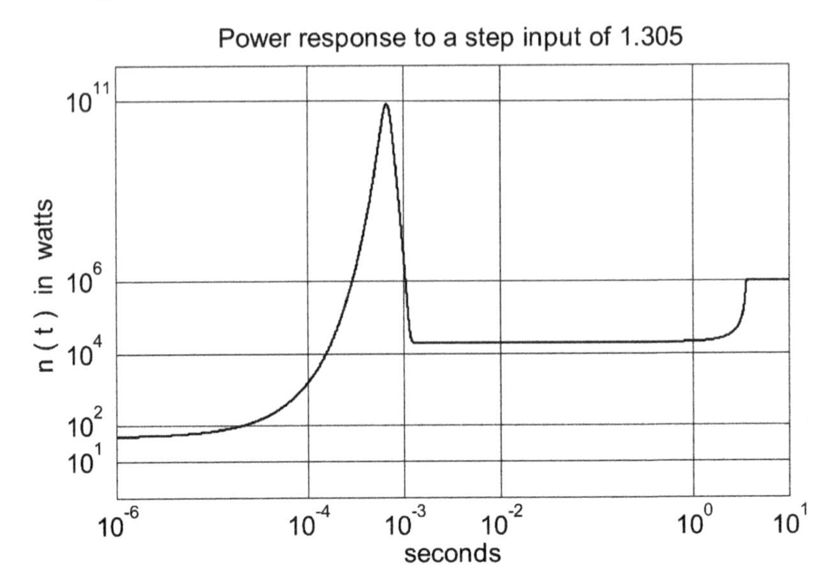

Fig. 2. Reactor response to an external reactivity step of 1.305

reactor shutdown every time the power ratio is greater than 1.00333889 on a 10 ms time period. This simple example illustrates the need of a control system for the ascent of power, by a proper withdrawal/insertion motion of the control rod, to attain the desired power in the shortest time possible, preventing at the same time a high overshoot and a period scram. The work by Pérez-Cruz [14] reports some disadvantages detected in previous control designs, which include the following: a) An input/output linearizing controller does not take into account, at least explicitly, the period scram; and b) The controllers based on Mamdani's rules have not been tested for neither a wide range of final powers nor a wide range of initial power levels. Thus, this work proposes a novel control scheme to overcome the above mentioned drawbacks.

4 Development of the Controller

State variable feedback [8], a first order numerical integration (predictive stage) [10], and fuzzy logic [20] are combined in such a way that the disadvantages of every single technique are attenuated by the combined action, obtaining in consequence a relatively simple and sufficiently fast controller. To the best of the authors' knowledge, no similar control strategy has been proposed in the literature for the power control of nuclear research reactors.

The design procedure started with the analytical definition, based on the mathematical model of the reactor, of the scaling factor in steady state between the external reactivity $\rho_{ext}(t)$ and the reactor power $n(t)$. Then, using the theory of state space, a state feedback control law was proposed.

The resulting closed-loop control system, although eliminates the overshoot; it renders an excessively slow neutron power response. The solution to this problem was given by incorporating a stage that predicts the required feedback gain k_3. The predictor, derived from the theory of first order numerical integration, produces very good results during the first stage of the ascent of power. However, the neutron power presents an abnormal behaviour (including irregular oscillations) when it approaches the desired reference level. To cope with this problem, a Mamdani fuzzy stage was added to regulate the predictor action. Details are given in the next subsections. The results obtained include a fast response and independence of the wide variety of potential operating conditions -something not easy and even impossible to obtain with other procedures.

4.1 Mathematical Model

A rigorous description of a nuclear reactor behavior implies the use of neutron transport and/or diffusion equations. These equations consider both space and time as independent variables, and are not very useful for control purposes [11]. However, under some assumptions such as independence of space and time functions, and zero net reactivity ($\rho_{ext}(t)+\rho_{\text{int}}(t) = 0$) under near critical operation of the reactor, it is possible to simplify the reactor's dynamics to what is known as the point kinetic equations [11]. This results in a model that is very adequate for designing and simulating closed-loop control systems.

Hereafter, we use the following notation: $x_1(t) = n\ (t)$, $x_2(t) = C\ (t)$, $x_3(t) = \rho_{int}\ (t)$ and $u\ (t) = \rho_{ext}\ (t)$. Also, $n_0 = x_1(0)$ will represent the initial power value (between 1 W and 50 W) when $u = 0$, and R will represent the desired steady-state power level. In principle, $R = x_1(\infty)$. Thus, the point kinetic equations of the reactor are:

$$\dot{x}_1(t) = \left(\frac{u(t) + x_3(t) - \beta}{\Lambda}\right) x_1(t) + \lambda x_2(t) \tag{2}$$

$$\dot{x}_2(t) = \frac{\beta}{\Lambda} x_1(t) - \lambda x_2(t) \tag{3}$$

$$\dot{x}_3(t) = -\alpha K\left(x_1(t) - n_0\right) - \gamma x_3(t) \tag{4}$$

The constant parameters $\alpha, \beta, \lambda, \Lambda, \gamma$, and K are determined considering the composition of the fuel (^{235}U as the fissile material), its enrichment, and the geometry of the reactor and core. The nominal values of these parameters that best correlate the model with the physical system [19] are shown in Table 1.

For fast neutron power changes, the simulation of the point kinetics is better accomplished with the ode45 Matlab function, which is based on the Runge-Kutta-Fehlberg method. On the other hand, for gradual changes of the neutron power response, it is possible and faster to use a fixed-step numerical method. A 4th-order Runge-Kutta with a step of 10 ms has given satisfactory results.

Table 1. Parameters of the reactor's point kinetic equations

Parameter	Nominal value	Units
α	0.01359875	°C^{-1}
β	6.433* 10^{-3}	
λ	0.4024	S^{-1}
Λ	38	μs
γ	0.2	s^{-1}
K	1/5.21045* 10^{4}	°C/(W · s)

4.2 Steady State Open-loop Scaling Factor, g

An open-loop control design consists of finding the steady-state relation between the external reactivity (input signal) $u\ (t)$ and the power $x_1(t)$. Thus, the magnitude of external reactivity to be inserted in the reactor core to produce a given R will be directly computed. In the case of linear systems with the transfer function $G(s)$, such a relation or scaling factor g is the reciprocal of the DC system gain [8], that is:

$$g = s \overset{Lim}{\longrightarrow} 0 \ \frac{1}{G(s)} \tag{5}$$

For the case of the nonlinear model of the reactor, the problem consists of determining the existence of such a factor g between $u\ (\infty)$, R and n_0 such that:

$$u(\infty) = g\,(R - n_0) \tag{6}$$

The open-loop control scheme is shown in Fig. 3.

Considering a zero rate of change with respect to time of $x_1(t)$, $x_2(t)$, and $x_3(t)$ as t $\longrightarrow \infty$, then (2) to (4) can be written as follows:

$$0 = \left(\frac{u(\infty) + x_3(\infty) - \beta}{\Lambda}\right) x_1(\infty) + \lambda x_2(\infty) \tag{7}$$

$$0 = \frac{\beta}{\Lambda} x_1(\infty) - \lambda x_2(\infty) \tag{8}$$

$$0 = -\alpha K\,(x_1(\infty) - n_0) - \gamma x_3(\infty) \tag{9}$$

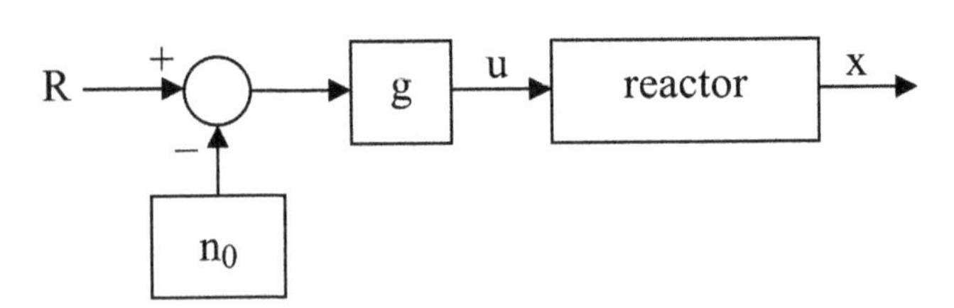

Fig. 3. Open-loop control scheme

The addition of (7) and (8) results in:

$$0 = \frac{u(\infty) + x_3(\infty)}{\Lambda} x_1(\infty) \tag{10}$$

or

$$x_3(\infty) = -u(\infty) \tag{11}$$

By substituting (11) into (9), and considering that the steady-state neutron power has attained the demanded level $x_1(\infty) = R$, then:

$$u(\infty) = \frac{\alpha K}{\gamma}(R - n_0) \tag{12}$$

From Fig. 3 and (12), the scaling factor is given as:

$$g = \frac{\alpha K}{\gamma} \tag{13}$$

4.3 Constant State Feedback

Given the efficiency of state feedback [8] for controlling linear systems, a similar scheme is designed for the nonlinear model. Considering $x_0 = (x_{1,0}\ x_{2,0}\ x_{3,0})^T$, where $x_{1,0} = n_0$, $x_{2,0} = \frac{\beta}{\lambda\Lambda} n_0$, and $x_{3,0} = 0$, the proposed control law is:

$$u = g'(R - n_0) - k_1(x_1 - n_0) - k_2\left(x_2 - \frac{\beta}{\lambda\Lambda} n_0\right) - k_3 x_3 \tag{14}$$

where g' is now the closed loop scaling factor in steady state between the desired power R, the initial power n_0, and the external reactivity $\rho_{ext}(t)$, considering state feedback. A block diagram of the resulting controller (scaling and state feedback) is shown in Fig. 4.

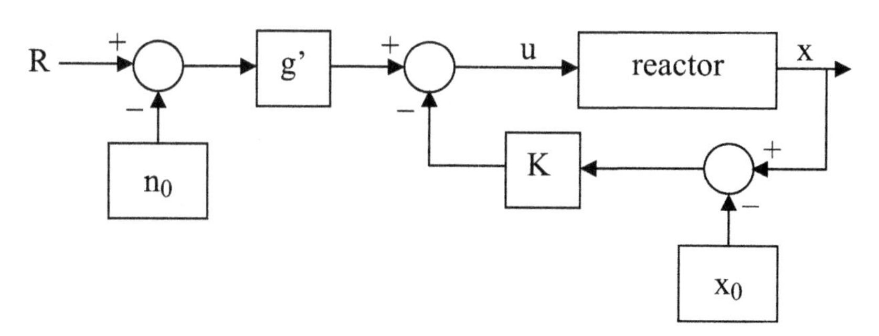

Fig. 4. Block diagram of the controller with scaling and state feedback

Steady State Closed-loop Scaling Factor

The control input (14) when $t \longrightarrow \infty$ is given by:

$$u(\infty) = g'\left(R - n_0\right) - k_1\left(x_1(\infty) - n_0\right) - k_2\left(x_2(\infty) - \frac{\beta}{\lambda\Lambda}n_0\right) - k_3 x_3(\infty) \quad (15)$$

Considering that: (a) $x_3(\infty) = -u(\infty)$ (from Eq. (11)), (b) $x_2(\infty) = \frac{\beta}{\lambda\Lambda}x_1(\infty)$ (from (8)), (c) the steady-state neutron power has attained the demanded level R, that is $x_1(\infty) = R$, and (d) $u(\infty) = \frac{\alpha K}{\gamma}\left(R - n_0\right)$ (from (9) and (11) and steady neutron power), then (15) is rewritten as:

$$(1 - k_3)\frac{\alpha K}{\gamma}\left(R - n_0\right) = \left(g' - k_1 - k_2\frac{\beta}{\lambda\Lambda}\right)\left(R - n_0\right) \quad (16)$$

from which the closed-loop scaling factor g', which now includes the effect of k_1, k_2 and k_3, is given by:

$$g' = k_1 + \frac{\beta}{\lambda\Lambda}k_2 + \frac{\alpha K}{\gamma}(1 - k_3) \quad (17)$$

Equation (15) reduces to (13) when $k_1 = 0$, $k_2 = 0$ and $k_3 = 0$.

Effect of the Independent Variation of k_1, k_2 and k_3 on the Reactor Response

This section deals with the feedback effect on the reactor's neutron power response. In order to systematically study the effect of the feedback, in particular with respect to the minimization of the overshoot, it is convenient to independently consider each gain k_1, k_2 and k_3. Thus, when different gain values are assigned to k_1, both k_2 and k_3 are kept null. Values from 1^*10^{-10} to 0.001 were assigned to k_1. Although the overshoot is practically eliminated, an observed abrupt change in power would lead to the automatic shutdown of the reactor. When values between 1^*10^{-12} and 1^*10^{-7} were assigned to k_2, oscillations and relatively large overshoots on the neutron power were observed.

Table 2. Overshoot of the neutron power response to different values of feedback gain k_3 when $R = 1\ MW$

k_3	Overshoot
0.1	7.64^*10^{6}%
0.9	7.76^*10^{5}%
0.99	2.21^*10^{4}%
0.994	1.77^*10^{3}%
0.997	1.58^*10^{1}%
0.999	7.45^*10^{-3}%

Finally, as k_3 takes on values from 0.1 to 0.9992, the overshoot is significantly reduced till practically being eliminated (see Table 2). Moreover, the ascent of power occurs smoothly as k_3 tends to a value of 0.999.

5 First Order Predictor for Feedback Gain k_3

Although a value of $k_3 = 0.992$ practically eliminates the overshoot, it does not avoid the occurrence of a reactor period scram, in particular during the first seconds of operation. An inadmissible solution would consist in the assignment of a value of 0.999999 to k_3. This value would prevent the occurrence of the scram, but it would dramatically slow down the dynamic response. For instance, departing from an initial power value, n_0, of 1 W, in 1000 s a risible increment of 0.08 W would occur in the neutron power. It is clear that the value of k_3 should change as the power increases. This raises the question of determining an adequate manner to carry out such a modification. Given the wide range of values for n_0 and R, the proposed solution consists in estimating the value of $n(t_i + h)$ from the current value $n(t_i)$ using the first-order Euler method [5]. Then, the value of k_3 that satisfies inequality (1) is computed. It can be shown [14] that a first-order expression for predicting k_3 is given by:

$$k_3 = 1 + \frac{\lambda\Lambda\frac{x_2[n]}{x_1[n]} - \beta - \frac{\Lambda}{h}\left(e^{\frac{h}{T_k}} - 1\right)}{\frac{\alpha K}{\gamma}\left(R - n_0\right) + x_3[n]} \tag{18}$$

As long as $T_k > T$, the value of k_3 given by (18) always satisfies inequality (1). Normally, $T = 3$ s. The incorporation of the predictive stage is shown in Fig. 5. During the first seconds of operation, the predictor action leads to a fast power response, independently of the values of n_0 and R and without period scram –something that is difficult or even impossible to obtain with other control approaches. However, not everything goes fine and smooth; a

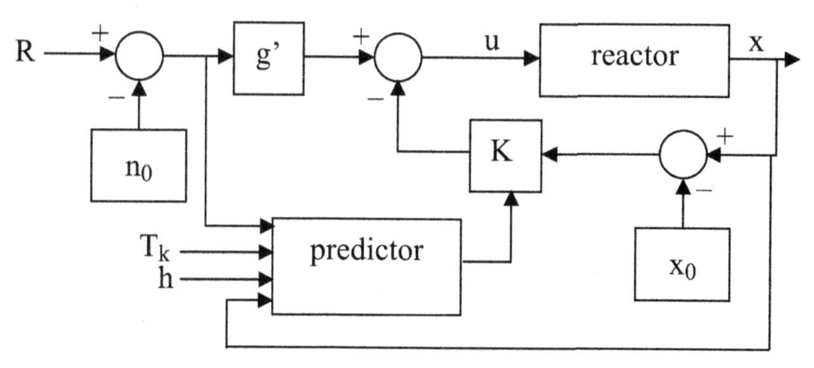

Fig. 5. Block diagram of the closed-loop system that combines scaling factor, state feedback, and prediction

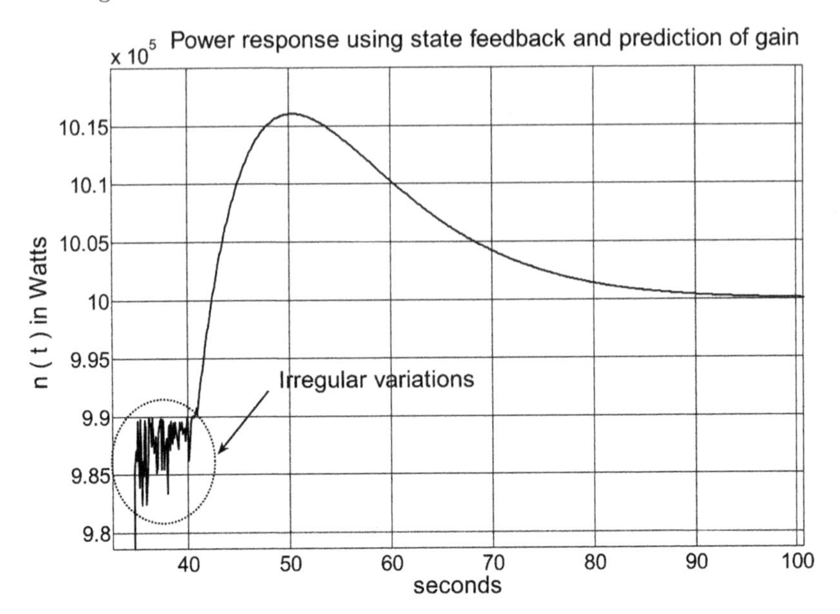

Fig. 6. Irregular behavior of power response near the set point

problem is observed when the power approaches the desired level R: some irregular variations occur. This behavior is depicted in Fig. 6.

6 Fuzzy Regulator of State Gain Predictor

In order to keep the advantages provided by the predictor but without the negative effects just mentioned, it is now necessary to regulate the predictor response, especially when the neutron power $n\ (t)$ approaches the desired level R. To accomplish this task, a Mamdani type fuzzy block is developed and incorporated into the controller [20]. Let $n\ (t) = x_1(t)$ be the input to the fuzzy stage and U the universe of discourse of $x_1(t)$ such that $U = [n_0,\ 2R]$. Eight fuzzy sets, A_1 to A_8, are defined on U. In order to maintain the adaptability of the algorithm to different initial and final power levels, the limits of U have been defined in terms of n_0 and R. The membership functions of fuzzy sets A_1 to A_8 are defined next. The membership function of the left-hand set A_1 is described as:

$$\mu_{A_1}(x_1) = \begin{cases} 1 & , \quad \mathbf{x_1 \in [n_0, 0.2\ R]} \\ \frac{-\mathbf{x_1 + 0.35\ R}}{\mathbf{0.15\ R}} & , \quad \mathbf{x_1 \in [n_0, 0.2\ R, 0.35\ R]} \\ 0 & , \quad \mathbf{U - (n_0, 0.5\ R)} \end{cases} \tag{19}$$

The fuzzy sets A_2 to A_7 all have triangular shapes and their membership functions are defined by:

$$\mu_{A_i}(x_1) = \begin{cases} \frac{x_1 - p_i R}{0.1\ R} & , \quad x_1 \in [p_i R, (p_i + 0.1)R] \\ \frac{-x_1 + q_i R}{0.1\ R} & , \quad x_1 \in ((q_i - 0.1)R, q_i R] \\ 0 & , \quad x_1 \in U - (p_i R, q_i R) \end{cases} \tag{20}$$

where $i = 2, \ldots, 7$. The values of the parameters p_i and q_i are given in Table 3.

Finally, the membership function of the right-hand set A_8 is described as:

$$\mu_{A_8}(x_1) = \begin{cases} \frac{x_1 - 0.85\ R}{0.1\ R} & , \quad x_1 \in [0.85\ R, 0.95\ R] \\ 1 & , \quad x_1 \in (0.95\ R, 2.0\ R] \\ 0 & , \quad x_1 \in U - (0.85\ R, 2.0\ R) \end{cases} \tag{21}$$

The distribution of the fuzzy sets A_1 to A_8 on U is shown in Fig. 7.

Let T_k be the output of the fuzzy regulator stage and V its universe of discourse such that $V = [3.1s,\ 200s]$. On V, the following eight singleton fuzzy sets are defined: $T_1 = 1/3.1$, $T_2 = 1/3.8$, $T_3 = 1/4.6$, $T_4 = 1/8$, $T_5 = 1/14$, $T_6 = 1/30$, $T_7 = 1/70$ and $T_8 = 1/150$, where the values 3.1, 3.8, 4.6, 8, 14, 30, 70, and 150 on V are given in seconds. Then, eight fuzzy rules are defined, where each rule associates each fuzzy set A_i to its corresponding singleton T_i as follows:

$$If\ x_1\ is\ A_i\ then\ T_k\ is\ T_i \tag{22}$$

The crisp output (singleton defuzzification) is simply the weighted average of $T_1, T_2, \ldots, T_8$, that is:

$$T_k = \sum_{i=1}^{8} T_i \mu_{Ai}(x_1) \Big/ \sum_{i=1}^{8} \mu_{Ai}(x_1) \tag{23}$$

The idea behind these rules is the following: The value of the period parameter $\boldsymbol{T}_k$ to be used by the predictor is determined by the neutron power level $\boldsymbol{x}_1$ in every control cycle. Thus, at low power levels the period $\boldsymbol{T}_k$ to be used

Table 3. Values of the parameters that define fuzzy sets A_2 to A_7

i	p_i	q_i
2	0.25	0.45
3	0.35	0.55
4	0.45	0.65
5	0.55	0.75
6	0.65	0.85
7	0.75	0.95

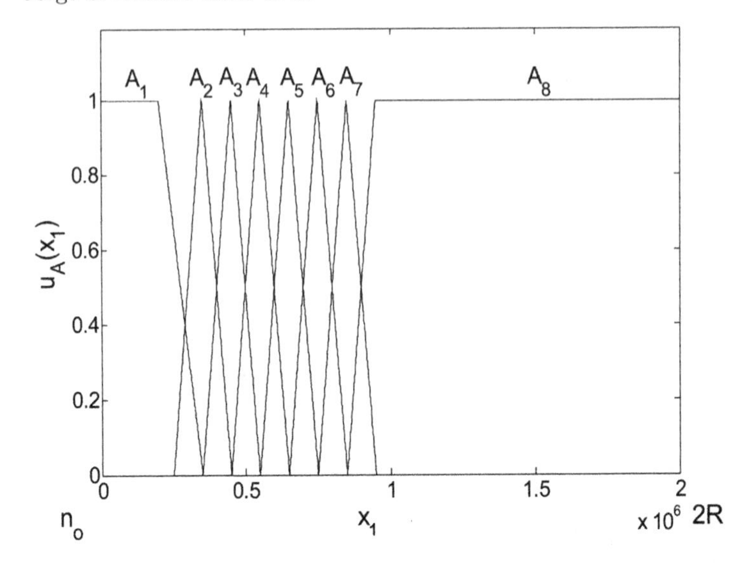

Fig. 7. Membership functions of the fuzzy sets associated to $x_1(t)$

is small. The resulting gain $\boldsymbol{k}_3$ estimated by the predictor will cause a rapid increase of the reactor power. On the contrary, at high power levels the period $\boldsymbol{T}_k$ to be used is large. The value of $\boldsymbol{k}_3$ estimated by the predictor will cause the reactor power to increase slowly.

In addition, the predictor is activated when $T_k < 149\ s$; otherwise, the new value of k_3 will be equal to the value obtained in the previous iteration. The complete control scheme is shown in Fig. 8.

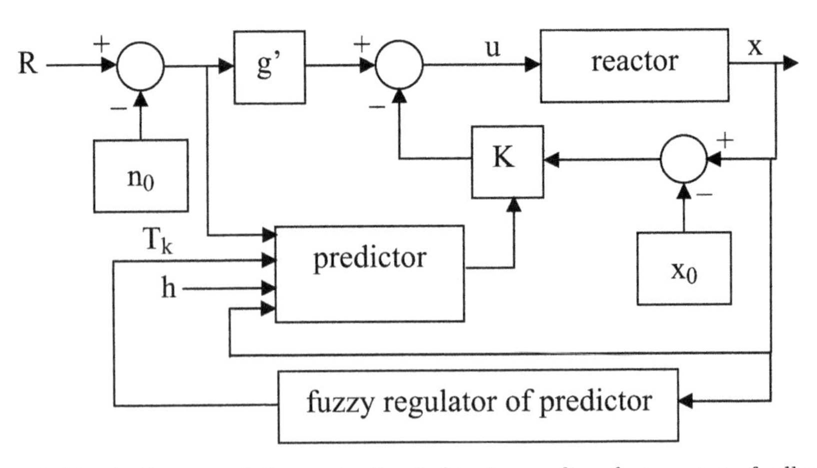

Fig. 8. Block diagram of the controller fashioning scaling factor, state feedback, prediction, and fuzzy regulator

7 Experimental Results

For the case $n_0 = 50\,W$ and $R = 1\,MW$, the proposed controller attains the desired power level in about 40 s, maintaining the instantaneous period above the 3-second safety limit at all times, thus preventing the reactor period scram. Simulation results of the closed-loop system for different values of R starting at the same initial power $n_0 = 1\,W$ are shown in Fig. 9.

Conversely, the response curves of the neutron power for different values of the initial reactor power n_0 and one desired power level $R = 1\,MW$ are shown in Fig. 10.

Inequality (1) is satisfied in all cases. The effectiveness of the control scheme was tested for the extreme case (although unreal): $n_0 = 10\,\mu W$ and $R = 1\,TW$. Once again the pattern obtained was the same: the desired power was attained in short time, the instantaneous period was maintained above the 3-second limit, no irregular oscillations occurred near the desired power level, and the response did not present overshoot.

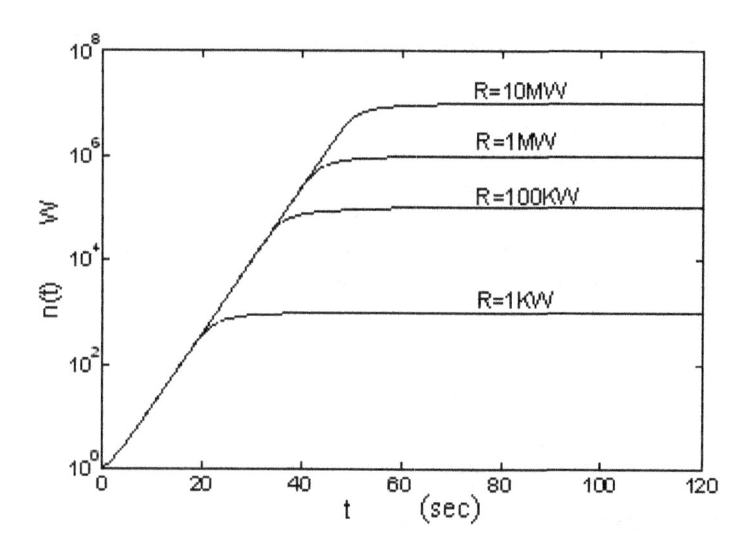

Fig. 9. Closed-loop system response for different values of R starting at $n_0 = 1\,W$

8 Conclusions

The proposed control scheme attains any desired R in the range from 100 W to 2.988 MW, starting at any n_0 in the range from 1 W to 50 W. The time required has always been less compared to other schemes, and no overshoot is observed in the neutron power response. It is worth-mentioning the homogeneous characteristic of the neutron power response. This characteristic can

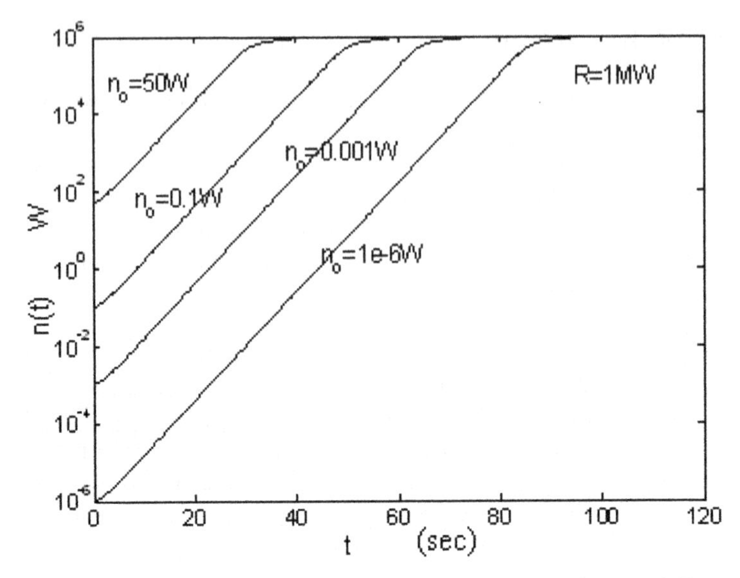

Fig. 10. Closed-loop system response for different values of n_0 and $R = 1\ MW$

be defined as a shape pattern of the time response that is properly scaled in time and magnitude depending on the initial and final conditions. Moreover, the resulting control algorithm is very compact and fast, which makes it suitable for real-time applications. Certain degree of robustness is provided by the fuzzy logic based stage. The algorithm can be easily codified in assembler language or any high-level computing languages, and allows modifications, in a simple manner, of the reactor period lower limit.

References

1. Benítez-Read J.S., López-Callejas R., Pacheco-Sotelo J.O., Longoria-Gándara L.C. (2002), Exact aggregation and defuzzification applied to a neutron fuzzy controller. *Soft Computing, Multimedia and Image Processing* 13, 183–188.
2. Bernard J.A. (1988), Evaluation of period-generated control laws for the time-optimal control of reactor power. *IEEE Trans. on Nuclear Science*, February 35 (1), 888–893.
3. Bernard J.A., Lanning D.D., Ray A. (1984), Digital control of power transients in a nuclear reactor. *IEEE Trans. on Nuclear Science*, February 31 (1), 701–705.
4. Bubak M. and Moscinski J. (1983), A fuzzy-logic approach to HTR nuclear power plant model control. *Annals of Nuclear Energy.* UK, 10 (9), 467–471.
5. Chapra S. and Canale R. (1988), "Numerical Methods for Engineers". Mc GrawHill.
6. Cottingham W.N. and Greenwood D.A. (2001), An Introduction to Nuclear Physics, 2nd Ed. Cambridge University Press.
7. DeGroot M.N. (1968), TRIGA Mark III Reactor: Instrumentation Maintenance Handbook, Document No. GA-8585. Gulf General Atomic, Inc.

8. Dorf R.C. and Bishop R.H. (1995), Modern Control Systems, 7th Ed. Addison-Wesley.
9. Edwards R.M., Garcia H.E., Turso J.A., Chavez C.M., Abdennour A.B., Weng C.K., Ku C.C., Ray A., Lee K.Y. (1992), Advanced Control Research at the Pennsylvania State University. Advanced digital computers, controls, and automation technologies for power plants: Proceedings, Electric Power Research Inst., Palo Alto, CA, USA, August, pp 26.1–26.10.
10. Gerald C.F. and Wheatley P.O. (1994), Applied Numerical Analysis, 5th Ed. Addison-Wesley.
11. Hetrick, D.L. (1971), Dynamics of Nuclear Reactors. The University of Chicago Press.
12. Marquez H.J. (2003), Nonlinear Control Systems: Analysis and Design. Wiley-Interscience. 2003.
13. Moon B.S. (1993), Fuzzy logic controllers for the nuclear power plants: simulation experiences. Hungarian Korean Symposium on Nuclear Energy, pp 237–250, Balatonfuered, Hungary, 30 March – 2 April.
14. Pérez-Cruz, J.H. (2004), Design of an adaptive feedback controller for power regulation in a TRIGA Mark III reactor (Spanish), M.Sc. Thesis; Instituto Tecnológico de Toluca.
15. Ruan D. (2001), Implementation of adaptive fuzzy control for a real time control demo-model. *Real-Time Systems* 21, 219–239.
16. Ruan D. (2003), Initial experiments on fuzzy control for nuclear reactor operations at the BR1, *Nuclear Technology* 143 (2), 227–240.
17. Sarabia-Guajardo O. (2003), *Nuclear Contact*, Official Newsletter of the National Institute for Nuclear Research of Mexico, No. 33, December.
18. Vélez-Díaz D. and Benítez-Read J.S. (1995), Study of the Behaviour of a TRIGA Reactor Point Kinetic Model, Based on Simulations. (In Spanish), Instituto Nacional de Investigaciones Nucleares, México, Technical report IT.ET.A-9514.
19. Viais-Juárez, J. (1994), Computation of the fundamental parameters of the Mexican TRIGA Mark III reactor to study its dynamic behavior (Spanish), B.Sc. Thesis; Universidad Nacional Autónoma de México.
20. Wang, L. (1996), A Course in Fuzzy Systems and Control. Prentice Hall.
21. Weng C.K., Edwards R.M., Ray A. (1994), Robust wide-range control of nuclear reactors by using the feedforward-feedback concept. *Nuclear Science and Engineering*, July 117 (3), 177–185.

The Fusion of Genetic Algorithms and Fuzzy Classification for Transient Identification

Enrico Zio and Piero Baraldi

Abstract. The basis for transient identification in complex engineering systems and processes is that different faults and anomalies lead to different patterns of evolution of the monitored variables. Two main steps need to be carried out in order to effectively perform the identification: i) the selection of the features carrying information relevant for the identification; ii) the classification of the dynamic patterns into the different transient types. This chapter illustrates: i) the combination of genetic algorithms and Fuzzy K-Nearest Neighbors classification for the feature selection; ii) the combination of a supervised, evolutionary algorithm and fuzzy-possibilistic clustering for the classification. An example of application of the proposed approach to pattern classification is given with reference to the classification of simulated nuclear transients in the feedwater system of a Boiling Water Reactor.

1 Introduction

In this Chapter, the issue of fault diagnosis in safety-critical components and systems, such as the nuclear ones, is framed as a pattern classification problem. The basis for the classification is that different faults and anomalies lead to different patterns of evolution of the involved process variables.

An approach to transient classification based on genetic algorithms [1] and possibilistic clustering [2,3] is adopted. The classification is performed by the following two tasks:

1. the selection of the features relevant for the classification;
2. the classification of the dynamic patterns into different transient types.

The first step of feature selection is particularly important since irrelevant or noisy features unnecessarily increase the complexity of the diagnostic problem and can degrade modeling performance [4]. Moreover, in modern industrial plants hundreds of parameters are monitored for operation and safety reasons so that expert judgment alone cannot effectively drive the feature selection.

The technique proposed to carry out this task combines a genetic algorithm search [1] with a Fuzzy K-Nearest Neighbors (FKNN) classification algorithm [5]. The performance achieved by the latter is used as criterion for comparing the different feature subsets searched by the former.

The second task of dynamic pattern classification is tackled within a possibilistic clustering approach. Given the complexity and variety of cluster shapes and dimensions which can be expected in the transient classification, an approach based on different Mahalanobis metrics for each cluster is embraced [6, 7]. The a priori known information regarding the true classes which the available patterns belong to is exploited in the classification construction phase to select, by means of a supervised evolutionary algorithm, the optimal Mahalanobis metrics. To avoid misclassifications of unknown transients during operation, the incoming transient patterns are processed by a possibilistic algorithm which is able to filter out unknown plant conditions.

An example of application of the approach is given with respect to the classification of simulated transients in the feedwater system of a nuclear Boiling Water Reactor.

2 Feature Selection

The diagnostic task may be viewed as a problem of partitioning of objects (the measured data patterns) into classes (the faults). From a mathematical point of view, a classifier is a mapping function $\Phi(\cdot)$ which assigns an object $\mathbf{x}$ in an h-dimensional domain $\Omega \subset \Re^h$ to a given class i. If one knew the exact expression of $\Phi(\cdot)$, the question of which features of $\mathbf{x}$ to use would not be of interest. In fact, in such situation adding features does not decrease the accuracy of the classifier, and hence restricting to a subset of features is never advised. However, as it is often the case in engineering, it is not possible to determine the exact analytical expression of the function $\Phi(\cdot)$ due to the complexity of the systems under analysis. Hence, one resorts to empirical classification techniques in which the classifier is built through a process based on a set of classification example pairs $\{(\mathbf{x}, i)\}$, each one constituted by a pattern $\mathbf{x}$ in the feature space labelled by the corresponding class i.

In practice, the number of measured features is quite large. At least various reasons call for a reduction of this number for use in the classification model. First of all, irrelevant, non informative features result in a classification model which is not robust [8, 9]. Second, when the model handles many features, a large number of observation data are required to properly span the high-dimensional feature space for accurate multivariable interpolation [8, 10]. Third, by eliminating unimportant features the cost and time of collecting the data and developing the classifier can be reduced [8, 10]. Finally, reducing the number of selected features permits to achieve more transparent and easily interpretable models [4].

Given a labelled dataset, the objective of feature selection is that of finding a subset of the features such that the classifier built on these features classifies the available data with the highest possible accuracy [4].

Notice that different classification algorithms may require different feature subsets to achieve highest accuracy, so that a universal optimal feature subset does not exist.

Finally, other requirements are often added to the feature selection objective of maximum classification accuracy, e.g. the reduction of the number of features for reducing the computational and data storage burdens.

2.1 An Overview on Feature Selection Techniques

Feature selection involves conducting a search for an optimal feature subset in the space of possible features. The inclusion or not of a feature in the subset can be encoded in terms of a binary variable which takes value 1 or 0, respectively. For n features, the size of the binary vector search space is 2^n. Thus, an exhaustive search is impractical unless n is small.

Each feature subset selected during the search must be evaluated with respect to the given objective functions, e.g. classification performance and number of features.

Several methods of feature selection have been proposed. They are usually classified into two categories: filter and wrapper methods [4].
In filter methods, the feature selector algorithm is independent of the specific algorithm used in the classification and it is used as a filter to discard irrelevant features, a priori of the construction of the classifier. A numerical evaluation function is used to compare the feature subsets proposed by the search algorithm. The subset with the highest value of the evaluation function is the final feature set which feeds the algorithm for the classification.

The evaluation functions are usually of two types:

- those which apply distance metrics in the feature space to measure the separability between classes, e.g. the classifiability evaluation function [11].
- those based on the rationale that a good feature subset contains features highly correlated with the class label but uncorrelated with the other features of the subset, e.g. those based on the concept of mutual information [12].

Contrary to filter methods, in wrapper methods the feature selector behaves as a "wrapper" around the specific algorithm used to construct the classifier whose performance is used to compare the different feature subsets [4].

The filter approach is generally computationally more efficient than the wrapper one because for each feature subset of trial, the computation of an evaluation function is less time consuming than the development of a complete classification model. Indeed, a high number of feature subsets are tested during the search for the optimal and the time consumption of a wrapper approach depends mainly from the time necessary for the development of the classifier and the subsequent classification of the patterns to test its performance. Hence, for many practical applications the wrapper approach is feasible only if the classifier is a fast-computing algorithm, e.g., the K-Nearest Neighbor (KNN) [13] algorithm or its fuzzy extension (FKNN) [5]. On the other hand, wrapper approaches are more performing than the filter ones since the former ensure the selection of the features more suitable for the specific classification algorithm used, whereas the latter totally ignore the effects of the selected feature subspace on the performance of the classifier that will actually be used.

With respect to the search algorithms, three approaches are commonly adopted: complete, heuristic and probabilistic [4].

In the complete approach, the properties of a pre-defined evaluation function are used to prune the feature space to a manageable size, thus avoiding that the complete search is also exhaustive [14]. Only some evaluation functions give rise to a search that guarantees the optimum feature subset selection without being exhaustive.

The heuristic approach does not guarantee that the best feature subset is achieved, but is less time consuming than the complete one and may be employed in combination with any evaluation function [11]. At present, the most employed heuristic methods are greedy search strategies such as the sequential forward selection (SFS) or the sequential backward elimination (SBE) "hill climbing" methods, which iteratively add or subtract features and at each iteration the evaluation function is evaluated. The forward selection refers to a search that begins with no features and at each step a feature is added to the subspace; on the contrary, the backward elimination refers to a search that begins with the n-dimensional feature set and at each step a feature is removed. At each step, the choice of which feature to add or remove is driven by its effect on the classifiability function in the direction of climbing towards its maximum value. The hill-climbing search is usually stopped when adding or removing new features does not increase the value of the classifiability function or when the number of features has reached a predefined threshold.

The hill climbing methods suffer from the so called "nesting effect": if the features added cannot be removed, a local minimum of the evaluation function may be found. To reduce this effect, it is possible to use the so called plus-l-take-away-r method (PTA) [4]. In this method, after l steps of the forward selection, r steps of the backward elimination are applied so as to allow escaping from local minima. Still, there is no guarantee of obtaining the absolute optimum.

The probabilistic approach is based on population-based methauristics guided by fittest solutions, such as the genetic algorithms, presented in this paper, or on methods like simulated annealing and tabu search algorithms [15]. In the next Section, a genetic algorithms-based method is introduced.

2.2 GA-based Feature Selection for Transient Classification

In this Chapter, the wrapper scheme is adopted and a Multi-Objective Genetic Algorithm (MOGA) for searching the optimal feature subset upon which to perform the diagnostics of nuclear transients is presented.

The total number of n-dimensional pre-labelled available data are partitioned into a set (hereafter denoted by A) to be used for the feature selection task and a separate set (hereafter denoted by A') to be used for validating the performance of the classifier resting upon the optimal feature subset selected.

The structure of the chromosome is straightforward [16]. For n features, the size of the chromosome is n bits and each bit of the chromosome is associated with a feature (Fig. 1): if the r-th bit equals 1, then the r-th feature is included in the subset and viceversa if the bit is 0. Thus, the number m of features in the subset is the total number of 1's in the chromosome.

In this view, the l-th chromosome represents a binary transformation vector V^l of dimension n, which operates on the n-th dimensional patterns of set A producing a modified set of m-dimensional patterns $B = V^l(A)$ (Fig. 2). Note that contrary to other GA applications, in this case of feature selection the binary chromosome does not encode real-valued control factors: the information regarding the features presence or absence in the optimal subset for classification is included in the bits themselves so that no decoding is necessary.

Two objective functions (fitness) are used for evaluating and comparing the feature subsets during the search: the fraction of patterns correctly classified (recognition rate) by a Fuzzy K-Nearest Neighbors classifier (to be maximized) and the number m of features forming the subsets (to be minimized).

With respect to these two objective functions, the feature subsets are compared in terms of dominance and the optimization leads to the identification of the so called Pareto optimal set or front of non-dominated feature subsets [1,17].

The efficiency of the search depends on the ability to maintain genetic diversity through the generations so as to arrive at a population of individuals which uniformly represent the real nondominated solutions of the Pareto set [1]. This can be achieved by resorting to niching techniques, e.g. sharing [1, 18], which apply a "controlled niched pressure" to spread out the

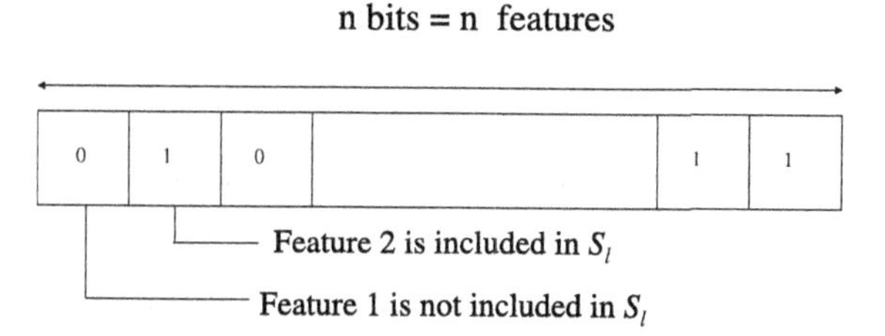

Fig. 1. n-dimensional binary chromosome

population in the search space so that convergence is shared on different niches of the Pareto front which is thus evenly covered.

For the FKNN evaluation of the classification performance fitness function associated to a given feature subset S_l, with corresponding binary transformation vector V_l, the labelled patterns constituting the transformed dataset $B = V_l(A)$ are randomly subdivided into a set B_1 containing 75% of the data which are used for the classifier construction and a tuning set B_2 of 25% of the data, which are used to compute the performance of the classifier in terms of its accuracy. By trial and error, a number $K = 5$ of neighbors has been found to produce good classification results. The obtained fuzzy partition of the tuning dataset B_2 $\{\mu_i(\mathbf{x}_k)\}$, where $0 \leq \mu_i(\mathbf{x}_k) \leq 1$ is the membership function of pattern $\mathbf{x}_k$ to class i, is converted into a hard partition assigning each pattern to the class with highest membership value.

The subdivision of the available patterns in training and tuning sets (B_1 and B_2, respectively) is randomly repeated 10 times (10 cross-validation) and for each tuning set B_2 the accuracy of the FKNN classifier operating on the proposed subset of features S_l is evaluated in terms of the recognition rate

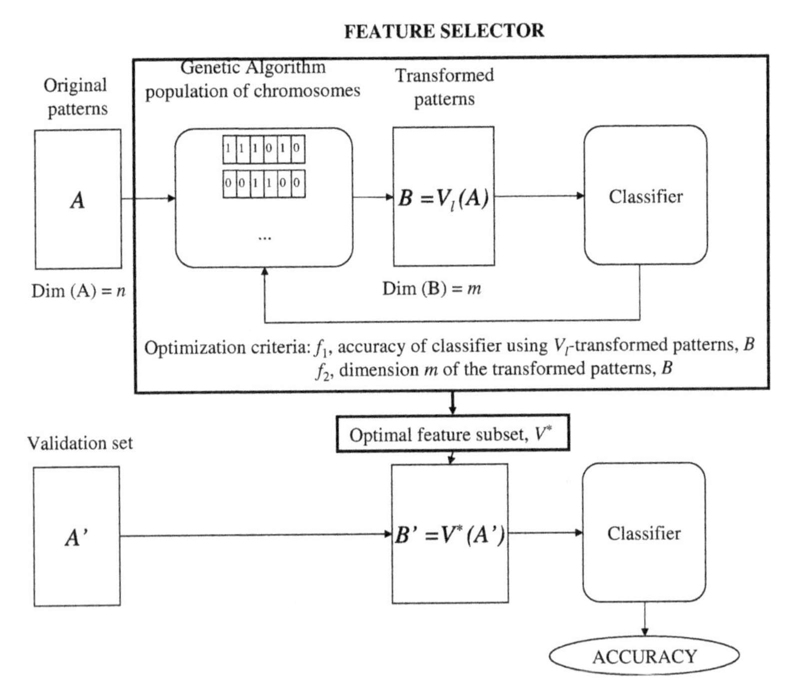

Fig. 2. GA-based feature selection using classification accuracy and number m of selected features as optimization criteria. Each binary chromosome from the GA population transforms the original patterns to a reduced feature space and passes them to the classifier. The objective function values of the chromosome are the classification accuracy attained on the transformed patterns and their dimension m

(the fraction of tuning patterns of B_2 correctly classified) [4]: then, the mean recognition rate is calculated and sent back to the GA as the fitness value of the transformation chromosome V_l used to produce the transformed set of patterns B.

At convergence, a family of non-dominated chromosomes (feature subsets) with different trade-offs of classification performance (FKNN recognition rate) and complexity (number of features) are obtained. Based on these results, an informed choice can be made on the features to be actually monitored for the diagnostic task considering also practical issues of costs, ease of data acquisition and data interpretability. Eventually, the analyst must select the preferred solution according to some subjective preference values.

Finally, the pre-constructed validation set A', separate from the training and tuning sets (B_1 and B_2) used for the feature selection task, is processed through the optimal transformation vector V^* and the corresponding classifier to verify the classification accuracy on a set B' of new patterns (never used during the feature selection process) specified in terms of the optimal selected feature subset encoded by the transformation vector V^* (Fig. 2). This validation procedure is of paramount importance for safety applications in critical technologies such as the nuclear one considered in this work.

3 Transient Identification

In the previous Section, a wrapper scheme for selecting the features relevant for the classification task has been devised by combining a MOGA search with a FKNN classification. The former effectively searches the large feature space whereas the latter offers the needed fast computing characteristics.

For increased accuracy and reliability in the classification, more refined soft computing techniques may be adopted [19]. In the following, an approach to transient identification based on pattern classification by possibilistic clustering [3] is presented as an extension of fuzzy clustering.

3.1 Fuzzy and Possibilistic Clustering

Fuzzy clustering algorithms have been widely studied and applied in various substantive areas such as taxonomy, medicine, geology, business, engineering, image processing and others. A general classification of these algorithms is offered in [20] in terms of three categories: fuzzy clustering based on fuzzy relations, fuzzy clustering based on the minimization of an objective function and the class of nonparametric classifiers based on the generalized fuzzy K-nearest neighbors rule. The interested reader is referred to [20] for a detailed discussion of the three categories and an extensive literature review of works in the field.

The approach illustrated in the following paragraphs falls under the second of the above categories. The fuzzy clustering is performed by minimizing an objective function under the following constraints on the membership functions μ_{ik} of pattern $\mathbf{x}_k$ in cluster i [20, 21]:

$$0 \leq \mu_{ik} \leq 1; \qquad i = 1, ..., c; \quad k = 1, ..., N \tag{1}$$

$$\sum_{i=1}^{c} \mu_{ik} = 1; \qquad k = 1, ..., N \tag{2}$$

In particular, the 'probabilistic' constraint (1), that the memberships of a given pattern must sum up to 1, is a generalization of the condition which ensures that in a 'hard' (crisp) partition a pattern is a member of one class only and avoids the trivial solution of all memberships equal to 0. As a result of this constraint, the membership of a pattern to a cluster depends on the memberships to all other clusters; geometrically speaking, it depends on where the pattern is located with respect to not only that cluster but also to the others. Hence, in the framework of fuzzy clustering the membership functions take the meaning of degrees of sharing, i.e. they measure how much a pattern belongs to a cluster relatively to the others.
Under these conditions, two major drawbacks arise [3]:

1. The constrained memberships cannot distinguish between 'equal evidence' and 'ignorance' or, in other words, between 'equally likely' and 'unknown' membership to a cluster.
2. Since most distance functions used in fuzzy clustering are geometric in nature, noisy patterns, which typically lie far from the clusters, can drastically influence the estimates of the clusters prototypes and, hence, the final partition and the resulting classification.

In this situation, an 'unknown', atypical pattern not belonging to any cluster would still belong more to one cluster than to the others, relatively speaking, and thus it may receive high membership values to some clusters even if it lies far from all clusters in the feature space.

On the contrary, in the diagnostic practice it is required that unknown, atypical patterns be recognized as such, i.e. bear low membership to all clusters. In this respect, thus, the 'conservation of total membership' constraint (2) is too restrictive since it gives rise to relative membership values, dependent on the number of clusters.

To overcome the above limitations, the clustering problem can be recast into the framework of possibility theory [3, 22, 23]. In this interpretation, the membership function μ_{ik} represents the degree of compatibility of the pattern $\mathbf{x}_k$ with the prototypical member $\mathbf{v}_i$ of cluster i. If the classes represented by the clusters are thought of as a set of fuzzy sets defined over the Universe of Discourse (UOD), then there should be no constraint on the sum of the memberships. The only constraint is that the membership values do represent degrees of compatibility, or possibility, i.e., they must lie in [0,1].

This is achieved by substituting the fuzzy clustering constraints (1-2) with the following [2]:

$$0 \leq \mu_{ik} \leq 1; \qquad i = 1, ..., c; \quad k = 1, ..., N \tag{3}$$

$$\max_{i} \mu_{ik} > 0 \qquad k = 1, ..., N \tag{4}$$

where constraint (4) simply ensures that the set of fuzzy clusters covers the entire UOD. A possibilistic partition derived under these constraints defines a set of distinct, uncoupled possibilistic distributions (and the corresponding fuzzy subsets) over the UOD [2].

3.2 The Supervised Evolutionary Possibilistic Clustering Algorithm for Classification

In this Section, a supervised evolutionary possibilistic clustering algorithm is developed to perform the diagnostic identification of transients.

The traditional, unsupervised possibilistic algorithm based on a Euclidean metric to measure compatibility leads to spherical clusters that rarely are adequate to represent the data partition in practice [7]. A significant improvement in classification performance is achieved by considering a different Mahalanobis metric for each cluster, thus obtaining different ellipsoidal shapes and orientations of the clusters that more adequately fit the a priori known data partition [6, 7].

The information on the membership of the available patterns $\mathbf{x}_k, k = 1, ..., N$, to the c a priori known classes, can be used to supervise the algorithm for finding the optimal Mahalanobis metrics such as to achieve geometric clusters as close as possible to the a priori known physical classes. Correspondingly, the possibilistic clustering algorithm is said to be constructed through an iterative procedure of 'training' based on a set of available patterns, pre-labeled with their possibilistic memberships to the a priori classes. The procedure for the optimization of the metrics is carried out via an evolutionary procedure, presented in the literature within a supervised fuzzy clustering scheme [7] and further extended to diagnostic applications [6]. Here, the procedure is employed within the possibilistic clustering scheme.

To this purpose, the distance between the set Γ_i^t (t = true) of memberships of the N available patterns to the a priori known class i and the corresponding set Γ_i of the possibilistic memberships to cluster $i = 1, ..., c$, is computed by:

$$D(\Gamma_i^t, \Gamma_i) = \sum_{k=1}^{N} \frac{|\mu_{ik}^t - \mu_{ik}|}{N} \tag{5}$$

where $0 \leq \mu_{ik}^t \leq 1$ is the a priori known (possibilistic) membership of the k-th pattern to the i-th physical class and $0 \leq \mu_{ik} \leq 1$ is the possibilistic membership to the corresponding geometric cluster in the feature space.

The target of the supervised optimization is the minimization of the distance $D(\Gamma^t, \Gamma^*)$ between the a priori known physical class partition $\Gamma^t \equiv (\Gamma_1^t, \Gamma_2^t, ..., \Gamma_c^t)$ and the obtained geometric possibilistic cluster partition $\Gamma^* \equiv (\Gamma_1^*, \Gamma_2^*, ..., \Gamma_c^*)$:

$$D(\Gamma_i^t, \Gamma_i^*) = \sum_{i=1}^{c} \frac{D(\Gamma_i^t, \Gamma_i^*)}{c} = \sum_{i=1}^{c} \sum_{k=1}^{N} \frac{|\mu_{ik}^t - \mu_{ik}^*|}{N \cdot c} \tag{6}$$

The optimal membership function, $\mu_{ik}^*, i = 1, 2, ..., c, k = 1, 2, ..., N$ can be computed accordingly to the possibilistic clustering algorithm [2] in which the distance $s_{ik} = s_i(\mathbf{x}_k, \mathbf{v}_i^*)$ between the pattern $\mathbf{x}_k$ and the optimal cluster center $\mathbf{v}_i^*$ is computed by:

$$s_i(\mathbf{x}_k, \mathbf{v}_i^*) = (\mathbf{x}_k - \mathbf{v}_i^*)^T \underline{\underline{M_i}} (\mathbf{x}_k - \mathbf{v}_i^*) \tag{7}$$

$\underline{\underline{M_i}}$ being the metric for the cluster i proposed by the evolutionary supervised procedure and T denoting the transpose operator. The overall iterative training scheme can be summarized as follows (Fig. 3):

1. At the first iteration ($\tau = 1$), initialize the metrics of all the c clusters to the Euclidean metrics, i.e. $\underline{\underline{M_i}} = \underline{\underline{I}}, i = 1, 2, ..., c$ where $\underline{\underline{I}}$ is the identity matrix.
2. At the generic iteration step τ, run the possibilistic clustering algorithm [2] to partition the N training data into c clusters, based on the current metrics $\underline{\underline{M_i}}(\tau)$.
3. Compute the distance $D(\Gamma^t, \Gamma^*(\tau))$ between the a priori known physical classes and the geometric possibilistic clusters obtained in step 2. At the first iteration ($\tau = 1$) initialize the best distance D^+ to $D(\Gamma^t, \Gamma^*(1))$, D_i^+ to $D(\Gamma_i^t, \Gamma_i^*(1))$ and the best metrics $\underline{\underline{M_i^+}}$ to $\underline{\underline{M_i}}(\tau)$ and go to step 5.

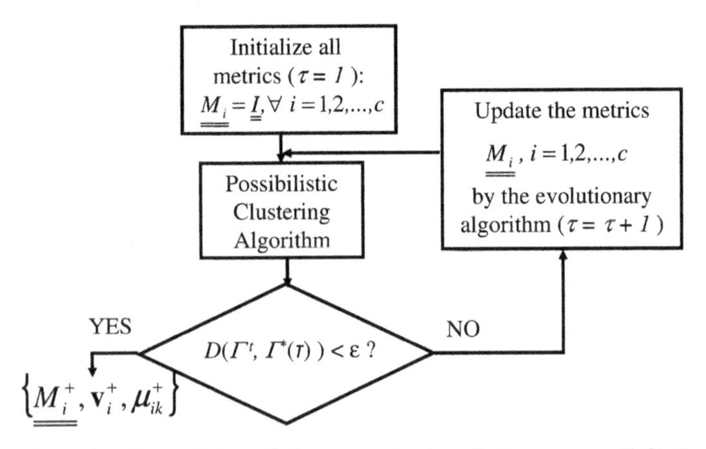

Fig. 3. Scheme for the training of the supervised evolutionary possibilistic classifier

4. Increment τ by 1. Update each matrix $\underline{\underline{M_i^+}}$ by exploiting its unique decomposition into Cholesky factors [7], $\underline{\underline{M_i^+}} = \underline{\underline{G_i^+}}^T \underline{\underline{G_i^+}}$, where $\underline{\underline{G_i^+}}$ is a lower triangular matrix with positive entries on the main diagonal. More precisely, at iteration τ, the entries $g^i_{l_1,l_2}(\tau)$ of the Cholesky factor $\underline{\underline{G_i}}(\tau)$ are updated as follows:

$$g^i_{l_1,l_2}(\tau) = g^{i+}_{l_1,l_2} + N^i_{l_1,l_2}(0, \delta_i^+) \quad \text{if } l_1 \leq l_2 \tag{8}$$

$$g^i_{l_1,l_2}(\tau) = \max\left(10^{-5}, g^{i+}_{l_1,l_2} + N^i_{l_1,l_2}(0, \delta_i^+)\right) \quad \text{if } l_1 = l_2 \tag{9}$$

where $\delta_i^+ = \alpha D_i^+$, α is a parameter that controls the size of the random step of modification of the Cholesky factor entries $g^{i+}_{l_1,l_2}$, $N^i_{l_1,l_2}(0, \delta)$ denotes a Gaussian noise with mean 0 and standard deviation δ, and eq.(9) ensures that all entries in the main diagonal of the matrices $\underline{\underline{G_i}}(\tau)$ are positive numbers and so $\underline{\underline{M_i}}(\tau)$ are definite positive distance matrices. Notice that the elements of the i-th Mahalanobis matrix are updated proportionally to the distance between the i-th a priori known class and the i-th found cluster. In this way, only the matrices of those clusters which are not satisfactory for the classification purpose are modified.
5. Return to step 2.

The overall structure of the algorithm is depicted in Fig. 3. The closed external loop iterates until an acceptable clustering of the training data is found, i.e. until the obtained possibilistic partition $\Gamma^*(\tau)$ is "close" to the a priori known partition Γ^t, where "close" means that the distance $D(\Gamma^t, \Gamma^*(\tau))$ is smaller than a defined threshold ϵ.

At convergence, the supervised evolutionary possibilistic clustering algorithm provides the c optimal metrics $\underline{\underline{M_i^+}}$ with respect to the classification task, the possibilistic cluster centers $\mathbf{v}_i^+$ and the possibilistic membership values μ_{ik}^+ of the patterns $\mathbf{x}_k$, $k = 1, ..., N$ to the clusters $i = 1, 2, ..., c$.

When fed with a new pattern $\mathbf{x}$, the classification algorithm provides the values of the membership functions $\mu_i^*(\mathbf{x}), i = 1, 2, ..., c$ to the possibilistic clusters. These values give the degree of compatibility or "typicality" of $\mathbf{x}$ to the c clusters. In practice, three situations may arise (Fig. 4):

1. $\mathbf{x}$ does not belong to any cluster with enough membership, i.e. all the membership values $\mu_i^*(\mathbf{x})$ are below a given threshold ϵ_c (degree of confidence): this means that $\mathbf{x}$ is an atypical pattern with respect to the labelled data set used for the construction of the classification algorithm;
2. at least two membership values are above the threshold ϵ_c: $\mathbf{x}$ is thus ambiguous. In this case, the ambiguity must be regarded as equal "evidence", i.e. the pattern is typical of more than one class and thus cannot be assigned to a class with enough confidence. This situation occurs if $\mathbf{x}$ is at the boundary between two classes;
3. $\mathbf{x}$ belongs only to a cluster with a membership value greater than the threshold ϵ_c: in this case, it is assigned to the corresponding class.

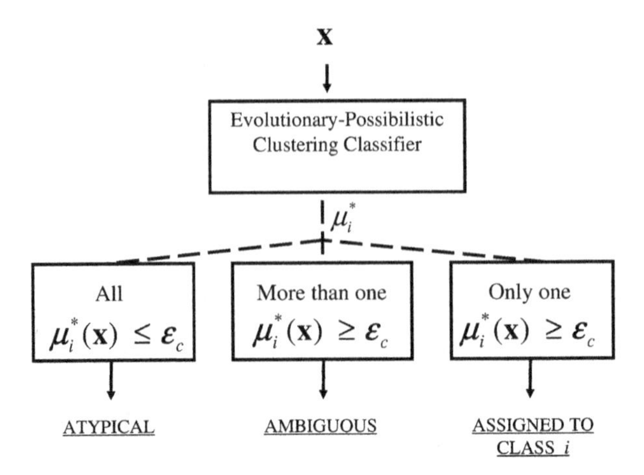

Fig. 4. Classification of pattern $\mathbf{x}$

4 Classification of Transients in the Feedwater System of a Boiling Water Reactor

As an example of application, the identification of a predefined set of faults in a Boiling Water Reactor (BWR) is considered. Transients corresponding to the faults have been simulated by the HAMBO simulator of the Forsmark 3 BWR plant in Sweden [24]. Figure 5 shows a sketch of the system [24]. The considered faults occur in the section of the feedwater system where the feedwater is preheated from 169 ^{0}C to 214 ^{0}C in two parallel lines of high-pressure preheaters while going from the feedwater tank to the reactor. Process experts have identified a set of 18 faults that are generally hard to detect for an operator and that produce efficiency losses if undetected [25]. Two power levels, 50% and 80% of full power, have been considered.

The faults may be divided into three categories:

- F1-F5,F7 regard the line 1 of the feedwater system.
- F11-F15,F17 regard the line 2 of the feedwater system.
- F6,F8,F9,F10,F16,F18 regard both lines.

Data coming from the simulations of five transients were available for each of the 18 faults, for varying degrees of leakage and valve closures and with step and ramp changes at different leak sizes. 363 plant parameters had been recorded with a sampling frequency of 1 Hz. All transients start after 60 seconds of steady state operation. From the analysis developed in [11], it is clear that faults 6, 10, 16 have no significant consequence on the plant measured parameters because the size of the leakage considered is too small. Hence, these faults were not considered further. Moreover, given that the ramp changes cause variations of the parameters later than the step changes, only the three step changes for each fault were considered.

For the fault classification purpose of this work, the number of parameters was reduced to 123 by combining redundant measurements of the same physical quantity and by eliminating those parameters linearly correlated or carrying basically the same time evolution in all the different faults.

Compact wavelet features were then extracted from the 123 measured signals by Haar wavelet decomposition from a sliding window on the actual signal time-series [26]. The selected wavelet features [27] are: the mean residual signal taken at the highest, i.e. coarsest, scale and the minimum and maximum wavelet coefficients over all the scales. The rationale behind this choice is that the first wavelet feature captures the general trend of the signal across the windows in a compact way, being very much related to the average signal value within the analysis window, while the minimum and maximum wavelet coefficients capture important variations in the signal within a single window which would otherwise be severely smoothed out by the compression process. In particular, the maximum coefficient reflects negative trends, step changes and the negative component of spikes, whereas the minimum coefficient reflects the positive trends, step changes and positive components of spikes. The window size is selected so as to correspond to wavelet dyadic decomposition values (i.e. powers of 2) and consecutive windows are chosen with a slight overlap to avoid missing features that might be present at the window borders. Because of its ability of continuously applying the wavelet transform on a sliding window, and since the transform is used as a pre-processing step for the final

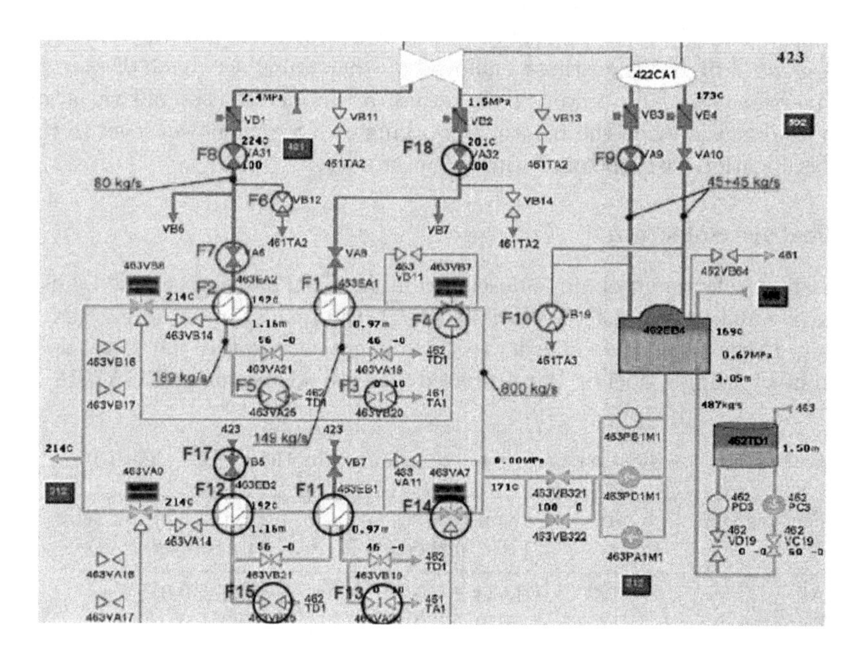

Fig. 5. Sketch of the feedwater system [24]

transient classification, this technique has been named Wavelet On-Line Pre-processing (WOLP) [27].

In a compromise between a high level of transient compression and an acceptable resolution, time windows which are 16 patterns long, with an overlap of 6 seconds, have been chosen. Consequently, the evolution of anyone of the wavelet features in a given transient from t= 58s to t=133s is summarized in 7 points each one representing the segment of dynamics in one time-window. Thus, the application of the WOLP pre-processing on the 123 original plant measured signals generates 369 wavelet coefficients, increasing from 2^{123} to 2^{369} the dimension of the search space from which the optimal subset of features relevant for the fault classification task is to be selected.

The transients at 50% of full power were used for the feature selection task, whereas the transients at 80% of full power were left out for validation of the resulting classifier.

For comparison purposes, Table 1 reports the mean recognition rate and the standard deviation obtained by the classifier based on all the 123 available features and on the feature subset selected by the experts of the plant process. Notice that there is a significant improvement in the classification performance with the reduced feature subset selected by the experts compared with the feature subset formed by all the available 123 features, both considering the test set and the validation set. This confirms the advantage of using a feature selection technique to support the classification task, otherwise based on the large number of available features. Finally, notice that the recognition rates (rr) achieved on the validation set obtained from transients at 80% power level are in general higher than those achieved on the tuning set made of patterns from transients at 50% power. This is due to the fact that signal variations are more relevant when the reactor is working at a higher power level so that the classification task becomes somewhat simpler.

4.1 Feature Selection

The feature selection technique described in Sect. 2.2 has been applied. Given the large number of possible solutions (2^{369}), the task of maintaining genetic diversity in the population in order to explore more accurately the search space is sought by using a Niched Pareto-based Genetic Algorithm (NPGA) [18,28]

Table 1. Classification performances achieved by the FKNN algorithm

Feature subset	m	rr on test set (50% power level)	rr on validation set (80% power level)
All	123	0.534 ± 0.035	0.647 ± 0.055
Expert-selected	17	0.679 ± 0.034	0.789 ± 0.029

with a large population size ($n_p = 200$) and a high probability of mutation ($p_m = 0.008$). In a single run, the NPGA identifies a family of non-dominated solutions with different classification performance (FKNN mean recognition rate) vs. complexity (number of features) trade-offs.

Figure 6 shows the Pareto front and the final population found by the NPGA at convergence. The niching "pressure" applied by the equivalence class sharing method succeeds in spreading the population out along the Pareto optimal front: indeed, the NPGA Pareto solutions cover from $m = 0$ to $m = 22$ with only individuals with $m = 8, 9, 15$ not present.

The analyst has to take the final decision on the features subset to be actually used for the classification, depending on his/her favouring high classification accuracy or low number of features. Thus, the closure of the problem must still rely on techniques of decision analysis such as utility theory, multi-attribute value theory or fuzzy decision making, to name a few.

In the present case, the optimization results for the non-dominated subset with 7 features show a good compromise between high classification accuracy and low number of features (pointing arrow in Fig. 6).

A set of patterns taken from transients at 80% power level, never employed during the feature selection phase, has then been used to cross-validate the performance of the FKNN classifier resting upon the selected 7-features subset. The resulting recognition rate of 0.9190 ± 0.0301 is significantly better than that obtained by using all the 123 features (0.647 ± 0.055) and by using the feature selected by the plant expert (0.789 ± 0.029)(Table 1).

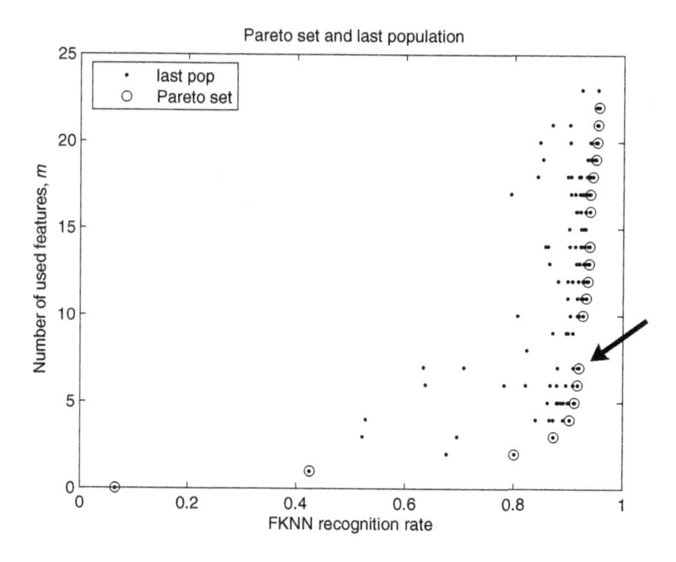

Fig. 6. Pareto front and final population found by the NPGA at convergence

4.2 Transient Classification

The possibilistic clustering technique illustrated in Sect. 3 has been applied to the problem of classifying the transients generated by the six faults F1,F2,F3,F4,F5 and F7 that regard line 1 of the feedwater system of Fig. 5.

Five features are used for the transient classification, extracted from the 7 features identified as optimal compromise in the previous Sect. 4.1. In particular, the 3 signals that regard line 1 of the the feedwater system, one signal that regards line 2 and one signal that regards the common part of the system have been chosen for the classification of the transients generated by the identified faults. The remaining 2 signals not considered regard line 2 of the feedwater system and are basically constant in the considered faults. Table 2 reports the five signals selected and Fig. 7 their behaviour for the 6 faults at 80% of full power.

The possibilistic classifier is built on patterns taken every 6 seconds from t=80s to t=200s from simulated transients of each fault type, with the plant at 80% of full power.

After the evolutionary supervised training of the possibilistic classifier, its performance has been tested using patterns taken every second from t = 0s to t = 300s from both the training transients and from an unknown transient caused by F13. Figure 8 shows the obtained transient classification as time progresses. Considering a degree of confidence $\epsilon_c = 0.7$, the results are quite satisfactory, even though at the beginning of the transients the possibilistic classifier assigns the steady state patterns of the first 60 seconds to the class of fault F2 albeit with low membership. This is explained by the fact that for transients of class F2 there are no significant effects on the relevant signals of Table 2 so that understandably the steady state may be confused with a fault of class F2 (Fig. 7).

Table 3 reports the time necessary for the possibilistic classifier to assign the transients to the right class with a membership value greater than $\epsilon_c = 0.7$. Notice that class F3 is recognized only 6 seconds after the beginning of the transient at t=60s while for the recognition of class F7 it is necessary to wait 34 seconds. This is due to the fact that the two most sensible signals for F7,

Table 2. Input signals of the classification model

Feature number as reported in [16]	Signal name	Unit
77	drain temperature before valve VB3 in line 1	^{o}C
160	water level in tank TD1	m
195	feedwater temperature after preheater EA2 in line 1	^{o}C
241	feedwater temperature after preheater EB2 in line 2	^{o}C
320	position level of the control valve for preheater EA1	%

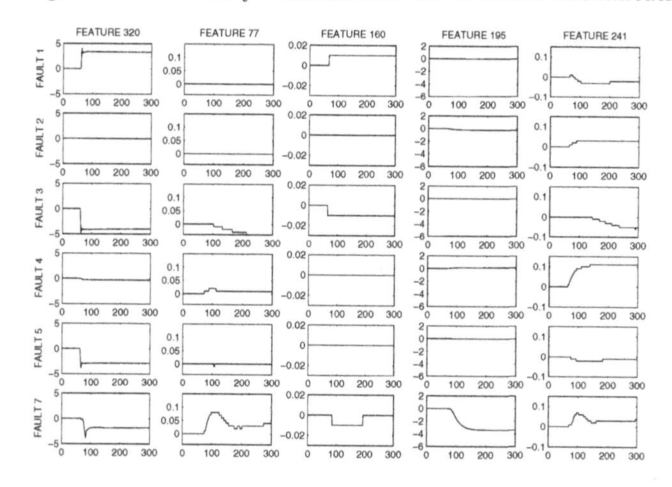

Fig. 7. Behaviour of the features at 80% of full power level

320 and 195, start departing from their steady state at t=82s and t=78s, respectively, reaching significant variations after 30s from the beginning of the transient at t = 60s (Fig. 7).

Also, note that the possibilistic classifier is able to assign to the right class the foreseen transients at times well beyond the temporal domain of 200s

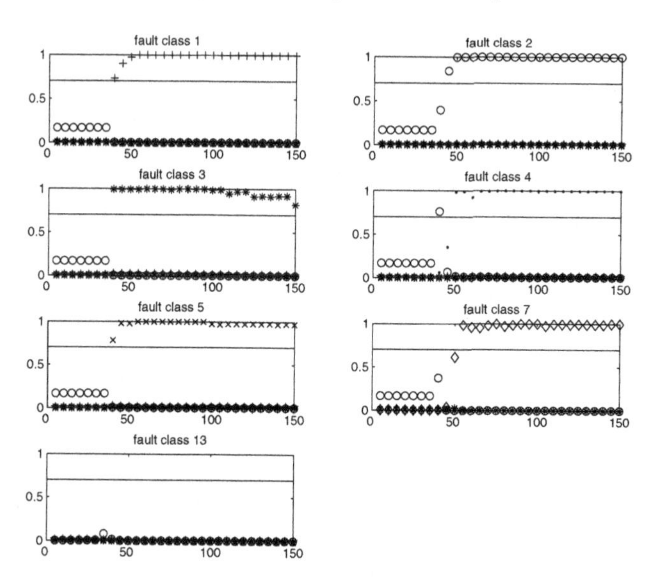

Fig. 8. Time profiles of the pattern assignment to the different classes: (+) membership to class F1, (o) membership to class F2, (*) membership to class F3, (.) membership to class F4, (x) membership to class F5 and ($\diamond$) membership to class F7. The horizontal solid line represents the degree of confidence ϵ_c here set to 0.7

Table 3. Time at which the transient is assigned to the right class (the transient begins at t=60s)

Class of the test transient	Time (s) at which there is the first assignment with membership value $> \epsilon_c = 0.7$
1	69
2	75
3	66
4	87
5	68
7	94

considered in the construction phase of the algorithm, due to the increased significance of the signals as the transients continue evolving away from their initial steady state.

Finally, the algorithm is very efficient in filtering out the patterns of the unknown fault F13 as atypical by assigning them very low membership values to all classes (Fig. 8, bottom). This is particularly important in diagnostic problems, given the impossibility to identify and explicitly enumerate a priori all transient scenarios possibly occurring in a system.

The identification as unknown of those transient conditions that have not been foreseen when building and tuning the diagnostic system, possibly allows to avoid taking wrong protection and emergency actions as a consequence of an incorrect diagnosis [19]. This may be important since an incorrect inference by the diagnostic system during an emergency condition risks to confuse and mislead the operator, with the potential of producing catastrophic consequences upon taking incorrect actions. Hence, the importance of classifying as unknown the transients that are not within the utilization domain of the classification algorithm.

5 Conclusions

The reliable and timely diagnosis of transients is a matter of paramount importance for the safety and operation of modern industrial plants. In this Chapter, this issue has been framed as a pattern classification problem.

First of all, the problem of discerning among the several measured plant parameters, those features to be used for early transient diagnosis has been tackled. This is a crucial issue to be resolved for the application of advanced diagnostic techniques to complex plants where hundreds of parameters are measured. A wrapper approach which combines a genetic algorithm search with a Fuzzy K-Nearest Neighbors (FKNN) classification algorithm has been embraced. The key advantages of the proposed methodology are that the 'wrapping' search measures explicitly the goodness of the considered feature subsets on the performance of the 'wrapped' classification algorithm (in our case the FKNN) and that the GA search is immune from the "nesting effect" characteristic of "hill climbing" methods, so that the probability of getting stuck in local optima is significantly lower.

Then, a supervised, evolutionary possibilistic clustering algorithm has been presented for more accurate and reliable classification. A supervised evolutionary algorithm finds a Mahalanobis metric for each possibilistic cluster which is optimal with respect to the classification of an available set of labelled patterns.

An example of application of the proposed approach to pattern classification has been given with reference to the classification of simulated nuclear transients in the feedwater system of a Boiling Water Reactor. Using five parameters identified as relevant by the feature selection algorithm, the possibilistic classifier correctly classifies the foreseen plant transients, while filtering out the patterns of an unknown fault type by assigning them very low membership values to all classes.

Acknowledgments

The authors wish to thank Drs. Paolo Fantoni and Davide Roverso of the IFE, Halden Reactor Project for providing the transient simulation data and Davide Mercurio and Nicola Pedroni for their contribution in developing the work.

The authors acknowledge that the work presented in this chapter has been partially funded by the Italian Ministry of University and Research (MIUR), within the project PRIN 2005093881-004 (years 2005-2006) and by the Institutt For Energiteknikk (IFE)-OECD-Halden Reactor Project (HRP), Norway.

References

1. D. E. Goldberg, Genetic Algorithms in Search, Optimization and Machine Learning, Addison-Wesley Longman Publishing Co., Inc., 1989.
2. R. Krishnapuram, J. Keller, The possibilistic c-means algorithm: insights and recommendations, Fuzzy Systems, IEEE Transaction 1 (2) (1993) 98–110.
3. D. Dubois, H. Prade, Possibility Theory: An Approach to Computerized Processing of Uncertainty, Plenum Press, New York, 1988.
4. R. Kohavi, G. H. John, Wrappers for feature subset selection, Artificial Intelligence 97 (1-2) (1997) 273–324.
5. J. Keller, M. Gray, J. Givens, A fuzzy k-nearest neighbor algorithm, IEEE Trans.Syst.,Man,Cybern. SMC-15 (4) (1985) 580–585.
6. E. Zio, P. Baraldi, Identification of nuclear transients via optimized fuzzy clustering, Annals of Nuclear Energy 32 (2005) 1068–1080.
7. B. Yuan, G. Klir, J. Swan-Stone, Evolutionary fuzzy c-means clustering algorithm, Proc. Fourth IEEE International Conference on Fuzzy Systems.

8. M. Na, Y. R. Sim, K. H. Park, Power Plant Surveillance and Diagnostics, Springer, 2002, Ch. Failure detection using a fuzzy neural network with an automatic input selection algorithm.
9. A. Verikas, M. Bacauskiene, Feature selection with neural networks, Pattern Recogn. Lett. 23 (11) (2002) 1323–1335.
10. C. Emmanoulidis, A. Hunter, J. MacIntyre, C. Cox, Selecting features in neurofuzzy modelling by multi-objective genetic algorithms, in: Proc. of ICANN99. 9th International Conference on Artificial Neural Networks, Vol. 2, Edinburgh. UK, 1999, pp. 749–754.
11. E. Zio, P. Baraldi, D. Roverso, An extended classifiability index for feature selection in nuclear transients, Annals of Nuclear Energy 32 (15).
12. D. Huang, T. W. Chow, Effective feature selection scheme using mutual information, Neurocomputing 63 (2005) 325–343.
13. B. Duran, P. Odell, Cluster Analysis: A survey, Springer-Verlag, 1974.
14. P. Narendra, K. Fukunaga, A branch and bound algorithm for feature subset selection, IEEE Trans. Comput C-26 (1977) 917–922.
15. H. Zhang, G. Sun, Feature selection using tabu search method, Pattern Recognition 35 (2002) 701–711.
16. E. Zio, P. Baraldi, N. Pedroni, Selecting features for nuclear transients classification by means of genetic algorithms, IEEE transactions on nuclear science 53 (3).
17. T. T. Y. Sawaragy, H. Nakayama, Theory of multiobjective optimization, Academic Press, Orlando, Florida,, 1985.
18. J. Horn, N. Nafpliotis, D. E. Goldberg, A Niched Pareto Genetic Algorithm for Multiobjective Optimization, in: Proc. of the First IEEE Conference on Evolutionary Computation, 1994.
19. J. Reifman, Survey of artificial intelligence methods for detection and identification of component faults in nuclear power plants, Nuclear Technology 119 (1997) 76–97.
20. M. Yang, A survey of fuzzy clustering, Mathematical and Computer Modelling 18 (11) (1993) 1–16.
21. J. Bezdek, Pattern Recognition with Fuzzy Objective Function Algorithms, Plenum Press, 1981.
22. G. J. Klir, T. A. Folger, Fuzzy sets, uncertainty, and information, Prentice-Hall, Inc., Upper Saddle River, NJ, USA, 1987.
23. P. F. Fantoni, Experiences and applications of PEANO for online monitoring in power plants, Progress in Nuclear Energy 46 (2005) 206–225.
24. E. Puska, S. Norrman, 3-d core studies for hambo simulator, in: Proc. of Enlarged Halden Programme Group Meeting, 2002.
25. D. Roverso, Fault diagnosis with the aladdin transient classifier, in: Proc. of System Diagnosis and Prognosis: Security and Condition Monitoring Issues III, AeroSense2003, 2003.
26. G. Strang, T. Nguyen, Wavelets and Filter Banks, Wellesley-Cambridge Press, 1996.
27. D. Roverso, Soft computing tools for transient classification, Inf. Sci. 127 (3-4) (2000) 137–156.
28. E. Zio, P. Baraldi, N. Pedroni, A niched pareto genetic algorithm for selecting features for nuclear transients classification, in: D. Ruan, P. D'hondt, P. Fantoni, M. D. Cock, M. Nachtegael, E. E. Kerre (Eds.), Applied Artificial Intelligence: Proceeding of the 7th International Flins Conference, 2006.

The Role of Fuzziness in Decision Making

Javier Montero, Victoria López and Daniel Gómez

Abstract. In this paper we discuss the true objective of fuzzy decision making models. In particular, we stress that fuzzy models should focus their attention on decision processes, instead of referring to the final output of a decision, which is a crisp act. In fact, while Probability Theory can properly model crisp acts, this is not the case for fuzzy decisions. But in fact human beings manage poorly defined arguments and alternatives, and what we usually call a decision is always ill defined, although we still expect they should produce consistent acts as their consequence. These acts, still being consistent with the previous fuzzy decision, are somewhat unpredictable, being extremely dependant on the specific circumstances at the moment an act is required by decision makers. Therefore, in many cases pursuing consistency of these subsequent crisp acts may be misleading. The main issue should be in principle to check consistency between those acts and the true decision behind them, usually poorly formalized. But this objective may sometimes be unrealistic, since there may be few chances to repeat the experience. We should also focus our attention on the arguments that led us to such a poorly defined decision. This should be a relevant role of fuzziness in decision making, viewed as a decision support problem (acts are supported by a fuzzy decision which is supported in fuzzy arguments).

Key words: Decision making, preference representation, probabilistic and fuzzy uncertainties.

1 Introduction

Human beings are often viewed as decision makers, and in fact a great effort is being made within Management and Social Sciences in order to develop decision making tools, so that human decisions can be understood and rationalized. A standard approach to these tools assumes the existence of a preference representation model, to be applied to a set of alternatives.

For example, within Social Sciences we can stress the historical impact of Arrow's impossibility theorem [5], which proved that given a group of rational individuals there is no possible amalgamation of their opinions, being simultaneously ethical and consistent. Many similar impossibility theorems have followed since then (see, e.g. [28]), producing a certain consternation to classical democratic ideals (but see [11, 43] for alternative possibility theorems). In fact, many political systems can be explained by this theorem: people putting too much stress on rationality seem to evolve towards dictatorial-like positions (the best for society is to leave decisions in their hands, of course), people putting too much stress on ethical issues are sometimes accused of slowing down economy, the number of political parties are quite often artificially reduced to very limited options (Arrow's impossibility theorem does not apply for only two alternatives), or individual freedom is reduced by means of certain terminology that induces certain undeclared assumptions into people's mind (Arrow's impossibility theorem can be avoided if everybody thinks in a left-center-right scheme, see [11]). Of course, every piece in each impossibility theorem is being precisely defined, from the ethical issues to the particular meaning of rationality or consistency of individual and social preferences, which represent the corner stone of the model. We must then point out that a preference relation in Arrow's model is understood as a mapping

$$\mu : X \times X \rightarrow \{0, 1\}$$

where X represents the set of alternatives, and $\mu(x, y) = 1$ if, and only if, alternative $x \in X$ is being considered as strictly better than alternative $y \in Y$ (standard consistency conditions require this preference relation to define a linear order on the set of alternatives). Note that in this model indifference between different alternatives is not considered: transitivity for indifference cannot be accepted, see e.g. [32, 41]). Notice also that this approach implies that the set of alternatives X has been precisely defined in advance.

Multi-criteria decision making models, for example, acknowledge the complexity of the problems we are faced with. Reality can be viewed as a multi-dimensional structure, where several criteria explain different aspects. But as ethics and rationality are contradictory in a group decision making context, those criteria may also be contradictory, and we can only search for some kind of equilibrium, since we cannot get everything (cheap things used to be low quality, and high quality used to be expensive, but perhaps we can find a good enough alternative that is not too expensive). Again, classical multi-criteria models are based upon analogous *binary* preference mappings, but applied to each individual criteria point of view. The above condition of consistency is, in this context, translated into a linear ordering assumption for each criteria, although incomparability is quite often natural in this context (given two alternatives $x, y \in X$, neither $\mu(x, y) = 1$ nor $\mu(y, x) = 1$ hold, so a decision maker does not establish any comparison between these two alternatives). Defining those underlying criteria is sometimes the key issue (see [45] but also [46]).

Fuzzy Sets [47] allow the possibility of introducing degrees of preference, in such a way that alternative decision making models can be built from fuzzy preference relations [48], i.e. mappings

$$\mu : X \times X \rightarrow [0, 1]$$

where $\mu(x, y)$ can be understood as the degree to which alternative x is strictly better than alternative y (see [19, 32, 36], where the whole preference structure is considered). Within this context we can avoid Arrow's paradox and propose a trivial meaning of such a discouraging result: *nothing is perfect.* But of course we can do our best, outside such an extreme binary approach that declares unacceptable anything that is not perfect. As pointed out in [32], Arrow's paradox is a consequence of a hidden hypothesis in his model (the Aristotelic logic) that considers *unacceptable* any potential situation not assuring *complete rationality* (see also [17, 31, 33] for alternative approaches). Note that we still assume that the set of alternatives is already defined (see [44]), and they are defined as crisp. Nevertheless, allowing degrees of preference should remind us that any model (mathematical or linguistic) will always be a simplification of reality, which is extremely complex. Within decision making this is sometimes so clear that we forget about optimization procedures and we focus our attention on developing decision making aid tools [13, 42].

In this paper, we want to stress that a decision making aid approach is the natural approach within fuzzy sets, and that putting too much stress on modelling decision making can be misleading if decisions are identified with acts, as already pointed out in [34]. As a consequence, we demand more effort in two main issues: fuzzy representation techniques [23] and fuzzy algorithmic verification [30].

2 Decision Making Within Probabilistic Uncertainty

Most students still today only know of one logic, the Aristotelic logic. It is not the one in existence, but it is the only one they know of. This situation confuses people, since they might think that any argument not following the Aristotelic logic is not logical, and therefore rejectable (from a psychological and social point of view, it would be desirable if there could be a more educative message included in children's text books, or at least a reference to some alternative logic, showing in this way the term *logic* does not have only one meaning).

In some way, an analogous situation appears with uncertainty. During many years, the only model for uncertainty being taught at schools and universities has been Probability. Hence, some people may think that uncertainty is equivalent to Probability. Nevertheless, we think it is quite easy to convince people that the meaning of *uncertainty* is not fixed: probabilistic uncertainty refers to the uncertainty regarding a certain event that may happen or not, but in any case such an event either happens or not. Events in Probability Theory are always crisp. Then, in order to check if such an event happens or not, we need the existence of an appropriate experiment, which should allow us to decide if such an event either happens or not. Hence, again we find that the Aristotelic logic is also behind Probability Theory.

The hard criticism from Bayesian researchers did not come from a defence of the existence of a unique kind of uncertainty, but from the useness of fuzziness in order to model uncertainty and its lack of *observability* in an operational sense (see [35] for a review). While Frequency Probability Theories of Probability demand the existence of randomness as part of the real word, Subjective Probability Theories claim that Probability is just the way by which we explain our decisions. And Bayesian Probability Theory claims that this is the only thing we should be doing in order to build a scientific theory: we should build models from facts, which will be observed, verifiable data and acts. In this way, initial information is crisp in nature. Hence they conclude there is no need for an alternative model of uncertainty. Note that initially, researchers on Probability assumed that there is no room for uncertainty in Science, which pursues a perfect knowledge.

If decision making refers to a set of crisp alternatives under consideration, fuzzy models may be considered in fact useless, at least in its initial stage (we can still introduce here the argument of simplifying reality so it can be managed by decision makers, but such an argument is subject to comparison with alternative approximation models).

3 Decision Making within Fuzzy Uncertainty

Probability and Fuzziness deal with uncertainty about information. But both theories should be combined in order to get a better understanding of reality, since many circumstances, present in human decision making, are poorly defined. Recent relevance of linguistic issues in the so-called *soft* sciences is recognizing this. The key question we bring here is the following: should Decision Making be a main objective of Fuzzy Sets research? Decision Making based on (random, whatever that means) facts indeed is a cornerstone of Probability, but should Decision Making play a similar role within Fuzziness?

Let us note that fuzzy uncertainty used to be associated with words representing concepts, not acts. Most examples supporting fuzzy uncertainty refer to information given in terms of linguistic terms, while most examples of probabilistic uncertainty refer to information given in terms of acts (data produced by an experiment). Each of us does realize that all the information we have, no matter its nature, has some influence on the decisions we make. But fuzziness cannot be properly applied to acts, because they are crisp (their linguistic description can be fuzzy, but acts are always crisp).

Some time ago, one of the authors was walking on Shattuck Avenue, Berkeley, and he noticed a homeless man sitting up ahead on the pavement. This author made what he thought was a *decision*: to help this person. But this decision was extremely poorly defined, very far from becoming an act. Giving a particular amount of money is an act, and it depends on many circumstances (some of them are basically random, as the total amount of money each of us has in our pockets and how it is distributed in bank notes and coins). Once this author had decided to help the person, the author faced the problem of specifying what he could do in order to help. But again the alternatives crossing his mind while still walking towards him were ill-defined: "*spend some time talking to him*", "*take him to some restaurant*", "take him to have a shower somewhere, etc.) No matter which *alternative* he could choose, it was not well enough defined, and another set of less ill-defined alternatives was needed to specify the previous election. The fact is that this analysis (decision process) consumed so much time that the author reached the man without choosing any particular act, so at the very last moment he decided to give him "some spare change", which was neither a well-defined decision either. He finally decided to pick up, at random, a couple of coins from his pocket, and they happened to be a quarter (25 cents of a U.S. dollar) and a dime (10 cents). This was the observable and verifiable act. Probabilistic uncertainty properly works at the final stage of his inner decision process. But is this final lottery really important in terms of human decision making? The fact is that the final set of crisp alternatives appears after a hierarchical ill-defined decision process, and it may even be the case that the final act has never been a proper alternative (we cannot spend unlimited time to fix alternatives into a crisp framework). More importantly, in most cases human beings make their decisions without waiting to have perfectly defined alternatives, but they expect that details will be fixed when the decision is implemented. Each step in the inner decision process is about a personal attitude which is usually not well defined, but they conform to the proper human decision problem (deciding being good "now" is perhaps the main human decision we can make, but it is far from being *observable*).

Fuzziness does refer to Decision Making, but mainly to that logical process leading to the definition of the particular framework where a particular act will be produced (perhaps not properly chosen). Coming back to Arrow's paradox, the most important issue in democracy is not the way we vote for alternatives, but how we reach the final set of alternatives. Fuzziness is fully present in the previous negotiation process, which should be characterized by a sincere and creative search for new proposals, where all participants should actively collaborate. Of course we cannot keep negotiating for ever, so at one moment we have to vote. But in a crisp world of alternatives there is no real negotiation, but bargaining. In fact, we must realize that a certain vagueness is essential in the first stages of negotiation, if we really pursue an agreement, most probably to be reached by means of a new alternative that appears during a joint creative negotiation process. Any negotiation starts from open or soft alternatives, not only due to obvious game advantages (see, e.g. [39]), but because non precise language establishes unexpected links that should never be avoided in the presence of a conflict (the process of creating new alternatives, as with any creative activity, cannot be fully explained from a systematic and rational point of view).

Meanwhile our experiments deal with crisp events, checking the final output (act) of a (fuzzy) decision process may be partially irrelevant and perhaps totally unrealistic within a crisp context. We can check that a certain desired event is produced for example 95 out of 100 times when we apply certain decision procedures, because of the existence of a crisp experiment assuring this kind of results. But how can we check that we have decided "to be good" when there is no such crisp experiment? Giving a coin to a homeless man does not imply that one has decided "to be good". In addition, how can we check a decision that will be made only once, where it is impossible that we can build up anything related to a measure of the frequency of success? The only thing we can be sure of is that there were serious supporting arguments for that unique decision and for the acts that resulted from that decision.

The fact is that quite often in our lives the only thing we can check is that our analysis has been made in the right way according to the information we are able to get. Science, for example, mainly refers to a systematic analysis of reality: we cannot be sure that a theory is true, because we can only check that predictions from that theory are not inconsistent with observations, but even in case of no observations, we can still argue that the methodology supporting that theory has been scientific.

And looking to the consequences of our decisions, quite often the only thing we should be pursuing is a better knowledge of reality: it has been proven in Medicine [6, 7] that the part of the brain in charge of analyzing the decision making problem is located in a different area from the part of the brain in charge of making the final decision, i.e., ordering an act (certain brain damage produces, as a consequence, rational individuals still able to develop detailed analysis of the different available options they have, but not emotionally able to choose one of those options). In fact, the brain structures responsible for emotion and reasoning must cooperate in order to make decisions [27, 29], by by-passing consistency difficulties (lack of information, excessive information, apparent contradictions, time limitations, etc.)

Organizing information is quite often more important than obtaining additional information. In order to build up good decision procedures we should first be sure that the problem has been stated in the correct terms, and this implies that we understand to some extent the problem we are facing with. It is only on this assumption that we can build a rational solution. Without rationality we can only measure *how lucky we are*, and perhaps we can even conclude that there is enough statistical support to think that we have special *intuition* or *authority* (people may follow our suggestions without requiring any supporting argument on our part). It is the supporting argument the one mainly subject to rationality, and the relevant consistency is the consistency between our acts and our decisions or purposes, not between our acts. Some acts may appear inconsistent but they can be justified in terms of an individual decision maker (not understanding another's behavior does not mean this behavior is not consistent). Moreover, acknowledging complexity in decision making is more related to the existence of contradictory arguments than to the difficulties in modelling the problem or the difficulties in obtaining the optimal solution when its existence is assured.

We should not expect that complex problems have a *direct* solution. In decision making, for example, we could even say that a *complex* problem is, by definition, one not allowing an optimal solution. A problem may be *difficult* to represent, but complexity comes as soon as we realize that there is no best solution. We should expect complex problems to be approached by means of *ad hoc* algorithms we are forced to design. The problem appears within a fuzzy framework: how the reliability of this algorithm can be checked with standard methods, if the algorithm is based upon fuzzy concepts or its results are fuzzy? A statistical analysis of results may not work, due to fuzziness in results, and the only available verification procedures require crisp algorithms.

The world of (crisp) acts is not the natural framework for Fuzzy Sets. But the decisions we make always involve the natural uncertainty attached to words, concepts or complexity. The same applies to decision aiding, where the main objective should be to allow a better understanding of the decision making problem (in other words, to improve our knowledge of the problem we are facing with).

Hence, two issues should become more relevant than the analysis of the consistency of those crisp *observable* acts being the consequence of a decision process:

- On one hand, the problem of checking the consistency (inner logic) of the decision analysis that justifies our decision processes, and the verification of the correspondence between the algorithm we are considering and our previously declared objectives.
- On the other hand, the problem of information treatment towards a useful visualization and representation of decision making problems, with the declared objective of allowing a better understanding of the problem, which will allow the definition of possible alternatives.

4 Verification Procedures

Verification of a decision process should be in some way analogous to the problem of verification in software, taking into account the natural presence of fuzziness present within any decision process.

Verification processes require a logical specification of the prerequisites and results. These verification mechanisms are studied for many different paradigms such as imperative [26], functional [24] or functional-logic languages [14]. According to the characteristics of the program and the properties to be verified, different techniques can be taken into account. The main techniques usually applied to verify properties of the systems are model-checking [8], theorem proving [38] and testing [37], plus different combinations of them and other formal methods such as abstract interpretation [16], for example. The point of view of model checking requires the verification of that a program, formalized as a transition system, and satisfying a given temporal logic formula. Model checking is a very efficient technique to verify such temporal formulas from an initial state (the point at which our computation begins). Theorem proving requires proving a given logical formula over a system specified as a program. Then we can prove many theorems satisfied by such a system. There exists many theorem provers that can be distinguished by the language in which systems are specified: some examples are Isabelle [38], Coq [9] or PVS [40]. Testing is another technique used in general for very large systems in which other approaches cannot give a result within a reasonable time (it is also used to speed up the decisions over the specification of a system).

From the point of view of the formalization of systems and their verification, the classical approach is that of Hoare [26], where the specification of the system is done by a pair of first order logical (FOL) formulas and the verification of the imperative system uses Hoare's deduction rules. For imperative systems this approach is not enough to deal with other programs that can be executed in parallel. For these systems the specification considered is a temporal logic that reflects the idea of the evolution of the system. But systems nowadays are more complex and they not only evolve in time but in space. For such a purpose programs are specified using spatial-temporal logics. But specification requisites are quite inexact to put them into a crisp formalism, so for that reason the possibility of specification of properties of programs using a fuzzy logic approach was introduced [30].

Of course, we should also take into account that framework is not constant in real life, so that logic we consider for verification should be subject to evolution (see, e.g. [15]).

5 Fuzzy Visualization Techniques

Presenting complex information in a comprehensible and helpful way is another key issue. The objective of decision aiding can be fulfilled when a decision maker *visualizes* and understands the tremendous complexity of the problem, perhaps suggesting some particular strategy after a careful sequence of small decisions. We should be able to develop some kind of descriptive fuzzy representations, analogous to those existing in statistics, but showing degrees of membership. This fuzzy approach should allow a conceptual representation, rather closer to reality than a crisp approach, which forces artificial partitions. We should be able to paint fuzzy sets, with the limitation of the three-band visual spectrum.

In a pure decision making context, for example, decomposition techniques such as the one initiated in [22, 23], which pursue a certain compact representation of preferences in terms of some possible linearly ordered underlying criteria. A classical (crisp) dimension analysis [45] applied to the sequence of α-cuts allows an informative *dimension function*, which in some way shows the preference structure (dimension for each α-cut is obtained by means of a key algorithm presented in [46]). Decision maker inconsistency can be explained in this way, in terms of a multicriteria approach.

Another field of special interest is classification. In fact, most human decisions are in fact a package of possible alternatives with key common characteristics: we choose one of these classes and leave the context to fix details about the particular act we produce.

In [20, 21], e.g. a particular coloring algorithm is provided in order to get possible homogeneous regions in remotely sensing images, to be considered as candidates to become a class (perhaps associated with a concept like "forest" or "urban area", e.g.). In particular, possible regions are presented in the form of a tree, by means of the sequential application of a basic binary coloring procedure that allows a first approach to a picture in terms of regions, taking into consideration the behavior of adjacent pixels. In this way a structured partition of the image is offered, to be considered for example as a posterior classification process. These possible classes can be derived, for example, from any existing fuzzy classification technique, e.g. fuzzy C-means (see [10]). Alternatively, we can consider the approach proposed in [2, 3] (see also [1, 4]), now taking advantage of the neighborhood information connected to those regions we have obtained. Such a segmentation can be viewed as a preprocessing tool in order to reduce the amount of information needed to explain the image, allowing faster posterior studies.

In [20, 21], e.g. a digital image is divided into connected regions, taking into account the surrounding behavior of each information unit (pixel). Local miss behavior is in this way bypassed, explaining the image in terms of few homogeneous regions, each one being described in terms of the aggregated value of *most* of its pixels. Each region will be a serious candidate to become a class. But two disjointed homogeneous regions are still not guaranteed to belong to the same class even if the properties of the pixels they contain are highly similar; and two disjointed homogeneous regions can be assigned to the same class even if the pixels in one region are very different from pixels in the other. Classification is rather more complex than segmentation, although segmentation represents a standard pre-processing technique for a better understanding of the image under study. It is in this sense that we stress the role of fuzzy methodologies, in order to provide an accurate useful visualization of the problem.

Coloring techniques should play a key role in the future, in order to help decision makers capture a global view of complex images, including those where regions do not show clear boundaries. Without appropriate representation techniques, without a manageable visualization of the available information, decision makers may look at some sophisticated models as *black boxes*, and therefore reject them in practice.

6 Final Comments: Knowledge and Decision Making

In this paper we maintain that since Probability Theory models crisp events, such a model can be properly applied to a decision making problem with well defined alternatives. But most human decisions refer to a sequence of sets of ill-defined alternatives, where information cannot be associated with those experiments considered in Probability Theory. Hence, a direct objective for fuzzy researchers should be to develop appropriate procedures in order to assure consistency of the decision process itself. On the other hand, since the information is complex in nature, we do need more descriptive tools for summarizing information for decision makers. In this sense, we postulate that fuzziness is more relevant in designing consistent (fuzzy) classification models for knowledge acquisition rather than developing (crisp) decision making tools about acts. As pointed out above, human acts are subject to complex influences, including those of a random nature, but not only those. In this sense, consistency of human acts with respect to other human acts can sometimes be considered unrealistic. We should of course look for the consistency of our standard decisions (mostly ill-defined), but with respect to their supporting arguments and with respect to their consequences (acts following those decisions). An intermediate approach is centering on linguistic models (see, e.g. [25]), while they represent concepts.

There are many different possible rationalities we can consider when dealing with fuzzy preferences (see, e.g. [18], where it was also pointed out that standard fuzzy consistency definitions are still crisp). Most of them are related in some way to transitivity or acyclicity, but such a concept can be excessive at a knowledge organizational level, closer to a classification problem [2] than to a decision making problem, if this decision making problem is understood as a choice problem between possible acts (see also [12]).

Fuzzy approaches should focuss attention on those decision processes allocated to a knowledge level, where the decision output is much closer to a fuzzy class rather than to an act. Calvin (see cartoon below) may desire a simplified world where things are simply true or false, but the fact is that the world is indeed complex. The key picture is the third one, where Calvin looks thoughtful and shocked at reality. Calvin will finally answer "yes" or "no" at random. But of course his relevant decision is neither to answer "yes" or "no". His relevant decision is neither to throw a coin and give an answer depending if the upper side of the coin is a "head" or "tail". His relevant decision is perhaps "I do not know" or even "I do not care". Such a decision is not observable, but in the end Calvin has to produce a single, crisp act. This act is subject to many contextual influences and it is absolutely irrelevant from a decision making point of view. The important stage is at the previous level, the true human decision level, closer to knowledge and classification problems, where fuzzy logic plays a key role. Human decision making is the natural framework for Fuzzy Set Theory. But our decisions should not be confused with our acts.

Fig. 1. Calvin and Hobbes, by B. Watterson (reproduction authorized by Universal Press Syndicate)

Acknowledgement. This work has been supported by the National Science Foundation of Spain, grant number MTM2005-08982-C04-01.

References

1. A. Amo, D. Gómez, J. Montero and G. Biging (2001): Relevance and redundancy in fuzzy classification systems. *Mathware and Soft Computing* 8:203–216.
2. A. Amo, J. Montero, G. Biging and V. Cutello (2004): Fuzzy classification systems. *European Journal of Operational Research* 156:459–507.
3. A. Amo, J. Montero and V. Cutello (1999): On the principles of fuzzy classification. *Proc. NAFIPS Conference.* IEEE Press, Piscataway, NJ; pp. 675–679.
4. A. Amo, J. Montero, A. Fernàndez, M. López, J. Tordesillas and G. Biging (2002): Spectral fuzzy classification, an application. *IEEE Trans on Systems, Man and Cybernetics, Part C,* 32:42–48.
5. K.J. Arrow (1951, 1964): *Social Choice and Individual Values.* Wiley, New York.
6. A. Bechara, H. Damasio and A.R. Damasio (2003): Role of the amygdala in decision-making. *Annals of the New York Academy of Sciences* 985, 356–369.
7. A. Bechara, D. Tranel and H. Damasio (2000): Characterization of the decision-making deficit of patients with ventromedial prefrontal cortex lesions. *Brain* 123, 2189–2202.
8. B. Bérard, M. Bidoit, A. Finkel, F. Laroussinie, A. Petit, L. Petrucci and Ph. Shnobelen (2001): *Systems and Software Verification: Model-Checking Techniques and Tools.* Springer .
9. Y. Bertot and P. Castéran (2004): *Interactive Theorem Proving and Program Development Coq'Art: The Calculus of Inductive Constructions.* Springer.
10. J.C. Bezdek and S. K. Pal (1992): *Fuzzy Models for Pattern Recognition.* IEEE Press, New York.
11. D. Black (1958): *The Theory of Commitees and Elections.* Cambridge U.P., Cambridge.
12. P. Bonissone (1997): Soft computing, the convergence of emerging reasoning technologies, *Soft Computing* 1:6–18.
13. J.P. Brans (1998): The DGS Prometee procedure, *Journal of Decision Systems* 7:283–307.
14. J.M. Cleva, J. Leach and F.J. López-Fraguas (2004): A logic programming approach to the verification of functional-logic programs. *Proc. Principles and Practice of Declarative Programming.* ACM Press; pp. 9–19.

15. C.A. Coello (2006): Evolutionary multiobjective optimization, a historical view of the field, *IEEE Computational Intelligence Magazine* 1:28–36.
16. P. Cousot and R. Cousot (1999): Refining model checking by abstract interpretation. *Automated Software Engineering Journal* 6:69–95.
17. V. Cutello and J. Montero (1993): A characterization of rational amalgamation operations. *International Journal of Approximate Reasoning* 8:325–344.
18. V. Cutello and J. Montero (1994): Fuzzy rationality measures. *Fuzzy Sets and Systems* 62:39–44.
19. J. Fodor and M. Roubens (1994): *Fuzzy Preference Modelling and Multicriteria Decision Support*. Kluwer, Dorcrecht.
20. D. Gómez, J. Montero and J. Yáñez (2006): A coloring algorithm for image classification. *Information Sciences* 176:3645–3657.
21. D. Gómez, J. Montero, J. Yáñez and C. Poidomani (2007): A graph coloring approach for image segmentation. *OMEGA-Int.J. Management Science* 35: 273–183.
22. J. González-Pachón, D. Gómez, J. Montero and J. Yánez (2003): Soft dimension theory. *Fuzzy set and Systems* 137:137–149.
23. J. González-Pachón, D. Gómez, J. Montero and J. Yánez (2003): Searching for the dimension of binary valued preference relations. *International Journal of Approximate Reasoning* 33:133–157.
24. M.J.C. Gordon and T.F. Melham (1993): *Introduction to HOL*. Cambridge University Press.
25. F. Herrera, L. Martínez and P.J. Sánchez (2005): Managing non-homogeneous information in group decision making. *European Journal of Operational Research* 166, 115–132.
26. C.A.R. Hoare (1969): An axiomatic basis for computer programming. *Comm. ACM* 12:89–100.
27. M. Hsu, M. Bahtt, R. Adolfs, D. Tranel and C.F. Camarer (2005): Neural systems responding to degrees of uncertainty in human decision-making. *Science* 310, 1680–1683.
28. J.S. Kelly (1975): *Arrow Impossibility Theorems*. Academic Press, New York.
29. J. Kounios, J.L. Frymiare, E.M. Bowden, J.I. Fleck, K. Subramaniam, T.B. Parrish and M. Jung-Beeman (in press): The prepared mind, neural activity prior to problem presentation predicts subsequent solution by sudden insight. *Psycological Science.*
30. V. López, J.M. Cleva and J. Montero (2006): A functional tool for fuzzy first order logic evaluation. In D. Ruan *et al.*, eds.: *Applied Artificial Intelligence, World Scientific,* New Jersey; pp. 19–26.
31. J. Montero (1985): A note on Fung-Fu's theorem, *Fuzzy Sets and Systems* 13:259–269.
32. J. Montero (1987): Arrow's theorem under fuzzy rationality, *Behavioral Science* 32:267–273.
33. J. Montero (1987): Social welfare functions in a fuzzy environment, *Kybernetes* 16:241–245.
34. J. Montero (2003): Classifiers and decision makers. In D. Ruan *et al.*, eds.: *Applied Computational Intelligence*. World Scientific, Singapore; pp. 19–24.
35. J. Montero and M. Mendel (1998): Crisp acts, fuzzy decisions. In S. Barro *et al.*, eds.: *Advances in Fuzzy Logic*. Universidad de Santiago de Compostela; pp. 219–238.

36. J. Montero, J. Tejada and V. Cutello (1997): A general model for deriving preference structures from data. *European Journal of Operational Research* 98: 98–110.
37. G.J. Myers (1979): *The Art of Software Testing.* Wiley.
38. T. Nipkow, L.C. Paulson and M. Wenzel (2002): *Isabelle/HOL - A Proof Assistant for Higer-Order Logic.* Springer.
39. G. Owen (1982): *Game Theory.* Academic Press.
40. S. Owre, N. Shankar, J.M. Rushby and D.W.J. Stringer-Calvert (2001): *PVS System Guide.* SRI International.
41. P.K. Pattanaik (1971): *Voting and Collecive Choice.* Cambridge University Press.
42. B. Roy (1993): Decision sciences or decision aid sciences, *European Journal of Operational Research* 66:184–203.
43. A.K. Sen (1970): *Collective Choice and Social Welfare.* Holden-Day, San Francisco.
44. G. Shafer (1986): Savage revisited. *Statistical Science* 1:463–501.
45. W.T. Trotter (1992): *Dimension Theory.* The Johns Hopkins University Press, Baltimore and London.
46. J. Yáñez and J. Montero (1999): A poset dimension algorithm, *Journal of Algorithms* 30:185–208.
47. L.A. Zadeh (1965): Fuzzy Sets, *Information and Control* 8:338–353.
48. L.A. Zadeh (1971): Similarity relations and fuzzy orderings. *Informations Sciences* 3:177–200.

Fuzzy Linear Bilevel Optimization: Solution Concepts, Approaches and Applications

Guangquan Zhang, Jie Lu and Tharam Dillon

Abstract. Bilevel programming provides a means of supporting two level non-cooperative decision-making. When a decision maker at the upper level (the leader) attempts to optimize an objective, the decision maker at the lower level (the follower) tries to find an optimized strategy according to each of the possible decisions made by the leader. A bilevel decision model is normally based on experts' understanding of possible choices made by decision makers at both levels. The parameters, either in the objective functions or constraints of the leader or the follower in a bilevel decision model, are therefore hard to characterize by precise values. Hence this study proposes a fuzzy parameter linear bilevel programming model and its solution concept. It then develops three approaches to solve the proposed fuzzy linear bilevel programming problems by applying fuzzy set techniques. Finally, a numerical example and a case study illustrate the applications of the proposed three approaches.

1 Introduction

A bilevel programming (BLP) problem can be viewed as a static version of the non-cooperative, two-player (decision maker) game [24]. The decision maker at the upper level is termed as the leader, and the lower level, the follower. In a BLP problem, the control for decision factors is partitioned amongst the decision makers who seek to optimize their individual objective functions [1]. Perfect information is assumed so that both the leader and the follower know the objective and feasible choices available to the other. The problem is said to be a 'static game' which implies that each decision maker has only one move. The leader attempts to optimize his/her objective function but he/she must anticipate all possible responses of the follower [13]. The follower observes the leader's decision and then responds to it in a way that is personally optimal. Because the set of feasible choices available to either decision maker is interdependent, the leader's decision affects both the follower's payoff and allowable actions, and vice versa. For example, consider a logistic company making decision on how to use commission as a means for its distributors to improve its product sale volume. The company, as the leader, attempts maximizing its benefit of product sale through offering a most suitable commission to its distributors. For each of the possible commission strategies made by the company, the distributors, as the follower, will response on product sale volume which is based on the maximized benefit obtained through product sale. Therefore, in such a BLP problem which is described by a bilevel optimization model, a subset of the decision variables (such as 'commission' in the example) is constrained to be a solution of a given optimization problem parameterized by the remaining variables (such as 'sale volume') [2, 4]. In mathematical terms, a BLP problem consists of finding a solution to the upper level problem

$$\begin{cases} \max\limits_{y_0} & F(y_0, y_1, \cdots, y_m) \\ \text{subject to}: & AY \leq 0 \end{cases} \tag{1.1}$$

where $y_i (i = 1, 2, \cdots, m)$, for each value of y_0, is the solution of the lower level problem:

$$\begin{cases} \max\limits_{y_i} & f_(y_0, y_1, \cdots, y_m) \\ \text{subject to}: & B_i Y \leq 0 \end{cases} \tag{1.2}$$

where $Y = (y_0, y_1, \cdots, y_m)^T$ and $A, B_i (i = 1, 2, \cdots, m)$ are matrixes.

The majority of BLP research has centered on the linear version of the problem. A linear BLP problem with a finite optimal solution shares the important property that at least one optimal (global) solution is attained at an extreme point of the constraint region. This result was first established by Candler and Townsley [8] for a linear BLP problem with no upper level constraints and with unique lower level solutions. Later Bard [3], Bialas and Karwan [7] proved this result under the assumption that the constraint region is bounded. The result for the case where upper level constraints exist has been established by Savard [18] under no particular assumptions. Based on these results, there have been nearly two dozen algorithms proposed for solving linear BLP problems [3, 4, 6–10, 23]. Kuhn-Tucker approach [5–7], Kth-best approach [7, 8], and Branch-and-bound approach [3, 5, 11] are three popular approaches of them.

The establishment of a BLP model assumes that both the leader and the follower have perfect information about the objects, constraints and feasible choices of the other. However, in real situations, the informatics, described by quantitative or qualitative data, obtained is often uncertain. Let us still consider a logistic company example mentioned above. The company can only estimate various feasible choices taken and various costs spent by its distributors through survey-based understanding of the distributors. Therefore, in establishing a bilevel model for the decision problem, related parameters of functions of these estimated choices and constraints are hard to describe with precise values. Noting this, it would certainly be more appropriate to interpret the experts' understanding of the parameters as fuzzy numerical data which can be represented by means of fuzzy sets [25]. Furthermore, a decision support approach should have the ability to process both precise and imprecise (it can be qualitative, at varying levels of precision) data in a decision model, and transform it into decision makers' options and judgments as optimal solutions. Existing bilevel optimization approaches mainly suppose the situation in which the objective functions and constraints of a bilevel model are characterized with precise parameters. This limits to a certain degree the applications of bilevel decision techniques in uncertain environments. Bilevel linear programming in which the parameters are characterized by fuzzy numbers is called fuzzy linear bilevel programming (FLBLP).

The FLBLP problem was well researched by Shih et al. [22], Sakawa et al. [17] and Lai [13]. Shih et al [22] and Lai [13] have proposed a solution concept which is different from the concept of the Stackelberg solution [24] for linear BLP. Their approach is based on the idea that the follower optimizes an objective function, taking a goal of the leader into consideration. Therefore, the follower solves a fuzzy optimization problem with a constraint imposed by the degree of satisfaction of the goal of the leader. As indicated by Sakawa [17] there is a possibility that the approach gives a final solution that is undesirable because of inconsistency between the fuzzy goals of the objective function and the decision variables. Sakawa et al. [16,17] formulated the linear bilevel fractional programming problems with fuzzy parameters which are characterized by fuzzy numbers. A related approach was proposed to derive a satisfactory solution by updating the satisfactory degrees of the leader with consideration of the overall satisfactory balance between both levels. The approach uses a λ-level set of fuzzy numbers as the universe of a fuzzy parameters defined as the ordinary λ set in which the degree of the membership function exceeds the λ-level. Each fuzzy parameter is therefore assigned a value from the λ-level's corresponding close interval rather than the end points of the projection interval of the membership function of the fuzzy parameter.

The approaches developed in the study first transform the single objective function into an infinite number of objective functions which are generated by two end-points of each of the infinite λ-level sets. The classical bilevel programming algorithms are then applied to solve these infinite objective functions. The objectives obtained are then substituted by the model to give an optimal solution for the original model. The solution obtained by using our approach still keeps the order-relationship of the fuzzy numbers introduced. It can fully represent the expected solution of the FLBLP model with original values of the fuzzy parameters. Therefore the solution is more reliable as it fully reflects the solution features of an original fuzzy bilevel decision problem.

Following the introduction, Sect. 2 reviews a solution concept of linear BLP and related algorithms for solving linear BLP problems. Three fuzzy-number based FLBLP approaches, called fuzzy Kuhn-Tucker, fuzzy Kth-best and fuzzy Branch-and-bound approaches, for solving FLBLP problems are presented in Sect. 3. A numeral example and a case study are shown in Sect. 4 for illustrating the proposed approaches. Conclusions and further study are discussed in Sect. 5.

2 Fuzzy Linear Bilevel Programming Problems

This section presents a FLBLP model, related solution concepts and theorems for solving a FLBLP problem. Consider the following FLBLP problem:

For $x \in X \subset R^n, y \in Y \subset R^m, F, f : X \times Y \longrightarrow F^*(R)$,

$$\min_{x \in X} F(x,y) = \tilde{c}_1 x + \tilde{d}_1 y \tag{2.1a}$$

$$\text{subject to } \tilde{A}_1 x + \tilde{B}_1 y \preceq \tilde{b}_1 \tag{2.1b}$$

$$\min_{y \in Y} f(x,y) = \tilde{c}_2 x + \tilde{d}_2 y \tag{2.1c}$$

$$\text{subject to } \tilde{A}_2 x + \tilde{B}_2 y \preceq \tilde{b}_2 \tag{2.1d}$$

where $\tilde{c}_1, \tilde{c}_2 \in F^*(R^n), \tilde{d}_1, \tilde{d}_2 \in F^*(R^m)$, $\tilde{b}_1 \in F^*(R^p), \tilde{b}_2 \in F^*(R^q)$, $\tilde{A}_1 = (\tilde{a}_{ij})_{p \times n}$, $\tilde{a}_{ij} \in F^*(R), \tilde{B}_1 = \left(\tilde{b}_{ij}\right)_{p \times m}$, $\tilde{b}_{ij} \in F^*(R), \tilde{A}_2 = (\tilde{e}_{ij})_{q \times n}$, $\tilde{e}_{ij} \in F^*(R)$, $\tilde{B}_2 = (\tilde{s}_{ij})_{q \times m}$, $\tilde{s}_{ij} \in F^*(R)$.

Associated with the FLBLP problem, we consider the following multi-objective linear bilevel programming (MLBLP) problem:

For $x \in X \subset R^n, y \in Y \subset R^m, F, f : X \times Y \longrightarrow F^*(R)$,

$$\min_{x \in X} \ (F(x,y))_\lambda^L = c_{1\lambda}^L x + d_{1\lambda}^L y, \quad \lambda \in [0,\ 1] \tag{2.2a}$$

$$\min_{x \in X} \ (F(x,y))_\lambda^R = c_{1\lambda}^R x + d_{1\lambda}^R y, \quad \lambda \in [0,\ 1]$$

$$\text{subject to } A_{1\lambda}^L x + B_{1\lambda}^L y \leqq b_{1\lambda}^L,\ A_{1\lambda}^R x + B_{1\lambda}^R y \leqq b_{1\lambda}^R, \quad \lambda \in [0,\ 1] \tag{2.2b}$$

$$\min_{y \in Y} (f(x,y))_\lambda^L = c_{2\lambda}^L x + d_{2\lambda}^L y, \quad \lambda \in [0,\ 1] \tag{2.2c}$$

$$\min_{y \in Y} (f(x,y))_\lambda^R = c_{2\lambda}^R x + d_{2\lambda}^R y, \quad \lambda \in [0,\ 1]$$

$$\text{subject to } A_{2\lambda}^L x + B_{2\lambda}^L y \leqq b_{2\lambda}^L,\ A_{2\lambda}^R x + B_{2\lambda}^R y \leqq b_{2\lambda}^R, \quad \lambda \in [0,\ 1] \tag{2.2d}$$

where $c_{1\lambda}^L, c_{1\lambda}^R, c_{2\lambda}^L, c_{2\lambda}^R \in R^n$, $d_{1\lambda}^L, d_{1\lambda}^R, d_{2\lambda}^L, d_{2\lambda}^R \in R^m$, $b_{1\lambda}^L, b_{1\lambda}^R \in R^p$, $b_{2\lambda}^R \in R^q, A_{1\lambda}^L = \left(a_{ij\lambda}^L\right), A_{1\lambda}^R = \left(a_{ij\lambda}^R\right) \in R^{p \times n}, B_{1\lambda}^L = \left(b_{ij\lambda}^L\right), B_{1\lambda}^R = \left(b_{ij\lambda}^R\right)$ $\in R^{p \times m}$ $A_{2\lambda}^L = \left(e_{ij\lambda}^L\right), A_{2\lambda}^R = \left(e_{ij_\lambda}^R\right) \in R^{q \times n}, B_{2\lambda}^L = \left(s_{ij\lambda}^L\right), B_{2\lambda}^R = \left(s_{ij_\lambda}^R\right)$ $\in R^{q \times m}$.

The following definition provides a solution concept for FLBLP problems:

Definition 2.1 [20]

(a) The constraint region of the linear BLP problems is given by

$$S = \{(x,y) : x \in X, y \in Y, A_1 x + B_1 y \leqq b_1, A_2 x + B_2 y \leqq b_2\}$$

(b) The feasible set for the follower for each fixed $x \in X$, $S(x)$ is given by

$$S(x) = \{y \in Y : B_2 y \leqq b_2 - A_2 x\}$$

(c) The projection of S onto the leader's decision space is

$$S(X) = \{x \in X : \exists y \in Y, A_1 x + B_1 y \leqq b_1, A_2 x + B_2 y \leqq b_2\}$$

(d) The follower's rational reaction set for $x \in S(X)$ is

$$P(x) = \{y \in Y : y \in \arg\min[(f(x, \hat{y})) : \hat{y} \in S(x)]\}$$

where

$$\arg\min[f(x, \hat{y}) : \hat{y} \in S(x)] = \{y \in S(x) : (f(x, y)) \leqq (f(x, \hat{y})), \hat{y} \in S(x)\}$$

(e) The inducible region is

$$IR = \{(x, y) : (x, y) \in S, y \in P(x)\}$$

The rational reaction set *P(x)* defines the response while the inducible region *IR* represents the set over which the leader may optimize his/her objective. Thus in terms of the above notations, the linear BLP problem can be written as

$$\min\{F(x, y) : (x, y) \in IR\}.$$

Theorem 2.1 [20] If S is nonempty and compact there exists an optimal solution for the linear BLP problem.

Proof: Since S is nonempty, there exists a point $(x^*, y^*) \in S$. Then, we have

$$x^* \in S(X) \neq \phi,$$

by Definition 2.1(b). Consequently, we have

$$S(x^*) \neq \phi,$$

by Definition 2.1(c). Because S is compact and Definition 2.1(d), we have

$$\begin{aligned} P(x^*) &= \{y \in Y : y \in \arg\min[f(x^*, \hat{y}) : \hat{y} \in S(x^*)]\} \\ &= \{y \in Y : y \in \{y \in S(x^*) : f(x^*, y) \preceq f(x^*, \hat{y}), \hat{y} \in S(x^*)\}\} \neq \phi. \end{aligned}$$

Hence, there exist $y_0 \in P(x^*)$ such that $(x^*, y_0) \in S$. Therefore, we have

$$IR = \{(x, y) : (x, y) \in S, y \in P(x)\} \neq \phi,$$

by Definition 2.1(e). Because we are minimizing a linear function $F(x, y) = c_1 x + d_1 y$ over *IR*, which is nonempty and bounded, a Pareto optimal solution to the linear BLP problem must exist. So the proof is complete. QED

Theorem 2.2 [26] Let (x^*, y^*) be the solution of the MLBLP problem (2.2). Then it is also a solution of the FLBLP problem defined by (2.1).

Proof. This can be proved by the definition of the order relationship of fuzzy numbers. QED

Lemma 2.1 [26] If there is (x^*, y^*) such that $cx+dy \geqq cx^*+dy^*$, $c_0^L x+d_0^L y \geqq c_0^L x^*+d_0^L y^*$ and $c_0^R x+d_0^R y \geqq c_0^R x^*+d_0^R y^*$, for any (x,y) and isosceles triangle fuzzy numbers $\tilde{c}$ and $\tilde{d}$, then

$$c_\lambda^L x + d_\lambda^L y \geqq c_\lambda^L x^* + d_\lambda^L y^*,$$
$$c_\lambda^R x + d_\lambda^R y \geqq c_\lambda^R x^* + d_\lambda^R y^*,$$

for any $\lambda \in (0,1)$, where c and d are the centre of $\tilde{c}$ and $\tilde{d}$ respectively.

Proof. As λ-section of isosceles triangle fuzzy numbers $\tilde{c}$ and $\tilde{d}$ are

$$c_\lambda^L = c_0^L(1-\lambda) + c\lambda \text{ and } c_\lambda^R = c_0^R(1-\lambda) + c\lambda$$
$$d_\lambda^L = d_0^L(1-\lambda) + d\lambda \text{ and } d_\lambda^R = d_0^R(1-\lambda) + d\lambda.$$

Therefore, we have

$$\begin{aligned} c_\lambda^L x + d_\lambda^L y &= c_0^L(1-\lambda)x + c\lambda x + d_0^L(1-\lambda)y + d\lambda y \\ &= (c_0^L x + d_0^L y)(1-\lambda) + (cx+dy)\lambda \\ &\geqq (c_0^L x^* + d_0^L y^*)(1-\lambda) + (cx^*+dy^*)\lambda \\ &= c_\lambda^L x^* + d_\lambda^L y^*, \end{aligned}$$

from $cx+dy \geqq cx^*+dy^*$ and $c_0^L x + d_0^L y \geqq c_0^L x^* + d_0^L y^*$. We can prove $c_\lambda^R x + d_\lambda^R y \geqq c_\lambda^R x^* + d_\lambda^R y^*$ using a similar approach.

Theorem 2.3. For $x \in X \subset R^n, y \in Y \subset R^m$, if all the fuzzy coefficients $\tilde{a}_{ij}, \tilde{b}_{ij}, \tilde{e}_{ij}, \tilde{s}_{ij}, \tilde{c}_i$ and $\tilde{d}_i$ have triangle membership functions of the (FLBLP) problem (2.1), then these are given by

$$\mu_{\tilde{z}}(t) = \begin{cases} 0 & t < z_0^L \\ \frac{t - z_0^L}{z - z_0^L} & z_0^L \leqq t < z \\ \frac{-t + z_0^R}{z_0^R - z} & z \leqq t < z_0^R \\ 0 & z_0^R \leqq t \end{cases}, \tag{2.3}$$

where $\tilde{z}$ denotes $\tilde{a}_{ij}, \tilde{b}_{ij}, \tilde{e}_{ij}, \tilde{s}_{ij}, \tilde{c}_i$ and $\tilde{d}_i$ and z are the centre of $\tilde{z}$ respectively. Then, the solution of the problem (2.1) $(x^*, y^*) \in R^n \times R^m$ must satisfy:

$$\min_{x \in X} (F(x,y)) = c_1 x + d_1 y,$$
$$\min_{x \in X} (F(x,y))_0^L = {c_{10}}^L x + {d_{10}}^L y, \tag{2.4a}$$
$$\min_{x \in X} (F(x,y))_0^R = {c_{10}}^R x + {d_{10}}^R y,$$
$$\text{subject to } A_1 x + B_1 y \leqq b_1,$$
$${A_{10}}^L x + {B_{10}}^L y \leqq {b_{10}}^L, \tag{2.4b}$$
$${A_{10}}^R x + {B_{10}}^R y \leqq {b_{10}}^R,$$
$$\min_{y \in Y} (f(x,y)) = c_2 x + d_2 y,$$
$$\min_{y \in Y} (f(x,y))_0^L = {c_{20}}^L x + {d_{20}}^L y, \tag{2.4c}$$
$$\min_{y \in Y} (f(x,y))_\lambda^R = {c_{20}}^R x + {d_{20}}^R y,$$
$$\text{subject to } A_2 x + B_2 y \leqq b_2,$$
$${A_{20}}^L x + {B_{20}}^L y \leqq {b_{20}}^L, \tag{2.4d}$$
$${A_{20}}^R x + {B_{20}}^R y \leqq {b_{20}}^R.$$

Proof. From Lemma 2.1, if (x^*, y^*) satisfies (2.4a) and (2.4c), then it satisfies (2.2a) and (2.2c). Then we need only prove, if (x^*, y^*) satisfies (2.4b) and (2.4d), then it satisfies (2.2b) and (2.2d). In fact, for any $\lambda \in (0,1)$, ${a_{ij}}_\lambda^L = a_{ij}\lambda + {a_{ij}}_0^L(1-\lambda)$, ${b_{ij}}_\lambda^L = b_{ij}\lambda + {b_{ij}}_0^L(1-\lambda)$ and ${b_1}_\lambda^L = b_1\lambda + {b_1}_0^L(1-\lambda)$, we have

$$
\begin{aligned}
&{A_1}_\lambda^L x^* + {B_1}_\lambda^L y^* = (a_{ij}\lambda^L)x^* + ({b_{ij}}_\lambda^L)y^* \\
&= \left(a_{ij}\lambda + {a_{ij}}_0^L(1-\lambda)\right)x^* + \left(b_{ij}\lambda + {b_{ij}}_0^L(1-\lambda)\right)y^* \\
&= (a_{ij})\,x^*\lambda + \left({a_{ij}}_0^L\right)x^*(1-\lambda) + (b_{ij})\,y^*\lambda + \left({b_{ij}}_0^L\right)y^*(1-\lambda) \\
&= ((a_{ij})\,x^* + (b_{ij})\,y^*)\,\lambda + \left(\left({a_{ij}}_0^L\right)x^* + \left({b_{ij}}_0^L\right)y^*\right)(1-\lambda) \\
&= (A_1 x^* + B_1 y^*)\,\lambda + \left({A_{10}}^L x^* + {B_{10}}^L y^*\right)(1-\lambda) \leqq b_1\lambda + {b_{10}}^L(1-\lambda) = {b_1}_\lambda^L,
\end{aligned}
$$

from (2.4b). Similarly, we can prove

$$
\begin{aligned}
{A_1}_\lambda^R x^* + {B_1}_\lambda^R y^* &\leqq {b_1}_\lambda^R, \\
{A_2}_\lambda^L x^* + {B_2}_\lambda^L y^* &\leqq {b_2}_\lambda^L, \\
{A_2}_\lambda^R x^* + {B_2}_\lambda^R y^* &\leqq {b_2}_\lambda^R,
\end{aligned}
$$

for any $\lambda \in (0,1)$ from (2.4b) and (2.4d). The proof is complete. QED

These definitions and theorems will be used for the development of the following three FLBLP approaches.

3 Three Approaches for Solving Fuzzy Linear Bilevel Programming Problems

This section presents three approaches for solving FLBLP problems, namely, (1) the fuzzy Kuhn-Tucker approach, (2) the fuzzy Kth-best approach and (3) the fuzzy Branch-and-bound approach

3.1 The Fuzzy Kuhn-Tucker Approach

The Kuhn-Tucker approach is proposed and discussed by [5–7]. The principle of the Kuhn-Tucker approach is that the upper level's problem with constraints involves optimality conditions of the lower level's problem solution. That is, to first replace the follower's decision problem with its Kuhn-Tucker conditions, and append the resultant system to the leader's decision problem. The BLP therefore becomes a standard mathematical programming problem where all the constraints are linear except one. It can be solved by using a normal mathematical programming method. For a FLBLP problem, we first apply Theorem 2.3 to transform it to a MLBLP problem. The original Kuhn-Tucker approach is then used to solve the transformed model. Finally, the objective values, of both the leader and the follower, described by fuzzy numbers are obtained. The approach is presented in the proof of the following Theorem 3.3.

Theorem 3.1 [26] A necessary and sufficient condition that (x^*, y^*) solves the FLBLP problem (3.1) with triangle fuzzy numbers is that there exist (row) vectors u^*, v^*, and w^* such that $(x^*, y^*, u^*, v^*, w^*)$ is the solution to:

$$\min_{x\in X} (F(x,y)) = (c_1 x + d_1 y) + \left(c_{10}{}^L x + d_{10}{}^L y\right) + \left(c_{10}{}^R x + d_{10}{}^R y\right) \tag{3.1a}$$

$$\begin{aligned} \text{subject to } & A_1 x + B_1 y \leqq b_1, \\ & A_{10}{}^L x + B_{10}{}^L y \leqq b_{10}{}^L, \\ & A_{10}{}^R x + B_{10}{}^R y \leqq b_{10}{}^R, \end{aligned} \tag{3.1b}$$

$$\begin{aligned} & A_2 x + B_2 y \leqq b_2, \\ & A_{20}{}^L x + B_{20}{}^L y \leqq b_{20}{}^L, \\ & A_{20}{}^R x + B_{20}{}^R y \leqq b_{20}{}^R. \end{aligned} \tag{3.1c}$$

$$\begin{aligned} & u_1 B_1 + u_2 B_{10}{}^L + u_3 B_{10}{}^R + v_1 B_2 + v_2 B_{20}{}^L + v_3 B_{20}{}^R - w \\ & \quad = -\left(d_2 + d_{20}{}^L + d_{20}{}^R\right) \end{aligned} \tag{3.1d}$$

$$\begin{aligned} & u_1 (b_1 - A_1 x - B_1 y) + u_2 \left(b_{10}{}^L - A_{10}{}^L x - b_{10}{}^L y\right) \\ & \quad + u_3 \left(b_{10}{}^R - A_{10}{}^R x - B_{10}{}^R y\right) + v_1 (b_2 - A_2 x - B_2 y) \\ & \quad + v_2 \left(b_{20}{}^L - A_{20}{}^L x - B_{20}{}^L y\right) \\ & \quad + v_3 \left(b_{20}{}^R - A_{20}{}^R x - B_{20}{}^R y\right) + wy = 0 \end{aligned} \tag{3.1e}$$

$$x \geq 0, y \geq 0, u \geq 0, v \geq 0, w \geq 0. \tag{3.1f}$$

Proof: (1) From Theorem 2.3, we know that we need to only solve the problem (2.4). In fact, to solve the problem (2.4), we can use the method of weighting [15] to this problem, such that it is the following problem:

$$\min_{x\in X}(F(x,y)) = (c_1x + d_1y) + \left(c_{10}{}^L x + d_{10}{}^L y\right) + \left(c_{10}{}^R x + d_{10}{}^R y\right) \quad (3.2a)$$

$$\text{subject to } A_1x + B_1y \leqq b_1,$$

$$A_{10}{}^L x + B_{10}{}^L y \leqq b_{10}{}^L, \quad (3.2b)$$

$$A_{10}{}^R x + B_{10}{}^R y \leqq b_{10}{}^R,$$

$$\min_{y\in Y}(f(x,y)) = c_2x + d_2y + c_{20}{}^L x + d_{20}{}^L y + c_{20}{}^R x + d_{20}{}^R y \quad (3.2c)$$

$$\text{subject to } A_2x + B_2y \leqq b_2,$$

$$A_{20}{}^L x + B_{20}{}^L y \leqq b_{20}{}^L, \quad (3.2d)$$

$$A_{20}{}^R x + B_{20}{}^R y \leqq b_{20}{}^R.$$

Therefore, the linear BLP problem can be written as

$$\min\{F(x,y) : (x,y) \in IR\} \quad (3.3)$$

We can prove (3.3) is equivalent to the following linear BLP programming by Definition 2.1.

$$\min_{x\in X}(F(x,y)) = (c_1x + d_1y) + \left(c_{10}{}^L x + d_{10}{}^L y\right) + \left(c_{10}{}^R x + d_{10}{}^R y\right) \quad (3.4a)$$

$$\text{subject to } A_1x + B_1y \leqq b_1,$$

$$A_{10}{}^L x + B_{10}{}^L y \leqq b_{10}{}^L, \quad (3.4b)$$

$$A_{10}{}^R x + B_{10}{}^R y \leqq b_{10}{}^R,$$

$$A_2x + B_2y \leqq b_2,$$

$$A_{20}{}^L x + B_{20}{}^L y \leqq b_{20}{}^L,$$

$$A_{20}{}^R x + B_{20}{}^R y \leqq b_{20}{}^R.$$

$$\min_{y\in Y}(f(x,y)) = c_2x + d_2y + c_{20}{}^L x + d_{20}{}^L y + c_{20}{}^R x + d_{20}{}^R y$$

$$\text{subject to } A_1x + B_1y \leqq b_1, \quad (3.4c)$$

$$A_{10}{}^L x + B_{10}{}^L y \leqq b_{10}{}^L,$$

$$A_{10}{}^R x + B_{10}{}^R y \leqq b_{10}{}^R,$$

$$A_2x + B_2y \leqq b_2,$$

$$A_{20}{}^L x + B_{20}{}^L y \leqq b_{20}{}^L, \quad (3.4d)$$

$$A_{20}{}^R x + B_{20}{}^R y \leqq b_{20}{}^R.$$

(2) Necessity is obvious from (3.4).
(3) Sufficiency. If (x^*, y^*) is the optimal solution of (3.2), we need to show that there exist (row) vectors u_1^*, u_2^*, u_3^*, v_1^*, v_2^*, v_3^* and w^* such that $(x^*, y^*, u_1^*, u_2^*, u_3^*, v_1^*, v_2^*, v_3^*, w^*)$ to solve (3.1). Going one step farther, we only need to prove that there exist vectors u_1^*, u_2^*, u_3^*, v_1^*, v_2^*, v_3^* and w^* such that $(x^*, y^*, u_1^*, u_2^*, u_3^*, v_1^*, v_2^*, v_3^*, w^*)$ satisfy:

$$\begin{aligned}
& u_1 B_1 + u_2 {B_{10}}^L + u_3 {B_{10}}^R + v_1 B_2 + v_2 {B_{20}}^L + v_3 {B_{20}}^R - w \\
& \quad = -(d_2 + {d_{20}}^L + {d_{20}}^R) && \text{(3.5a)} \\
& u_1(b_1 - A_1 x - B_1 y) = 0 && \text{(3.5b)} \\
& u_2({b_{10}}^L - {A_{10}}^L x - {B_{10}}^L y) = 0 && \text{(3.5c)} \\
& u_3({b_{10}}^R - {A_{10}}^R x - {B_{10}}^R y) = 0 && \text{(3.5d)} \\
& v_1(b_2 - A_2 x - B_2 y) = 0 && \text{(3.5e)} \\
& v_2({b_{20}}^L - {A_{20}}^L x - {B_{20}}^L y) = 0 && \text{(3.5f)} \\
& v_3({b_{20}}^R - {A_{20}}^R x - {B_{20}}^R y) = 0 && \text{(3.5g)} \\
& wy = 0, && \text{(3.5h)}
\end{aligned}$$

where $u_1, u_2, u_3 \in R^p$, $v_1, v_2, v_3 \in R^q$, $w \in R^m$ and they are not negative variables.
Because (x^*, y^*) is the optimal solution of (3.2), we have

$$(x^*, y^*) \in IR,$$

by (3.3). Thus we have

$$y^* \in P(x^*),$$

by Definition 2.1(e). y^* is the optimal solution to the following problem

$$\min(f(x^*, y) : y \in S(x^*)), \tag{3.6}$$

by Definition 2.1(d). Rewrite (3.6) as follows

$$\begin{aligned}
& \min f(x, y) \\
& \text{subject to } y \in S(x) \\
& \qquad\qquad\quad x = x^*.
\end{aligned}$$

From Definition 2.1(b), we have

$$\begin{aligned}
& \min_{y \in Y} (f(x, y)) = c_2 x + d_2 y + {c_{20}}^L x + {d_{20}}^L y + {c_{20}}^R x + {d_{20}}^R y && \text{(3.7a)} \\
& \text{subject to } A_1 x + B_1 y \leqq b_1, && \text{(3.7b)} \\
& \qquad {A_{10}}^L x + {B_{10}}^L y \leqq {b_{10}}^L, && \text{(3.7c)} \\
& \qquad {A_{10}}^R x + {B_{10}}^R y \leqq {b_{10}}^R, && \text{(3.7d)}
\end{aligned}$$

$$A_2x + B_2y \leqq b_2, \quad (3.7e)$$
$$A_{20}{}^Lx + B_{20}{}^Ly \leqq b_{20}{}^L, \quad (3.7f)$$
$$A_{20}{}^Rx + B_{20}{}^Ry \leqq b_{20}{}^R. \quad (3.7g)$$
$$x = x^* \quad (3.7h)$$
$$y \geqq 0. \quad (3.7i)$$

To simplify (3.7), we can have

$$\min g(y) =(d_2 + d_{20}{}^L + d_{20}{}^R)y \quad (3.8a)$$
$$\text{subject to } - B_1y \geqq -(b_1 - A_1x^*), \quad (3.8b)$$
$$- B_{10}{}^Ly \geqq -(b_{10}{}^L - A_{10}{}^Lx^*), \quad (3.8c)$$
$$- B_{10}{}^Ry \geqq -(b_{10}{}^R - A_{10}{}^Rx^*), \quad (3.8d)$$
$$- B_2y \geqq -(b_2 - A_2x^*), \quad (3.8e)$$
$$- B_{20}{}^Ly \geqq -(b_{20}{}^L - A_{20}{}^Lx^*), \quad (3.8f)$$
$$- B_{20}{}^Ry \geqq -(b_{20}{}^R - A_{20}{}^Rx^*), \quad (3.8g)$$
$$y \geqq 0. \quad (3.8h)$$

Let we note

$$B = \begin{pmatrix} B_1 \\ B_{10}{}^L \\ B_{10}{}^R \\ B_2 \\ B_{20}{}^L \\ B_{20}{}^R \end{pmatrix}, \quad A = \begin{pmatrix} A_1 \\ A_{10}{}^L \\ A_{10}{}^R \\ A_2 \\ A_{20}{}^L \\ A_{20}{}^R \end{pmatrix}, \quad \text{and} \quad b = \begin{pmatrix} b_1 \\ b_{10}{}^L \\ b_{10}{}^R \\ b_2 \\ b_{20}{}^L \\ b_{20}{}^R \end{pmatrix}. \quad (3.9)$$

We rewrite (3.8) by using (3.9) and we get

$$\min g(y) = (d_2 + d_{20}{}^L + d_{20}{}^R)y \quad (3.10a)$$
$$\text{subject to } - By \geqq -(b - Ax^*) \quad (3.10b)$$
$$y \geqq 0. \quad (3.10c)$$

Now we see that y^* is the optimal solution of (3.10) which is a linear programming problem. By Proposition 2 from [20], there exists vector λ^*, μ^*, such that (y^*, λ^*, μ^*) satisfy the system below

$$\lambda B - \mu = -(d_2 + d_{20}{}^L + d_{20}{}^R) \quad (3.11a)$$
$$- By + (b - Ax^*) \geqq 0 \quad (3.11b)$$
$$\lambda(-By + (b - Ax^*)) = 0 \quad (3.11c)$$
$$\mu y = 0, \quad (3.11d)$$

where $\lambda \in R^{3p+3q}$ and $\mu \in R^m$.

Let $u_1, u_2, u_3 \in R^p, v_1, v_2, v_3 \in R^q$ and $w \in R^m$ and define

$$\lambda = (u_1,\ u_2,\ u_3, v_1,\ v_2, v_3)$$
$$w = \mu.$$

Thus we have $(x^*, y^*, u_1^*, u_2^*, u_3^*, v_1^*, v_2^*, v_3^*, w^*)$ that satisfy (3.5). Our proof is completed. QED

Theorem 3.3 shows that the most direct approach to solve (2.1) is to solve the equivalent mathematical program given in (3.1). One advantage of the approach is that it allows a more robust model to be solved without introducing any new computational difficulty.

3.2 The Fuzzy Branch-and-Bound Approach

The core of the Branch-and-bound approach is the Branch-and-bound algorithm which can deal with the complementary constraints arising from the Kuhn-Tucker conditions. The algorithm requires additional $m + q$ variables and the explicit satisfaction of complementary slackness. Here, m is the number of the follower's variables and q is the number of the follower's constraints. In our previous work [19] an extended Branch-and-bound algorithm was proposed which requires $p + m + q$ variables, where p is the number of leader's constraints. Based on the result, similar to the fuzzy Kuhn-tucker approach, the fuzzy Branch-and-bound approach first transforms a FLBLP to a MLBLP problem. It then solves the MLBLP problem by using the extended Branch-and-bound algorithm. Finally, the objectives, of both the leader and the follower, described by fuzzy values are obtained. The detailed work process of the approach is shown as follows [29].

Write all the inequalities (except of the leader's variables) of (2.1) as $g_i(x, y) \geq 0, i = 1, \ldots, p + q + m$, and note that complementary slackness simply means $u_i g_i(x, y) = 0 (i = 1, \ldots, p + q + m)$. We suppress the complementarity's term and solve the resulting linear sub-problem. At each iteration stage, the approach checks if (3.1e) is satisfied. If it is satisfied, the corresponding point is in the inducible region and hence a potential solution to (2.1) is obtained. Otherwise, a branch and bound scheme is used to implicitly examine all the combinations of complementarities slackness. A flowchart of the approach is displayed in Fig. 1.

Let $W = \{1, \ldots, p + q + m\}$ be the index set for the terms in (3.1e), $\bar{F}$ be the incumbent upper bound on the leader's objective function. At the kth level of the search tree we define a subset of indices $W_k \subset W$, and a path P_k corresponding to an assignment of either $u_i = 0$ or $g_i = 0$ for $i \in W_k$. Let

$$S_k^+ = \{i : i \in W_k, u_i = 0\}$$
$$S_k^- = \{i : i \in W_k, g_i = 0\}$$
$$S_k^0 = \{i : i \notin W_k\}.$$

For $i \in S_k^0$, the variables u_i or g_i are free to assume any nonnegative value in the solution of (3.1) with (3.1e) omitted, so complementary slackness will not necessarily be satisfied. The approach can be accomplished with the following procedure.

Step 1 To transform problem (2.1) into the problem (2.4) by using Theorem 2.3.

Step 2 To transform problem (2.4) into the following linear bilevel programming problem (3.1) by using the method of weighting [15].

Step 3 To solve the problem (3.1).

Step 3.0 (Initialization) Set $k = 0$, $S_k^+ = \phi$, $S_k^- = \phi$, $S_k^0 = \{1, \ldots, p+q+m\}$ and $\bar{F} = \infty$.

Step 3.1 (Iteration k) Set $u_i = 0$ for $i \in S_k^+$ and $g_i = 0$ for $i \in S_k^-$. Attempt to solve (3.1) without (3.1e). If the resultant problem is infeasible, go to Step 3.5; otherwise, put $k \leftarrow k+1$ and label the solution (x^k, y^k, u^k).

Step 3.2 (Fathoming) If $F(x^k, y^k) \geq \bar{F}$, go to Step 3.5.

Step 3.3 (Branching) if $u_i^k g_i(x^k, y^k) = 0$, $i = 1, \ldots, p+q+m$, go to Step 3.4. Otherwise, select i for which $u_i^k g_i(x^k, y^k) \neq 0$ is the largest

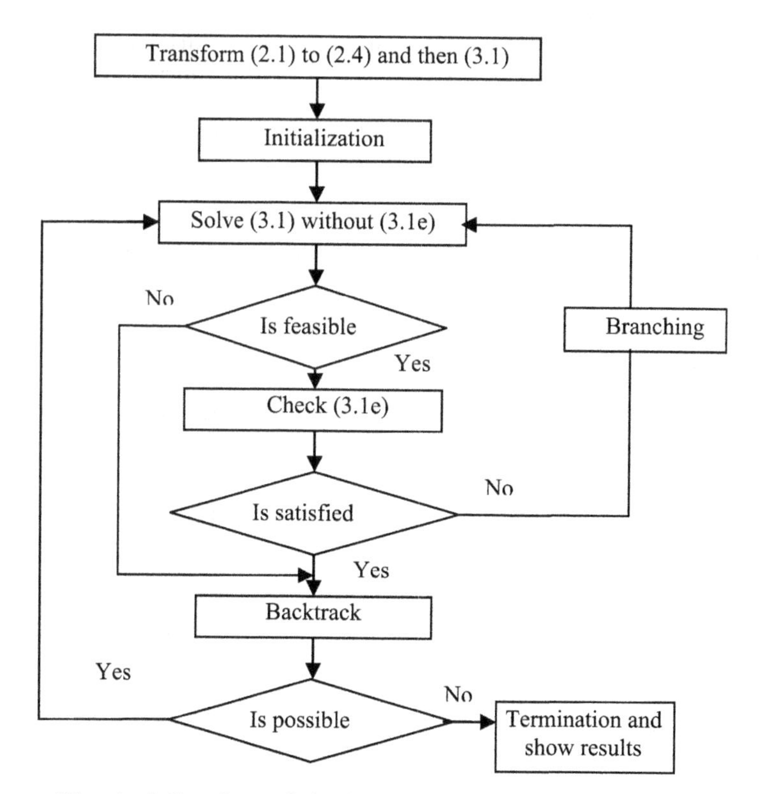

Fig. 1. A flowchart of the fuzzy branch and bound approach

and label it i_1. Put $S_k^+ \leftarrow S_k^+ \cup \{i_1\}$, $S_k^0 \leftarrow S_k^0 \backslash \{i_1\}$, $S_k^- \leftarrow S_k^-$, append i_1 to P_k, and go to Step 3.1.

Step 3.4 (Updating) $\bar{F} \leftarrow F(x^k, y^k)$.

Step 3.5 (Backtracking) If no live node exists, go to Step 3.6. Otherwise branch to the newest live vertex and update S_k^+, S_k^-, S_k^0 and P_k as discussed below. Go to Step 3.1.

Step 3.6 (Termination) If $\bar{F} = \infty$, there are no feasible solution to (2.1). Otherwise, declare the feasible point associated with $\bar{F}$ the optimal solution to (2.1).

Step 4 Show the result of problem (2.1).

3.3 The Fuzzy Kth-best Approach

The Kth-best approach uses vertex enumeration. The solution search procedure of the method starts from a point which is an optimal solution to the problem of the upper level and checks whether it is also an optimal solution to the problem of the lower level or not. If the first point is not a Stackeberg solution, the procedure continues to examine the second best solution to the problem of the upper level and so on. The Kuhn-Tucker approach is used by Bialas and Karwan [7] in their parametric complementary pivot algorithm. So the core of the approach is to compute global solutions of linear BLP problems by enumerating the extreme points of the constraint region. The proposed fuzzy Kth-best approach will use the original Kth-best approach after transforming a FLBLP to a MLBLP problem. The following theorems and corollaries will be used to prove the proposed fuzzy Kth-best approach.

Theorem 3.2 [27] The inducible region can be written equivalently as a piecewise linear equality constraint comprised of supporting hyperplanes of S.

Proof. Let us begin by writing the inducible region of Definition 2.1(e) explicitly as follower:

$$\begin{aligned} IR &= \{(x, y) : (x, y) \in S, d_2 y \\ &= \min[d_2 \tilde{y} : B_1 \tilde{y} \leq b_1 - A_1 x, B_2 \tilde{y} \leq b_2 - A_2 x, \tilde{y} \geq 0]\}. \end{aligned}$$

Now define

$$Q(x) = \min\{d_2 y : B_1 y \leq b_1 - A_1 x, B_2 y \leq b_2 - A_2 x, y \geq 0\}. \tag{3.12}$$

$$B = \begin{pmatrix} B_1 \\ B_2 \end{pmatrix}, \quad b = \begin{pmatrix} b_1 \\ b_2 \end{pmatrix}, \quad A = \begin{pmatrix} A_1 \\ A_2 \end{pmatrix}.$$

We rewrite (3.12) as follows

$$Q(x) = \min\{d_2 y : By \leq b - Ax, y \geq 0\}. \tag{3.13}$$

For each value of $x \in S(X)$, the resulting feasible region to problem (2.1) is nonempty and compact. Thus $Q(x)$, which is a linear program parameterized in x, always has a solution. From duality theory we get

$$\max\{u(Ax - b) : uB \geq -d_2, u \geq 0\}, \tag{3.14}$$

which has the same optimal value as (3.12) at the solution u^*. Let $u^1, \ldots, u^s$ be a listing of all the vertices of the constraint region of (3.14) given by $U = \{u : uB \geq -d_2, u \geq 0\}$. Because we know that a solution to (3.14) occurs at a vertex of U, we get the equivalent problem

$$\max\{u^j(Ax - b) : u^j \in \{u^1, \ldots, u^s\}\}, \tag{3.15}$$

which demonstrates that $Q(x)$ is a piecewise linear function. Rewriting IR as

$$IR = \{(x, y) \in S : Q(x) - d_2 y = 0\}, \tag{3.16}$$

yields the desired result. QED

Corollary 3.1 [28] The linear BLP problem (2.1) is equivalent to minimizing F over a feasible region comprised of a piecewise linear equality constraint.

Proof. By (2.3) and Theorem 2.3, we have the desired result.

The function $Q(x)$ defined by (2.4) is convex and continuous. In general, because we are minimizing a linear function $F = c_1 x + d_1 y$ over IR, and because F is bounded below S by, say, $\min\{c_1 x + d_1 y : (x, y) \in S\}$, the corollary can be concluded. QED

Corollary 3.2 [28] A solution for the linear BLP problem occurs at a vertex of IR.

Proof. A linear BLP programming can be written as (2.3). Since $F = c_1 x + d_1 y$ is linear, if a solution exists, one must occur at a vertex of IR. The proof is completed. QED

Theorem 3.3 [28] The solution (x^*, y^*) of the linear BLP problem occurs at a vertex of S.

Proof. Let $(x^1, y^1), \ldots, (x^r, y^r)$ be the distinct vertices of S. Since any point in S can be written as a convex combination of these vertices, let $(x^*, y^*) = \sum_{i=1}^{\bar{r}} \alpha_i (x^i, y^i)$, where $\sum_{i=1}^{\bar{r}} \alpha_i = 1, \alpha_i \geq 0, i = 1, \ldots, \bar{r}$ and $\bar{r} \leq r$. It must be shown that $\bar{r} = 1$. To see this let us write the constraints to (2.1) at (x^*, y^*) in their piecewise linear form (3.16).

$$\begin{aligned} 0 &= Q(x^*) - d_2 y^* \\ &= Q(\sum_i \alpha_i x^i) - d_2(\sum_i \alpha_i y^i) \\ &\leq \sum_i \alpha_i Q(x^i) - \sum_i \alpha_i d_2 y^i \qquad \text{by convexity of } Q(x) \\ &= \sum_i \alpha_i (Q(x^i) - d_2 y^i). \end{aligned}$$

But by Definition 2.1,

$$Q(x^i) = \min_{y \in S(x^i)} d_2 y \leq d_2 y^i.$$

Therefore, $Q(x^i) - d_2 y^i \leq 0, i = 1, \ldots, \bar{r}$. Noting that $\alpha_i \geq 0, i = 1, \ldots, \bar{r}$, the equality in the preceding expression must hold or else a contradiction would result in the sequence above. Consequently, $Q(x^i) - d_2 y^i = 0$ for all i. This implies that $(x^i, y^i) \in IR$, $i = 1, \ldots, \bar{r}$ and (x^*, y^*) can be written as a convex combination of points in IR. Because (x^*, y^*) is a vertex of IR, a contradiction results unless $\bar{r} = 1$. QED

Corollary 3.3 [28] If x is an extreme point of IR, it is an extreme point of S.

Proof. We can prove it by using Theorem 3.3. QED

Theorem 3.3 and Corollary 3.3 have provided theoretical foundation for the fuzzy Kth-best approach. It means that by searching extreme points on the constraint region S, we can efficiently find an optimal solution for a linear BLP problem. According to the objective function of the upper level, we order all the extreme points on S in descending order, and select the first extreme point to check if it is on the inducible region IR. If yes, the current extreme point is the optimal solution. Otherwise, continue the process.

More specifically, let $(x_{[1]}, y_{[1]}), \ldots, (x_{[N]}, y_{[N]})$ denote the N ordered extreme points to the linear programming problem

$$\min\{c_1 x + d_1 y : (x, y) \in S\}, \tag{3.17}$$

such that $c_1 x_{[i]} + d_1 y_{[i]} \leq c_1 x_{[i+1]} + d_1 y_{[i+1]}, i = 1, \ldots, N-1$.

Let $\tilde{y}$ denote the optimal solution to the following problem

$$\min(f(x_{[i]}, y) : y \in S(x_{[i]})). \tag{3.18}$$

We only need to find the smallest i ($i \in \{1, \ldots, N\}$) under which $y_{[i]} = \tilde{y}$.

Let us write (3.18) as follows

$$\begin{aligned} &\min f(x, y) \\ &\text{subject to } y \in S(x) \\ &x = x_{[i]}. \end{aligned}$$

From Definition 2.1(a) and (c), we have

$$\min f(x, y) = c_2 x + d_2 y \tag{3.19a}$$
$$\text{subject to } A_1 x + B_1 y \leq b_1 \tag{3.19b}$$
$$A_2 x + B_2 y \leq b_2 \tag{3.19c}$$
$$x = x_{[i]} \tag{3.19d}$$
$$y \geq 0. \tag{3.19e}$$

The solution is equivalent to selecting one of the ordered extreme points $(x_{[i]}, y_{[i]})$, and then solve (3.19) to obtain the optimal solution $\tilde{y}$. If $\tilde{y} = y_{[i]}$, $(x_{[i]}, y_{[i]})$ is the global optimum to (2.1). Otherwise, check the next extreme point.

This approach can be accomplished with the following procedure.

Step 1 Transform problem (2.1) into the problem (2.4) by using Theorem 2.3.
Step 2 Transform problem (2.4) into the following linear bilevel programming problem (3.1) by using the method of weighting [15].
Step 3 Solve the problem (3.1).
Step 3.1 Put $i \leftarrow 1$. Solve (3.17) with the simplex method to obtain the optimal solution $(x_{[1]}, y_{[1]})$. Let $W = \{(x_{[1]}, y_{[1]})\}$ and $T = \phi$. Go to Step 2.
Step 3.2 Solve (3.19) with the bounded simplex method. Let $\tilde{y}$ denote the optimal solution to (3.19). If $\tilde{y} = y_{[i]}$, stop; $(x_{[i]}, y_{[i]})$ is the global optimum to (2.1) with $K^* = i$. Otherwise, go to Step 3.
Step 3.3 Let $W_{[i]}$ denote the set of adjacent extreme points of $(x_{[i]}, y_{[i]})$ such that $(x, y) \in W_{[i]}$ implies $c_1 x + d_1 y \geq c_1 x_{[i]} + d_1 y_{[i]}$. Let $T = T \cup \{(x_{[i]}, y_{[i]})\}$ and $W = (W \cup W_{[i]}) \backslash T$. Go to Step 4.
Step 3.4 Set $i \leftarrow i + 1$ and choose $(x_{[i]}, y_{[i]})$ so that

$$c_1 x_{[i]} + d_1 y_{[i]} = \min\{c_1 x + d_1 y : (x, y) \in W\}.$$

Go back to Step 2.
Step 4 Show the result of problem (2.1).

In principle, the fuzzy Kuhn-tucker approach provides a principle to deal with FLBLP problem. The fuzzy Kth-best and the fuzzy Brand-and-bound approaches are developed based on this principle. The two approaches can be described by steps and therefore are easily applied in real decision making processes.

4 Numerical Examples and Case-based Applications

This section presents a numeral example and a case based application for illustrating the proposed three FLBLP approaches.

4.1 A Numerical Example for the Fuzzy Kuhn-Tucker Approach

Consider the following FLBLP problem with $x, y \in R$, and $X = \{x \geq 0\}$, $Y = \{y \geq 0\}$,

$$\min_{x \in X} F(x, y) = \tilde{1}x - \tilde{2}y \tag{4.1a}$$

$$\text{subject to} - \tilde{1}x + \tilde{3}y \leq \tilde{4} \tag{4.1b}$$

$$\min_{y \in Y} f_1(x, y) = \tilde{1}x + \tilde{1}y \tag{4.1c}$$

$$\text{subject to } \tilde{1}x - \tilde{1}y \leq \tilde{0} \tag{4.1d}$$
$$-\tilde{1}x - \tilde{1}y \leq \tilde{0} \tag{4.1e}$$

where $\mu_{\tilde{1}}(t) = \begin{cases} 0 & t < 0 \\ t & 0 \leqq t < 1 \\ 2-t & 1 \leqq t < 2' \\ 0 & 2 \leqq t \end{cases}$ $\mu_{\tilde{2}}(t) = \begin{cases} 0 & t < 1 \\ t-1 & 1 \leqq t < 2 \\ 3-t & 2 \leqq t < 3' \\ 0 & 3 \leqq t \end{cases}$

$\mu_{\tilde{3}}(t) = \begin{cases} 0 & t < 2 \\ t-2 & 2 \leqq t < 3 \\ 4-t & 3 \leqq t < 4' \\ 0 & 4 \leqq t \end{cases}$ $\mu_{\tilde{4}}(t) = \begin{cases} 0 & t < 3 \\ t-3 & 3 \leqq t < 4 \\ 5-t & 4 \leqq t < 5' \\ 0 & 5 \leqq t \end{cases}$

$\mu_{\tilde{0}}(t) = \begin{cases} 0 & t < -1 \\ t+1 & -1 \leqq t < 0 \\ 1-t & 0 \leqq t < 1 \\ 0 & 1 \leqq t \end{cases}$

We use the fuzzy Kuhn-Tucker approach to solve the FLBLP problem.

Step 1. The problem (4.1) is transformed into the following MLBLP problem by using Theorem 2.3

$$\begin{aligned}
&\min_{x \in X} (F(x,y))_c = 1x - 2y \\
&\min_{x \in X} (F(x,y))_0^L = 0x - 3y \\
&\min_{x \in X} (F(x,y))_0^R = 2x - 1y \\
&\text{subject to } -1x + 3y \leq 4 \\
&\qquad -2x + 2y \leq 3 \\
&\qquad 0x + 4y \leq 5 \\
&\qquad \min_{y \in Y} (f(x,y))_c = 1x + 1y \\
&\qquad \min_{y \in Y} (f(x,y))_0^L = 0x + 0y \\
&\qquad \min_{y \in Y} (f(x,y))_0^R = 2x + 2y \\
&\qquad \text{subject to } 1x - 1y \leq 0 \\
&\qquad\qquad 0x - 2y \leq -1 \\
&\qquad\qquad 2x - 0y \leq 1 \\
&\qquad\qquad -1x - 1y \leq 0
\end{aligned}$$

$$0x - 0y \leq 0$$
$$-2x - 2y \leq -1.$$

Step 2. The problem is transformed into the following linear BLP problem by using a weighting method [15].

$$\min_{x \in X} F(x,y) = 3x - 6y$$
$$\text{subject to } -1x + 3y \leq 4$$
$$-2x + 2y \leq 3$$
$$0x + 4y \leq 5$$
$$\min_{y \in Y} f(x,y) = 3x + 3y$$
$$\text{subject to } 1x - 1y \leq 0$$
$$0x - 2y \leq -1$$
$$2x - 0y \leq 1$$
$$-1x - 1y \leq 0$$
$$-2x - 2y \leq -1.$$
$$0x - 0y \leq 1-$$

Step 3. Solve this linear BLP problem by using the extended Kuhn-Tucker approach [20]

$$\min_{x \in X} F(x,y) = 3x - 6y$$
$$\text{subject to } -1x + 3y \leq 4$$

$$0x + 4y \leq 5$$
$$1x - 1y \leq 0$$
$$0x - 2y \leq -1$$
$$2x - 0y \leq 1$$
$$-1x - 1y \leq 0$$
$$-2x - 2y \leq -1$$
$$0x - 0y \leq 1$$
$$3u_1 + 2u_2 + 4u_3 - u_4 - 2u_5 - 0u_6 - u_7 - 2u_8 - 0u_9 - u_{10} = -3$$
$$u_1(4 + 1x - 3y) + u_2(3 + 2x - 2y) + u_3(5 - 4y) + u_4(-x + y)$$
$$+ u_5(-1 + 2y) + u_6(1 - 2x) + u_7(x + y) + u_8(-1 + 2x + 2y) + u_9 + u_{10}y$$
$$= 0 \; x \geq 0, y \geq 0, u_1 \geq 0, \ldots, u_{10} \geq 0.$$

Step 4. The result is

$$\begin{aligned}
&\min_{x\in X}(F(x,y))_c = 1x - 2y = -1\\
&\min_{x\in X}(F(x,y))_0^L = 0x - 3y = -1.3\\
&\min_{x\in X}(F(x,y))_0^R = 2x - 1y = -0.5\\
&\min_{y\in Y}(f(x,y))_c = 0.5\\
&\min_{y\in Y}(f(x,y))_0^L = 0\\
&\min_{y\in Y}(f(x,y))_0^R = 1\\
&x = 0, y = 0.5
\end{aligned}$$

Consequently, we have the solution of the problem (4.1)

$$\begin{aligned}
&\min_{x\in X} F(x,y) = \tilde{1}x - \tilde{2}y = \tilde{c}\\
&\min_{y\in Y} f_1(x,y) = \tilde{1}x + \tilde{1}y = \tilde{d}
\end{aligned}$$

and

$$x = 0, y = 0.5,$$

where

$$\mu_{\tilde{c}}(t) = \begin{cases} 0 & t < -1.3 \\ \frac{t+1.3}{0.3} & -1.3 \le t < -1 \\ \frac{-0.5-t}{0.5} & -1 \le t < -0.5 \\ 0 & -0.5 \le t \end{cases}, \quad \mu_{\tilde{d}}(t) = \begin{cases} 0 & t < 0 \\ \frac{t}{0.5} & 0 \le t < 0.5 \\ \frac{1-t}{0.5} & 0.5 \le t < 1 \\ 0 & 1 \le t \end{cases}.$$

4.2 A Numerical Example for the Fuzzy Branch-and-Bound Approach

We still use the example (4.1) shown in Sect. 4.1 for illustrate the fuzzy Branch-and-bound approach.

Step 1. Same as the Step 1 of the fuzzy Kuhn-Tucker approach presented in Sect. 4.1.

Step 2. Same as the Step 2 of the fuzzy Kuhn-Tucker approach presented in Sect. 4.1.

Step 3. According to the extended Branch-and-bound approach [19] we have

$$g_1(x,y) = 4 - (-1x + 3y) \geq 0$$
$$g_2(x,y) = 3 - (-2x + 2y) \geq 0$$
$$g_3(x,y) = 5 - (0x + 4y) \geq 0$$
$$g_4(x,y) = -(1x - 1y) \geq 0$$
$$g_5(x,y) = -1 - (0x - 2y) \geq 0$$
$$g_6(x,y) = 1 - (2x - 0y) \geq 0$$
$$g_7(x,y) = 1x + 1y \geq 0$$
$$g_8(x,y) = -1 - (-2x - 2y) \geq 0$$
$$g_9(x,y) = y \geq 0$$

and also have

$$\begin{aligned}
\min_{x \in X} F(x,y) &= 3x - 6y \\
\text{subject to} \quad & -1x + 3y \leq 4 \\
& -2x + 2y \leq 3 \\
& 0x + 4y \leq 5 \\
& 1x - 1y \leq 0 \\
& 0x - 2y \leq -1 \\
& 2x - 0y \leq 1 \\
& -1x - 1y \leq 0 \\
& -2x - 2y \leq -1 \\
& 3u_1 + 2u_2 + 4u_3 - u_4 - 2u_5 - 0u_6 - u_7 - 2u_8 - u_9 = -3 \\
& \sum_{i=1}^{9} u_i g_i(x,y) = 0 \\
& x \geq 0, y \geq 0, u_1 \geq 0, \ldots, u_9 \geq 0.
\end{aligned}$$

We get the following linear programming problem with one condition.

$$\begin{aligned}
\min_{x \in X} F(x,y) &= 3x - 6y \\
\text{subject to} \quad & -1x + 3y \leq 4 \\
& -2x + 2y \leq 3 \\
& 0x + 4y \leq 5 \\
& 1x - 1y \leq 0 \\
& 0x - 2y \leq -1 \\
& 2x - 0y \leq 1
\end{aligned}$$

$$
\begin{aligned}
&\quad - 1x - 1y \leq 0 \\
&- 2x - 2y \leq -1 \\
&\qquad 3u_1 + 2u_2 + 4u_3 - u_4 - 2u_5 - 0u_6 - u_7 - 2u_8 - u_9 = -3 \\
&\qquad x \geq 0, y \geq 0, u_1 \geq 0, \ldots, u_9 \geq 0.
\end{aligned}
$$

At each iteration stage, the following condition is checked.

$$\sum_{i=1}^{9} u_i g_i(x, y) = 0.$$

As shown in Fig. 2, the algorithm finds a feasible solution to the Kuhn-Tucker representation with the complementary slackness conditions omitted and proceeds to Step 3.3. The current point, $x^1 = 0$, $y^1 = 1.25$, $u^1 = (0,0,0,3,0,0,0,0,0)$, with $F(x^1, y^1) = -7.5$ does not satisfy complementarities so a branching variable is selected (u_4) and the index sets are updated, giving $S_1^+ = \{4\}$, $S_1^- = \phi$, $S_1^0 = \{1,2,3,5,6,7,8,9\}$ and $P_1 = \{4\}$. In the next four iterations, the algorithm branches on u_5, u_7, u_8 and u_9, respectively. Now, five levels down in the tree, the current subproblem at Step 3.1 turns out to be infeasible so the algorithm goes to Step 3.5 and backtracks. The index sets are $S_5^+ = \{4,5,7,8\}$, $S_5^- = \{9\}$, $S_5^0 = \{1,2,3,6\}$ and $P_5 = \{4,5,7,8,\underline{9}\}$. Go to Step 3.1, and algorithm turns out to be infeasible so the algorithm goes to Step 3.5 and backtracks. The index sets are $S_6^+ = \{4,5,7\}$, $S_6^- = \{8\}$, $S_6^0 = \{1,2,3,6,9\}$ and $P_6 = \{4,5,7,\underline{8}\}$. Go to Step 3.1, and continue the process. Finally, at Step 3.4, $\bar{F} = -3$. The algorithm backtracks at Step 3.5 and updates the sets, $S_9^+ = \phi$, $S_9^- = \{4\}$, $S_9^0 = \{1,2,3,5,6,7,8,9\}$ and $P_9 = \{\underline{4}\}$. Returning to Step 3.1, another feasible solution is found, but at Step 3.2, the value of the leader's objective function is greater than the incumbent upper bound, so the algorithm goes to Step 3.5 and backtracks. However, no live vertices exist. We have found an optimal solution, occurring at the point $(x^*, y^*) = (0, 0.5)$, $(u^*) = (0,0,0,3,0,0,0,0,0)$ with $F^* = -3$ and $f^* = 1.5$. By examining the above procedure, the optimal solution occurs at $(x^*, y^*) = (0, 0.5)$ with $F_1^* = -1,$, $F_2^* = -1.5$ $F_3^* = -0.5$, $f_1^* = 0.5$ and $f_2^* = 1$.

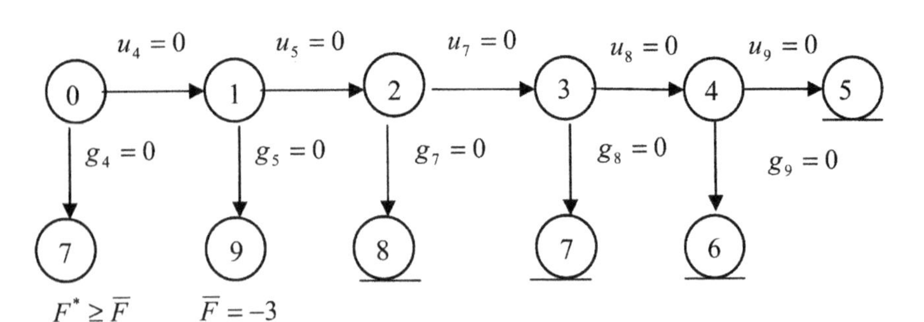

Fig. 2. Search tree

Step 4. We have the same solution as shown in Step 4 of Sect. 4.1 using fuzzy Kuhn-Tucker approach.

4.3 A Case Study of Traffic Management for the Fuzzy Kth-best Approach

In general, a FLBLP problem involves two issues. The first issue is to establish a FLBLP model including determination of variables, objective functions and constraints for both the leader and the follower. The second issue is to use a suitable approach to solving the problem. This section shows a road network with hypothetical problem parameters and how the proposed fuzzy Kth-best approach is used for solving it.

The improvement of a road network can be performed through capacity expansion, traffic signals synchronization or vehicle guidance systems [12,14]. The road management committee (the leader) is assumed to control these decision variables. The committee's decision can influence directly or indirectly the travel choices of the road network users. Let x denote the decision vector of the road management committee, X the set of feasible decision variables, y the decision vector of the follower, and $c_i(x, y)$ the travel delay along a link i. The road management committee's main objective is to minimize, over the set X, the system travel cost $\sum_i a_i c_i(x, y) = F(x, y)$. Network users seek to minimize their travel delays $\min f(x,\ y)$. The committee also seeks its minimized travel delays. However, the committee is interested in minimizing total travel time for all kinds of users, while the user group only wants to optimize its own travel time. The committee knows the objectives and constraints of the road network users. The parameters such as a_i are set by some experts and by their understanding of road-network users' reactions for any road control policy. Obviously, it is hard to characterize human understanding using precise values into these parameters. This is a typical fuzzy bilevel decision making case. In order to easily show the application of the proposed fuzzy Kth-best approach, the road-network decision problem is established by simplifying it into the following FLBLP model. $\min_{x \in X}\ F(x, y) = -\tilde{1}x + \tilde{2}y + \tilde{3}$ is a simplified objective function of road management committee, and $\min_{y \in Y}\ f_1(x, y) = \tilde{1}x - \tilde{1}y + \tilde{2}$ is a simplified objective function of road network users, with $x \in R^1,\ y \in R^1$, and $X = \{x \geq 0\},\ Y = \{y \geq 0\}$,

$$\min_{x \in X} F(x, y) = -\tilde{1}x + \tilde{2}y + \tilde{3} \tag{4.2a}$$

$$\text{subject to } -\tilde{1}x + \tilde{3}y \underline{\underline{\prec}} \tilde{4} \tag{4.2b}$$

$$\min_{y \in Y} f_1(x, y) = \tilde{1}x - \tilde{1}y + \tilde{2} \tag{4.2c}$$

$$\text{subject to } \tilde{1}x - \tilde{1}y \underline{\underline{\prec}} \tilde{0} \tag{4.2d}$$

$$-\tilde{1}x - \tilde{1}y \underline{\underline{\prec}} \tilde{0} \tag{4.2e}$$

where

$$\mu_{\tilde{1}}(t)=\begin{cases}0 & t<0\\ t & 0\leqq t<1\\ 2-t & 1\leqq t<2\\ 0 & 2\leqq t\end{cases},\ \mu_{\tilde{2}}(t)=\begin{cases}0 & t<1\\ t-1 & 1\leqq t<2\\ 3-t & 2\leqq t<3\\ 0 & 3\leqq t\end{cases},$$

$$\mu_{\tilde{3}}(t)=\begin{cases}0 & t<2\\ t-2 & 2\leqq t<3\\ 4-t & 3\leqq t<4\\ 0 & 4\leqq t\end{cases},$$

$$\mu_{\tilde{4}}(t)=\begin{cases}0 & t<3\\ t-3 & 3\leqq t<4\\ 5-t & 4\leqq t<5\\ 0 & 5\leqq t\end{cases},\ \mu_{\tilde{0}}(t)=\begin{cases}0 & t<-1\\ t+1 & -1\leqq t<0\\ 1-t & 0\leqq t<1\\ 0 & 1\leqq t\end{cases}.$$

Step 1. The problem (4.2) is transformed into the following problem by using Theorem 2.3.

$$\begin{aligned}
&\min_{x\in X}\,(F(x,y))_c = -1x+2y+3\\
&\min_{x\in X}\,(F(x,y))_0^L = -2x+1y+2\\
&\min_{x\in X}\,(F(x,y))_0^R = 0x+3y+4\\
&\text{subject to } -1x+3y\le 4\\
&\qquad\qquad\quad -2x+2y\le 3\\
&\qquad\qquad\quad 0x+4y\le 5\\
&\qquad\qquad\quad \min_{y\in Y}\,(f(x,y))_c = 1x-1y+2\\
&\qquad\qquad\quad \min_{y\in Y}\,(f(x,y))_0^L = 0x-2y+1\\
&\qquad\qquad\quad \min_{y\in Y}\,(f(x,y))_0^R = 2x-0y+3\\
&\qquad\qquad\quad \text{subject to } 1x-1y\le 0\\
&\qquad\qquad\qquad\qquad\quad 0x-2y\le -1\\
&\qquad\qquad\qquad\qquad\quad 2x-0y\le 1\\
&\qquad\qquad\qquad\qquad\quad -1x-1y\le 0\\
&\qquad\qquad\qquad\qquad\quad 0x-0y\le 0\\
&\qquad\qquad\qquad\qquad\quad -2x-2y\le -1.
\end{aligned}$$

Step 2. The problem is then transformed into the following linear BLP problem by using a weighting method.

$$\begin{aligned}
\min_{x\in X} F(x,y) &= -3x+6y+9 \\
\text{subject to } & -1x+3y \le 4 \\
& -2x+2y \le 3 \\
& 0x+4y \le 5 \\
& \min_{y\in Y} f(x,y) = 3x-3y+6 \\
& \text{subject to } 1x-1y \le 0 \\
& \qquad 0x-2y \le -1 \\
& \qquad 2x-0y \le 1 \\
& \qquad -1x-1y \le 0 \\
& \qquad -2x-2y \le -1 \\
& \qquad 0x-0y \le 1
\end{aligned}$$

Step 3. According to the extended Kth-best approach proposed in [21], we rewrite above problem and get the following problem:

$$\begin{aligned}
\min F(x,y) &= -3x+6y+9 \\
\text{subject to } & -1x+3y \le 4 \\
& -2x+2y \le 3 \\
& 0x+4y \le 5 \\
& 1x-1y \le 0 \\
& 0x-2y \le -1 \\
& 2x-0y \le 1 \\
& -1x-1y \le 0 \\
& -2x-2y \le -1 \\
& x \ge 0, y \ge 0.
\end{aligned}$$

Let $i = 1$, and solve the above problem with the simplex method to obtain the optimal solution $(x_{[1]},\ y_{[1]}) = (0.5, 0.5)$. Let $W = \{(0.5, 0.5)\}$ and $T = \phi$. Go to Step 3.2.

Loop 1. By (3.19), we have

$$\begin{aligned}
\min f(x,y) &= 3x-3y+6 \\
\text{subject to } & -1x+3y \le 4 \\
& -2x+2y \le 3 \\
& 0x+4y \le 5 \\
& 1x-1y \le 0 \\
& 0x-2y \le -1 \\
& 2x-0y \le 1
\end{aligned}$$

$$
\begin{aligned}
&-1x - 1y \leq 0\\
&-2x - 2y \leq -1\\
&x = 0.5\\
&y \geq 0.
\end{aligned}
$$

Using the bounded simplex method, we have $\tilde{y} = 1.25$. Because of $\tilde{y} \neq y_{[i]}$, we go to Step 3.3 and have $W_{[i]} = \{(0.5,\ 1.25),\ (0,\ 0.5),\ (0.5,\ 0.5)\}$, $T = \{(0.5,\ 0.5)\}$ and $W = \{(0,\ 0.5), (0.5,\ 1.25)\}$, then go to Step 3.4. Update $i = 2$, and choose $(x_{[i]}, y_{[i]}) = (0, 0.5)$, then go back to Step 3.2.
Loop 2. By (3.19)

$$
\begin{aligned}
\min f(x, y) = {}& 3x - 3y + 6\\
\text{subject to } & -1x + 3y \leq 4\\
& -2x + 2y \leq 3\\
& 0x + 4y \leq 5\\
& 1x - 1y \leq 0\\
& 0x - 2y \leq -1\\
& 2x - 0y \leq 1\\
& -1x - 1y \leq 0\\
& -2x - 2y \leq -1\\
& x = 0\\
& y \geq 0.
\end{aligned}
$$

Using the bounded simplex method, we have $\tilde{y} = 1.25$. As $\tilde{y} \neq y_{[i]}$, go to Step 3.3, have $W_{[i]} = \{(0.5,\ 1.25),\ (0,\ 0.5),\ (0.1,\ 1.25)\}$, $T = \{(0.5, 0.5),\ (0, 0.5)\}$ and $W = \{(0, 1.25), (0.5, 1.25)\}$, then go to Step 3.4. Update $i = 3$, and choose $(x_{[i]}, y_{[i]}) = (0.5, 1.25)$, then go to Step 3.2.
Loop 3. By (3.19), we have

$$
\begin{aligned}
\min f(x, y) = {}& 3x - 3y + 6\\
\text{subject to } & -1x + 3y \leq 4\\
& -2x + 2y \leq 3\\
& 0x + 4y \leq 5\\
& 1x - 1y \leq 0\\
& 0x - 2y \leq -1\\
& 2x - 0y \leq 1\\
& -1x - 1y \leq 0\\
& -2x - 2y \leq -1\\
& x = 0.5\\
& y \geq 0.
\end{aligned}
$$

Using the bounded simplex method, we have $\tilde{y} = 1.25$. Because of $\tilde{y} = y_{[i]}$, we stop here. $(x_{[i]}, y_{[i]}) = (0.5, 1.25)$ is the global solution to this Example. By examining the above procedure, we found that the optimal solution occurs at the point $(x^*, y^*) = (0.5, 1.25)$ with $F_1^* = 5$, $F_2^* = 2.25$, $F_3^* = 7.75$, $f_1^* = 1.25$, $f_2^* = -1.5$ and $f_3^* = 4$.

Step 4. The result is

$$\min_{x \in X} (F(x,y))_c = -1x + 2y + 3 = 5$$

$$\min_{x \in X} (F(x,y))_0^L = -2x + 1y + 2 = 2.25$$

$$\min_{x \in X} (F(x,y))_0^R = 0x + 3y + 4 = 7.75$$

$$\min_{y \in Y} (f(x,y))_c = 1x - 1y + 2 = 1.25$$

$$\min_{y \in Y} (f(x,y))_0^L = 0x - 2y + 1 = -1.5$$

$$\min_{y \in Y} (f(x,y))_0^R = 2x - 0y + 3 = 4$$

$$x = 0.5, y = 1.25$$

Consequently, we have the solution of the problem (4.2): $\min_{x \in X} F(x,y) = -\tilde{1}x + \tilde{2}y + \tilde{3} = \tilde{c}$ is the road management committee's objective, $\min_{y \in Y} f_1(x,y) = \tilde{1}x - \tilde{1}y + \tilde{2} = \tilde{d}$ is the road network users' objective, and $x = 0.5$, $y = 1.25$ are the values of decision variables of the road committee and the road users respectively, where

$$\mu_{\tilde{c}}(t) = \begin{cases} 0 & t < 2.25 \\ \frac{t-2.25}{2.75} & 2.25 \leq t < 5 \\ \frac{7.75-t}{2.75} & 5 \leq t < 7.75 \\ 0 & 7.75 \leq t \end{cases}, \mu_{\tilde{d}}(t) = \begin{cases} 0 & t < -1.5 \\ \frac{t+1.5}{2.75} & -1.5 \leq t < 1.25 \\ \frac{4-t}{2.75} & 1.25 \leq t < 4 \\ 0 & 4 \leq t \end{cases}.$$

5 Conclusions

Many organizational decision units are within a two-level structure. The execution of their decisions is often sequential and the leader's decision can be affected by the responses of their follower. Uncertainty is involved in the model for such decision problems, and the solution approach has to account for this uncertainty. Fuzzy bilevel decision making is a common issue in organizational management. This chapter presents three FLBLP approaches for the fuzzy bilevel decision problems. A numerical example and a real-case study of a road-network problem illustrate the application of the proposed three fuzzy bilevel optimization approaches. Further study will involve the development of applications of these proposed approaches. A decision support system will then be further developed to applying the proposed techniques for supporting decision makers in their management activities.

Acknowledgment

This research is partially supported by Australian Research Council (ARC) under discovery grants DP0557154 and DP0559213.

References

1. Aiyoshi E and Shimizu K (1981) Hierarchical decentralized systems and its new solution by a barrier method. IEEE Transactions on Systems, Man, and Cybernetics 11: 444–449
2. Anandalingam G and Friesz T (1992) Hierarchical optimization: An introduction. Annals of Operations Research 34: 1–11
3. Bard J (1984) An investigation of the linear three level programming problem. IEEE Transactions on Systems, Man, and Cybernetics 14: 711–717
4. Bard J (1998) Practical Bilevel Optimization: Algorithms and Applications. Kluwer Academic Publishers.
5. Bard J and Falk J (1982) An explicit solution to the programming problem. Computers and Operations Research 9: 77–100
6. Bialas W and Karwan M (1978) Multilevel linear programming. Technical Report 78-1, State University of New York at Buffalo, Operations Research Program.
7. Bialas W and Karwan M (1984) Twolevel linear programming. Management Science 30: 1004–1020
8. Candler W and Townsley R (1982) A linear twolevel programming problem. Computers and Operations Research 9: 59–76
9. Chen Y, Florian M and Wu S (1992) A descent dual approach for linear bilevel programs. Technical Report CRT-866, Centre de Recherche sur les Transports.
10. Dempe S (1987) A simple algorithm for the linear bilevel programming problem. Optimization 18: 373–385
11. Hansen P, Jaumard B. and Savard G (1992) New branchandbound rules for linear bilevel programming. SIAM Journal on Scientific and Statistical Computing 13: 1194–1217
12. Leblanc L and Boyce D (1986) A bilevel programming algorithm for exact solution of the network design problem with useroptimal flows. Transportation Research 20: 259–265
13. Lai YJ (1996) Hierarchical optimization: a satisfactory solution, Fuzzy Sets and Systems 77: 321–335
14. Marcotte P (1986) Network design with congestion effects: a case of bilevel programming. Mathematical Programming 34: 142–162.
15. Sakawa M (1993) Fussy sets and interactive mulitobjective optimization. Plenum Press, New York
16. Sakawa M and Nishizaki I (1998) Interactive fuzzy programming for multilevel linear programming problems. Computers and Mathematics with Applications 36: 71–86
17. Sakawa M, Nishizaki I and Uemura Y (2000) Interactive fuzzy programming for multilevel linear programming problems with fuzzy parameters. Fuzzy Sets and Systems 109: 3–19

18. Savard G (1989) Contributions á la programmation mathématique á deux niveaux. PhD thesis, Université de Montréal, École Polytechnique
19. Shi C, Zhang G and Lu J (2004), An algorithm for linear bilevel programming problems, *Applied Computational Intelligence*, The 6th International FLINS Conference, Blankenberge, Belgium, 1–3, September, 2004, Published by *World Scientific*, 304–307
20. Shi C, Lu J and Zhang G (2005) An extended Kuhn-Tucker approach for linear bilevel programming. Applied Mathematics and Computation 162:51–63
21. Shi C, Lu J and Zhang G (2004) An extended Kth-best approach for linear bilevel programming. Applied Mathematics and Computation 164: 843–855
22. Shih HS, Lai YJ and Lee ES (1996) Fuzzy approach for multi-level programming problems. Computers & Operations Research 23: 73–91
23. White D and Anandalingam G (1993) A penalty function approach for solving bilevel linear programs. Journal of Global Optimization 3: 397–419
24. Von Stackelberg H (1952) The theory of the market economy. Oxford University Press, Oxford
25. Zadeh LA (1965) Fuzzy sets. Inform & Control 8: 338–353
26. Zhang G and Lu J (2004) The definition of optimal solution and an extended Kuhn-Tucker approach for fuzzy linear bilevel programming. The IEEE Computational Intelligence Bulletin 5:1–7
27. Zhang G, Lu J, Steele R and Shi C (2005) An Extended Kth-best approach for fuzzy linear bilevel problems. 10th International Conference on Fuzzy Theory and Technology, July 21–26, 2005, Salt Lake City, USA
28. Zhang G and Lu J (2005) Model and approach of fuzzy bilevel decision making for logistics planning problem. Journal of Enterprise Information Management (In Press)
29. Zhang G, Lu J and Dillon T (2006) An extended branch-and-bound algorithm for fuzzy parameter linear bilevel programming. Accepted by 7th International FLINS Conference, Genoa, Italy, 29–31, August 2006.

Fuzzy Predictive Earth Analysis Constrained by Heuristics Applied to Stratigraphic Modeling

Jeffrey D. Warren, Robert V. Demicco and Louis R. Bartek

Abstract. A three-dimensional computer simulation was developed to examine the sensitivity of stratigraphic evolution of a continental margin to variable sediment influx, subsidence and sea level. The preliminary results of this model, created using MATLAB and referred to as fuzzyPEACH (Predictive Earth Analysis Constrained by Heuristics), are presented in this chapter. Fuzzy logic provides a method to quantify subjective variables with a collection of linguistic-based rules and fuzzy sets. To date, few geological investigations have utilized fuzzy logic to simulate complex, geologically reasonable stratigraphic responses to variable depositional systems. Fuzzy logic was chosen as a modeling platform because it is easy to use, adaptable, and a computationally efficient alternative to traditional, mathematically complex models. FuzzyPEACH uses a series of five fuzzy inference systems that rely on 15 rules and 47 fuzzy sets to control grain size, sediment distribution, fluvial avulsions, compaction of sediment, and isostatic (load-driven) subsidence. User-defined variables include margin physiography (width, depth of shelf-slope break), sedimentation rates, sea level, tectonic subsidence, duration of model and number of time steps. Graphical tools allow output visualization via strike- and dip-oriented stratigraphic cross sections, isopach (sediment thickness) maps, and structure maps. The locations and time-step intervals for these tools are user-defined. Preliminary comparisons of model output to an extensive seismic dataset from the East China Sea continental margin help validate the ability of fuzzy logic to recreate complex depositional conditions and advance further the idea that fuzzy logic is well suited not only for stratigraphic analysis but geological applications in general.

1 Introduction

An effective geologic model is one that expands on descriptions by logical or mathematical relations to synthesize and/or predict a geologic system [1]. The goal of this investigation was to determine whether a set of general, non-mathematical rules could be developed to describe the generic stratigraphic response of a continental margin to changes in eustasy (global sea level change), tectonics, sediment influx and margin physiography. The fuzzy sets and linguistic rules of fuzzy logic provide the framework to test this approach.

The fuzzy simulation described here is able to run numerous permutations of sea level, tectonic subsidence and sediment influx (although users can also change margin geometry as well as simulate the response to various margin physiographies by merely changing sea level magnitude and periodicity).

Studies of the stratigraphic response of low-gradient continental margins, especially those with a deep shelf-slope break (i.e. not exposed subaerially during sea level lowstands) are rare. Such margins still exhibit the shelf-slope-rise morphology typical of most passive continental margins [2], but major incision does not occur at the margin's edge during sea level lowstands. The response of an unconfined fluvial system across a low gradient (especially one that is lower than the fluvial gradient) is deposition and avulsion throughout the basin with little sedimentary bypass beyond the shelf-slope break. These geologic responses are contrary to the traditional sequence stratigraphic theory, which predicts incision and sedimentary bypass across a continental margin during sea level lowstand. Quantitative models have been developed to better understand the depositional processes within low-gradient foreland basins [3–5]. Furthermore, the fluvial response to a low gradient has been discussed theoretically [6] and the fluvial response (latrally extensive, unincised, braided sand sheets) has been verified empirically using flume experiments [7, 8]. However, computational simulations of the stratigraphic evolution of low-gradient continental margins remain undeveloped.

A three-dimensional computer simulation was created to examine the sensitivity of stratigraphic evolution on a passive continental margin with respect to sea level, sediment supply and subsidence. The model uses the MATLAB programming language and the software's Fuzzy Logic Toolbox and is referred to as fuzzyPEACH (Predictive Earth Analysis Constrained by Heuristics). Here, the term "heuristics" represents general rules that simply describe complicated processes within geologic systems. The heuristics driving this particular geologic simulation are controlled by a series of 21 rules contained within five fuzzy inference systems (FISs). Sediment distribution within the fluvial system (i.e. grain size and sediment volume) is governed by three rules contained within a single FIS. Deltaic deposition is controlled with three rules and one FIS. Three additional FISs govern fluvial avulsion, sediment compaction and isostatic flexure (crustal subsidence sediment loading).

Fuzzy logic was used in this investigation to address the stratigraphic evolution of continental margins for three reasons. First, the lack of detailed investigations of low-gradient continental margins creates a knowledge gap for common variables needed for the complicated differential equations used in traditional depositional models. For example, the delta progradation and basin filling model DELTA2 requires input of values of seasonal concentrations of suspended sediment, removal-rate constants, bulk densities of sediments, seasonal velocity of river at river mouth, seasonal dimension of the river mouth (width and depth), and maximum depth of river plume [9]. The sediment transport simulator SEDTRANS96 considers over 100 variables [10]. The SedFlux model simulates and predicts the geomorphic evolution of a continental margin. It uses a suite of nine separate models to address sediment discharge, deltaic plain sedimentation and erosion, bedload dumping on a tidal flat, fallout from river mouth plumes, storm re-suspension and transport, sediment stability and failure, erosion and deposition by turbidity currents, sedimentation from debris flows, flexural subsidence, tectonic motion ad faulting, sediment compaction and sea level fluctuations [11].

In comparison, "geometric" models do not simulate the actual flow processes responsible for the sediment distribution [12]. Geometric models distribute sediments beneath specified geometric profiles of equilibrium based on empirical or observational data to approximate the gross-scale results of sedimentation as geometries, which in turn creates a stratigraphic framework [1, 13]. Numerous geometric stratigraphic models have been developed in both two and three dimensions [12, 14–24]. Geometric modeling is based on generalized rules expressed mathematically (as equilbrium profiles; slopes of coastal plains, shelves, or basin floors; angles of repose for stratal slumping and/or erosion). At first, this approach appears similar to fuzzy logic because of the generalization of expected geometries that strata are expected to achieve. However, stratigraphic simulations using fuzzy logic are vastly different. The fuzzyPEACH model presented in this chapter is governed by general rules on how a depositional system operates (i.e. thickness and grain size of sediment with respect to distance from fluvial system), but the general rules do not pre-determine or govern stratal geometry in any way whatsoever; The stratigraphic response is based on available accomodation space (the area available for sediment to be deposited).

The second reason fuzzy logic was chosen as a modeling platform for this study is that a FIS can be assembled with general concepts and expert knowledge about a geologic system (or, for that matter, any complex system) without relying on hundreds of variables, non-linear relationships, and complicated differential equations. It is not suggested that fuzzy logic modeling can, or should, replace traditional techniques. However, fuzzy applications can certainly complement existing analytical tools. For example, fuzzy logic has the ability to handle complicated, nonlinear variables that cannot be, or have not yet been, defined mathematically. Fuzzy logic may also narrow potential values needed for traditional model input to a field of few. Mathematical models may still be used, but the computational efficiency and ease of fuzzy logic modification has the potential to accelerate the process of variable selection by discarding values that are inaccurate or not important.

Finally, because fuzzy logic applications require no special knowledge apart from basic mathematical logic, the fuzzyPEACH (and fuzzy applications in general) can be understood and widely applied by the geologic community rather than being limited to modeling specialists. For the time being, the use of fuzzy logic within the geological sciences is not widespread, although fuzzy techniques have proven successful in the simulation of stratigraphic development [25–33]. The fuzzy logic simulations described in this chapter build on these earlier stratal simulations.

2 Geological Examples of Fuzzy Variables

Application of fuzzy logic to the geological sciences has increased dramatically during the past few decades. For example, one (and subsequently the first) paper on the subject was published during the 1970s [34], 151 in the 1980s, and 640 in the 1990s. Complete journal volumes [35] and books focusing on specific geological applications of fuzzy logic were published only recently [36–39]. A comprehensive literature review of many of these earth science-specific studies was presented by Demicco [40], and the general applicability of fuzzy techniques to the field of geological science is addressed by Demicco [40,41], Fang [42,43] and Nordlund [26,27].

Fuzzy logic is well suited for a broad array of geological concepts and data. The inherent complexity of geological systems and the knowledge of their processes are many times too incomplete for quantitative modeling with mathematical equations [42]. Many geological phenomena are described or classified using rigid boundaries that do not correspond to the conceptual notion of gradational transitions along a continuum. In the examples presented below, this type of "pigeon holing" is eliminated with fuzzy sets. Fuzzy logic is able to exploit the highly descriptive nature of geological phenomena and the subjective nature of the majority of geological data [26,27].

2.1 Sediment Classification

Sediment grain size classification is the most common geological example used to illustrate the difference between crisp and fuzzy sets [26,27,30,41,44]. Using the traditional Udden-Wentworth classification scheme [45], a grain diameter of 1.999 mm is classified as coarse sand and a grain diameter of 2.001 is classified as gravel. Nordlund [26] pointed out that such a rigid classification does not correspond to the conceptual notion of a continuum between arbitrarily defined grain-sized classes (Fig. 1A). For example, a fuzzy set reflects the similarity in grain diameters of 1.999 and 2.001 mm by assigning similar, if not identical, degrees of membership in both sand and gravel (i.e. both diameters express virtually the same degree of "sand-ness" and "gravel-ness"). Although differences of the interpretation of grain size classification are observable in Fig. 1B-D, these variations do not detract from the overall ability of fuzzy sets to provide greater accuracy than traditional logic. The categorization of other grain size descriptors (e.g. roundness, shape, sorting) arguably is achieved with greater accuracy using fractional membership along a continuum.

2.2 Fluvial Classification

Fluvial systems classically are subdivided into three general channel patterns: straight, meander and braided [46]. The axes of Fig. 2 illustrate the relative relationship between straight, meandering and braided channel morphologies as well as many of the controlling morphological variables, including sediment

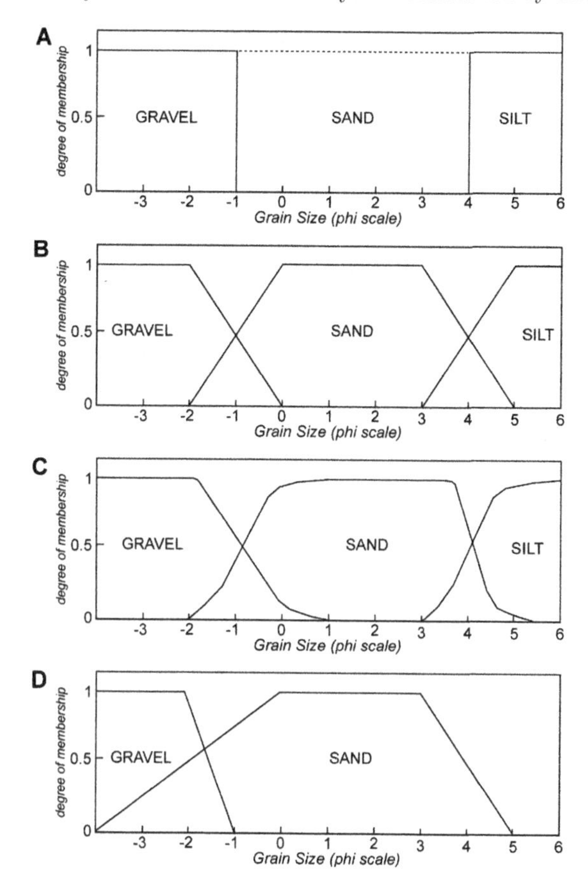

Fig. 1. Sediment grain size classification sets based on traditional (crisp) logic (A) and fuzzy logic (B, C, D) [adapted from 45, 26, 44, and 41, respectively]

load (suspended, mixed, bed), sediment size (small versus large), flow velocity (low versus high) and stream power (low versus high). In the absence of exact units of measurement, information is still conveyed in linguistic terms such as "high" and "low" noted along the axes. Fuzzy sets defining the three classical fluvial patterns, and their relationship to sediment size, sediment load, flow velocity and stream power, are presented in Fig. 3.

In addition to the ability to delineate each of these three channel classifications, fuzzy sets and fuzzy logic can also be used to describe the transition between each major morphological category. Early research suggested that abrupt thresholds exist between straight, meandering and braided fluvial systems [47–49]. Recent studies have suggested gradual transitions [50–54]. Minor modifications to the shape and slope of the three MFs (one for each major channel pattern) contained within a single fuzzy set can accommodate the range of morphological transitions between the "gradual" and "abrupt" end

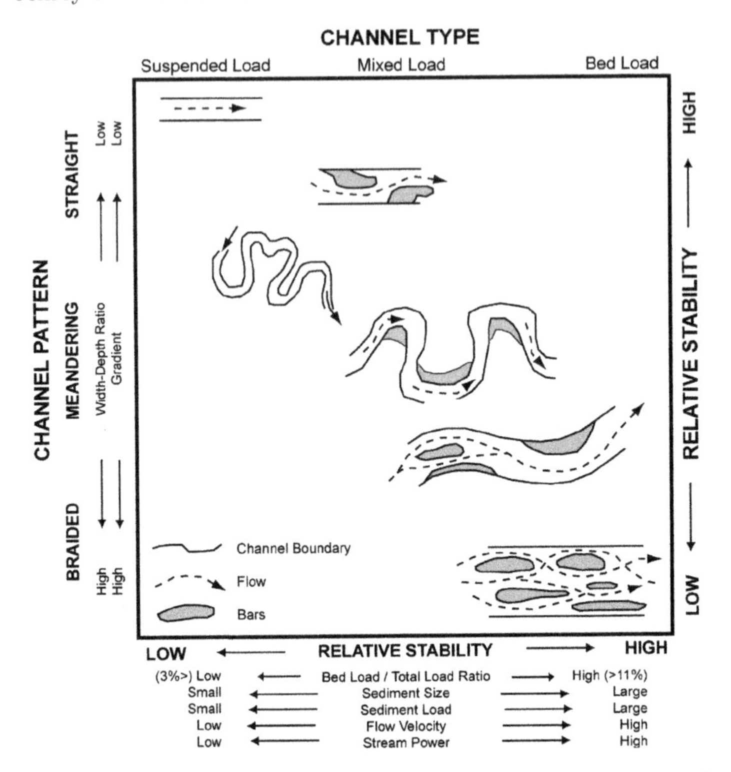

Fig. 2. Classification of channel morphology based on pattern type and sediment load [49]

members (Fig. 3). However, more important than understanding whether the transition between channel morphologies is abrupt or gradual is the derivation of relationships between channel morphology and the processes of deposition and erosion that control it. Schumm [55] and Orton and Reading [56] presented some general links between fluvial variables (e.g. slope, discharge, sediment load, velocity, etc.) and channel morphology. Based on these process-response investigations, it is not difficult to create FISs since the variables controlling channel morphology are defined easily by fuzzy sets. The manner in which these variables interact with each other is the basis for general rules that can describe the system and, hence, drive the FIS.

Some classifications in the literature are, unwittingly, already in a fuzzy format (although the investigation had nothing to do with fuzzy logic). For example, a figure presented in Dalrymple et al. [57] describes the relative nature of active processes defining an estuary (Fig. 4). The transitional relationships between fluvial and marine systems is described more accurately with fuzzy sets rather than crisply defined boundaries. The nature of the curves in Fig. 4 is analogous to a MF and, by default, fuzzy logic. The values on the y-axis are along a continuum between 0 and 100% (representing none

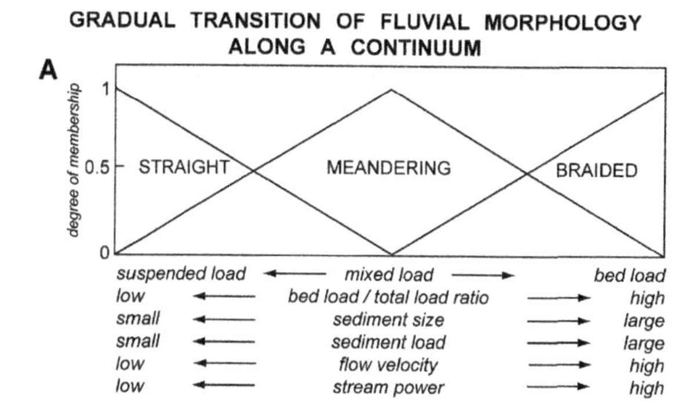

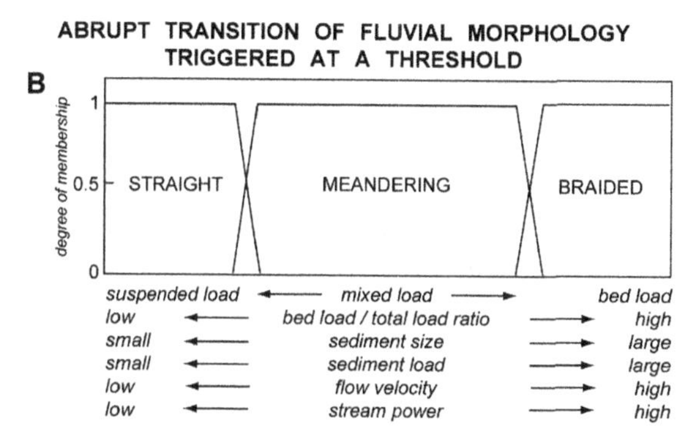

Fig. 3. Channel classification based on fuzzy sets indicating A) a gradual morphological transition and B) a morphological transition that is more abrupt

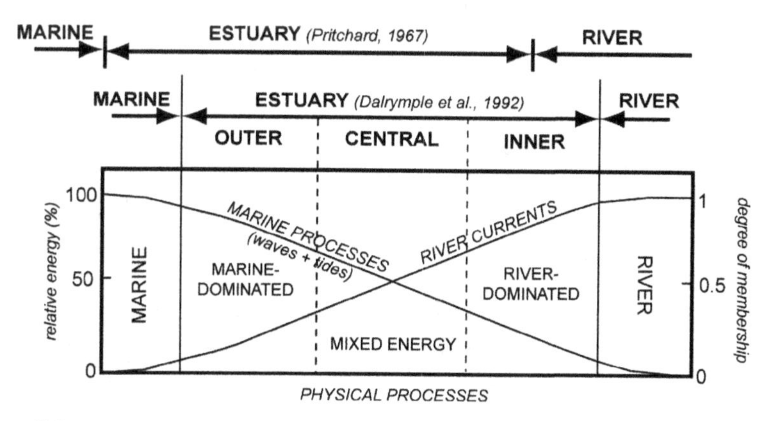

Fig. 4. Schematic distribution of the physical processes operating within estuaries and the resulting zonation similar in form and function to fuzzy sets [57]

and all, respectively). The domain is a relative, unit-less axis that defines the transitional environment between fluvial and marine processes that define an estuary. Two MFs define the river-dominated and marine-dominated processes, and an estuary is defined as a combination of the two overlapping processes (and two overlapping fuzzy sets).

2.3 Deltaic Classification

The ternary, process-based delta classification has three end-member variables: sediment input, wave energy flux, tidal energy flux [58]. This classification also has been adapted to include grain size (Fig. 5) [56]. The categories "mixed mud and silt", "fine sand" and "gravelly sand" illustrate the fuzzy nature of gradual transitions between grain-size categories defined linguistically, based on relative (rather than absolute) values. Dalrymple et al. [57] used a similar ternary classification scheme to place not only deltas but coastal systems in general into the framework of a continuum and expanded these deltaic descriptions linguistically (e.g. low, moderate, high, as well as low to moderate and extremely high) based on tidal, wave and fluvial-influenced processes [57]. Similarly, the linguistic approach to deltas and related depositional systems (e.g. drainage basin, fluvial parameters, shoreline and marine processes) developed by Orton and Reading [56] creates the general rules that can be assembled into a FIS (although the data were not presented in the context of fuzzy logic).

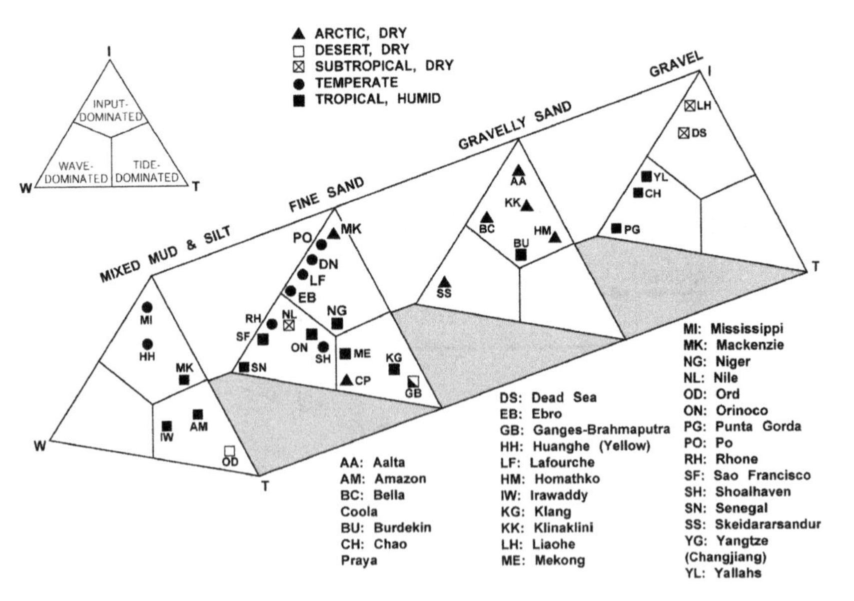

Fig. 5. Deltaic classification adapted from [56]. The continuums between the apices on each ternary diagram [58], as well as between each individual ternary diagram, are well suited for fuzzy logic classifications

3 Stratigraphic Modeling Using Fuzzy Logic

As of 2007, there were only three known fuzzy applications developed to model sediment deposition and stratigraphic evolution. Each of these three stratigraphic simulations is summarized in this chapter to offer an overview of fuzzy stratigraphic modeling and provide evidence that a "fuzzy" approach offers an alternative, yet complementary, method to traditional mathematical models. While there are only minor differences to the overall mechanics of how these three models operate as compared to the fuzzyPEACH simulation presented in this chapter, there is a significant difference in the amount of data used to constrain and interpret modeling output. However, these models, especially FUZZIM, the work of Demicco [33], Demicco and Klir [30] and Warren et al. [59, 60], contributed to the development of fuzzyPEACH.

The first stratigraphic simulator to be developed was FUZZIM, a Macintosh-based program developed in C/C++ [25–28]. FUZZIM simulated sedimentation rates (stratal thickness), grain size and erosion rates across idealized ramp margins and was able to consider both clastic and carbonate systems. The initial simulator used the following conditions: 1) a set of 10 fuzzy rules, 2) a 25 km × 25 km model grid (10 km shelf with no antecedent topography/bathymetry, shelf-slope break < 50 m below sea level at final time step), 3) a run time of 70 ky (ky = 1,000 years) and 4) a sinusoidal sea level curve with an amplitude of 30 m and a frequency of 20 ky [26]. Tectonic subsidence was implemented at a rate of 1 mm/yr (maximum rate located at distal margin with a linear decrease to zero at a landward hinge point located 10 km outside the modeled universe). Nordlund [27] modified FUZZIM to simulate the Miocene carbonate platform of Mallorca, Spain with the following conditions: 1) a set of 12 fuzzy rules that controlled deposition (grain size and distribution), erosion and carbonate production, 2) a model grid < 10 km square, 3) a run time of 130 ky and 4) a sea level curve inferred from field observations (max amplitude of sea level curve 100 m below present with a frequency of 100 ky). Subsidence and compaction variables were not reported for this model. Nordlund [28] provided a general tutorial of FUZZIM with the following conditions: 1) a set of five fuzzy rules, 2) a model grid of 40 km × 80 km, 3) a run time of 100 ky (5 ky time steps) and 4) a sinusoidal sea level curve with an amplitude of 50 m and a frequency of 100 ky. Regional tectonic subsidence was defined using a model of a titled plane defined by three separate subsidence curves (one for each of three geographically defined reference points). Compaction and load-driven subsidence (simple Airy isostasy) were included using separate fuzzy systems. FUZZIM simulated deposition and erosion in both subaerial and submarine environments and limited the submarine erosion to simulated gravity deposits affecting strata exceeding a slope of 10° [27, 28]. Visualization of output for FUZZIM simulations included sediment distribution maps for each time step and dip-oriented cross-sections.

A second set of models developed by Demicco [33] and Demicco and Klir [30] using MATLAB and its associated Fuzzy Logic Toolbox simulated clastic, carbonate and evaporite depositional systems in three dimensions. Visualization of output included sediment type superimposed on the topography generated for each time step, synthetic cross-sections (strike- and dip-oriented) through the final thickness of the deposits, and synthetic stratigraphic columns for predetermined locations in the simulation. The first model in the suite used four rules to govern depositional environments in Death Valley, CA (freshwater lake, playa mud flat, salt pan, saline lake) with respect to precipitation and temperature. Conditions of the model included: 1) a 15 km × 65 km basin, 2) a run time of 191 ky, 3) initial basin floor based on present topography and 4) a subsidence component that varied between 0.2 and 1 mm/yr. The fuzzy systems governing basin floor sediments were calibrated with core data. Erosion, compaction and isostatic flexure were not incorporated into this model.

Another fuzzy model in the Demicco and Klir suite used 19 rules to determine carbonate production, erosion, and lithology to simulate the past 10,000 years of tidal flat deposition on Western Andros Island (western side of the Great Bahama Bank) [30]. Conditions of the model included: 1) a 150 km × 300 km ramp margin, 2) a run time of 101 ky (100-yr time steps), 3) a data-based sea level curve and a 4) bathymetry/topography simplified from the literature. The fuzzy logic systems were calibrated with cores and maps. Tectonic subsidence, compaction and isostatic flexure were not incorporated into this model. Both of these models were tuned to reproduce the deposition conditions observed in data from Death Valley and Western Andros Island.

The Demicco and Klir approach also contained a simplified, hypothetical delta and floodplain system simulator [30]. The river system, adjacent levee, crevasse-splay systems and the simple deltaic dispersive cone were modeled by fuzzy logic systems based on the rules of Nordlund (1996). Simulation conditions included: 1) a 125 km × 125 km ramp margin approximately 60 km wide, 2) a run time of 50 ky (200-yr time steps), 3) a simple sinusoidal sea level oscillation (amplitude of 10 m and a frequency of 20 ky) and 4) random upstream avulsions of the river system. Tectonic (thermal) subsidence rates (maximum subsidence rate of approximately 3 mm/yr in center of model decreased to zero towards the edges) remained constant (specific to spatial location) for the duration of the simulation. This simulation did not incorporate erosion, compaction and isostatic flexure. Later modifications of the model included 17 rules specifically applied to the Southwest Pass of the Mississippi River Delta complex [33]. Additional revisions included additional fuzzy rules to simulate bedload transport, suspended sediment plumes, variable wave regimes and long-shore drift at the river mouth. Isostatic compensation (subsidence due to sediment loading) was incorporated into this model, but erosion and compaction were not.

The third group of fuzzy stratigraphic simulations, referred to as FUZZYREEF, was developed by Parcel [29, 31, 32] as a Windows-based program created in C/C + +. FUZZYREEF modeled depositional facies distribution and productivity rates on a carbonate platform using an example of microbial reef development on a Jurassic carbonate ramp from the US Gulf coast (Smackover Formation). Simulated conditions included: 1) a ramp margin less than 5 km × 5 km (initial topography determined from 3D seismic data), 2) a run time of 4 million years (100 ky time steps) and 3) eustatic curves with variable amplitudes between 0 and 200 m. Carbonate productivity and facies distribution were determined by linguistic rules based on climate (arid, temperate, humid), latitude (low, mid, high), water energy (low, mid, high), slope (low, mid, high), and hardground location (soft, firm, hard). Although three FISs were identified, the individual rules driving each FIS were not reported. Subsidence parameters (thermal, loading and compaction) were included in the simulation but implemented using traditional mathematical equations rather than fuzzy logic.

4 Justification for Model Development

The development of fuzzyPEACH occurred in two phases. The first phase of the study assessed the ability of fuzzy logic to model complex stratigraphy under various depositional conditions on an idealized, passive continental margin. The second phase of this investigation used fuzzyPEACH to simulate the physical conditions associated with low-gradient margins. For the sake of brevity, only a general description of the model as it relates to the study's second phase is presented here to illustrate its utility. A detailed discussion of the results is documented in Warren [61].

The typical gradient of a passive continental margin is $< 0.5°$ [62] but generally $> 0.05°$ [6]. Numerous quantitative models have simulated the stratigraphic evolution of a passive margin and have been validated with examples that span the geologic record [1, 16, 17]. Low-gradient margins ($< 0.02°$), on the other hand, have received minimal attention by the modeling community. These margins still exhibit the shelf-slope-rise morphology typical of most passive continental margins [2], but the stratigraphic evolution is different. The response of an unconfined fluvial system across a low gradient (especially one that is lower than the fluvial gradient) is deposition and avulsion throughout the basin with little sedimentary bypass beyond the shelf-slope break. Furthermore, there is no incision at the shelf-slope break that, because of its depth, remains submerged during lowstands of sea level. Margins with these characteristics do not conform to the sequence stratigraphic model.

In order to better understand the depositional processes and stratigraphic response of low-gradient systems, a broader investigation of the East China Sea (ECS) continental margin acquired a high-resolution seismic dataset (14,000 km over a 300,000 km^2 area) on the East China Sea (ECS) continental margin to investigate the stratal architecture on a low-gradient margin with a deep shelf-slope break [63–70]. The combination of these data, in addition to data from previous studies, provides an understanding of the geologic conditions that affected relative sea level, eustasy, tectonic subsidence, sediment influx and margin physiography during the late Quaternary [71–74]. The extensive nature of the ECS data, and the general knowledge gained from it, provide many of the constraints needed to assess the accuracy of fuzzyPEACH. Constraining simulations with these data not only aids in model validation, but also provides the opportunity to explore different scenarios of stratigraphic sensitivity under numerous scenarios of sea level, tectonic subsidence, sediment influx and margin physiography.

5 How the FuzzyPEACH Works

The fuzzyPEACH model uses a combination of triangular and trapezoidal MFs in each of its five FISs. These simple function shapes were chosen because of their successful application in previous stratigraphic models [26–33]. The shape of a MF can vary, and the process of choosing a MF has been addressed by multiple authors [75–78]. Lotfi and Tsoi [75] suggested bell-shaped MFs (Gaussian or Cauchy curves) as better choices than triangular MFs when designing a fuzzy system. The effect of MF shape on fuzzy logic simulations of depositional processes and stratigraphic response has not been addressed in detail on a large-scale, and further research seems warranted. However, preliminary sensitivity testing that replaced triangular MFs with trapezoidal and bell-shaped MFs showed variations in fuzzyPEACH output to be negligible. For example, the deltaic deposition FIS consistently returned sediment volumes with a difference of less than 10% (average $< 5\%$) when bell-shaped MFs replaced triangular MFs. Trapezoidal MFs returned results with a difference no higher than 5% (average $< 2\%$) than those generated with triangular MFs. Many times, there was no difference in output regardless of MF shape.

User-defined variables not defined and/or controlled by fuzzy logic include model duration, length of individual time steps, margin physiography (width and gradient), sedimentation rate, sea level (frequency, amplitude and overall shape of curve), and tectonic subsidence. The general assumptions of the overall simulation are those of the sequence stratigraphic model [79]: 1) a passive margin, 2) constant rate of tectonic subsidence at any given location on the margin each time step, 3) a basinward increase in subsidence, 4) a sediment supply that remained constant during each time step and 5) curvilinear sea level trends. The general processes of the fuzzyPEACH are presented as a flow chart in Fig. 6.

5.1 Defining Geological Variables for the FuzzyPEACH

Model Duration. For the purpose of the second phase of this investigation this investigation, fuzzyPEACH was designed to simulate conditions on a low-gradient continental margin (0.017°) over a period of approximately 200 ky (although model duration is a user-defined variable). This period was quantized into 500-year time steps, an increment that resolves the avulsion

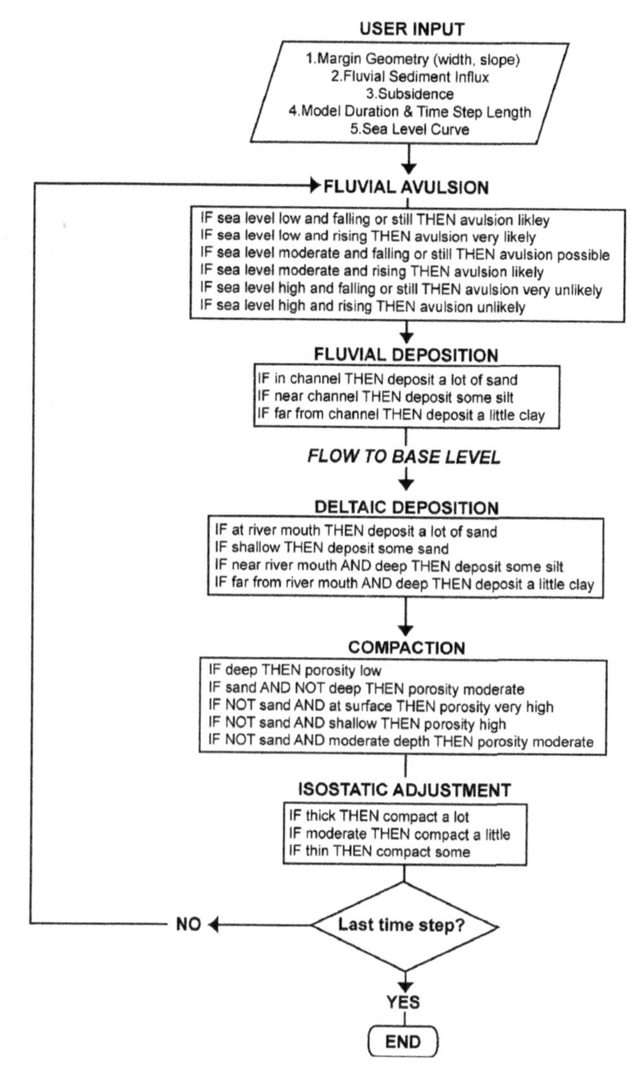

Fig. 6. Flow chart identifying the major processes and general operation of stratigraphic simulations using fuzzyPEACH

frequency of highly avulsive river systems (e.g. Yellow, Mississippi, Po) [80]. (The simulation of fluvial avulsions is discussed in greater detail below.)

Margin Physiography. The simulated margin geometry at the first time step includes a flat margin (lacking antecedent topography) with a length (shore parallel) of 600 km and a shelf width (shore normal) of 500 km (Fig. 7). The shelf-slope break is shore parallel at 150 m below modern sea level and creates a shelf gradient of 0.017°. The synthetic margin is spatially referenced to a 600 km × 600 km horizontal grid of $1-\text{km}^2$ cells (360,000 total cells). This

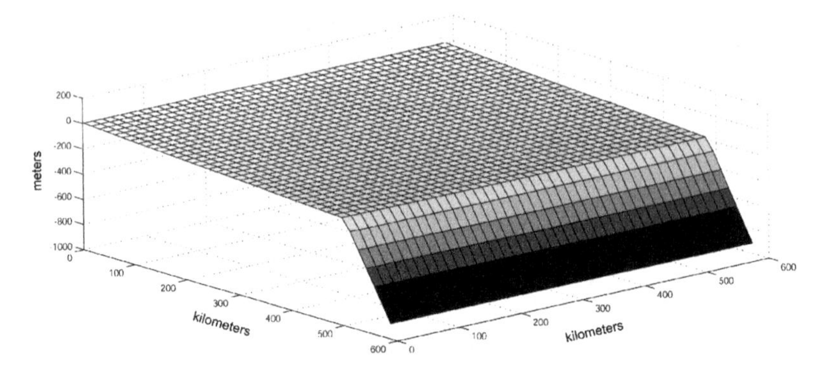

Fig. 7. The synthetic continental margin as it appears at the initiation of a simulation

idealized margin physiography was established based on the geometry of margins sharing similar unincised fluvial architecture. Widths of these margins are 500 km (northwestern Australia) [81], 100 km (southeastern New Zealand) [82] and 500 km (ECS) [65–67]. Depths of shelf-slope breaks are 113 m (northwestern Australia), 150 m (southeastern New Zealand) and 150 m (ECS). Offshore northern Java is a ramp margin (gradient = 0.03°) without a defined shelf-slope break. The gradient of the other margins is 0.1° (northwestern Australia), 0.1° (southeastern New Zealand) and 0.013° (ECS). Similar physiographic data (i.e. gradient and depth of shelf-slope break) for margins and basins throughout the Phanerozoic (the last 545 million years) are rare in the literature.

Subsidence. Tectonic subsidence within fuzzyPEACH is user-defined based on the assumed relationship to thermal contraction [83]. For the simulations presented here, subsidence rates varied between 0 and greater than 4 mm/yr based on values associated with modern margins. Rates of subsidence are well established for numerous continental margins, including the northern Gulf of Mexico from south Texas to western Louisiana (up to 5 mm/yr) [84], the northern Gulf of Mexico at the mouth of the Mississippi River (up to 8 mm/yr) [85], the Ganges River delta plain in the Bengal Basin (up to 5.5 mm/yr) [86], the Canterbury Plains in New Zealand (up to 2 mm/yr) [87] and the ECS near the mouth of the Yangtze River (up to 4.4 mm/yr) [88]. However, these rates reflect total subsidence and do not separate tectonics from compaction and isostatic flexure. Because compaction and flexural isostasy associated with sediment loading have considerable effect on sequence architecture [18, 20], an attempt was made in this investigation to de-couple these two components from tectonic subsidence. Tectonic (thermal) subsidence is held constant for every time step and compaction and isostatic compensation from sediment loading are controlled separately by fuzzy logic.

Sediment Influx. FuzzyPEACH uses a single fluvial system to deliver sediment to the simulated margin. The sedimentation rate is a user-defined variable held constant for each time step. Rates varied in this investigation from between 1×10^8 and 5×10^8 tons/yr and were based on values expected from modern fluvial systems (the average of the world's top 25 rivers is about 2.43×10^8 tons/yr) [89, 90]. Simulated fluvial sedimentation is tied to a mass balance algorithm embedded within fuzzyPEACH to ensure the volume of sediment deposited in both the fluvial and deltaic systems is equal to the sediment influx for each time step. This is accomplished by using a mass-to-volume conversion algorithm that assumes a clastic-dominated, silica-rich sediment load (density of quartz = $\rho_{\mathrm{qtz}} = 2643\,\mathrm{kg/m^3}$) and an average porosity (Φ) of 50% (although ρ and Φ can be user-defined variables with fuzzyPEACH). The average porosity was chosen here as a moderate value between unconsolidated sands and sandstones that can range in porosity from 28% to 46% [91–94] and silts and clays that commonly have Φ values between 60% and 70% [93, 95, 96].

Fluvial influx is a user-defined variable, and sensitivity testing used sedimentation rates spanning 1×10^8 to 5×10^8 tons/yr. These rates are similar to the largest of the world's modern river systems [89, 90]. An assumption of constant sediment influx one of the general assumptions of the sequence stratigraphic model and, therefore, not uncommon. However, of all the depositional variables used in fuzzyPEACH simulations, sediment influx is certainly the one with the highest level of uncertainty. Based on variable precipitation rates and climatic shifts associated with the high-frequency glacio-eustatic signal throughout the Pleistocene, constant sedimentation rates were unlikely. Future fuzzyPEACH modifications will consider variable sediment supply controlled by an additional FIS and based on parameters such as sea level (ocean currents, fluvial gradient) and precipitation (suspended load, runoff, discharge). Like all other FISs in fuzzyPEACH, rules and sets could be altered easily to test multiple hypotheses in relatively short order (i.e. higher or lower sediment supply with high precipitation, higher or lower sediment supply with submerged versus exposed margin). Based on relative relationships, some potential rules defining such an FIS could be:

- IF precipitation high THEN sed rate high
- IF precipitation low THEN sed rate low
- IF climate warm and humid THEN foliage dense AND basin erosion low
- IF climate cold and arid THEN foliage sparse AND basin erosion high
- IF sea level high THEN fluvial gradient higher AND erosion
- IF sea level low THEN fluvial gradient lower AND deposition

Eustatic Sea Level. FuzzyPEACH relies upon prescribed eustatic curves to drive simulations. During this investigation, idealized sinusoids of multiple periodicities (100 ky, 40 ky, 20 ky) were used to simulate Milankovitch-band frequencies observed throughout the Quaternary [97, 98]. Functions of different frequencies and magnitudes were convolved to simulate more complex eustatic change (e.g. parasequences). In addition to sea level curves that were either theoretical and/or idealized, actual sea level conditions were based on SPECMAP (Mapping Spectral Variability in Global Climate Project) data [99,100].

5.2 Fuzzy Variables, Fuzzy Sets, and Fuzzy Inference Systems

Five individual FISs accomplish the following tasks: 1) determine relative grain size (coarse to fine) and distribution (thickness and lateral extent) of fluvial sediment, 2) control fluvial avulsion, 3) determine relative grain size of deltaic sediment (coarse to fine) and distribute deltaic sediment (thickness and lateral extent), 4) control compaction based on porosity and depth and 5) control subsidence via stratigraphic thickness and isostatic compensation. Together, the five FISs and their components (e.g. rules, MFs, domain values of fuzzy sets, etc.) are the engine that drives the fuzzy portion of the model and, therefore, remained constant throughout the entire investigation. The user-defined variables (i.e. margin geometry, sediment influx, sea level, tectonic subsidence) interact with the logic rules and fuzzy sets within each FIS to determine sediment type and thickness for each time step at each of the 360,000 1-km-square cells that define the idealized continental margin (i.e. 2 variables $\times$ $360,000$ cells $\times$ 390 time steps).

During each time step, the simulation of sediment distribution and deposition throughout the fluvial system, flood plain, delta and continental shelf is controlled by rules, or heuristics, based on basic geologic principles. Five distinct Takagai-Sugeno type FISs apply 21 rules and their logic operators (AND, OR, NOT) to a collection of fuzzy sets. (Each FIS is outlined in Fig. 8) Two types of FISs, Mamdani [101] and Takagi-Sugeno [102], commonly are used in fuzzy logic applications to create an aggregate of the appropriate MFs through a single "defuzzified" variable. The difference between the Mamdani and Takagi-Sugeno FIS is the averaging process that calculates this variable. Summarily, the Mamdani FIS integrates across the entire domain of the final output fuzzy set, and the Takagi-Sugeno FIS employs only the weighted average of a few points to create a single spike. A more detailed discussion of the Takagi-Sugeno and Mamdani FIS is illustrated by Demicco [33,41] using mutliple geologic examples, including the determination of carbonate production, carbonate exposure indices and the determination of paleoclimates. The computational efficiency of the Takagi-Sugeno process is well suited for the more than 280 million data points generated during each fuzzyPEACH stratigraphic simulation.

Previous simulations of fluvio-deltaic deposition using fuzzy sets and fuzzy logic were critical to this investigation [25–28, 30, 33]. The geological engine driving fuzzyPEACH, as well as earlier fuzzy stratigraphic models reviewed in this chapter, is based on a set of general rules governing deposition (grain size and volume) within an idealized fluvio-deltaic system. The actual rules are presented below, but can be summarized as follows: 1) the coarsest sediment is deposited proximal to both the fluvial channel and the mouth of the river at the delta, 2) the highest sediment volume is also deposited proximal to

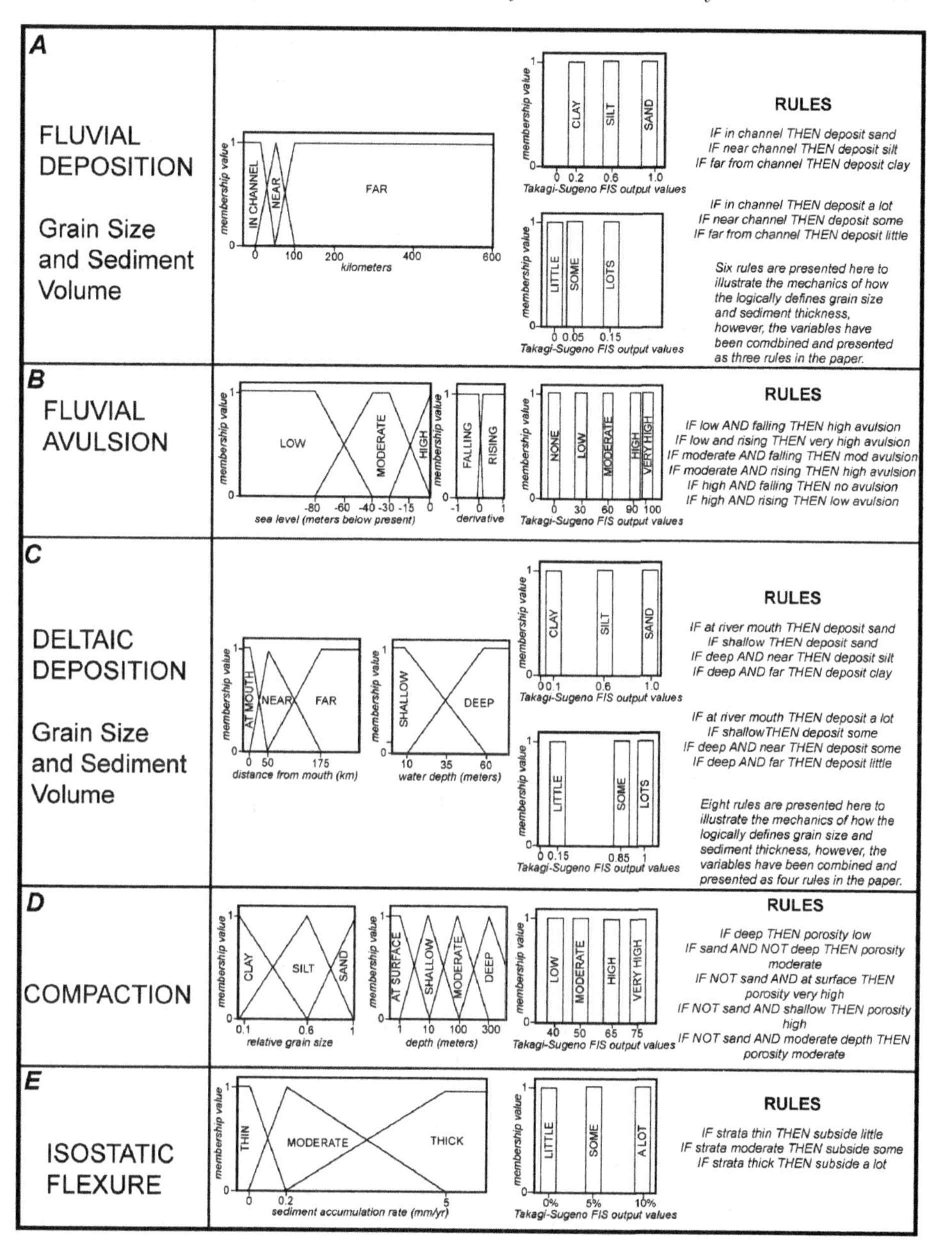

Fig. 8. The five fuzzy inference systems included in fuzzyPEACH that simulate fluvial deposition, fluvial avulsion, deltaic deposition, compaction and isostatic flexure

the fluvial channel and river mouth, 3) sediment becomes more fine-grained and 4) the sediment volume decreases as distance from the channel axis and river mouth increase. Therefore, in relative terms, the coarsest sediments and thickest strata are deposited in the channel and at the head of the delta, while the finest-grained sediments and thinnest strata are deposited the farthest from the channel on the floodplain or farthest from the delta on the continental shelf. Variation of sediment grain size and stratal thickness between these spatial end members (i.e. proximal versus distal relative to the channel or delta head) are calculated along the continuum defined by fuzzy sets. The values and justifications for these variables, fuzzy sets and linguistic rules associated with each FIS are discussed in detail below. Variables were tuned during sensitivity testing in order to ensure that reasonable stratal architectures were produced.

Fluvial Deposition. At the beginning of each simulation, a single river flows from the upstream end of the model toward adjacent cells with lower elevations. During rising sea level, the river mouth backtracks up the river course. During falling sea level or stillstands, the river seeks out the lowest adjacent cell in front of it until reaching zero elevation (base level). Fluvial deposition is simulated during every time step for every cell above base level, and a single FIS (Fig. 8A) uses three rules to determine grain size and sediment volume relative to the location of the channel.

- IF in channel THEN a lot of sand
- IF near channel THEN some silt
- IF far from channel THEN little clay

The left-hand side of the rule (the premise) defines the distance from the channel. The right-hand side of the rule (the conclusion) determines the amount and type of sediment deposited per time step at each grid cell. The three variables in these rules (i.e. distance from source, sediment volume, grain size) contain multiple fuzzy sets, and the boundary of each set is determined by a MF. The geological data and/or justification(s) for these sets (domain variables and set boundaries) are discussed below.

The simulated fluvial channel in fuzzyPEACH is idealized and does not distinguish between the end-members of fluvial morphology (i.e. straight and braided; Fig. 2). Therefore, the widths of the channel axis are based on both modern and recent fluvial systems. The average widths of individual channels and channel braids are used to define the simulated channel width. The distance from the channel axis under the influence of the channel is based on floodplain widths. The distance relative to the channel is defined (in kilometers) by three fuzzy sets: 1) "in channel", 2) "near channel" and 3) "far from channel." The minimum channel width is set at 2 km ("in channel"). The main channel of the Mississippi River is between 1 and 2 km wide [85]. Paleo channels of similar widths of are preserved in the shallow strata offshore Java and are approximately the same width (2 km) [103]. The Yangtze River is also about 5 km wide just upstream from its estuary [104]. Individual channels within braided complexes are not as wide, but braids associated with these channels have similar widths. The entire braid is considered a site of active deposition based on the spatial and temporal resolution of fuzzyPEACH simulations. The braided Rangitata River on the Canterbury Plains (New Zealand) has an overall width between 2 and 5 km wide [105]. The bankfull widths of the braided Tana (Kenya) and Slims (Canada) rivers are 0.6 to 2 km and 0.3 to 1.8 km, respectively [106]. The fuzzy set "in channel" is defined as less than, or equal to, 5 km. This is a reasonable value based on the width of fluvial channels and channel complexes presented above.

The fuzzy set "near" encompasses the floodplain, the width of which is based on fluvial observations from various geologic settings. The floodplain associated with the modern Mississippi River averages 100 km wide from the Gulf of Mexico up into Missouri. This width has remained fairly constant throughout the Holocene [107]. Core data from the Yangtze coastal plain identify a floodplain width of at least 100 km throughout the Quaternary [108]. The floodplain from Cooper Creek (Australia) has a smaller, but still substantial, floodplain of 50 km [109]. The fuzzy set "near channel" is defined as less than or equal to 50 km. This is measured in both directions from the axis of the channel (i.e. a total width of 100 km). This is a reasonable value based on the floodplain data above. The fuzzy set "far" includes distances greater than 100 km from the channel axis. Deposition in fuzzyPEACH does not occur farther than 100 km from the channel.

Justified by the widths above, complete membership (MF = 1) in each of the fuzzy sets that define distance from channel occurs at = 5 km ("in channel"), 50 km ("near channel") and $\geq$ 100 km ("far from channel"). Partial membership (overlap in adjacent sets where MF $<$ 1) is defined along the continuum of values between 0 and 50 km (partial membership in both "in channel" and "near channel") as well as 50 km and 100 km (partial membership in both "near channel" and "far from channel"). Distances from the channel axis greater than 100 km are classified solely as "far."

Two outputs, volume of sediment and grain size, subdivide into multiple fuzzy sets. Volume is defined by three fuzzy sets ("little", "some", "a lot") whereas grain size categories correspond to the three basic grain-size fractions ("sand", "silt", "clay"). Volume of sediment deposited (stratal thickness in a 1-km-square grid) defines full membership (MF = 1) at 0.15 ("a lot"), 0.05 ("some"), and 0 ("little"). Partial membership (overlap in adjacent sets where MF $<$ 1) is defined along the continuum of values between 0.15 and 0.05 (partial membership in both "a lot" and "some") as well as 0.05 and 0 (partial membership in both "some" and "little"). The values are unit-less as well as dimensionless (i.e. they are scalable based on total sediment influx). In conjunction with the sedimentation mass balance algorithm described above, the output defines the depositional volume (stratal thickness) to each grid cell above base level at that particular time step. Sediment volumes are determined relative to the output ratios described above (e.g. 0.15 being the maximum thickness and 0 being the minimum and determining areas without deposition). Due to the unit-less nature of this FIS, sediment accumulation rates were calculated during sensitivity testing to ensure reasonable deposition was occurring. Sediment influx primarily affected the deltaic system, therefore, sediment accumulation in the fluvial system averaged about 1 mm/yr during high rates of sediment influx ($> 2 \times 10^8$ tons/yr). Simulated, in-channel sediment accumulation rates averaged 1.3 mm/yr. Similar fluvial accumulation is calculated from the Canterbury Plains (New Zealand) between 0.7 mm/yr and 1.7 mm/yr [82]. An average accumulation rate of 0.5 mm/yr is calculated from fluvial deposits deposited on the ECS margin between 125 and 175 ka (ka = 1,000 years ago) [66]. Floodplain accumulation rates from fuzzyPEACH simulations averaged 0.2 mm/yr. Rates similar to those in the model are observed in the floodplains of the Indus River (0.2 mm/yr) [110], the Tigris and Euphrates (0.2 mm/yr) [111], the Delaware River (as low as 0.1 mm/yr) [112] and the Wisconsin valley (0.35 mm/yr) [111].

Grain size is determined along a unit-less, relative continuum spanning coarse and fine grain sizes. The relative nature of these fuzzy grain size sets is dimensionless and scalable. Therefore, identical results could be achieved with grain size defined in relative terms as “coarse”, “medium” and “fine.” Full membership (MF = 1) at each fuzzy singleton occurs at 1.0 (sand), 0.6 (silt) and 0.2 (clay). Partial membership (overlap in adjacent sets where MF < 1) is defined along the continuum of values between 1.0 and 0.6 (partial membership in both “sand” and “silt”) as well as 0.6 and 0.2 (partial membership in both “silt” and “clay”). The mass balance algorithm embedded within the fuzzyPEACH ensures depositional volume of sediment deposited in both the fluvial and deltaic systems is equal to sediment influx for each time step.

Fluvial Avulsions. Fluvial avulsion is the relatively sudden shift to a new course on a floodplain or deltaic plain [113]. At the beginning of each 500-year time step, the fluvial system has the potential to avulse. The point of avulsion along the longitudinal axis of the river, as well as the direction of movement (right or left of the existing river), is programmed to be random. The new fluvial pathway establishes along the lowest topographic profile between the point of avulsion and sea level. Avulsions, when they do occur, are instantaneous and resolve to the model's 500-year time steps. This time window is similar to that of Törnqvist [114] who defined instantaneous avulsion as one that occurs within a period of 200 years (^{14}C dates limit resolution to ± 200 years). Avulsion data tabulated by Stouthamer and Berendsen [80] also validate the resolution of avulsion within 500-year time steps. For example, the lower Mississippi River and its delta complexes have avulsion frequencies that average 1380 and 1400 years, respectively. Other rivers experiencing frequent avulsion include the Rhine and Meuse rivers (945 years), the Saskatchewan River (675 years), the Yellow River (600 years) and the Po River (490 years).

The mechanics of fluvial avulsion, while relatively easy to describe in general terms, are complicated, and a governing set of mathematical equations does not exist. However, trends have been observed and described [8, 80, 113–121]. It is these general trends that form the foundation for the six linguistic rules controlling avulsion during each time step (Fig. 8B).

- IF sea level low AND falling THEN chance of avulsion high
- IF sea level low AND rising OR still THEN avulsion chance very high
- IF sea level moderate AND falling THEN avulsion chance moderate

- IF sea level moderate AND rising OR still THEN avulsion chance high
- IF sea level high AND falling THEN no chance of avulsion
- IF sea level high AND rising OR still THEN chance of avulsion low

The premise defines the position of sea level relative to present (i.e. modern sea level is considered high) and its direction of movement (rise or fall). The conclusion determines the likelihood the river will avulse during any given time step. The three variables of each rule (i.e. sea level, direction of sea level movement, chance of avulsion) contain multiple fuzzy sets. The boundary of each set is determined by a MF. The geological data and/or justification(s) for these sets (domain variables and set boundaries) are discussed below.

The elevation of sea level determines the degree of subaerial exposure of the margin that, in turn, determines the distance over which a river must flow to reach sea level. As fluvial systems become overextended, avulsion potential increases because an increasing length decreases fluvial gradient and results in an overall loss of power (i.e. the ability for the system to carry sediment) [117, 119]. The Yellow River illustrates the effect of overextension and power loss, although it is occurring at present during a period of slow sea level rise. It is the river's abundant sediment load (1.08×10^9 tons/yr) [89], and not sea level lowstand, that is responsible for delta progradation rates as high as 2 km/yr [117]. However, the result is the same: a lowered fluvial gradient (from either 1) a shelf or exposed margin with a gradient equal to or lower than the fluvial system, or 2) rapid progradation of the delta) leads to overextension and power loss. As these individual delta lobes rapidly build seaward, stream gradients substantially lower and cause an overall straightening and overextension of the fluvial system [119]. The resulting power loss causes channels to aggrade above the surrounding floodplain, and an eventual levee breach shifts the channel to a shorter, higher-gradient course to base level. To illustrate this point, nine avulsions of the Yellow River (nodal points 20–50 km updip from delta lobes) since 1855 are attributed to this cycle of overextension, aggradation, and avulsion [117]. An ancient example of fluvial overextension and high-frequency avulsion is the laterally extensive Mesa Rica Sandstone (Lower Cretaceous, New Mexico) [119].

The unconfined fluvial systems on the margins of Australia [81], New Zealand [82], Java [103] and eastern China [63–66] were deposited during subaerial exposure of the low-gradient margin during recent glacial maxima. The 100 ky periodicity of these extreme lowstands correlates to the eccentricity Milankovitch band [97]. The inferred eustatic signal during those times was less than 80 m below present. Since sea level is used as an input variable for the six rules in the avulsion FIS, the fuzzy classification considers sea level "low" for elevations greater than 80 m below present. The remaining sea level values are defined by two additional sets: "moderate" and "high." Complete membership (MF = 1) in each of these three fuzzy sets occurs at 0 m (modern sea level = "high"), 40 m below present ("moderate") and = 80 m below present ("low"). Partial membership (overlap in adjacent sets where MF < 1) is defined along the continuum of values between 0 and −40 m (partial membership in both "high" and "moderate") as well as −40 and −80 m (partial membership in both "moderate" and "low").

Two general trends relative to avulsion are defined by the direction of sea level movement. First, avulsion rates increase during rising sea level (and continue to increase as rate and magnitude of sea level rise increases) as a river aggrades to positively adjust its equilibrium profile [8, 115]. Second, avulsion rates decrease (and continue to decrease as rate and magnitude of sea level rise decreases) during falling sea level as fluvial sedimentation tries to keep pace and negatively adjust its equilibrium profile [8, 80, 115]. Two fuzzy sets define sea level movement: 1) falling and 2) rising. The derivative of the sea level curve for each time step determines slope, and complete membership ($MF = 1$) in each fuzzy set occurs for slopes $= -0.1$ ("falling") and $= 0.1$ ("rising"). Partial membership (overlap in adjacent sets where $MF < 1$) is defined along the continuum of values between −0.1 and 0.1 where sea level change rates are negligible (essentially stillstand).

In addition to sea level elevation, the direction of elevation change affects the stability of a fluvial system and also contributes to the rate of avulsion [8]. Rate of change also is important. Higher rates of sea level change make it more difficult for fluvial systems to keep an equilibrium profile. Faster rates of change during sea level rise increase the chance for avulsion (the fluvial system cannot reach equilibrium), and slower rates of change during falling sea level increase the chance for avulsion (sediments prograde and lower the fluvial gradient). During the past 195 ky, rates of sea level change were consistently high (up to 7 mm/yr). Therefore, rate was not included in the fuzzyPEACH simulations presented in this paper, but will be considered during future upgrades of the model.

The output of the avulsion FIS determines whether fluvial avulsions will occur during any given time step. This possibility for avulsion is defined by five fuzzy sets: 1) "none", 2) "low", 3) "moderate", 4) "high" and 5) "very high." Full membership ($MF = 1$) occurs at 0% ("none"), 30% ("low"), 60% ("moderate"), 90% ("high") and 100% ("very high"). Partial membership (overlap in adjacent sets where $MF < 1$) is defined along the continuum of values between 0 and 30% (membership in both "none" and "low"), 30 and 60% (membership in both "low" and "moderate"), 60 and 90% (membership in "moderate" and "high") as well as 90 and 100% (membership in "high" and "very high"). Based on this fuzzy approximation of probability, avulsions frequencies can be established at 500 years (100% chance at each time step), 1,000 years (50% chance), 4,000 years (25% chance), 8,000 years (12.5% chance), etc.

FuzzyPEACH uses output values as a proxy for avulsion probability because, even though membership functions are similar to probability density functions in form and function [26], fuzzy sets address imprecision from the absence of sharply defined criteria rather than the presence of random variables [122]. Fuzzy logic addresses the degree to which something is believed to be possible rather than the likelihood of an event to occur [123]. Dealing with these two forms of uncertainty (i.e. probability and possibility) has been widely discussed and is somewhat controversial [124–126]. For the purpose of fuzzyPEACH simulations, fluvial avulsion either occurs or it does not occur (crisp, traditional logic). However, the FIS bases the possibility of avulsion on geologically reasonable conditions easily defined and governed by linguistic terms. The domain variables (i.e. the values along the x-axis that represented percent chance of avulsion between 1 and 100) defining the fuzzy set boundaries were chosen arbitrarily between 1 and 100% to represent a logical subdivision of avulsion potential. After setting the boundaries to these fuzzy sets, numerous sensitivity tests during model development produced avulsion rates similar to natural systems (e.g. Rhine and Meuse, Saskatchewan, Yellow and Po rivers) [80]. Future versions of fuzzyPEACH will record a detailed avulsion history of each simulation, including the number of fluvial avulsions and when they occurred (time step, eustatic elevation, direction of sea level movement) to facilitate a more quantitative review.

Deltaic Deposition. Simulated deltaic deposition propagates seaward at the intersection of the river and the shoreline (base level). The delta FIS controls deposition for every grid cell below base level for any given time step with a single FIS (Fig. 8C). Four rules within this FIS determine grain size and sediment volume relative to the location of the river mouth and water depth. The second rule (below) simulates shoreface deposition by spreading a thin veneer of sand along the entire shoreline.

- IF at river mouth THEN deposit a lot of sand
- IF shallow THEN deposit some sand
- IF near river mouth AND deep THEN deposit some silt
- IF far from river mouth AND deep THEN deposit a little clay

The premise defines the distance from the river mouth and the water depth. The conclusion determines the amount and type of sediment deposited per time step at each grid cell. The four variables of the rules (i.e. distance from source, depth of water, sediment volume, grain size) contain multiple fuzzy sets, the boundary of each set being determined by a MF. The geological data and/or justification(s) for these sets (domain variables and set boundaries) are discussed below.

The distance from the river mouth is defined (in kilometers) by three fuzzy sets: 1) “at river mouth”, 2) “near river mouth” and 3) “far from river mouth.” These sets only deal with distance; the physical processes in the water column affecting deltaic deposition are addressed below. The 1-km grid spacing of the model, by default, sets “at river mouth” at 1 km. However, domain variables defining distance for the fuzzy sets “near river mouth” and “far from river mouth” are based on actual distances of the 30 largest modern deltas in the world in terms of area of delta plain [127]. This group of 30 deltas was broken up into two tiers (the top 15 = Tier I and the bottom 15 = Tier II). The average area of Tier I is about $4,500\,\mathrm{km}^2$. The radius of a delta with this area, assuming a semi-circular delta where area = $[(pR^2)/2]$, the radius is just over 50 m. The average area of the deltas from Tier II is about 48,000 km^2, and the radius is 175 km. An assumption is made here that the deltaic radii from these two groups are a good representation of average boundaries expected between major deltas in the modern record [56,127,128]. All deltas, on average, are expected to have a radius of at least 50 km and a maximum width no more than 175 km. These two values define the fuzzy sets "near river mouth" and "far from river mouth", where complete membership (MF = 1) 50 and 175 km, respectively. Therefore, complete membership (MF = 1) in the fuzzy set "far from river mouth" is set at 175 km. Values equal to, or greater than, 175 km were classified solely as "far from river mouth." Partial membership (overlap in adjacent sets where MF < 1) is defined along the continuum of values between 0 and 50 km (membership in both "at river mouth" and "near river mouth") as well as 50 and 175 km (membership in both "near river mouth" and "far from river mouth").

The second input variable, water depth, is defined (in meters) by two fuzzy sets: 1) "shallow" and 2) "deep." Water depth is determined by the general depth at which the physical processes affecting depositional processes are no longer affected by surface waves. The depth of fair-weather wave base has been reported as deep as 60 m for specific locations (Ebro Delta) [129] to as shallow as 10 m for modern, high-energy coasts in general [130]. Therefore, "shallow" was defined for water depths less than 10 m and "deep" for depths greater than 60 m. The range between 10 and 60 m was a combination of both "shallow" and "deep" to varying degrees with equal membership (MF = 0.5) at 35 m. In the ECS, the 10 m isobath defines the boundary of maximum turbidity [131] and the seaward boundary of the Yangtze depocenter [132]. Fluvial and coastal processes dominate the areas shoreward of this position and a combination of fluvial and marine processes occur seaward of the 10 m depth. The 30 m isobath defines the general boundary dividing coarse- and finer-grained sediment fractions near the mouth of the Yangtze River [133]. The grain size distribution is widespread and represents a potential boundary between depositional processes (e.g. where the influence of coastal currents and fluvial processes on deltaic deposition cease to dominate). Thus, the boundary between the fuzzy sets "shallow" and "deep" defined at 35 m are taken to be a reasonable proxy for the ECS to define the boundary between dominant processes affecting seafloor deposition.

The amount of wave energy at the shoreline depends mainly on the subaqueous profile of the delta (i.e. the flatter the slope, the greater the attenuation of deep water wave energy). The major effect of incoming wave energy impacts sediment distribution [127]. Wells and Coleman [127] correlated offshore delta slope with wave power at the shoreline for 18 major deltas of the world. The average wave energy of these deltas was approximately 2×10^7 Ergs/sec. Transposing this average wave energy from the vertical axis onto a best-fit line through the plot of Wells and Coleman [127] results in a slope value along the horizontal axis of 0.04°. Application of this slope to the distance defining "near river mouth" above (i.e. horizontal distance of 50 km) produces a water depth of 35 m, a value that falls halfway between 10 and 60 m (the depths of wave base presented above). Therefore, full membership (MF = 1) in the fuzzy set "shallow" occurs from 0 to −10 m and is the area of the delta dominated by fluvial and other coastal processes. Full membership (MF = 1) in the fuzzy set "deep" occurs in depths greater than 60 m. Partial membership (overlap in adjacent sets where MF < 1) is defined along the continuum of values between 10 and 60 m (membership in both "shallow" and "deep") where wave processes still affect deposition, but to a lesser degree at increasing depths. Hence, the environment is a combination of fluvial, coastal and marine processes. The crossing point of the two MFs occurs at 35 m (membership in both "shallow" and "deep" is 0.5).

Similar to the FIS controlling fluvial deposition, two outputs (volume of sediment and grain size) subdivide linguistically into multiple fuzzy sets. Volume is defined by three fuzzy sets ("little", "some", "a lot") whereas grain size categories correspond to the three basic grain-size fractions ("sand", "silt", "clay"). Full membership (MF = 1) occurs at 1 ("a lot"), 0.85 ("some") and 0.15 ("little"). Partial membership (overlap in adjacent sets where MF < 1) is defined along the continuum of values between 1 and 0.85 (partial membership in both "a lot" and "some") as well as 0.85 and 0.5 (partial membership in both "some" and "little"). The values are unit-less as well as dimensionless (i.e. they are scalable based on total sediment influx). In conjunction with the mass balance algorithm described above, the output defines the depositional volume (stratal thickness) to each grid cell below base level at that particular time step. The depositional volume is relative to the output ratios described above (e.g. 1 being the maximum thickness and 0.15 being the minimum and determining areas with minimal thickness). Due to the unit-less nature of this FIS, sediment accumulation rates were calculated during model development to ensure reasonable deposition was occurring. Sediment influx primarily affected the deltaic system (higher sedimentation rates correspond to higher discharge at the river mouth and higher deposition at the delta and surrounding continental margin.

Using a sedimentation rate of 5×10^8 tons/yr to illustrate (the average of the world's top 10 rivers) [89, 90], the sediment accumulation rate at the delta mouth was about 10 mm/yr. For comparison, the Nile River delta has a sediment accumulation rate of 10 mm/yr [110] and an annual sediment load of 1×10^8 tons/yr [56]. The Orinoco River has an average annual sedimentation rate of 2×10^8 tons/yr [89, 90] and a sediment accumulation rate associated with its delta of 5 to 6 mm/yr [134]. The Yangtze River has a sediment accumulation rate at its mouth between 10 and 54 mm/yr [135] and an annual sedimentation rate of 5×10^8 tons/yr. Sediment accumulation rates at the toe of the delta in fuzzyPEACH were also reviewed to ensure a reasonable value. Using a sediment influx of 5×10^8 tons/yr, simulations had accumulation rates between 1.5 and 2 mm/yr. Modern systems, such as the Rhone River, accumulates 6 mm/yr of sediment 45 km offshore (the radius of the delta) [56] from the river mouth [134]. The Nile River delta toe has an approximate sediment accumulation rate of 0.7 mm/yr [112]. Finally, sediment accumulation on the outer shelf was also calculated from from fuzzyPEACH model results (also using a sedimentation rate of 5×10^8 tons/yr) at a rate of 0.1 mm/yr. Sedimentation accumulation at the Bengal cone (Ganges prodelta) is 0.06 mm/yr [136]. The sediment load of the Ganges-Brahmaputra River is 1.7×10^9 tons/yr, the largest in the world [89, 90]. Sediment accumulation on the open shelf, 30 km basinward of the Rhone River delta toe, is 1 mm/yr [134]. The Yangtze River has an accumulation rate on the outer margin, about 300 km from the Yangtze mouth, of 0.3 mm/yr [74]. The sedimentation rates in the Atlantic and Pacific basins (abyssal) are, on average, 0.01 mm/yr [137].

Grain size is determined along a unit-less, relative continuum spanning coarse and fine grain sizes. The relative nature of these fuzzy grain size sets is dimensionless and scalable. Therefore, identical results could be achieved with grain size defined in relative terms as "coarse", "medium", and "fine." Full membership (MF = 1) at each fuzzy singleton occurs at 1.0 (sand), 0.6 (silt) and 0.2 (clay). Partial membership (overlap in adjacent sets where MF < 1) is defined along the continuum of values between 1.0 and 0.6 (partial membership in both "sand" and "silt") as well as 0.6 and 0.2 (partial membership in both "silt" and "clay"). Values in the output set were unit-less and relative to each other (rather than absolute) where coarsest grain size was 1 and the finest grain size was 0.1. The value for "clay" deposited on the subaerial coastal plain is "clay" (0.2). This fractional difference is visible in synthetic cross-sections and is assigned a different color and allows floodplain deposits to be distinguished from finer-grained pelagic muds on the shelf.

Compaction. In addition to a user-defined variable for tectonic subsidence, fuzzyPEACH also simulates sediment compaction. This calculation occurs at the end of every time step for the entire vertical sediment succession. Demicco [41] effectively used a similar technique using fuzzy logic to simulate the burial and compaction of carbonate mud. Five rules within a single FIS (Fig. 8D) determine the amount of compaction based on porosity and depth.

- IF deep THEN porosity low
- IF sand AND NOT deep THEN porosity moderate
- IF NOT sand AND at surface THEN porosity very high
- IF NOT sand AND shallow THEN porosity high
- IF NOT sand AND moderate depth THEN porosity moderate

The premise defines the vertical extent (height) of the sediment column and grain size. The conclusion determines the amount of porosity expected after compaction, and the three variables of the rules (i.e. depth, grain size, porosity) contain multiple fuzzy sets. The boundary of each set is determined by a MF. The geological data and/or justification(s) for these sets (domain variables and set boundaries) are discussed below.

Both grain size and sediment volume classifications are assigned to each grid cell at every time step by the two FISs controlling fluvial and deltaic deposition (i.e. output from the fluvial FIS and delta FIS become input for the compaction FIS). Therefore, fuzzy grain size values correspond to the relative, unit-less categories described above: "sand", "silt" and "clay." Full membership (MF = 1) occurs at 1.0 ("sand"), 0.6 ("silt"), and 0.1 ("clay"). Partial membership (overlap in adjacent sets where MF < 1) is defined along the continuum of values between 0.1 and 0.6 (partial membership in both "clay" and "silt") as well as between 0.6 and 1 (partial membership in both "silt" and "sand"). In addition to grain size, the depth of sediment is also used to simulate compaction based on the application of expected porosity trends [96]. Depth is calculated for the entire stratigraphic column that was deposited at the end of each time for every grid cell. This is a cumulative process. For example, at the end of the first time step, sediment depth (and its respective grain size value) is merely the height (in meters) of the sediment deposited during only that time step. Subsequently, at the end of the last time step (time step 390), the model analyzes the matrix of values (grain size and height of sediment column) deposited at each grid cell during each subsequent time step (in this case, 390 time steps). FuzzyPEACH then simulates compaction by assigning a porosity value to each cell for each of the 390 time steps. Using this approach, the porosity matrix is re-calculated at each time step and the model will not continue to decrease porosity and over-compact sediments.

Depth is divided (in meters) into four fuzzy sets: 1) "at surface", 2) "shallow", 3) "moderate" and 4) "deep." These values are based upon the suite of compaction curves presented by Baldwin and Butler [96], although multiple investigations of porosity and volume loss during compaction were reviewed for both sand [138–141] and argillaceous sediment [95, 96, 139]. The logarithmic scale of depth (meters) of the Baldwin and Butler curves establishes the domain of four fuzzy sets: 1) "at surface", 2) "shallow", 3) "moderate" and 4) "deep." Complete membership in each of these sets (MF = 1) occurs at = 1 m ("at surface"), 10 m ("shallow"), 100 m ("moderate") and 300 m ("deep"). Partial membership (overlap in adjacent sets where MF < 1) is defined along the continuum of values between 1 and 10 m (membership in both "at surface" and "shallow"), 10 and 100 m (membership in both "shallow" and "moderate") and 100 and 300 m (membership in both "moderate" and "deep"). During fuzzyPEACH simulations specific to this study, it was not necessary to continue the logarithmic trend for the fuzzy set "deep" (e.g. depth = 1,000 km). Model results were validated using the shallow stratigraphy of the last 195 ky on the ECS margin and the average thickness of these strata is 100 m. However, a simple modification to fuzzyPEACH could either modify the boundaries of the fuzzy set "deep" or add an additional fuzzy set "very deep" to handle depths = 1 km. For the purpose of this investigation, all depths greater than 300 m were considered to have full membership (MF = 1) in the fuzzy set "deep."

The output of the compaction FIS (i.e. percent porosity) is defined based on the grain size and depth of sediment deposited during each discrete time step. Based on the compaction curves presented by Baldwin and Butler [96], the values defining the fuzzy sets for porosity corresponded the fuzzy set boundaries used for depth. Following their compaction curves for finer-grained sediments, the porosity is defined as 75% at 1 m, 65% at 10 m and 50% at 100 m. The volume loss for coarser material on these compaction curves is 50%. This is based on sandstone data [142] rather than unconcsolidated material. Unpacked, sand-sized sediments typically have porosity values between 30 and 40% [92, 140, 141] although values have been reported as high as 45 to 50% [91, 94]. FuzzyPEACH is programmed to assume that, regardless of the coarse-grained porosities, there is no compaction of sand-sized material until depths of 100 km are reached. Based on the empirical data on compaction of St. Peter quartz sand (St. Peter Formation, Minnesota, Ordovician) [140,143], which show a porosity change of less than 2% at 10 MPa (1450 psi = 500 km in depth) [144], this is a reasonable assumption. From 100 to 300 m, all sediments along the compaction curves lose an additional 10% of porosity. Based on these trends, porosity is defined by four fuzzy sets ("low", "moderate", "high" and "very high"). Full membership (MF = 1) at each of these porosity values occurs at 40% ("low"), 50% ("moderate"), 65% ("high") and 75% ("very high"). Partial membership (MF < 1) is defined along the continuum of values between 40 and 50% (membership in both "low" and "moderate"), 50 and 65% (membership in "moderate" and "high") and 65 and 75% (membership in both "high" and "very high"). Values equal to or greater than 75% are classified solely as "very high." The model can easily be adapted to include additional fuzzy sets for porosities less than 30% (e.g. "very low").

Isostatic Flexure. It is difficult to separate compaction from isostatic loading when measuring subsidence (a point that is underscored by the paucity of such data in the literature). The effect of isostatic compensation is best observed at rivers with abundant sediment influx. A regional comparison of the coastline flanking the river makes the point that, while compaction is certainly occurring at the site of deposition, isostatic compensation is also occurring at, but not away from, the depocenter. For example, the Mississippi River delivers 2×10^8 tons/yr of sediment to the Gulf of Mexico [80, 89]. While numerous, independent analyses of eustatic rise show similar rates of change between 1 and 1.5 mm/yr [145–148], the average rate of subsidence for the Mississippi delta region is 9.4 mm/yr [85], or about 8 mm/yr corrected for eustasy. Subsidence drops, moving laterally away from Louisiana and away from high sediment influx, to about 0.5 mm/yr (Florida-Alabama border) and 0.1 to 5 mm/yr (central Texas) [84]. The same observation is made along the eastern coast of China. The Yangtze River delivers 4.8×10^8 tons/yr of sediment to the ECS margin. On the inner shelf near the river mouth, subsidence rates vary between 1.6 and 4.4 mm/yr [88]. On the outer margin, the rates drop to 0.3 mm/yr [74]. Along the coast, north and south of the Yangtze, regional uplift is reported as high as 3 mm/yr [149, 150].

De-coupling isostatic flexure from total basinal subsidence (i.e. tectonic or thermal) has been addressed in other investigations by the application of differential equations with techniques such as backstripping (i.e. using variables such as flexural rigidity, density of mantle, average stratal density, density of water column, gravity acceleration, and elastic thickness) [151–155]. Backstripping is usually applied to basins on a much larger scale (kilometers-thick strata representing millions of years) and not considered appropriate for the thin successions and small time-scale of interest in this investigation. Similar to the fuzzy stratigraphic model FUZZIM [25–28], fuzzyPEACH applies a simple hydrostatic approach to simulate isostatic flexure (the compensating mass density is determined only by the topographic load directly above it) [156, 157]. The quantification of values for each of the three subsidence components on various continental margins is lacking in the literature. However, some general ratios were observed using the regional seismic dataset from the ECS margin, combined with published values for sediment accumulation and total subsidence. The FIS rules were written based on these relationships. During each time step, vertical displacement related to isostasy is simulated using a single FIS (Fig. 8E) containing three rules to determine cumulative stratal thickness and vertical displacement.

- IF thin THEN subside little
- IF moderate THEN subside some
- IF thick THEN subside a lot

The premise defines the thickness of a stratal unit. The conclusion simulates isostatic compensation by applying downward displacement of the sediment column for each 1-km-square grid cell at the end of each time step (after compaction). The two variables of the rules (i.e. stratal thickness and vertical displacement) contain multiple fuzzy sets. The boundary of each set is determined by a MF. The geological data and/or justification(s) for these sets (domain variables and set boundaries) are discussed below.

The results of load-driven subsidence are observed in the shallow strata of regional seismic profiles of the ECS margin. Subsidence rates between 1.6 and 4.4 mm/yr on the inner margin near the mouth of the Yangtze River are roughly 10% of sedimentation rates (between 10 and 54 mm/yr) [88, 135]. The tectonic component of subsidence at this location is regarded as negligible due to its location near the tectonic hinge associated with regional uplift along the east coast of China [150, 158]. This range of subsidence is similar to other marginal basins receiving a high sediment input, including the northern Gulf of Mexico from south Texas to western Louisiana (up to 5 mm/yr) [84], the northern Gulf of Mexico at the mouth of the Mississippi River (up to 8 mm/yr) [85], the Ganges River delta plain in the Bengal Basin (up to 5.5 mm/yr) [86] and the Canterbury Plains in New Zealand (up to 2 mm/yr) [82]. On the middle and outer ECS margin, the subsidence rate is 0.3 mm/yr [74]. When corrected for tectonic subsidence (0.3 mm/yr – 0.1 mm/yr = 0.2 mm/yr), this value is a little more than 5% of the sedimentation rate of 3 mm/yr [135].

Elsewhere on the margin, sedimentation is negligible and isostatic loading is assumed to be zero.

Isostatic adjustment is simulated using the trends between sediment accumulation and isostatic adjustment observed on the ECS margin. These data, at least for this investigation, provide a range of coefficients ranging between 0 and 10% that determine the amount of vertical displacement, at the end of each time step, based on sediment accumulation rate (i.e. total thickness/time elapsed). Regional seismic profiles verify subsidence rates presented above by measuring total downward displacement in the strata relative to a structural datum. This method is independent of compaction. Therefore, after correcting for tectonic displacement, this method is used as a proxy for isostatic flexure (at least for depositional conditions analogous to the ECS margin) and is incorporated easily into fuzzyPEACH. Based on this proxy of isostatic adjustment, sediment accumulation (stratal thickness) defines sediment loading (vertical adjustment of sediment column). At the end of each time step, sediment accumulation is determined for that particular time step. Three fuzzy sets define (in millimeters) stratal thickness: 1) "thin", 2) "moderate" and 3) "thick." The ranges of ECS sediment accumulation (presented above) define these linguistic terms (i.e. sediment accumulation about 0.2 mm on outer margin and up to 5 mm on inner margin). Therefore, full membership (MF = 1) in each fuzzy set occurs at 0 mm ("thin"), 0.2 mm ("moderate") and 5 mm ("thick"). Partial membership (overlap in adjacent fuzzy sets where MF = 1) is defined along the continuum of values between 0 and 0.2 mm (membership in both "thin" and "moderate") as well as 0.2 and 5 mm (membership in both "moderate" and "thick"). Based on these thicknesses, fuzzyPEACH calculates the approximate percentage of expected vertical displacement by multiplying the sediment accumulation rate with the appropriate coefficient (in this case, = 10%). Simulated isostasy displaces the entire vertical sediment column at each of the model's 1-km-square grids. The percentage of downward adjustment is defined (in meters) using three fuzzy sets: 1) "subside little", 2) "subside some" and 3) "subside a lot." The values for these sets are defined by the rate of subsidence relative to sediment accumulation on the inner margin that, relative to the observations presented above, was higher on the inner margin (~ 10%), lower on the middle to outer margin proximal to the deltaic depocenter (~ 5%) and negligible on portions of the outer shelf distal to the deltaic depocenter (~ 0%). Therefore, full membership (MF = 1) in each fuzzy set is set similarly at 0% ("subside little"), 5% ("subside some") and 10% ("subside a lot"). Partial membership (overlap in adjacent fuzzy sets where MF = 1) is defined along the continuum of values between 0 and 5% (membership in both "subside little" and "subside some") as well as 5 and 10% (membership in both "subside some" and "subside a lot").

The determination of isostatic flexure was based on the general premise that higher sediment loads require higher amounts of isostatic compensation. This isostatic reponse of the earth's crust via subsidence has been observed on continental margins receiving a high sediment load, including the Ganges-Brahmaputra delta, the Mobile River delta and the Mississippi River [86, 159, 160]. The geometry of the ECS shelf and the Okinawa Trough exhibits the fundamental shelf-slope-rise pattern that is typical of most passive continental margins [2]. Many have classified the ECS as tectonically inactive throughout the Quaternary [161–163]. However, the abundance of sediment on the ECS margin (between 8 and 10 km thick) has been considered by others to be greater than sediment accumulation attributed to accommodation space associated with subsidence driven solely by compaction [164–166]. Consequently, others attributed subsidence to a combination of tectonic down-warping and isostatic compensation during all of the Cenozoic [88, 164, 167]. The general relationship for isostatic compensation is $S = [(\rho m - \rho w)/(\rho m - \rho s)] \cdot (D - d)$ where S = is the change if thickness of the sedimentary column to which the the crust responds isostatically, ρm is mantle density (3300 kg/m3), ρw is density of ocean water (1027 kg/m3), ρs is the density of the sediment load (2643 kg/m3), D is water depth before deposition and d is the water depth after deposition [168, 169]. This calculation overestimates the amount of subsidence observed on the ECS margin based on published rates and observed trends in the seismic data. Therefore, ranges for the fuzzy variables “subside a little”, “subside a lot”, “thick” and “thin” were defined using observational data rather than calculated estimates. Subsidence rates between 1.6 and 4.4 mm/yr on the inner margin near the mouth of the Yangtze River are roughly 10% of sedimentation rates (between 10 and 54 mm/yr) [88, 135]. The tectonic component of subsidence at this location is regarded as negligible due to its location near the tectonic hinge that is associated with regional uplift along the east coast of China [150, 158]. On the middle and outer ECS margin, the subsidence rate is 0.3 mm/yr [74]. When corrected for tectonic subsidence (0.3 mm/yr – 0.1 mm/yr = 0.2 mm/yr), this value is a little more than 5% of the sedimentation rate of 3 mm/yr [135]. Elsewhere on the margin, sedimentation is negligible and isostatic loading is assumed to be zero.

Incision. Because major fluvial incision on low-gradient margins is limited or lacking altogther, the initial version of fuzzyPEACH presented in this chapter does not incise. An initial erosion component has been developed as a separate FIS but is still under development. Similar to fluvial the FIS controlling fluvial deposition discussed below, there are three rules of the prototype FIS.

- IF in channel THEN erode a lot
- IF near channel THEN erode some
- IF far from channel THEN erode little

Preliminary results are promising, however, there are additional challenges to be overcome with erosion and re-distribution of sediment as well as fluvial gradients, cohesiveness of substrate and knickpoint migration (instead of wholesale erosion at all points along the fluvial axis, as depicted in the rules above).

Physical Oceanographic Conditions. Physical oceanographic conditions were not incorporated explicity into fuzzyPEACH. The high-energy environment of the present-day ECS is a good example of how sediment deposition and distribution are affected by a complex interaction of oceanic and tidal currents as well as frequent and intense storm events. There is no doubt that sediment distribution, deposition and stratigraphic architecture has been affected by these high-energy oceanographic processes. The general oceanic circulation pattern of the ECS and the adjoining Yellow Sea (YS) and Bohai Sea (BS) is driven by the warm (T = 20° to 27 °C) and highly saline (S = 33‰) Kuroshio western boundary current, its offspring (e.g. Taiwan Warm Current, Tsushima Current, Jiangsu Warm or Yellow Sea Warm Current, Shandong Coastal Current, Jiangsu or Yellow Sea Coastal Current and Changjiang Coastal Current) and a minor thermohaline component from the colder, sediment-laden freshwater discharge of the Yellow and Yangtze Rivers [170]. The ECS, YS and BS are geographically and hydrodynamically inseparable and are, therefore, considered one system [171]. Semidiurnal tidal currents between 20 cm/sec (weakest in BS) and 100 cm/sec (strongest near mouth of Yangtze River) are sufficiently strong in some areas to cause localized resuspension or bedload transport [133, 142]. Approximately 7% of the global dissipation of tidal energy presently occurs in the shallow ECS/YS/BS system and causes sea level fluctuations from 5 to 6 m in Taiwan and up to 11 m in Hangchow Bay southwest of Shanghai [172,173]. Without these physical processes, the simulation of the ECS margin did not remove sediment from the ECS system via coastal currents (as mentioned above) nor did it remove, rework and redistribute sediment around the margin. For this reason, features such as the tidal ridges observed on the modern seafloor, as well as deeper in the seismic record, were not simulated by fuzzyPEACH. Because these strata are relatively thin and a minor component of the shallow seismic record, the lack of simulated oceanographic conditions relative to conditions of sea level and sediment influx is considered to be of minor importance.

6 Results

Hundreds of fuzzyPEACH simulations were completed during the course of this investigation. For the sake of brevity, only a general description of the model results is presented here to illustrate its utility. The complete results, discussion, and conclusions derived from the results are documented in Warren [61]. The primary purpose of these simulations was to investigate the relationship between stratigraphic architecture and variables associated with relative sea level (eustasy and tectonics), sediment influx and margin physiography. These tests contributed to the general understanding of fuzzyPEACH's ability to produce reasonable stratigraphic architecture under various permutations of sea level, tectonic subsidence and sedimentation (Figs. 9 and 10).

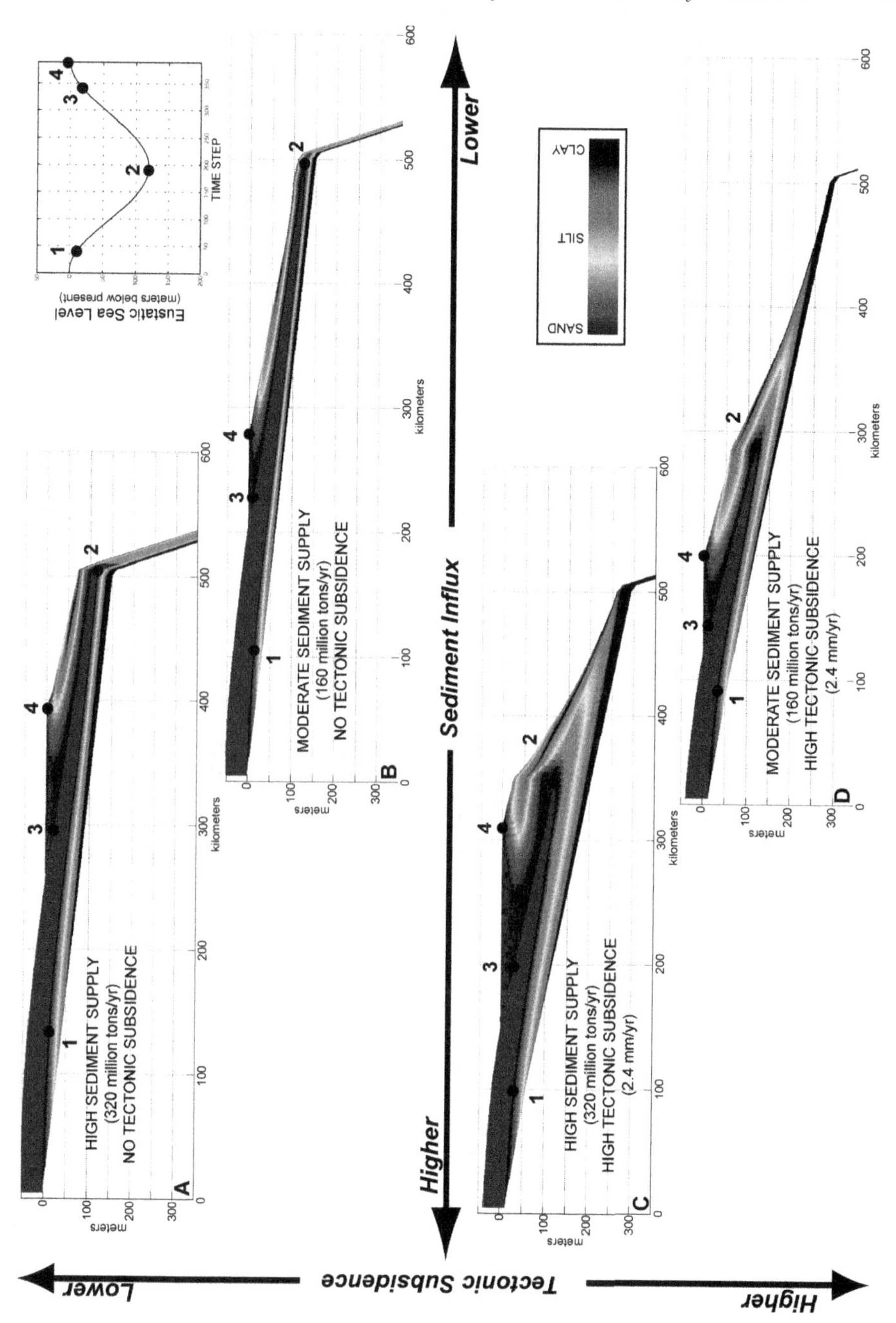

Fig. 9. Synthetic, dip-oriented stratigraphic cross-sections from the middle of the simulated continental margin. Strata deposited during a single, simple sinusoidal sea level cycle. Numbered positions on stratal sections correspond to numbered positions on sea level curve (inset upper right) and represent shoreline position. Four distinct simulations, each spanning 200 kg, are shown with respect to relative sediment influx (horizontal arrows) and relative tectonic subsidence (vertical arrows). Numbered positions on stratal cross sections correspond to numbered positions on sea level curve (upper right) and indicate location of shoreline at that time

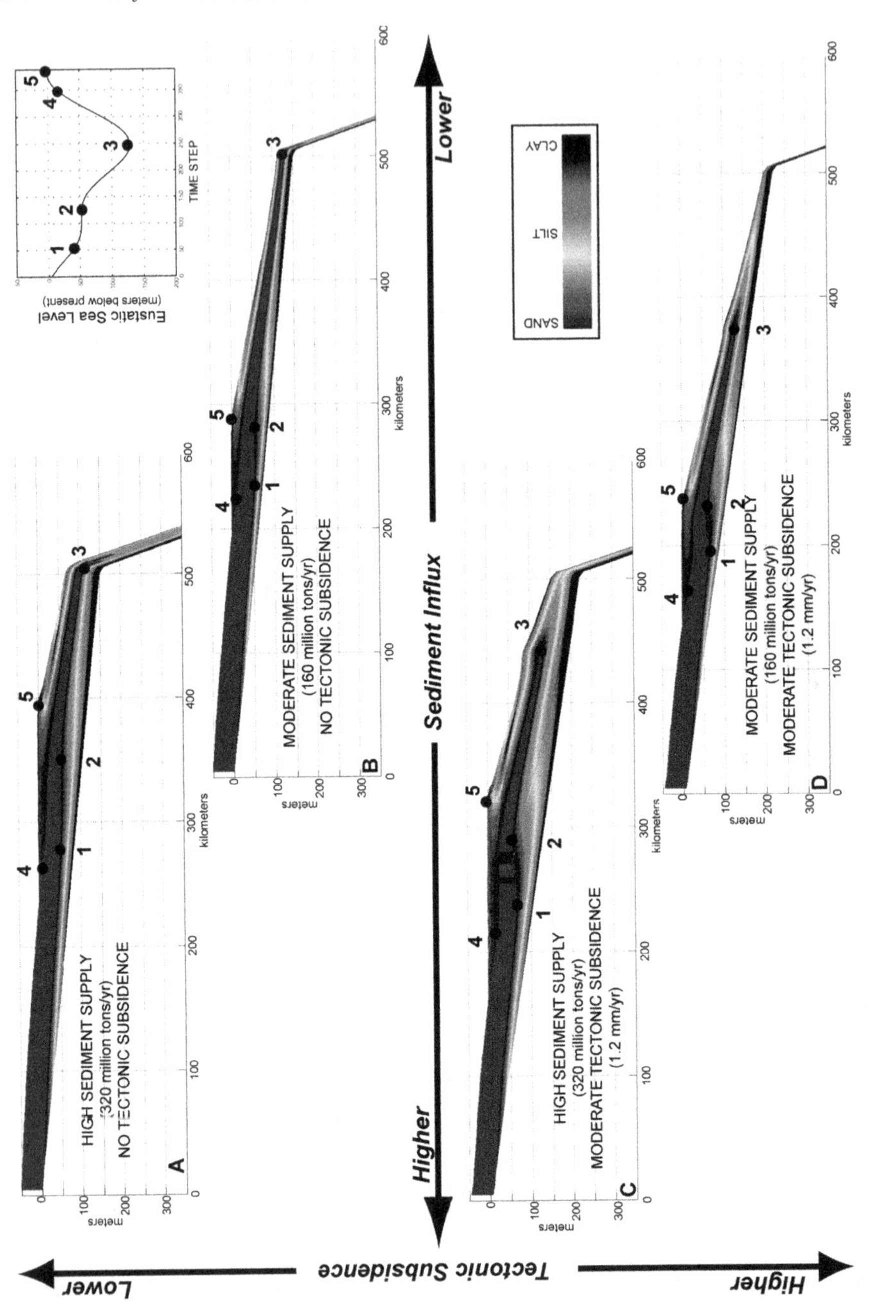

Fig. 10. Synthetic, dip-oriented stratigraphic cross-sections from the middle of the simulated continental margin. Strata deposited during a single, asymmetric sinusoidal sea level cycle. Numbered positions on stratal sections correspond to numbered positions on sea level curve (inset upper right) and represent shoreline position. Four distinct simulations, each spanning 200 kg, are shown with respect to relative sediment influx (horizontal arrows) and relative tectonic subsidence (vertical arrows). Numbered positions on stratal cross sections correspond to numbered positions on sea level curve (upper right) and indicate location of shoreline at that time

General stratigraphic trends from the sensitivity testing were consistent with the general concepts of both seismic stratigraphy [173–175] and sequence stratigraphy [6, 62, 79, 176–178]. Synthetic, dip-oriented stratigraphic cross-sections in Figures 9 and 10 provide an example of the simulated response of a continental margin to variable depositional conditions (elevations of sea level and rates of change, sediment influx, and tectonic subsidence).

Onlapping stratal patterns were formed during all sea level conditions (rises, falls and stillstands). Downstepping stratal patterns were products of forced regressions as falling sea level forced a basinward shift in the shoreline [179, 180]. A combination of aggradational and progradational strata as well as downlapping stratal patterns are associated with periods of slow sea level rise [179, 180]. Retrogradation (backstepping) patterns dominate periods of faster rates of sea level rise. The general stratal response to increasing eustatic periodicity was thinner strata and faster rates of shoreline translation across the margin. This resulted in higher progradation rates and lower aggradation during sea level fall (forced regression) as well as higher rates of backstepping and lower rates of progradation and aggradation during sea level rise. Therefore, normal regressions became less noticeable under increasing eustatic rate changes (but occurred as parasequences during eustatic slowdowns and stillstands). Exposure of the shelf-slope break during lowstands in sea level bypassed sediment from the exposed shelf basinward of the shelf-slope break and onto the continental slope. The initial fuzzyPEACH model did not simulate incision during subaerial exposure. However, subsequent versions of the model described here, but not presented in this chapter, tested the preliminary development of an erosion component. These fluvial "incisions" constrained avulsion and led to limited floodplain sedimentation. Further development of a dedicated FIS for erosion and incision will continue and be included in future versions of fuzzyPEACH.

In addition to sea level fluctuations, strata also responded to changes in tectonic subsidence and sedimentation rate. Stratal thickness as well as rate of basinward translation of the shoreline was proportional to sediment influx (i.e. higher sedimentation led to quicker rates of accommodation infill and basinward translation of the shoreline). Where sedimentation rates were high, shoreline regression occurred during periods of rising sea level. Increased tectonic subsidence also created thicker strata by increasing accommodation and, therefore, limiting the basinward movement of the shoreline. As sedimentation rates were steadily increased, strata steadily moved basinward and eventually deposited sediment beyond the shelf-slope break. As subsidence rates steadily increased, strata were deposited farther from the shelf-slope break and strata become much thicker than simulations with the same sediment influx and lower, or no, subsidence. Shoreline translation rates increased with higher sediment input but decreased with increased tectonic subsidence. These results are analogous to the stratal geometry typical of passive continental margins [181–183].

7 Discussion

Numerous conditions were, and can be, simulated with fuzzyPEACH, including the user-defined parameters for margin geometry, eustasy, tectonic subsidence, and sediment influx. A detailed discussion of these simulations is addressed in Warren (2006). Assignments of values to these variables are based on a combination of general geologic conditions (deltaic deposition, avulsion, compaction, subsidence) with a few being specific to low-gradient margins in general. The values used to constrain the fuzzy sets used in the five FISs driving the fuzzy logic engine of fuzzyPEACH were also established based primarily on general geologic conditions. Evaluation of model skill includes comparing the geologic variables used in fuzzyPEACH simulations to those observed, inferred or expected on the ECS margin.

This investigation occurred in two phases. The first phase of the study assessed the ability of fuzzy logic to model complex stratigraphy under various depositional conditions on idealized, passive continental margins (not at a low gradient). Fuzzy logic was chosen because it is simple to use and understand. Fuzzy logic provides the ability to quantify subjectivity by capturing the vagueness of linguistics terms, thus making it flexible and tolerant of imprecise data [33]. General concepts and expert knowledge assembled fuzzyPEACH, a MATLAB-based model that uses five FISs to simulate the complicated, nonlinear relationships of stratal evolution on a passive continental margin. No special knowledge, apart from basic mathematical logic, is required for this application.

Similar to previous fuzzy-logic-based stratigraphic simulations [25–33], the goal of fuzzyPEACH was to describe, in simplistic terms, the complexity of a fluvial system and its delta. The general instructions defining fluvial deposition are straightforward: the highest volume and coarsest sediments are deposited closest to the channel and the lowest volume and finest sediments are deposited farther away. The same concept is used for the delta (i.e. high volume and coarse sediment at the river mouth transitioning to low volume and fine sediment farther away). User-defined variables such as sediment influx, margin geometry and tectonic subsidence may be altered for any given simulation, but the rules, variables, fuzzy sets, and fuzzy set boundaries are "hard coded" and remained constant for all simulations. With the exception of fluvial avulsion, which is based on sea level elevation as a proxy for gradient, no special instructions are included within fuzzyPEACH to define how stratal geometry should be built during numerous permutations of eustasy, tectonic subsidence and sediment influx. Instead, stratal geometries are based solely on sediment distribution (how much, what type, and where) and available sediment accommodation. Accomodation is determined by relative sea level, eustasy, tectonic subsidence and sediment influx.

The second phase of this investigation used fuzzyPEACH to simulate the physical conditions associated with low-gradient margins. Phase II was justified because Phase I showed that the stratigraphic architecture generated during variable conditions of eustasy, tectonic subsidence, and sediment influx is consistent with the general concepts of both seismic stratigraphy and sequence stratigraphy. Low-gradient margins were studied because the traditional sequence stratigraphic model predicts incision at the shelf-slope break as well as sedimentary bypass across the continental margin during sea level lowstands. When the shelf-slope break of the margin remains submerged during lowstand, fluvial systems under low-gradient conditions are unconfined, laterally extensive, and do not bypass the majority of sediments beyond the shelf-slope break and into deeper water. Although these margins are rare in the modern and recent geologic record, examples include northeastern Australia [81], New Zealand (Canterbury Plains) [82], northern Java [103] and the ECS [63–67]. The evaluation of phase II used an extensive, regional dataset from the ECS margin that provided the constraints needed to assess the accuracy of fuzzyPEACH simulations and validates output.

The regional extent of the large seismic dataset from the ECS margin used in this study provides a unique set of observations that were not available to previous stratal models incorporating fuzzy logic. Interpretations of the stratigraphic architecture of the ECS margin, derived from the seismic data set, combined with numerous published studies, provide a good understanding of the depositional conditions under which the stratigraphy was formed. There appears to be a good correlation between fuzzyPEACH simulations constrained by these same conditions (i.e. margin physiography, eustatic curve, tectonic subsidence rate, sediment influx). Therefore, in areas with similar depositional conditions to the ECS, but with little or no data, strataigraphic trends can be inferred from fuzzyPEACH simulations.

In addition to using datasets from the ECS margin dataset and low-gradient margins in general, these stratigraphic simulations are applicable generally to the stratigraphic evolution of a wide range of depositional systems. The three-dimensional stratal geometries, stratal termination patterns, and rates of shoreline change produced by fuzzyPEACH are similar to those observed in natural systems. Therefore, understanding this complex stratigraphic architecture as it relates eustatic sea level, tectonic (thermal) subsidence and sediment influx has important implications. Without this understanding, it is difficult to relate these variables to outcrops and subsurface data (or lack thereof) throughout the geologic record.

8 Conclusions

The fuzzyPEACH is a three-dimensional, forward simulator using fuzzy logic to model the stratigraphic response continental margins. In preliminary applications, fuzzyPEACH simulated numerous scenarios of eustasy, tectonic subsidence and sediment influx onto a low-gradient continental margin. Although fuzzyPEACH allows the geometry to be user defined, the various rates and magnitudes of eustatic fluctuations may also be used as a proxy for the effect of margin physiography on stratal geometry (e.g. slow changes in sea level across a low-gradient margin can provide a similar shoreline migration rate of relatively fast sea level changes across a high-gradient margin). Fuzzy logic was chosen because it is simple yet powerful. General concepts and expert knowledge assembled a set of robust fuzzy logic inference systems that were able to describe complicated, nonlinear relationships found within geologic systems. FuzzyPEACH simulations collectively use only five FISs containing a total of 21 separate rules. These rules incorporate 15 variables and are defined by 47 fuzzy sets.

Numerous quantitative models have simulated the stratigraphic response of continental margins throughout the geologic record. These quantitative models, and sequence stratigraphy in general, predict incision and sedimentary bypass across a continental margin during sea level lowstand. On the other hand, low-gradient margins have not been addressed by the modeling community. While this justified the initial development of fuzzyPEACH, the results and observations of the fuzzy logic simulations generally are applicable to the stratigraphic evolution of a wide range of depositional systems. The three-dimensional stratal geometries, stratal termination patterns, and rates of shoreline change produced by fuzzyPEACH are similar to those observed in natural systems. Therefore, fuzzyPEACH is a potential tool to provide a better understanding of the complex stratigraphic architecture as it relates eustatic sea level, tectonic (thermal) subsidence and sediment influx.

Visual comparison of model output compares well with the stratigraphic trends observed in the ECS seismic dataset. The laterally extensive fluvial deposits from the ECS (older than 125 ka) were similar to fuzzyPEACH simulations with higher sediment influx (i.e. rates similar to the modern Yangtze River) and avulsion controlled by a set of fuzzy rules. The more defined deltaic lobes from the ECS (younger than 125 ka) shared trends simulated with moderate sediment influx (half the rate of the modern Yangtze River) and less frequent avulsions. Seismic data and simulation output support a hypothesis of variable sedimentation rates. The current sediment delivery method in fuzzyPEACH (i.e. constant for each time step) can be redesigned with additional fuzzy logic rules to handle variable rates of deposition by defining trends (where they exist) between sedimentation rate, climatic and physical conditions of the margin. The incised nature of the ECS central margin (incision did not occur at the submerged shelf-slope break) during the last ice age (24 to 12 ka), was not simulated by fuzzyPEACH. While incision was addressed partially during model development, continued modifications shall be considered for inclusion into future versions of the fuzzyPEACH.

Acknowledgments

The overall financial support for the regional East China Sea investigation, of which the fuzzy modeling is a subset, was provided by the Office of Naval Research (grants N00014-93-1-0921, N00014-96-0995, N00014-97-1-0382, N00014-99-1-0602, N00014-00-1-0275, N00014-01-1-0918 and

N00014-03-1-0190 awarded to LRB). The utilization of the Kingdom Suite seismic interpretation and visualization software was made possible by an additional grant to LRB from Seismic Micro-Technology, Inc. JDW was partially funded through the Department of Geological Science Martin Trust Fund and the Graduate School Dissertation Completion Fellowship, both at the University of North Carolina at Chapel Hill.

References

1. Watney WL, Rankey EC and Harbaugh J (1999) Perspectives on stratigraphic simulation models: Current approaches and future opportunities. In Harbaugh JW, Watney WL, Rankey EC, Slingerland R, Goldstein RH and Franseen EK (eds), Numerical experiemnts in stratigraphy: Recent advances in stratigraphic and sedimentologic computer simulations. SEPM Special Publication 62, Tulsa USA: 3–21
2. Heezen B, Tharp M and Ewing M. (1959) The floors of the oceans: I, The North Atlantic.Geological Society of America Special Paper 65
3. Garcia-Castellanos D, Fernandez M and Torne M (2002) Modeling the evolution of the Guadalquivir foreland basin (southern Spain). Tectonics 21: 1018–1034
4. Du Fornel E, Joseph P, Guillocheau F, Euzen T and Granjeon D (2003) Regional outcrop study and 3-D stratigraphic modeling ina foreland basin setting: The example of the Gres d'Annot turbidite formation (French Alps). AAPG Annual Meeting Expanded Abstracts 12: 44
5. Zhou D, Yu H-S, Xu H-H, Shi X-B and Chou Y-W (2003) Modeling of thermo-rheological structure of lithosphere under the foreland basin and mountain belt of Taiwan. Tectonophysics 374: 115–134
6. Posamentier HW and Allen GP (1999) Siliciclastic sequence stratigraphy – Concepts and applications. SEPM Concepts in Sedimentology and Paleontology 7, Tulsa USA
7. Wood LJ, Ethridge FG and Schumm SA (1993) An experimental study of the influence of subaqueous shelf angles on coastal plain and shelf deposits. In Weimer P and Posamentier HW (eds), Siliciclastic sequence stratigraphy: Recent developments and applications. AAPG Memoir 58, Tulsa USA: 381–391
8. Koss JE, Ethridge FG and Schumm SA (1994) An experimental study of the effects of baselevel change on fluvial, coastal plain and shelf systems. Journal of Sedimentary Research B64: 90–98
9. Syvitski JPM and Daughney S (1992) Delta2: Delta progradation and basin filling. Computers and Geosciences 18: 839–897
10. Li MZ and Amos CL (2001) SEDTRANS96: The upgraded and better calibrated sediment-transport model for continental shelves. Computers and Geosciences 27: 619–645
11. Syvitski JP, Pratson L and O'Grady D (1999) Stratigraphic predictions of continental margins for the US Navy. In Harbaugh JW, Watney WL, Rankey EC, Slingerland R, Goldstein RH and Franseen EK (eds), Numerical experiemnts in stratigraphy: Recent advances in stratigraphic and sedimentologic computer simulations. SEPM Special Publication 62, Tulsa USA: 219–236

12. Ross WC, Watts DE and May JA (1995) Insights from stratigraphic modeling: Mud-limited versus sand-limited depositional systems. AAPG Bulletin 79: 231–258
13. Cross TA and Harbaugh JW (1990) Quantitative dynamic stratigraphy: A workshop, a philosophy, a methodology. In Cross TA (ed), Quantitative dynamic stratigraphy, Prentice Hall, Englewood Cliffs NJ USA: 3–20
14. Jervey MT (1988) Quantitative geologic modeling of siliciclastic rock sequences and their seismic expression. In Wilgus K, Hastings BS, Kendall CGStC, Posamentier HW, Ross CA and Van Wagoner JC (eds), Sea-Level Changes: An Integrated Approach. SEPM Special Publication 42, Tulsa USA: 47–69
15. Pitman WC and Golovchenko X (1988) Sea-level changes and their effect on the stratigraphy of Atlantic-type margins. In Sheridan RE and Grow JA (eds), The Atlantic Continental Margin: US. Geological Society of America, Boulder USA: 429–436
16. Kendall CGStC, Moore P, Strobel J, Cannon R, Perlmutter M, Bezdek J and Biswas G (1991) Simulation of the sedimentary fill of basins. In Franseen EK, Watney WL, Kendall CGStC and Ross W. (eds), Sedimentary Modeling: Computer simulations and methods for improved parameter definition. Kansas Geological Survey Bulletin 223, Lawrence KS USA: 9–30
17. Kendall CGStC, Strobel J, Cannon R, Bezdek J and Biswas G (1991) The simulation of the sedimentary fill of basins. Journal of Geophysical Research 96 (B4): 6911–6929
18. Reynolds DJ, Steckler MS and Coakley BJ (1991) The role of the sediment load in sequence stratigraphy: The influence of flexural and isostatic compaction. Journal of Geophysical Research 96 (B4): 6931–6949
19. Nummedal D, Riley GW and Templet PL (1993) High-resolution sequence architecture: A chronostratigraphic model based on equilbrium profile studies. Special Publications of the International Association of Sedimentology 18: 55–68
20. Steckler MS, Reynolds DJ, Coakley BJ, Swift BA and Jarrard R (1993) Modeling passive margin sequence stratigraphy. In Posamentier HW, Summerhayes CP, Haq BU and Allen GP (eds), Sequence stratigraphy and facies associations. International Association of Sedimentologists Special Publication 18, Blackwell Scientific, Oxford UK: 19–41
21. Steckler MS (1999) High-resolution sequence stratigraphic modeling 1: The interplay of sedimentation, erosion, and subsidence. In Harbaugh JW, Watney WL, Rankey EC, Slingerland R, Goldstein RH and Franseen EK (eds), Numerical experiemnts in stratigraphy: Recent advances in stratigraphic and sedimentologic computer simulations. SEPM Special Publication 62, Tulsa USA: 139–149
22. Ritchie BD, Hardy S and Gawthorpe RL (1999) Three-dimensional numerical modeling of coarse-grained clastic deposition in sedimentary basins. Journal of Geophysical Research 104 B8: 17759–17780
23. Ritchie BD, Gawthorpe RL, Hardy S (2004) Three-dimensional numerical modeling of deltaic depositional sequences 1: Influence of the rate and magnitude of sea-level change. Journal of Sedimentary Research 74: 203–220
24. Ritchie BD, Gawthorpe RL and Hardy S (2004) Three-dimensional numerical modeling of deltaic depositional sequences 2: Influence of local controls. Journal of Sedimentary Research 74: 221–238

25. Nordlund U and Silfversparre M (1994) Fuzzy logic – a means for incorporating qualitative data in dynamic stratigraphic modeling. International Association for Mathematical Geology Annual Conference Proceedings: 265–266
26. Nordlund U (1996) Formalizing geological knowledge – with an example of modeling stratigraphy using fuzzy logic. Journal of Sedimentary Research 66: 689–698
27. Nordlund U (1999) Stratigraphic modeling using common-sense rules. In Harbaugh JW, Watney WL, Rankey EC, Slingerland R., Goldstein RH and Franseen EK (eds), Numerical experiemnts in stratigraphy: Recent advances in stratigraphic and sedimentologic computer simulations. SEPM Special Publication 62, Tulsa USA: 245–251
28. Nordlund U (1999) FUZZIM: Forward stratigraphic modeling made simple. Computers and Geosciences 25: 449–456
29. Parcell WC (2000) 3D computer simulation of carbonate depositional facies distribution and productivity rates using continuous set theory to mimic geologists' reasoning. Gulf Coast Association of Geological Societies Transactions 50: 439–449
30. Demicco RV and Klir GJ (2001) Stratigraphic simulations using fuzzy logic to model sediment dispersal. Journal of Petroleum Science and Engineering 31: 135–155
31. Parcell, WC (2003) Examining the definition and significance of the maximum flooding surface through fuzzy logic modeling. Gulf Coast Association of Geological Societies Transactions 53: 659–667
32. Parcell WC (2003) Evaluating the development of Upper Jurassic reefs in the Smackover Formation, eastern Gulf Coast, USA through fuzzy logic computer modeling. Journal of Sedimentary Research 73: 498–515
33. Demicco RV (2004) Applications of fuzzy logic to stratigraphic modeling. In Demicco RV and Klir GJ (eds), Fuzzy logic in geology. Elsevier, Amsterdam: 121–151
34. Chappaz RJ (1977) Application of the fuzzy sets theory to the interpretation of seismic sections. Geophysics 42: 1499
35. (2001) Journal of Petroleum Geology 24
36. Wong P, Aminzadeh F, Mikravesh M (eds) (2002) Soft computing for reservoir characterization and modeling. Studies in Fuzziness and Soft Computing 80, Physica-Verlag, Heidelberg, Germany
37. Nikravesh M, Aminzadeh F, Zadeh L (eds) (2003) Soft computing and intelligent data analysis in oil exploration. Developments in Petroleum Science 51, Elsevier, Amsterdam
38. Sandham W and Leggett M (2003) Geophysical applications of artificial neural networks and fuzzy logic. Kluwer Academic Publishers, Dordrecht
39. Demicco RV and Klir GJ (eds) (2004) Fuzzy logic in geology. Elsevier, Amsterdam
40. Demicco RV (2004) Fuzzy logic in geological Sciences: A literature review. In Demicco RV and Klir GJ (eds), Fuzzy logic in geology. Elsevier, Amsterdam
41. Demicco RV (2004) Fuzzy logic and earth science: An overview. In Demicco RV and Klir GJ (eds), Fuzzy logic in geology. Elsevier, Amsterdam
42. Fang JH (1987) Toward fuzzy expert systems in geology. Acta Geologica Taiwanica 25: 85–96
43. Fang JH (1997) Fuzzy logic and geology. Geotimes 42: 23–26

44. Saggaf, MM and Nebrija EL (2003) A fuzzy logic approach for the estimation of facies from wire-line logs. AAPG Bulletin 87: 1223–1240
45. Wentworth CK (1922) A scale of grade and class terms for clastic sediments. Journal of Geology 30: 377–392
46. Leopold LB and Wolman MG (1957) River Channel patterns: Braided, meandering and straight. Geological Survey Professional Paper 282-B, US Government Printing Office, Washington USA
47. Ackers P and Charlton FG (1970) Dimensional analysis of alluvial channels with special reference to meander length. Journal of Hydraulic Research 8: 287–316
48. Schumm SA and Khan HR (1972) Experimental study of channel patterns. GSA Bulletin 83: 1755–1770
49. Schumm SA (1981) Evolution and response of the fluvial system, sedimentologic implications. SEPM Special Publication 31: 19–29
50. Ferguson RI (1981) Channel Form and channel changes. In Lewin J (ed), British rivers. George, Allen and Unwin, London UK: 90–125
51. Knighton DA and Nanson GC (1993) Anastamosis and the continuum of channel pattern. Earth Surface Processes and Landforms 18: 613–625
52. Van den Berg JH (1995) Prediction of alluvial channel pattern of perennial rivers. Geomorphology 12: 259–279
53. Woolfe KJ and Balzary JR (1996) Fields in the spectrum of channel style. Sedimentology 43: 797–805
54. Heritage GL, Charlton ME and O'Regan S (2001) Morphological classification of fluvial environments: An investigation of the continuum of channel types. The Journal of Geology 109: 21–33
55. Schumm SA (1972) Fluvial paleochannels. In Rigby JK and Hamblin WK (eds), Recognition of ancient sedimentary environments. SEPM Special Publication 16, Tulsa, USA: 98–107
56. Orton GJ and Reading HG (1993) Variability of deltaic processes in terms of sediment supply, with particular emphasis on grain size. Sedimentology 40: 475–512
57. Dalrymple RW, Zaitlin BA and Boyd R. (1992) Estuarine facies models: Conceptual basis and stratigraphic implications. Journal of Sedimentary Petrology 62: 1130–1146
58. Galloway WE (1975) Process framework for describing the morphologic and stratigraphic evolution of deltaic depositional systems. In Broussard ML (ed), Deltas: Models for exploration. Houston Geological Society, Houston USA: 87–98
59. Warren JD, Bartek LR, Wang PP and Ramsey HN (2002) Fuzzy logic applied to the geological sciences: an emerging application for seismic interpretation. Proceedings from the Joint Conference on Information Science, March 9–14, Durham NC USA: 128–132
60. Warren JD, Bartek LR, Wang PP and Ramsey HR (2003) Relative sea-level modeling using fuzzy logic: Proceedings from the Joint Conference on Information Science, September 26–30, Research Triangle Park, NC, USA: 117–120
61. Warren JD (2006) The Sequence stratigraphy of the East China Sea continental margin. Unpublished doctoral dissertation, University of North Carolina, Chapel Hill USA

62. Van Wagoner JC, Mitchum RM, Campion KM and Rahmanian VD (eds) (1990) Siliciclastic sequence stratigraphy in well logs, cores, and outcrops: Concepts for high-resolution correlation of time and facies. AAPG Methods in Exploration Series 7, Tulsa USA
63. Bartek LB, Warren JD and Miller KL (2001) Perched lowstand stratigraphy on the East China Sea continental margin. AGU Chapman Conference on Formation of Sedimentary Strata on Continental Margins, June 17–19, Ponce, Puerto Rico
64. Bartek LR and Warren JD (2002) Concurrent incised and unincised valley systems and their geomorphic control during Quaternary lowstands on the East China Sea continental margin. SEPM Research Conference: Incised Valleys – Images and processes, August 18–23, Casper and Newcastle WY USA
65. Warren JD and Bartek LR (2002) The Metastable Fluvial Shelf System (MFSS): An alternative hypothesis to lowstand unincised fluvial bypass on the East China Sea continental margin. SEPM Research Conference: Incised Valleys – Images and processes, August 18–23, Casper and Newcastle WY USA
66. Warren JD and Bartek LR (2002) The Sequence Stratigraphy of the East China Sea: Where are the incised valleys? Sequence stratigraphic models for exploration and production: Evolving methodology, emerging models and applications history, GCSSEPM Foundation Bob F. Perkins Research Conference, December 8–11, Houston USA: 729–738
67. Warren JD, Bartek LR and Miller KL (2002) Stratigraphic architecture of the East China Sea continental margin: a case study of eustacy and sediment supply. AAPG Annual Meeting Program, March 10–14, Houston USA: A185
68. Wellner RW and Bartek LR (2003) The effect of sea level, climate, and shelf physiography on the development of incised valley complexes: a modern example from the East China Sea. Journal of Sedimentary Research 73: 926–940
69. Bartek LR, Wellner RW and Warren JD (2004) Climate change and shelf physiography – their role in shaping sequence architecture: An example from the East China Sea. AAPG Annual Meeting Program, April 18–21, Dallas USA: A185
70. Bartek, LR and Warren JD (2005) Comparison of the fill of high and low sediment supply incised valley systems. Abstracts Volume 14, AAPG Annual Convention, June 19–22, Calgary: A12
71. Saito Y, Katayama H, Ikehara K, Kato Y, Matsumoto E, Oguri K, Oda M and Yumoto M (1998) Transgressive and highstand systems tracts and post-glacial transgression, the East China Sea. Sedimentary Geology 122: 217–232
72. Liu ZX, Berne S, Saito Y, Lericolais G and Marsset,T (2000) Quaternary seismic stratigraphy and paleoenvironments on the continental shelf of the East China Sea. Journal of Asian Earth Sciences 18: 441–452
73. Saito Y, Yang Z, Hori K (2001) The Huanghe (Yellow River) and Changjiang (Yangtze River) deltas: A review on their characteristics, evolution and sediment discharge during the Holocene. Geomorphology 41: 219–231
74. Berne S, Vagner P, Guichard F, Lericolais G, Liu Z, Trentesaux A, Yin P and Yi HI (2002) Pleistocene forced regressions and tidal sand ridges in the East China Sea. Marine Geology 188: 293–315
75. Lotfi A and Tsoi AC (1994) Importance of membership functions: A comparative study on different learning methods for fuzzy inference systems. Proceedings of the Third IEEE Conference on Fuzzy Systems 3, IEEE World Congress on Computational Intelligence: 1791–1796

76. Rondeau L, Levrat E and Brémont J (1996) An analytical formulation of the influence of membership functions shape. Proceedings of the Fifth IEEE International Conference on Fuzzy Systems 2, Sept. 8–11, New Orleans USA: 1314–1319
77. Sancho-Royo A and Verdegay JL (1999) Methods for the construction of membership functions. International Journal of Intelligent Systems 14: 1213–1230
78. Klir GJ (2004) Fuzzy Logic: A specialized tutorial. In Demicco RV and Klir GJ (eds) Fuzzy logic in geology. Elsevier, Amsterdam: 11–61
79. Posamentier HW, Jervey MT and Vail PR (1988) Eustatic controls on clastic deposition I – conceptual framework. In Wilgus CK, Hastings BS, Kendall CGStC, Posamentier HW, Ross CA, Van Wagoner JC (eds), Sea-Level Changes: An Integrated Approach. SEPM Special Publication 42, Tulsa USA: 109–124
80. Stouthamer E and Berendsen HJA (2001) Avulsion frequency, avulsion duration, and interavulsion period of Holocene channel belts in the Rhine-Meuse delta, the Netherlands. Journal of Sedimentary Research 71: 589–598
81. Woolfe KJ, Larcombe P, Naish T and Purdon RG (1998) Lowstand rivers need not incise the shelf: An example from the Great Barrier Reef, Australia, with implications for sequence stratigraphic models. Geology 26: 75–78
82. Browne GH and Naish TR (2003) Facies development and sequence architecture of a late Quaternary fluvial-marine transition, Canterbury Plains and shelf, New Zealand: Implications for forced regressive deposits. Sedimentary Geology 158: 57–86
83. Watts AB, Karner GD and Steckler MS (1982) Lithospheric flexure and the evolution of sedimentary basins. Philosophical Transactions of the Royal Society of London, Mathematical and Physical Sciences 305: 249–281
84. Anderson JB, Rodriguez A, Abdulah KC, Fillon RF, Banfield LA, McKeown HA and Wellner JS (2004) Late Quaternary stratigraphic evolution of the northern Gulf of Mexico margin: A synthesis. In Anderson JB and Fillon RH (eds) Late Quaternary stratigraphic evolution of the northern Gulf of Mexico Margin, SEPM Special Publication no. 79, Tulsa USA: 1–23
85. Wells JT (1996) Subsidence, sea-level rise, and wetlands loss in the lower Mississippi River delta. In Milliman JD and Haq BU (eds) Sea-level rise and coastal subsidence: Causes, consequences, and strategies, Coastal Systems and Continental Margins 2, Kluwer Academic Publishers, Dordrecht: 281–311
86. Alam M (1996) Subsidence of the Ganges-Brahmaputra delta of Bangladesh and associated drainage, sedimentation and salinity problems. In Milliman JD and Haq BU (eds) Sea-level rise and coastal subsidence: Causes, consequences, and strategies, Coastal Systems and Continental Margins 2, Kluwer Academic Publishers, Dordrecht: 169–192
87. Wellman HW (1979) An uplift map for the South Island of New Zealand and a model for uplift of the Southern Alps. In Walcott RI and Cresswell MM (eds) The Origin of the Southern Alps, Royal Society of New Zealand Bulletin 18: 13–20
88. Stanley DJ and Chen Z (1993) Yangtze delta, eastern China: 1, Geometry and subsidence of Holocene depocenter. Marine Geology 112: 1–11
89. Milliman JD and Meade RH (1983) World-wide delivery of river sediments to the oceans. Journal of Geology 91: 1–21

90. Milliman JD and Syvitski JPM (1992) Geomoprhic/tectonic control of sediment discharge to the ocean: The importance of small mountainous rivers. Journal of Geology 100: 525–544
91. Beard DC and Weyl PK (1973) Influence of texture on porosity and permeability. AAPG Bulletin 57: 349–369
92. Domenico SN (1977) Elastic properties of unconsolidated sand reservoirs. Geophysics 42: 1339–1368
93. Giles MR, Indrelid SL and James DMD (1998) Compaction – the great unknown in basin modeling. In Düppenbecker,SJ and Iliffe JE (eds) Basin modelling: Practice and progress. Geological Society of London Special Publication 141: 15–43
94. Curry CW, Bennett RH, Hulbert MH, Curry KJ and Faas RW (2004) Comparative study of sand porosity and a technique for determining porosity of undisturbed marine sediment. Marine Georesources and Geotechnology 22: 231–252
95. Rieke HH III and Chilangarian GV (1974) Compaction of argillaceous sediments. Developments in Sedimentology16, Elsevier, Amsterdam
96. Baldwin B and Butler CO (1985) Compaction curves. AAPG Bulletin 69: 622–626
97. Imbrie J, Hayes JD, Martinson DG, McIntyre A, Mix AC, Morley JJ, Pisias NG, Prell WL and Shackleton NJ (1984) The orbital theory of Pleistocene climate: Support from a revised chronology of the marine δ^{18}O record. In Berger AL, Imbrie J, Hays J, Kukla G and Saltzman B (eds) Milankovitch and climate: Understanding the response to astronomical forcing part I. NATO ASI Series C, Mathematical and Physical Sciences 126, D. Reidel Publishing Company, Dordrecht: 269–305
98. Weedon GP (1991) The spectral analysis of stratigraphic time series. In Einsele G, Ricken W and Seilacher A (eds) Cycles and Events in Stratigraphy, Springer-Verlag, Berlin: 840–854
99. Winograd IJ, Szabo BJ, Coplen TB, Riggs AG (1988) A 250,000-year climatic record from Great Basin vein calcite: Implications for Milankovitch theory. Science 242: 1275–1280
100. Winograd IJ, Coplen TB, Landwehr JM, Riggs AC, Ludwig KR and Szabo BK (1992) Continuous 500,000-year climate record from vein calcite in devils hole Nevada. Science 258: 255–260
101. Mamdani EH and Assilian S (1975) An experiment in linguistic synthesis with fuzzy logic controller. International Journal of Man-Machine Studies 7: 1–13
102. Takagi T and Sugeno H (1985) Fuzzy identification of systems and its application for modeling and control. IEEE Transactions on Systems, Man and Cybernetics 15: 116–132
103. Posamentier HW (2001) Lowstand alluvial bypass systems: Incised vs. unincised. AAPG Bulletin 85: 1771–1793
104. Hori K, Saito Y, Zhao Q, Cheng X, Wang P, Sato Y and Li C (2001) Sedimentary facies of the tide-dominated paleo-Changjiang (Yangtze) estuary during the last transgression. Marine Geology 177: 331–351
105. Carson MA (1984) Observations on the meandering-braided river transition, the Canterbury Plains, New Zealand: Part one. New Zealand Geographer 40: 12–17
106. Miall AD (1977) A review of the braided-river depositional environment. Earth-Science Reviews 13: 1–62

107. Saucier RT (1994) Geomorphology and Quaternary geologic history of the Lower Mississippi Valley, volume 1. US Army Corps of Engineers Waterways Experiment Station Geotechnical Laboratory, Vicksburg MS USA
108. Baeteman C, Bodemans F, Govaert E, Huanzhong H and Jingxin L (1992) The Quaternary deposits of the Changjiang coastal plain (Shanghai area). Bulletin of the International Association of Engineering Geology 46: 7–23
109. Fagan SD and Nanson GC (2004) The morphology and formation of floodplain-surface channels, Cooper Creek, Australia. Geomorphology 60: 107–126
110. Kukal Z (1971) Geology of Recent Sediments. Academic Press, London
111. Bridge JS and Leeder MR (1979) A simulation model of alluvial stratigraphy. Sedimentology 26: 617–644
112. Schindel DE (1980) Microstratigraphic sampling and the limits of paleontologic resolution. Paleobiology 6: 408–426
113. Törnqvist TE and Bridge JS (2002) Spatial variation of overbank aggradation rate and its influence on avulsion frequency. Sedimentology 49: 891–905
114. Törnqvist TE (1994) Middle and late Holocene avulsion history of the River Rhine (Rhine-Meuse delta, Netherlands). Geology 22: 711–714
115. Schumm SA (1993) River response to base level changes: Implications for sequence stratigraphy. Journal of Geology 101: 279–294
116. Thorne J (1994) Constraints on riverine valley incision and the response to sea-level change based on fluid mechanics. In Dalrymple RW, Boyd R and Zaitlin BA (eds) Incised-valley systems: Origin and sedimentary sequences, SEPM Special Publication 51, Tulsa USA: 29–43
117. Van Gelder A, van den Berg JH, Cheng G and Xue C (1994) Overbank and channelfill deposits of the modern Yellow River delta. Sedimentary Geology 90: 293–305
118. Bryant M, Falk P and Paola C (1995) Experimental study of avulsion frequency and rate of deposition. Geology 23: 365–368
119. Holbrook J (1996) Complex fluvial response to low gradients at maximum regression: A genetic link between smooth sequence-boundary morphology and architecture of overlying sheet sandstone. Journal of Sedimentary Research 66: 713–722
120. Morozova GS and Smith ND (1999) Holocene avulsion history of the lower Saskatchewan fluvial system, Cumberland marshes, Saskatchewan-Manitoba, Canada. In Smith ND and Rogers J (eds) Fluvial Sedimentology VI. Special Publication 28 of the International Association of Sedimentologists, Blackwell Science, Oxford: 231–249
121. Ashworth PJ, Best JL, Jones M (2004) Relationship between sediment supply and avulsion frequency in braided rivers. Geology 32: 21–24
122. Zadeh LA (1965) Fuzzy sets. Information and Control 8: 338–353
123. Wolkenhauer O (1998) Possibility Theory with Applications to Data Analysis. Research Studies Press Ltd, Taunton UK
124. Laviolette M, Seaman JW Jr, Barrett JD and Woodall WH (1995) A probabilistic and statistical view of fuzzy methods. Technometrics 37: 249–261
125. Kandel A, Matins A and Pacheco R (1995) Discussion: On the very real distinction between fuzzy and statistical methods. Technometrics 37: 276–281
126. Zadeh LA (1995) Discussion: probability theory and fuzzy logic are complementary rather than competitive. Technometrics 37: 271–276

127. Wells JT and Coleman JM (1984) Deltaic morphology and sedimentology, with special reference to the Indus River Delta. In Haq BU and Milliman JD (eds) Marine Geology and Oceanography of Arabian Sea and Coastal Pakistan, Van Nostrand Reinhold Company, New York USA: 85–100
128. Coleman JM (1981) Deltas: Processes of Deposition and Models for Exploration. Burgess Publishing Company, CEPCO Division, Minneapolis USA
129. Puig P, Palanques A and Guillen J (2001) Near bottom suspended variability caused by storms and near-intertidal internal waves on the Ebro mid continental shelf (NW Mediterranean). Marine Geology 178: 81–93
130. Clifton EH (2000) Shoreface myths and misconceptions. AAPG Annual Meeting Extended Abstracts 2000: 29
131. Hori K, Saito Y, Zhao Q and Wang P (2002) Architecture and evolution of the tide-dominated Changjiang (Yangtze) River delta, China. Sedimentary Geology 146: 249–264
132. Chen Z, Song B, Wang Z and Cai Y (2000) Late Quaternary evolution of the sub-aqueous Yangtze Delta, China: Sedimentation, stratigraphy, palynology, and deformation. Marine Geology 162: 423–441
133. Milliman JD, Beardsley RC, Yang Z-S and Limeburner R (1985) Modern Huanghe-derived muds on the outer shelf of the East China Sea: Identification and potential transport mechanisms. Continental Shelf Research 4: 175–188
134. Lisitzin AP (1972) Sedimentation in the world ocean. SEPM Special Publication 17, Tulsa USA
135. DeMaster DJ, McKee BA, Nittrouer CA, Qian J and Cheng G (1985) Rates of sediment accumulation and particle reworking based on radiochemical measurements from continental shelf deposits in the East China Sea. Continental Shelf Research 4: 143–158
136. Moore DG, Curray JR, Raitt RW and Emmel FJ (1974) Stratigraphic-seismic section correlations and implications to Bengal Fan history. In von der Borch CC and Sclater JG (eds) Initial Reports of the Deep Sea Drilling Project, US Government Printing Office, Washington USA: 403–412
137. Dymond J and Lyle M (1994) Particle fluxes in the ocean and implications for sources and preservation of ocean sediments. Material Fluxes on the Surface of the Earth Studies in Geophysics series, National Research Council, Washington USA: 125–142
138. Athy LF (1930) Density, porosity, and compaction of sedimentary rocks. AAPG Bulletin 65: 2433–2436
139. Holbrook P (2002) The primary controls over sediment compaction. In Huffman AR and Bowers GL (eds) Pressure Regimes in Sedimentary Basins and their Prediction, AAPG Memoir 76, Tusla USA: 21–32
140. Karner SL, Chester FM, Kronenberg AK and Chester JS (2003) Subcritical compaction and yielding of granular quartz sand. Tectonophysics 377: 357–381
141. Karner SL, Chester JS, Chester FM, Kronenberg AK and Hajash A Jr (2005) Laboratory deformation of granular quartz sand: Implications for the burial of clastic rocks. AAPG Bulletin 89: 603–625
142. Sclater JG and Christie PAF (1980) Continental stretching: An explanation of the post-mid-Cretaceous subsidence of the central North Sea basin. Journal of Geophysical Research 85: 3711–3739
143. Borg I, Friedman M, Handin J, Higgs DV (1960) Experimental deformation of St. Peter Sand: A study of cataclastic flow. In Griggs D and Handin J (eds)

Rock Deformation (A Symposium), Geological Society of America Memoir 79, New York USA: 133–191
144. Maxwell JC (1960) Experiments on compaction and cementation of sand. In Griggs D and Handin J (eds) Rock Deformation (A Symposium), Geological Society of America Memoir 79, New York USA: 105–132
145. Gutenberg B (1941) Changes in sea-level, postglacial uplift, and mobility of the earth's interior. GSA Bulletin 52: 721–772
146. Fairbridge RW and Krebs OA (1962) Sea-level and the southern oscillation. Geophysical Journal, London 6: 532–545
147. Hicks SD (1978) An average geopotential sea-level series for the United States. Journal of Geophysical Research 83: 1377–1379
148. Gornitz V, Lebedeff S and Hansen J (1982) Global sea-level trend in the past century. Science 215: 1611–1614
149. Wang J and Wang P (1982) Relationship between sea-level changes and climatic fluctuations in East China since last Pleistocene. The Quaternary Research 21: 101–114 (in Japanese)
150. Congxian L, Gang C, Ming Y and Ping W (1991) The influence of suspended load on the sedimentation in the coastal zones and continental shelves of China. Marine Geology 96: 341–352
151. Holt WE and Stern TA (1991) Sediment loading on the Western Platform of New Zealand continent: Implcations for the strength of a continental margin. Earth and Planetary Science Letters 107: 523–538
152. Peper T (1993) Quantitative subsidence analysis of the Western Canada foreland basin with implications for short-term facies change. Tectonophysics 226: 301–318
153. Csato I, Cao S, Petersen K, Lerche I, Sullivan N and Lowrie A (1994) Basement motion and sediment loading: A quantitative study in northern Louisiana, Gulf of Mexico. AAPG Bulletin 78: 1453
154. Lavier LL, Steckler MS and Brigaud F (2000) An improved method for reconstructing the stratigraphy and bathymetry of continental margins: Applications to the Cenozoic tectonic and sedimentary history of the Congo margin. AAPG Bulletin 84: 923–939
155. Lee S-M, Kim J-W, Baag C-E (2003) 2-D flexural analysis of the Ulleung back-arc basin, East Sea (Sea of Japan). Terrestrial, Atmospheric and Oceanic Sciences 14: 431–444
156. Airy GB (1855) On the computation of the effect of the attraction of mountain-masses as disturbing the apparent astronomical latitude of stations in geodetic surveys. Philosophical Transactions of the Royal Society of London 145: 101–104
157. Li F, Dyt C and Griffiths C (2004) 3D modeling of flexural isostatic deformation. Computers and Geosciences 30: 1105–1115
158. Wang Y (1980) The coast of China. Geoscience Canda 7: 109–113
159. Fillon RH, Kohl B, Roberts HH (2004) Late Quaternary deposition and paleobathymetry at the shelf-slope transition, ancestral Mobile River delta complex, northeastern Gulf of Mexico. In Anderson JB and Fillon RB (eds) Late Quaternary stratigraphic evolution of the northern Gulf of Mexico margin, SEPM Special Publication 79, Tulsa USA: 111–141
160. Törnqvist TE, Blick SJ, van der Borg K, de Jong AFM and Greenberg J (2005) Using Holocene relative sea-level data for high-precision measurement

of tectonic subsidence rates in the Mississippi Delta. Abstracts with Programs 37 (5), GSA North-Central Section 39th Annual Meeting: 90

161. Desheng L (1984) Geologic evolution of petroliferous basins on continental shelf of China. AAPG Bulletin 68: 993–1003
162. Weiling X and Junying L (1989) Structural history of the East China Sea. China Earth Sciences 1: 59–73
163. Yu H-S (1991) East China Sea Basin revisited: Basin architecture and petroleum potential. Petroleum Geology of Taiwan 26: 33–44
164. Wang Y and Aubrey DG (1987) The characteristics of the China coastline. Continental Shelf Research 7 (4): 329–349
165. Shanshu W, Zuobin and Libin L (1990) Hydrocarbon accumulations on China's continental shelf. China Earth Sciences 1: 93–109
166. Yunshan Q and Fan L (1983) Study of influence of sediment loads discharged from the Hunaghe River on sedimentation in the Bohai Sea and the Hunaghai Sea. In Proceedings of International Symposium on Sedimentation on the Continental Shelf with Special Reference to the East China Sea, April 12–16, Hangzhou, China, China Ocean Press, Beijing: 83–92
167. Chen Z and Stanley DJ (1993) Yangtze delta, eastern China: 2, Late Quaternary subsidence and deformation. Marine Geology 112: 13–21
168. Bitzer K and Pflug R (1990) DEPO3D: A three-dimensional model for simulating clastic sedimentation and isostatic compensation in sedimentary basins. In Cross TA (ed) Quantitative Dynamic Stratigraphy, Prentice Hall, Englewood Cliffs NJ USA: 335–348
169. Turcotte DL and Schubert G (2002) Geodynamics. Cambridge University Press, UK
170. Yu H-S and Hong E (1992) Physiographic characteristics of the continental margin, northeast Taiwan. Terrestrial, Atmospheric and Oceanic Sciences 3: 419–434
171. Bingxian G and Hanli M (1982) A note on the circulation of the East China Sea. Chinese Journal of Oceanology and Limnology 1: 5–16
172. Choi BH (1980) A tidal model of the Yellow Sea and the eastern China Sea. Korea Ocean Research and Development Institute Report 80–82, Seoul
173. Payton CE (ed) (1977) Seismic stratigraphy – Applications to hydrocarbon exploration. AAPG Memoir 26, Tulsa USA
174. Brown LF Jr and Fisher WL (1980) Seismic stratigraphic interpretation and petroleum exploration. AAPG Continuing Education Course Note Series 16, Tulsa USA
175. Berg OR and Woolverton DG (eds) Seismic stratigraphy II – An integrated approach to hydrocarbon exploration. AAPG Memoir 26, Tulsa USA
176. Vail PR (1987) Seismic stratigraphic interpretation procedure. In Bally AW (ed) Atlas of Seismic Stratigraphy, AAPG Studies in Geology 27: 1–10
177. Posamentier HW and Vail PR (1988) Eustatic controls on clastic deposition II – sequence and systems tracts models. In Wilgus CK, Hastings BS, Kendall CGStC, Posamentier HW, Ross CA, and Van Wagoner JC (eds) Sea-Level Changes: An Integrated Approach. SEPM Special Publication 42, Tulsa USA: 125–154
178. Van Wagoner JC, Posamentier HW, Mitchum RM, Vail PR, Sarg JF, Loutit TS and Hardenbol J (1988) An overview of the fundamentals of sequence stratigraphy and key definitions. In Wilgus CK, Hastings BS, Kendall CGStC, Posa-

mentier HW, Ross CA and Van Wagoner JC (eds) Sea-Level Changes: An Integrated Approach, SEPM Special Publication 42, Tulsa USA: 39–46
179. Hunt D and Tucker ME (1992) Stranded parasequences and the forced regressive wedge systems tract: Deposition during base-level fall. Sedimentary Geology 81: 1–9
180. Catuneanu O (2006) Principles of Sequence Stratigraphy. Elsevier, Boston USA
181. Burk CA and Drake CL (eds) (1974) The geology of continental margins. Springer-Verlag, New York USA
182. Weimer P and Posamentier HW (eds) (1990) Siliciclastic sequence stratigraphy: Recent developments and applications. AAPG Memoir 58, Tulsa USA
183. Watkins JS, Zhiqiang F and McMillen KJ (eds) (1992) Geology and Geophysics of Continental Margins. AAPG Memoir 53, Tulsa USA

Fuzzy Logic for Modeling the Management of Technology

André Maïsseu and Benoît Maïsseu

Abstract. Managing the complexity and diversity of the parameters influencing consumer decisions, by relating them directly to corporate research and development policy, is one of the major components of an efficient economic policy. The quantification of a psychological variable implies subjectivity. Yet, the methods that have been developed to understand and analyze the behavior of consuming economic agents have become very relevant insofar as the requirements relative to minimum sample size and segmentation are fulfilled. It is then easy to formalize the consumer needs by fuzzy logic.

This article states how the use of fuzzy logic helps to directly connect consumer needs with the technology portfolio and thereby optimize the elucidation and implementation of a policy for the management of technology and technological resources. This short article merely attempts to be a starting point of a methodology based on fuzzy logic designed to optimize the management of technology.

1 Introduction

The formalization of consumer decision making processes according to the level of their income is a subject widely addressed in the economic and management literature, just like consumer behavior as a function of the ranking of their needs. Yet few studies have attempted a rapprochement between the technological content of the products and consumer behavior, because the essentially fluctuating nature of this behavior makes its formalization difficult.

Classic strategic analysis begins with consumer needs and product/market combinations, and work back to the technology. The estimation of the "value" attributed by consumers to technologies implies the formulation of relations between consumer needs and technology "qualities". But the methodologies followed for relating consumer needs with technologies, which are conventionally qualitative, preclude a quantification of these relations, the sole genuine foundation of an economic and financial analysis on which the entrepreneur can fully appreciate the resources that he will have to commit in an environment where the joint marketing and technology aspects absolutely must be clearly grasped.

The use of fuzzy mathematics, developed in the last forty years, appears to resolve this difficulty and help to elaborate the foundations of an efficient management of technology, in other words, one based, on the one hand, on consumer behavior, aspirations and income, and on the other, on the opportunities created by technological breakthroughs and the evolution of their complete costs. The model proposed below is developed according to a value chain stretching from the technologies to needs, and conversely, from needs to technologies, although the direction of the relationship introduces no bias, thereby fitting into one of the conventional "market pull" or "technology push" approaches.

2 Relations Between Technologies, Components, Functions and Characteristics

2.1 The "Technical Elementary Unit" Vector

Any product or services considered as a finite sum of components whose characteristics are accurately described in a specification, which the producer must meet to the letter, pursuing a "quality" approach that is widely described in the literature. The resultant of the characteristics of the product's components determines the characteristics of the product.

Using the concepts and the nomenclature of systemic analysis, any product, P, or service, is accordingly conceived as a system, a combination of parts, the subsystems forming a whole, a true aggregate of elements structured around one or more construction principles. This systemic concept of products covers the well-known concept of "product-system" that can be broken down into successively into subsets S^P, then into components C^P, and finally into elementary technical units ω^P

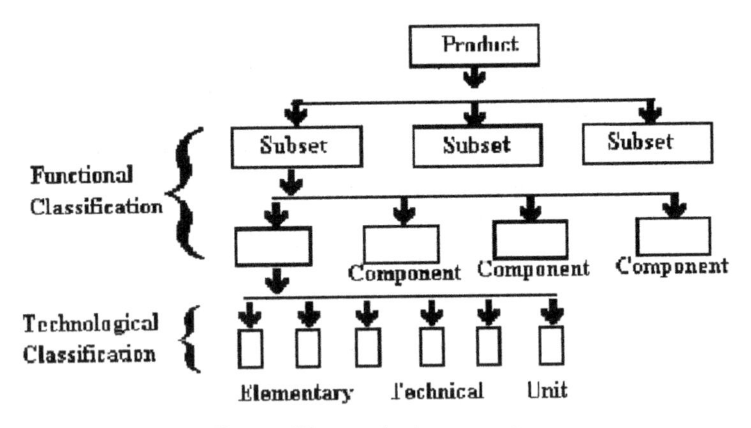

Fig. 1. The product as a system

Any elementary technical object ω^{P} belongs to two distinct categories: material elementary technical objects corresponding to the physical components of the product, and virtual technical objects corresponding to each of the productive operations necessary for the complete manufacture of the product, from design to marketing.

Let n be the total number of elementary technical objects ω^{P}, material and virtual, making of the product P, *each elementary technical object being represented by a scalar, its total production cost.* This must include the design and manufacturing stages, as well as the distribution costs of indirect operations involved in marketing the finished product.

Let $[\omega]$ be the set of all the elementary technical units. Let $\omega_{\mathrm{i}}{}^{\mathrm{P}}$ be an elementary technical object of this set, belonging to P, so that:

$$\mathrm{P} = \{\omega_1{}^{\mathrm{P}}, \omega_2{}^{\mathrm{P}}, \ldots, \omega_{\mathrm{i}}{}^{P}, \ldots, \omega_{\mathrm{n}}{}^{\mathrm{P}}\} = [\omega_{\mathrm{i}}{}^{\mathrm{P}}] \tag{1}$$

where n is the number of elementary technical units which belong to P (nok: n = m, l; see below for the definition of m and l).

A product can then be represented by its "elementary technical unit" vector:

$$\mathrm{P} = [\omega_{\mathrm{i}}{}^{\mathrm{P}}] \tag{2}$$

Any product can be described as being an ordered aggregate of the elementary technical units, components, and subsystems which make up the product.

2.2 The "Technology" Vector

By definition, any elementary technical unit ω_{i}. belongs to a technology Π_{j}

$$\omega_{\mathrm{i}}{}^{\mathrm{P}} \in \Pi_{\mathrm{j}}{}^{\mathrm{P}} \tag{3}$$

Any $\Pi_{\mathrm{j}}{}^{\mathrm{P}}$ can be represented by an elementary technical unit vector which is written

$$\Pi_{\mathrm{j}}{}^{\mathrm{P}} = \{\omega_{1,\mathrm{j}}{}^{\mathrm{P}}, \ldots, \omega_{\mathrm{i,j}}{}^{\mathrm{P}}, \ldots, \omega_{\mathrm{n,j}}{}^{\mathrm{P}}\} \tag{4}$$

where in $\omega_{\mathrm{i,j}}{}^{\mathrm{P}}$, i and j means the " i " elementary technical unit belongs to the " j " technology. In the "j" technology, some i can be nil.

Product P can then be defined by a partitioning into its constituent technologies j

$$\mathrm{P} = \{\Pi_1{}^{\mathrm{P}}, \Pi_2{}^{\mathrm{P}}, \ldots, \Pi_{\mathrm{j}}{}^{\mathrm{P}}, \ldots, \Pi_{\mathrm{m}}{}^{\mathrm{P}}\} = [\Pi_{\mathrm{j}}{}^{\mathrm{P}}] \tag{5}$$

where m is the number of technologies embodied in the product.

This can be called termed the "technology" vector: the product looked at as an aggregate of technical resources and technologies.

2.3 The "Components/Characteristics" Vector

The product can then be defined as a statement of particulars or a set [C] of components, which are well known as product specifications. This set [C] can be defined by the vector $[C_k{}^P]$.

$$P = \{C_1{}^P, C_2{}^P, \ldots, C_k{}^P, \ldots, C_l{}^P\} = [C_l{}^P] \tag{6}$$

where l is the number of components of the product.

Any $C_k{}^P$ can be represented by an elementary technical unit vector which is written as:

$$C_k{}^P = \{\omega_{k,1}{}^P, \ldots, \omega_{k,i}{}^P, \omega_{k,n}{}^P\} = [\omega_{k,i}{}^P] \tag{7}$$

where in $\omega_{i,k}{}^P$, i and k means the "i" elementary technical unit belongs to the "k" component. In the "k" component, some "i" can be nil

Each component C_k, corresponds to one and only one function F_k. Hence, a strictly univocal relation exists between the notions of component and of function.

$$C_k{}^P \equiv F_k{}^P \tag{8}$$

The product P can thus be defined according to a "function" effective that is strictly identical to the component vector:

$$P = \{F_1{}^P, F_2{}^P, \ldots, F_k{}^P, \ldots, F_1{}^P\} = [F_1{}^P] \tag{9}$$

2.4 The "Technology/Product" Matrix

A product P can thus be represented in three different vector forms:

- $P = [\Pi_j{}^P]$ vector of the associated technologies;
- $P = [C_k{}^P]$ vector of the components or its function vector $[F_k{}^P]$;
- $P = [\omega_i{}^P]$ vector of the elementary technical units.

It is readily demonstrated[1] that any product can be defined by its matrix $P = [\omega_{j,k}]$. The " i " elementary technical unit belongs to the "j" technology and the "k" component. The product can then be considered as a matrix of elementary technical units $[M_\omega]$ for transforming technologies $[\Pi]$ into characteristics given by its components or functions [C] or for transforming characteristics given by its components or functions [C] into technologies $[\Pi]$

Each column of the matrix $[M_\omega]$ corresponds to a technology incorporated in the product P, quantified by its cost. Each line of the matrix $[M_\omega]$ corresponds to a component or a function of the product P, also quantified by its cost.

Due to the way any elementary technical unit is valued, summation across the rows of the product matrix gives the cost of component C, summation down the columnsgives the quantification of the technology implemented in product P. Double summation across the rows and down the columns gives the total cost of product P.

$$P_{Tc} = \Sigma\, \Sigma\, \omega_{j,k} \tag{10}$$

The Technology/Product Matrix $[M_\omega]$ thus has the following form:

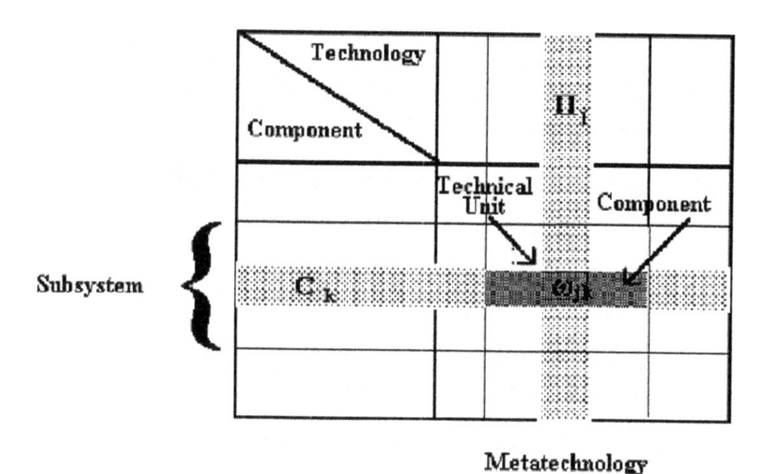

Fig. 2. The product matrix $[M_\omega]$

2.5 Cost Analysis of the "Technology/Product" Matrix

The technology/product matrix can be broken down according to the analysis of costs normally performed by cost accounting. Each type of cost will thus correspond to a specific matrix.

2.5.1 Definitions

Direct costs

Direct costs are directly proportional to production. They are related to the product cost without intermediate calculations, and can therefore be assigned directly thereto.

Indirect costs

Indirect costs are basically related to several products. They require an intermediate calculation (breakdown in analysis centers and posting of the cost of these centers to the various products concerned) before being allocated to the cost of a product.

Fixed costs

Fixed costs are independent of the level of activity, and do not vary with the volume of production or sale.

Variable costs: direct costs + indirect costs

Variable costs are formed of only the costs that vary according to the volume of activity (volume of production or sale).

Complete costs: variable cost + fixed cost

Complete cost represents the cost of a product or service in the final stage of production and marketing. It incorporates all the costs related to the group of objects, to the product or to the service concerned.

2.5.2 Breakdown of the "Technology/product" Matrix into Accounting Submatrixes

For each elementary technical object, it is thus possible to define:

itscost	–direct	] variable	} complete
	–indirect	]	
	–fixed		

The technology/product matrix can then be broken-down into a sum of three matrixes each representing a component of cost:

$$[\omega_{j,k}] = [\omega_{dir\ j,k}] + [\omega_{indir\ j,k}] + [\omega_{fixe\ j,k}] \tag{11}$$

where $\omega_{j,k}$ complete cost of the elementary technical object,
$\omega_{dir\ j,k}$ direct cost of the elementary technical object,
$\omega_{indir\ j,k}$ indirect cost of the elementary technical object,
$\omega_{fixe\ j,k}$ fixed cost of the elementary technical object.

3 Valorization of Technologies: The "Technology/Utility" Fuzzy Matrix

3.1 The Classic Model

Consider a first application (f) of the space of consumption Y on the representative space of the quantities of products (what Lancaster calls goods) produced or available X. This first application, the consumption function, has been the subject of a large number of analyses making it possible to introduce various constraints, including income.

If y is the consumption vector and x the products vector (in quantitative form), Lancaster described a matrix A representing the consumption function as a strictly univocal relation between products and consumption.

"We shall regard an individual product or a collection of products as a consumption activity and associate a scalar (the level of activity) with it. We shall assume the relationship between the level of activity k, y_k, and the products consumed in that activity to be both linear and objective, so that, if x_j is the jth commodity we have:

$$x_j =_{\Sigma} a_{jk^y k} \tag{12}$$

and the vector of total products required for a given activity vector is given by

$$x = Ay \tag{13}$$

Since the relationships are assumed objectives, the equations are assumed to hold for all individuals, the coefficients a_{jk} being determined by the intrinsic properties of the products themselves." [18][2] (Indexes are the Lancaster one's with no relation with the indexes of previous paragraphs)

3.2 The Lancaster Model

Lancaster was probably the first to consider the product as a sum of characteristics. Even if the validity of his model has been severely questioned because of the assumptions made on the efficient conditions of consumption, on the independence of utility with respect to the distribution of characteristics, and on the condition of equality between the number of characteristics and the number of products needed to solve the optimization problem, this model is of continuing interest on many grounds.

Lancaster introduces a second application, called the function of characteristics (g) of the space of products over the space of characteristics Z, which directly connects with the space of functions or components.

$$\mathrm{C_k}^{\mathrm{P}} \equiv \mathrm{F_k}^{\mathrm{P}} \equiv \mathrm{Z_k}^{\mathrm{P}} \tag{14}$$

As already described, the product can then be defined as a statement of particulars or on the set [F] of functions, which are well known as product specifications. This set [F] can be defined by the vector $[\mathrm{F_k}^{\mathrm{P}}]$ of its constituent functions.

$$\mathrm{P} = \{\mathrm{F_1}^{\mathrm{P}}, \mathrm{F_2}^{\mathrm{P}}, \ldots, \mathrm{F_k}^{\mathrm{P}}, \ldots, \mathrm{F_l}^{\mathrm{P}}\} = [\mathrm{F_k}^{\mathrm{P}}] \tag{15}$$

where l is the number of functions/components involved per product.

Each function/component is assigned a performance level, a characteristic or attribute determined by reference to the performance proposed by all the products on the market.

The product can best be described according to a "characteristics" vector:

$$\mathrm{P} = \{\mathrm{Z_1}^{\mathrm{P}}, \mathrm{Z_2}^{\mathrm{P}}, \ldots, \mathrm{Z_k}^{\mathrm{P}}, \ldots, \mathrm{Z_l}^{\mathrm{P}}\} = [\mathrm{Z_k}^{\mathrm{P}}] \tag{16}$$

In the general case, every product has several characteristics, and any given characteristic can belong to several products. The standard theory hence only appears in this logic as a special case, in which each product is associated with one and only one characteristic. "...we shall assume that each consumption activity produces a fixed vector of characteristics and that the relationship is again linear, so that if z_i is the amount of the ith characteristic

$$z_i =_\Sigma b_{ik} y_k \tag{17}$$

$$\mathrm{Z} = \mathrm{B} \bullet \mathrm{Y} \tag{18}$$

Again we shall assume that the coefficients b_{ik} are objectively determined - in principle, at least - for some arbitrary choice of the units of z_i"
given

$$z = \mathrm{BA}^{-1} x \tag{19}$$

3.3 Integration of Value/Price Concept

In his model, Lancaster so to speak substitutes the consumption of characteristics for the consumption of products. Utility is no longer based on the quantities consumed but on their characteristics. The utility function is no longer defined on the space of products but on the space of characteristics: "when characteristics, activities and products are equal. In this case, $U(z)$ can be written directly and unambiguously as a function of $u(x)$."

Yet Lancaster cannot introduce the utility concept directly in his model since this concept is not quantifiable, because ordinal and not cardinal. "We shall assume that the individual possesses an ordinal utility function on characteristics $U(z)$ and that he will choose a situation which maximizes $U(z)$." The obstacle of the ordinality of the utility function: "Unfortunately this is not the general case, and the relations are between vectors in spaces of different dimensions, can be easily removed by introducing the concept of fuzzy space and fuzzy vector, by considering the space of utilities as a fuzzy space. This makes it possible to solve the contested problem posed by Lancaster's analysis, for which consumer satisfaction is based on the intrinsic characteristics offered by the product and not on the satisfaction that would be derived from its use or its consumption, a simplification that is the target of many controversies throughout the economic literature: a product is considered as a solution devised to respond to a non-fixed fuzzy set of consumer aspirations. The use or consumption of a product enables consumers to satisfy a varied spectrum of needs, reflecting the concept of utility, of which the ranking is specific to each consumer.

To resolve this problem we introduce two new functions, one linking the quantities of products consumed to the satisfaction felt, their utility, itself a function of the characteristics of said products. The descriptive logic of the space of utilities is a fuzzy logic, as opposed to the logics which describe the space of products or of characteristics. The set of satisfactions is hence necessarily a fuzzy set, since the satisfactions derived from the "consumption" of a characteristic do not correspond to strict values but vary between intervals of values along scales within which each consumer situates his own need.

The present model is based on a systemic analysis of the product. The product is no longer considered as a unitary whole, but as a finite sum of

Table 1. Framework for evaluating the aspirations-value model

Characteristic (Attribute)	Weight	Attributed value ≡ Utility (Benefit)	Scoring	Evaluation
Z_1	P_1	U_1	$S_1 = P_1^* Z_1$	
Z_2	P_2	U_2	$S_2 = P_2^* Z_2$	$M = S_1 Ⓡ S_2 Ⓡ S_r Ⓡ \ldots Ⓡ S_n$
...	...	...	...	where * and Ⓡ are fuzzy operators.
Z_r	P_r	U_r	$S_r = P_r{}^* Z_r$	
...	...	...	...	
Z_n	P_n	U_n	$S_n = P_n{}^* Z_n$	

components whose characteristics are described accurately in a specification, which the constructor must follow to the letter, according to a "quality" approach widely described in the literature. The resultant of the characteristics of the product's components determines the characteristics of the product.

Kotler and Dubois (1989) developed a model in which consumers express their aspirations in terms of product characteristics. According to this model, the analysis and decision making procedure followed by consumers comprises four steps: the selection of the characteristics that represent the qualitative criterion for analysis, and the weighting the characteristics, which sets up the quantitative framework of the analysis, followed by the scoring of each product by these characteristics, and finally, the composition of an overall score for comparing the products with each other.

This scoring of the product characteristics by the use of fuzzy logic helps to assign a degree of appreciation between 1 and 0, to quantify the importance of each attribute or characteristic, then to quantify the consumer aspirations, and finally help to develop below the general relation associating the technologies and behavior of the economic agents, who are never "perfect", as presumed in conventional econometric analysis, but rather fluctuating, unpredictable and relatively precise. This method of using fuzzy logic satisfies the basic assumptions of marketing while adding the rigor that it needs, which cannot be provided by conventional mathematic tools based on conventional logic.

- The assumption of limited rationality as to whether the consumer is capable of calculating the utility of the product;
- the consumer's calculation is aimed to maximize the overall utility;
- differentiation of the products on an imperfect market is a prerequisite to the purchasing decision;
- the purchasing decision is based on a valorization of preferences.

Each consumer attributes a utility value to each of the product's characteristic, a value which depends on his tastes, his personal history, his mood and its variability, etc. a utility value of which the resultant will be the determining factor, concomitant with the consumer's income and the purchasing decision.

For a given consumer, the total utility of product P is defined as the resultant of the respective utilities of each of the product's descriptive functions. Thus where U^P is the total utility of product P, we have

$$U^P = \{\alpha_1 Z_1{}^P, \alpha_2 Z_2{}^P, \ldots, \alpha_k Z_k{}^P, \ldots, \alpha_1 Z_1{}^P\} = [\alpha_k Z_k{}^P] \tag{20}$$

where α_k is the utility of the characteristics $Z_k{}^P$ but also of the component/function $C_k{}^P/F_k{}^P$. Each coefficient is a number, between 0 and 1, measuring the utility of said function for the consumer. The function vectors each assigned their utility coefficient α_k define a fuzzy subset of the set of utilities describing the needs, aspirations and wishes of the consumers. The vector $[\alpha_k Z_k{}^P]$ is called the "utility" vector of the product.

If necessary, this first approach can be refined by incorporating the concept of "benefit" frequently used in marketing analysis, the positioning based only on the characteristics which can only be accepted, according to [20] if the performance of the products is similar, and if the products and attributes are too numerous. For complex markets, like the market for automobiles, computers, etc. consumers try to simplify the available information. They can achieve this by constructing a classification of a higher level of aggregation, by substituting other variables for the concept of characteristics.

The so-called *Means-Ends* typology is aimed to find the links between the tangible attributes and the intangible attributes describing the aspirations of consumers, such as "the benefits, not directly visible and resulting directly from the use of the product, for example, comfortable, easy ... and the values that are rather cognitive and relatively stable and which are related to consumer motivations, such as safety, contentment, etc." [20], or "estimated by others", "self-esteem", "enthusiasm", "amusement and pleasure in life", "security", "contentment", etc.

The utilities, benefits or values, as from now "value" will be used instead of "benefit", belong respectively to different socio-psychological constructs, in a network of extremely complex relations. Hofstede et al. (1998) proposed and validated a quantitative approach for gathering *Means-Ends* pattern data. This "Pattern Technique" approach, based on polling methods, uses two matrixes, a characteristic-utilities matrix and a utilities-values matrix, which are quantified by the conventional method of probabilities, which would appear to be better suited to a quantification based on the use of fuzzy logic. The characteristics, utilities and values variables are interrelated by unclear and non-independent relations.

Two complementary matrixes are accordingly used to refine the "utility" vector defined above:

Table 2. Attribute-Utility/Benefit fuzzy matrix g_{st} varies between 0 and 1

Characteristic Utility	Z_1	Z_2	...	Z_s	...	Z_v
U_1	g_{11}	g_{21}	...	g_{s1}	...	g_{v1}
U_2	g_{12}	g_{22}	...	g_{s2}	...	g_{v2}
...	...	...	......	...	......	...
U_t	g_{1t}	G_{2t}	...	g_{st}	...	g_{vt}
...	...	...	......	...	......	...
U_w	g_{1w}	g_{2w}	...	g_{sw}	...	g_{vw}

Table 3. Utility/Benefit-Value fuzzy matrix h_{tr} varies between 0 and 1

Utility Value	U_1	U_2	...	U_t	...	U_w
V_1	h_{11}	h_{21}	...	h_{t1}	...	h_{w1}
V_2	h_{12}	h_{22}	...	h_{t2}	...	h_{w2}
...	...	...	......	...	......	...
V_r	h_{1r}	h_{2r}	...	H_{tr}	...	h_{wr}
...	...	...	......	...	......	...
V_x	h_{1y}	h_{2y}	...	H_{ty}	...	h_{wy}

The existence of these quantitative characteristics-utilities relations in the quantitative modeling of means-technologies relations, helps to elucidate the fuzzy relationship existing between the characteristics and values that they provide to consumers, although these relations are only valid for the body of consumers to which the socio-psychological analysis has been addressed.

If β is the coefficient taking account of the corrections attributed to each elementary utility by the sampling subjected to the behavior analysis, the utility vector becomes

$$U^P = \{\alpha\beta_1 Z_1{}^P, \alpha\beta_2 Z_2{}^P, \ldots, \alpha\beta_k Z_k{}^P, \ldots, \alpha\beta_l Z_l{}^P\} = [\alpha\beta_l Z_l{}^P] \qquad (21)$$

the summation of which corresponds to the total value attributed by the sampling of consumers to the product, which can reasonably be considered to be equal to the price of the product.

This provides us with a second matrix defining the product $[U^P{}_{\alpha\beta}]$, based on its valorization, and of which the summation on the lines and columns is directly linked to the selling price of the product, margins included.

$$P_{Pt} = \Sigma\Sigma\ \alpha\beta\, Z_{j,k} \qquad (22)$$

We can then determine the relationship existing between the value attributed to the product by the consumer, i.e. its price, to the technologies it incorporates, and compare it to their costs to determine the margin released by each technology.

Price/Value—y→ utility—h→ characteristics/functions/components —g→ technologies
Price—y→ U—h→ Z/F/C—g→ Π

The comparison of the utility matrix with the sub-matrixes derived from the accounting analysis helps to automatically determine the margin obtained by each technology incorporated in the product. The effective quantification or the economic and financial profitability of each technology becomes immediate.

4 Modeling Consumer Needs

The product can also be considered as a partioning worked out with a view to responding to a non-fixed, fuzzy set of aspirations from the market. Product P can then be defined by its ability to meet a set of consumer needs.

4.1 Attitudes–Motivations Fuzzy Matrix (or Fuzzy Relation)

Consumer behavior is underpinned by a hierarchy of psychological constructs that coincide, conflict with or complement another hierarchy of product characteristics, and which must, at least theoretically, meet the consumer's aspirations. The evaluation and purchasing behavior results from this interaction between the psychological constructs of consumers and the attributes of products.

"The behavior iceberg" shown above is frequently alluded to in the literature to explain consumer behavior. It attempts to provide the psychological framework of consumer behavior, the purchase of any product being positioned at

Psychological Constructs	←→	Product
↓		↓
Opinion	←→	Characteristics
↓		↓
Attitude	←→	Utilities
↓		↓
Motivation	←→	Value
↓		↓
	Consumer Personality	

Fig. 3. The Behavior Iceberg

any level of the psychological constructs, opinions, attitudes and motivations, in which projected attributes, benefits and values on their "product" counterpart.

This makes it possible to class consumer behavior according to two different levels, attitudes and motivations. In the case of the automobile, for example, all the attitudes could comprise needs such as Speed, Comfort, Safety, Fuel Consumption, Maneuverability, Pollution, Maintenance, Roominess, Esthetics and price. The inventory of motivations is generally described eight elements already discussed above in the means-ends chart: "esteem from others", "self-esteem", "enthusiasm", "amusement and pleasure in life", "safety", "contentment", "feeling of accomplishment", and "sense of belonging". These two sets are interconnected by fuzzy relations.
With $a_{\alpha\beta}: \ M \bullet \Theta \longrightarrow [0,1]$

The value $a_{\alpha\beta}$ represents the score of a variable with respect to another.

Human behavior can be described using fuzzy logic. Russell and Fehr (1994) demonstrated that the concept of anger is a fuzzy subset containing several sub-categories (fury, discontent), each possessing their own degree of belonging. Chaplin, John, and Goldberg (1988) found that psychological traits, like anxious behavior, or psychic states, the feeling of anxiety, are fuzzy subsets. For [19], "Anything that exists, exists in a certain quantity anything that exists in a certain quantity, can be measured". In this analytical framework, [3] every attitude can be measured by a parameter weighted by a degree of belonging.

Consumer needs are described in the form of a multicriteria composition that lends itself readily to quantification (measurement and digitization), and generalization (weighting and calculation).

Hence the definition of product archetypes as a response to an aggregate of needs of economic agents

$$A = B^A = \{B_1{}^A, B_2Z_2{}^A, \ldots, B_kZ_k{}^A, \ldots, B_uZ_u{}^A\} = [B_uZ_u{}^A] \tag{23}$$

Table 4. Attitude-Motivation fuzzy matrix

Attitudes Motivations	Θ_1	Θ_2	...	Θ_β	...	Θ_n
M_1	a_{11}	a_{12}	...	$a_{1\beta}$	...	a_{1n}
M_2	a_{21}	a_{22}	...	$a_{2\beta}$	...	a_{2n}
...	...	...	...	...	......	...
M_α	$A_{\alpha 1}$	$a_{\alpha 2}$	...	$A_{\alpha\beta}$	...	$a_{\alpha n}$
...	...	...	...	...	......	...
M_m	A_{m1}	a_{m2}	...	$A_{m\beta}$	...	a_{mn}

4.2 Relationships Between Products and Archetypes

Relations between product categories and products themselves are conventionally perceived as simple, because in this simplistic approach, any product can be positioned in a single category. If we take the automobile as an example, a vehicle model is presumed to belong to a single category "sports", "family", "luxury sedan", etc. This conception has obviously been challenged. Many researchers recently proposed the use of fuzzy logic to introduce the multi-parameterization of any product according to several categories (Varki et al. 1997; [19], every product thereby appearing as the best economic compromise of archetypes ideally satisfying consumer needs. A product which, according to conventional analysis, can only belong to a single archetype, can now be described according to its degree of belonging to several archetypes.

Let A be a set of archetypes of the various client "needs" in terms of cars (archetypes of cars or needs below), with a limit, for example, to a number of a archetypes, as a function of the fineness of the definition to be applied to them. "The Chevrolet Métro 1998, with a retail price of $10,725, fuel consumption of 39 miles per gallon (mpg), and a 79 hp engine, is considered as the "economic vehicle" archetype [22,23], while the Ford Contour ($14,995, 24 mpg, 125 hp), would be the result of compromise between "family" archetype and "economic vehicle" archetype.

$$\mathrm{A} = \{\mathrm{archetype_i}\}_{\mathrm{i}\in[1,\alpha]}$$

Example (automobile):
X_E denotes each example reference set associated with the theoretical reference set:

$$A_E = \{Family, CityDweller, Sports, Luxury, 4WD\}$$

A complete list of these archetypes would naturally comprise all the generic types of vehicles such as coupes, convertibles, SUV (sport utility vehicle), people carriers, station wagons, as well as utility vehicles such as the whole variety of vans, pick-ups, and, in addition, for each of these categories, the various segmentations by level of range ($\mathrm{I}_{1/2}$, $\mathrm{M}_{1/2}$, $\mathrm{S}_{1/2}$).

Any actual vehicle can be defined as a fuzzy subset of all the vehicle archetypes.

$$\forall \text{vehicle} \in \mathrm{V} \longrightarrow \quad \text{vehicle} \subset_{\text{fuzzy}} \mathrm{A}$$
$$\text{vehicle} = \sum_{a\in A} f_{v\acute{e}hicule}(a)/a$$

Example:
A Clio Privilege 85hp can be written as a fuzzy subset of A or of this reference set of simplified archetypes A_E.

For A as the reference set:

$$Clio\ privilege85hp \subset \text{fuzzy A}$$
$$Clio\ privilege85hp = \sum_{a \in A} f_{Clio3\ portes1,4l16vprivilége}(a)/a$$

Where A_E is the reference set, so that the following score is obtained for the Clio fuzzy subset (the coefficients are merely examples):

$$\begin{aligned} &Clio\ privilege85hp = 0.2/Family + 0.8/Citydweller + \\ &0.4/Sports + 0.3/Luxury + 0.1/4wd \end{aligned} \tag{24}$$

4.3 Relationships Between Needs and Archetypes

The third product definition proposed in this work is that of an aggregate of components/functions/characteristics designed to satisfy a body of needs of economic agents. It is formed by the composition of the archetype/product and needs/archetypes matrixes.

Let P be a set of needs to which the automobiles respond, like "drive" or "impress the neighbors".

$$\mathrm{P} = \{\mathrm{needs_i}\}_{\mathrm{i} \in [1,\pi]} \tag{25}$$

Example (case of the automobile):

$$\begin{aligned} &P_E = \{Power,\ Safety,\ Roominess,\ Comfort,\ Maneuverability,\ Cost, \\ &Environment\} \end{aligned}$$

By market service and analyses, like the one described above, conducted on representative samplings, the importance of each need for each product archetype can be quantified, synthesizing the ideal needs expressed by the economic agents concerning the product. Each product is evaluated in terms of needs in the form of a fuzzy subset representing the ideal product, an archetype, that may be sometimes be quite different from economic reality. This makes it possible to associate each product in the form subset of P, described by the "product/technology" matrix defined in Sect. 2.4, with an archetype, the ideal product dreamed of by the economic agents, express in the form of a fuzzy subset of A.

Thus real products V, (where V denotes all real vehicles for which a mean expression of their offer has been obtained in terms ofutilities/benefits) is associated with a fuzzy subset A describing the utilities/benefits that the clients appear to seek:

$$\forall v \in V \qquad v \subset_{fuzzy} A \tag{26}$$

$$v = \sum_{a \in A} f_v(a)/a \tag{27}$$

$$\text{desired benefits(v)} \subset_{fuzzy} P \tag{28}$$

$$\text{desired benefits(v)} = \sum_{p \in P} f_{desired\,benefits(v)}(p)/p \tag{29}$$

Many alternatives are then available for determining the interrelation coefficients of this fuzzy relation. In the limited framework of this article, an attempt has been made to express, for each elementary archetype, the associated fuzzy subset of P. This requires defining a distance. The distance used can take the following euclidean form:

$$\forall v_1, v_2 \subset_{fuzzy} A \tag{30}$$

$$v_1 = \sum_{a \in A} f_{v_1}(a)/a \tag{31}$$

$$v_2 = \sum_{a \in A} f_{v_2}(a)/a \tag{32}$$

$$d(v_1, v_2) = \sqrt{\sum_{a \in A} (f_{v1}(a) - f_{v2}(a))^2} \tag{33}$$

This is the **archetypical distance** between two vehicles. Each archetype of A is denoted a_i. a_i is a classic singleton of A, but is also expressed in the form of fuzzy subset of A with the coefficient 1 for itself and coefficients reduced to zero for all the other elements of A. $a_i \in A$ and also $a_i \subset_{fuzzy} A$

The distance defined above can obviously be applied to it. Each archetype a_i is associated with the services that it provides expressed in the form of a fuzzy subset of P: Π_i.

$$\forall i \in [1, \alpha] \; \Pi_i \;\; \subset_{fuzzy} P$$

$$\Pi_i = \sum_{p \in P} f\Pi_i(p)/p$$

For any classic element P (singleton) of P, the associated coefficient $f_{\Pi_i}(p)$ is calculated as follows:

$$f_{\Pi_i}(p) = \frac{\sum_{v \in V} \left(\frac{f_v(p)}{1 + d^2(v, a_i)} \right)}{\sum_{v \in V} \left(\frac{1}{1 + d^2(v, a_i)} \right)} \tag{34}$$

It constitutes a weighted mean of the importance of the service considered for the archetype a_i in relation to the "proximity" of each vehicle to the archetype concerned.

Moreover, each coefficient can also be seen as the cardinality of the intersection of two fuzzy subsets of V: Ui and Wp.

Ui is the fuzzy subset of V of the vehicles of which the definition in terms of archetypes is close to that of the archetype a_i with:

$$\mathrm{U}_i = \sum_{v \in V} f_{U_i}(v)/v \quad \text{with} \quad f_{U_i}(v) = \frac{\frac{1}{(1+d^2(v,a_i))}}{\sum_{v \in V} \left(\frac{1}{1+d^2(v,a_i)} \right)} \tag{35}$$

And Wp is the fuzzy subset of V of the vehicles for which the need p is important.

$$W_\mathrm{P} = \sum_{v \in V} f_{W_p}(v)/v \qquad \text{with} \qquad f_{W_p}(v) = f_v(p) \tag{36}$$

The intersection is determined in this case by using the probabilistic (product) standard instead of the Zadeh (minimum) standard.

Finally, the fuzzy equation between A and P is thus defined as follows:

$$\forall \mathrm{i} \in [1, \alpha] \text{ and } \forall \mathrm{p} \in \mathrm{P} \qquad a_i \in \mathrm{A} \text{ and } f_{R_1}(a_i, p) = f_{\prod_i}(p)$$

Example:

For example, with the purchase of a Clio 3 doors 85hp privilege described as the fuzzy subset of A_E denoted:

0.2/Family + 0.8/City dweller + 0.4/Sports + 0.3/Luxury + 0.1/4WD,

The definition of this purchase in terms of archetypes of vehicles will be compared to the preferences that the client has expressed. In this case, in the order, the client's preferences are as follows: Maneuverability, Costs, Power, Environment, Comfort, Safety, Roominess. These preferences can be translated by the following fuzzy subset:

desired services (Clio 3 doors 85hp privilege) = 0.8/Power + 0.4/Saftey + 0.2/Roominess + 0.6/Comfort + 1/Manuevrability + 0.9/Cost + 0.7/ Environment.

4.4 "Archetype-Product" and "Archetype-Technology" Fuzzy Matrixes

The fuzzy matrix describing the relations between archetypes and products, for a given product range, is obtained

$$\mathrm{C}_\gamma \delta : \ \mathrm{AxP} \longrightarrow [0,1]$$

Note that it is obviously possible to replace C, representing the component $C_k{}^P$, of the product P, by F^u its utility function.

Each product is thus designed by the entrepreneur according to a composition of functions of belonging to archetypes, which he wants to coincide

Table 5. Archetype-Product fuzzy matrix

Products / Archetypes	P_1	P_2	...	P_δ	...	P_q
A_1	C_{11}	C_{12}	...	$C_{1\delta}$	...	C_{1q}
A_2	C_{21}	C_{22}	...	$C_{2\delta}$	...	C_{2q}
...	...	...	...	...	...	...
A_γ	$C_{\gamma 1}$	$C_{\gamma 2}$	...	$C_{\gamma\delta}$	...	$C_{\gamma q}$
...	...	...	...	...	...	...
A_n	C_{n1}	C_{n2}	...	$C_{n\delta}$	...	C_{nq}

with consumer aspirations. A perfect coincidence means commercial success. Straying too far away would mean commercial failure.

Since every product can be described, as discussed above, by its technology vector, the transition from the fuzzy matrix describing the relations between archetypes and technology is automatic, this matrix being strictly descriptive of the marketing and technology strategies pursued by the entrepreneur whose product range serves as the medium for composing the relations.

Table 6. Archetype-Technology fuzzy matrix

Technology / Archetype	Π_1	Π_2	...	Π_j	...	Π_m
A_1	D_{11}	D_{12}	...	D_{1j}	...	D_{1m}
A_2	D_{21}	D_{22}	...	D_{2j}	...	D_{2m}
...	...	...	...	...	...	...
A_γ	$D_{\gamma 1}$	$D_{\gamma 2}$	...	$D_{\gamma j}$	...	$D_{\gamma m}$
...	...	...	...	...	...	...
A_n	D_{n1}	D_{n2}	...	D_{nj}	...	D_{pm}

Note that it is obviously possible to replace C, representing the component $C_k{}^P$, of the product P, by F^u its utility function.

Consider a "fuzzy" set of consumer needs:

$$\Omega_B = \{B_1, \ldots B_i, \ldots B_m\}$$

where B_i, is the expressed or latent need of a consumer

Example:

• "fast" car	Power	: B_1
• "big" car	Size	: B_2
• "maneuverable" car	Maneuverability	: B_3
• "safe" car	Safe	: B_4
• "pleasurable" car	Contentment	: B_5

$$V = \{\mu_1/B_1, \ldots \mu_i/B_i \ldots \mu_m/B_m\}$$

with: $0 < \mu_B(x) \leq 1$ ($\mu_B : \Omega_B \longrightarrow [0, 1]$)
A was defined above as the set of ideal models. The archetypes

$$A = \{a_1, \ldots a_i, \ldots a_n\}$$

where a_i is an "ideal model", an archetype.
Example:

• "town" car	City dweller
• "sports" car	Sport
• "family" car	Station wagon
• "gran turismo" car	Top end

Table 7. "Car Need/Archetypes" fuzzy matrix

B_i	V_1	Sport	V_n
...		...	
Power		1.0	
Volume		0.3	
Safety		0.5	
Pleasure		0.72	
...		...	

The example described where the above matrix, a sports car is 1.0 "faster", 0.3 "bigger", 0.5 "safe", 0.7 "pleasurable"

- A "small" car is a car between 3.3 and 4.8 meters long.
- A "medium" car is a car between 4.5 and 5.8 meters long.
- A "big" car is a car between 5.5 and 6.9 meters long

The size of an archetype in this example is thus a composition of the "small car" and "medium car" and "big car" attributes

The "ideal" size λ_{Tsport} of the sport car is the result of the defuzzyfication operation.

The "ideal" size of the sports car archetype is defined by the fuzzy equation:

$$B_i = \{0/3;\ 0.1/3.5;\ 0.1/4.5;\ 0.6/4.8;\ 0.6/5.7;\ 0.2/5.8;\ 0.2/6.7;\ 0/6.9\} \tag{37}$$

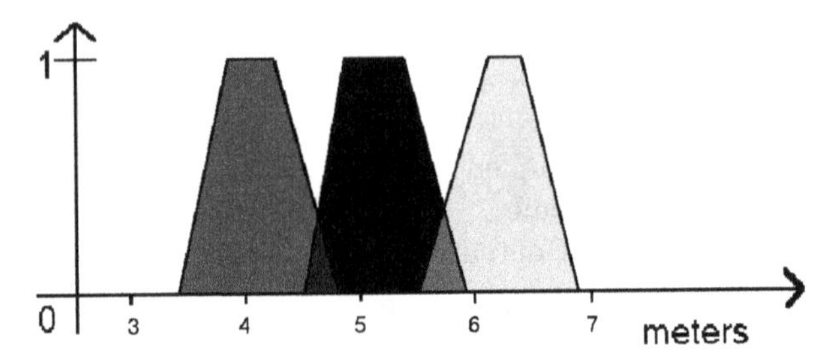

Fig. 4. Small, medium and big cars

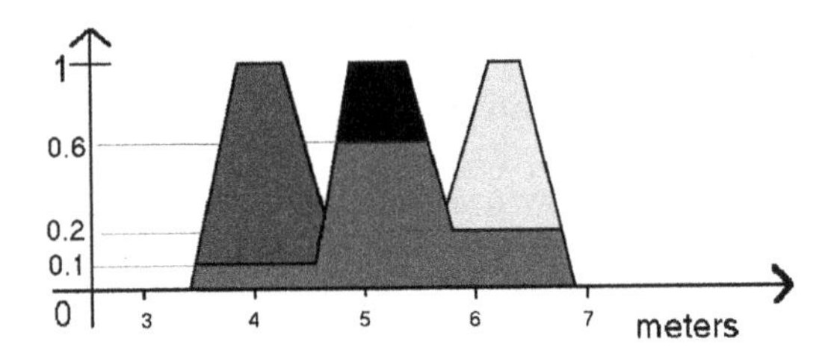

Fig. 5. Ideal size for a sport car

The size of the 307, is given by the fuzzy composition of the size of the archetypes by the degree of belonging e of the 306 to the archetypes.

In the example described by the above matrix, the 306 is 0.8 city dweller, 0.5 sports, "big", 0.1 station wagon, 0.3 top end.

The fuzzy relation between needs and products is given by the composition of the two fuzzy matrixes (Ω_B, Ω_A) and (Ω_A, Ω_P)

$$T(\Omega_B, \Omega_A) \otimes T(\Omega_A, \Omega_P)$$
$$P_j = \{\mu_1/B_{1j}, \dots \mu_i/B_{ij} \dots \mu_m/B_{mj}\} \tag{38}$$

where: $0 < \mu_B(x) \leq 1 \quad (\mu_B : \Omega_B \longrightarrow [0,1])$

Table 8. "Peugeot Archetype/Product" fuzzy matrix

Aj	**107**	**207**	**307**	**407**	**807**	...
...			...			
City dweller			**0.8**			
Sport			**0.5**			
Station Wagon			**0.1**			
Top end			**0.3**			
...			...			

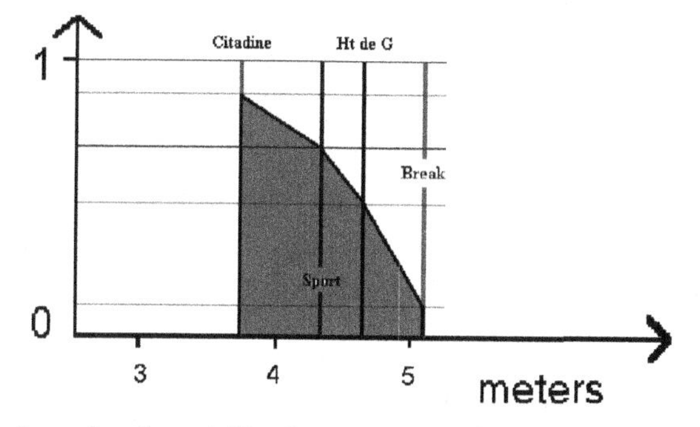

Fig. 6. Defuzzyfication yields the theoretical value of the size of the 306

It is obviously easy to relate the needs expressed by the economic agents to their typology (gender, age, habitat, profession, education, family status, and above all, income). These developments are not presented in this article despite their importance. The corresponding fuzzy matrix is represented below by [R]

5 Modeling the Management of Technological Resources

5.1 Various Definitions of the Product

The product P can be considered successively as:

- an aggregate of technology, $P = f(\Pi)$
- an aggregate of components, functions or characteristics $P = g'(C) = g''(F) = g'''(Z)$
- an aggregate of utility, $P = h\ (U)$
- an aggregate of archetypes $P = x\ (A)$
- an aggregate of needs $P = y\ (B)$

Fuzzy relations exist between the technologies incorporated in the product, and the needs of the economic agents underlying a technology management policy that takes account of the marketing analysis:

$$\text{Needs} = [B] \bullet [A] \bullet [C] \bullet [\Pi]$$

Fuzzy relations exist between these technologies incorporated in the product, and the needs of the economic agents, and hence their typology and particularly in their income (this point has only been touched on here), as well as between the technologies and the overall utility of the product permitting the utilization of the technologies and to the margins that they release by exploiting the breakdown of the technology/product matrix into accounting sub-matrixes (this point has only been touched on here):

$$\text{Consumer} = [R] \bullet [B] \bullet [A] \bullet [C] \bullet [\Pi]$$

$$\text{Price} = [U^P][C][\Pi]$$

Example: automotive industry
A ={family, city dweller, sports, luxury}
The car is a fuzzy subset of A:
0/family +0.7/city dweller +0.5/sports +0.5/luxury
B ={power, safety, roominess, comfort, maneuverability}

The car is also described by a fuzzy subset of B which can be obtained as the image of the subset of A by the following fuzzy relation:

	power	safety	roominess	comfort	maneuverability
Family	0.3	1	1	0.5	0
city dweller	0.4	0.2	0	0.4	1
Sports	1	0.8	0	0.5	0.2
Luxury	0.5	0.5	0.5	1	0.5

C ={engine, suspension, brakes, body, onboard electronics control electronics, (injection, ABS)}

	engine	suspension	brakes	body	Onboard electronics	Control electronics
Power	1	0	0	0.3	0	0.2
Safety	0.3	0.7	1	1	0	0.4
Roominess	0	0	0	1	0	0
Comfort	0.2	1	0.1	0.8	1	0.3
maneuverability	0.8	0.2	0.5	1	0	0

The car desired by the consumer can hence also be obtained as a fuzzy subset of C by using the fuzzy relation represented by the above table.

5.2 Incorporation of Technological Innovation

Technological innovation can be defined simply as equivalent to the gap between the product matrix $[M]_1$ before the incorporation of the innovation and the product matrix $[M]_2$ after its insertion. Technological innovation I, is hence equal to

$$I = \Delta P = [M_\omega]_2 - -[M_\omega]_1 \tag{39}$$

$$I = \delta[M_\omega]/\delta F, \text{so} - \text{called "market pull" innovation} \tag{40}$$

$$I = \delta[M_\omega]/\delta\pi, \text{so} - \text{called "technology push" innovation} \tag{41}$$

This definition of innovation can be resumed by applying it to each of the other matrixes [technologies/archetype], [technologies/utility], [technologies/needs] described in this article, which all describe the product.

One of the main advantages of fuzzy relations is that they are intrinsically "bidirectional", or rather, that they inherently have no direction contrary to applications or functions, and can be composed. For example, it is possible to compose the various relations established above to obtain a direct relation between the vehicle archetypes and technologies in order to identify the technologies that should be privileged when a manufacturer seeks reinforcement in any specific field.

Based on these matrixes, on their derivatives, and on their relations, it is accordingly possible to evaluate a very large body of relevant information for setting out an efficient technology management policy, like, for example, calculating which vehicle archetypes are the most affected by a fluctuation of a descriptive criteria of consumers (income, change in marital status, etc.).

Resuming the example of the automobile, for example, one can calculate the set of vehicles for which safety is not important, with the negation of the image by R_1 of the singleton of P {Safety}. {Safety} is also written in the form of a fuzzy subset: 0/Power + 1/Safety + 0/Roominess + 0/Comfort + 0/Maneuverability + 0/Cost + 0/Environment. The Zadeh negation enables us to determine the vehicles most popular with candidates for suicide: those for which power is important but safety is not, by means of an intersection calculation with the Zadeh standard.

The small example below is also useful for clarifying the opportunities offered by this model to make an important contribution to the management of technological resources and of innovation.

The gap between the size of the archetype and the product provides leeway for customer satisfaction (figure 7). This gap can be measured by variations in:

1) Composition of the archetypes mix describing the product;
2) Ranking of coefficients, describing change in consumers, moods, consumer sampling, etc. This induces a gap in components and technologies, giving the relation between the shift in consumer requirements and technological innovation.

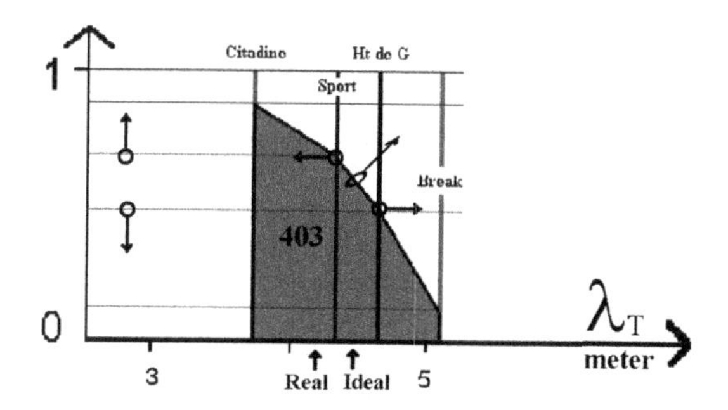

Fig. 7. Innovations possible changes for the size of the "403" car

6 Conclusions

The valorization of technologies by the market happens to be one of the key factors of industrial policies. However, the construction of relations between technologies and the market, conventionally done by an empirical method is unsatisfactory. By drawing on fuzzy logic and matrix calculation, formalized relations can be established between the market, of which the definition is limited in this work in terms of consumer needs and technologies, between technologies and economic and financial analysis, between technologies and technological innovation.

In this model, a product may be described as several elaborate solutions to fixed sets of technologies, to fuzzy sets of characteristics and of consumer requirements. It should also be analyzed as a resultant of technical resources. Synthesis of these many concepts involves defining the product as a combination of technical resources brought into play and meeting one or more requirements in a given market, constrained by the consumer's income.

The introduction of a mathematical definition of technological innovation, that can assume as many forms as definitions of the product presented in this work, sets the stage for an efficient management of innovation, of technology and of technological resources, based on reliable econometric and financial calculations, because founded on the corporation's accounting data.

Notes

1. Maïsseu, A; Le Duff, R. (1991) "*Management technologique*", Paris, Sirey
2. Lancaster K., (1966) "*A New Approach to Consumer Theory*", Journal of Political Economy, 74, p 135

References

1. Bergmann, A., et Uwamungu B., "Encadrement et comportement", Paris, Edition ESKA, 1997.
2. Bouchon-Meunier, B., (1993) "La Logique floue", Paris Presses Universitaires de France
3. Brabet, J., (1988) "Faut-il encore parler d'approche qualitative et d'approche quantitative", Recherche et application en marketing, vol 11, n° 1, pp.75–89
4. Calori, R., (1998) "Philosophizing on strategic management models", Organization Studies, Berlin
5. Gervais, M., (1995) "Stratégie de l'entreprise", Economica
6. Hutchinson, M., O. "The use of fuzzy logic in business decision-making", Derivatives Quarterly, New York, Summer 1998;
7. Kumar, V., K., A., Ganesh, L., S., (1999) "Fuzzy operations and Petri nets: Techniques for resource substitution in projects", Project Management Journal, volume 30, issue 3, p13–22, Sylva, Sep 1999

8. Lancaster K., (1966) "A New Approach to Consumer Theory", Journal of Political Economy, 74, pp 132–57
9. Maïsseu A., and Le Duff, R., (1988) "L'Anti-Déclin, ou les mutations technologiques maîtrisées" ESF, Paris
10. Maïsseu, A., (1990) "A new tool for strategy analysis : the technology-product matrix", Conference on Interregional Cooperation in Europe, Madrid, Spain, 23–24 oct 1990
11. Maïsseu, A., (1991) "Technogénèses de l'innovation et stratégie entrepreneuriales", 1er Congrès franco-espagnol de Gestion et d'Economie de l'Entreprise, Reus, Spain, 23–25 oct 1991
12. Maïsseu, A., (1992) Rev Europea de Direccion y Economica de la Empresa, vol 1, n°1, avril 1992, pp 167–175
13. Maïsseu, A., (1999) "Matrice technologie – produit" Encyclopédie de la gestion et du management, Dalloz, Paris, pp 805–806
14. Maïsseu, A., and Maïsseu, B., (2000) "The extension of the Lancaster model to the management of technology", IAMOT 2000, Miami, USA, 20–25 février 2000
15. Maïsseu, A. (2001) "Modelling the management of technological resources.", IAMOT 2001, Lausanne, Switzerland, 19–22 mars 2001
16. Maïsseu, A. (2002) "The application of fuzzy logic to Management", Polar Lights Conference, Saint Petersburg, Russia, January 2002
17. Rosa, J., A., Porac, J., F., Runser-Spanjol, J., Saxon, M., S., (1999) "Sociocognitive dynamics in a product market", Journal of Marketing, volume63, p64–77, New York
18. Tong-Tong, J., R., (1995) "Logique floue", Paris, Hermès
19. Varki, S., Bruce, C., Roland, P. R., (1997) "Defuzzification: Criteria and Classification". Fuzzy sets and systems 108(2): 159–178.
20. Vriens, M., (2000) "Linking Attributes, Benefits, And Consumer Values", Marketing Research: A Magazine of Management & Applications, Fall, 2000.
21. Bagnoli, C., Smith, H., C., "Fuzzy logic: The new paradigm for decision making", Real Estate Issues, volume 22, issue 2, p35–41, Chicago, Aug 1997;
22. Viswanathan, M, Mark, B., Shantanu, D., Childers, D., (1996) "Does a single response category in a scale completely capture a Response?" Psychology and Marketing 13(5): 457–479
23. Viswanathan, M, Childers, D., (1999) "Understanding how product attributes influence product categorization: development and validation of fuzzy set-based measures of gradedness in product categories" Journal of Marketing Research, 36: 75–94

Index